AF595060

Pharmaceutical Residues in the Environment

Pharmaceutical Residues in the Environment

Editor

Jolanta Kumirska

MDPI • Basel • Beijing • Wuhan • Barcelona • Belgrade • Manchester • Tokyo • Cluj • Tianjin

Editor
Jolanta Kumirska
University of Gdansk,
Faculty of Chemistry,
Department of Environmental Analysis
Poland

Editorial Office
MDPI
St. Alban-Anlage 66
4052 Basel, Switzerland

This is a reprint of articles from the Special Issue published online in the open access journal *Molecules* (ISSN 1420-3049) (available at: https://www.mdpi.com/journal/molecules/special_issues/pharmaceutical_residues_environment).

For citation purposes, cite each article independently as indicated on the article page online and as indicated below:

LastName, A.A.; LastName, B.B.; LastName, C.C. Article Title. *Journal Name* **Year**, *Article Number*, Page Range.

ISBN 978-3-03943-485-5 (Hbk)
ISBN 978-3-03943-486-2 (PDF)

Contents

About the Editor . . . vii

Jolanta Kumirska
Special Issue "Pharmaceutical Residues in the Environment"
Reprinted from: *Molecules* **2020**, *25*, 2941, doi:10.3390/molecules25122941 . . . 1

Joanna Giebułtowicz, Grzegorz Nałęcz-Jawecki, Monika Harnisz, Dawid Kucharski, Ewa Korzeniewska and Grażyna Płaza
Environmental Risk and Risk of Resistance Selection Due to Antimicrobials' Occurrence in Two Polish Wastewater Treatment Plants and Receiving Surface Water
Reprinted from: *Molecules* **2020**, *25*, 1470, doi:10.3390/molecules25061470 . . . 5

R. Guedes-Alonso, S. Montesdeoca-Esponda, J. Pacheco-Juárez, Z. Sosa-Ferrera and J. J. Santana-Rodríguez
A Survey of the Presence of Pharmaceutical Residues in Wastewaters. Evaluation of Their Removal Using Conventional and Natural Treatment Procedures
Reprinted from: *Molecules* **2020**, *25*, 1639, doi:10.3390/molecules25071639 . . . 27

Shuai Zhang, Yu-Xiang Lu, Jia-Jie Zhang, Shuai Liu, Hai-Liang Song and Xiao-Li Yang
Constructed Wetland Revealed Efficient Sulfamethoxazole Removal but Enhanced the Spread of Antibiotic Resistance Genes
Reprinted from: *Molecules* **2020**, *25*, 834, doi:10.3390/molecules25040834 . . . 45

Daniel Wolecki, Magda Caban, Magdalena Pazda, Piotr Stepnowski and Jolanta Kumirska
Evaluation of the Possibility of Using Hydroponic Cultivations for the Removal of Pharmaceuticals and Endocrine Disrupting Compounds in Municipal Sewage Treatment Plants
Reprinted from: *Molecules* **2020**, *25*, 162, doi:10.3390/molecules25010162 . . . 57

Grzegorz Nałęcz-Jawecki, Milena Wawryniuk, Joanna Giebułtowicz, Adam Olkowski and Agata Drobniewska
Influence o f S elected A ntidepressants o n t he C iliated P rotozoan *S pirostomum ambiguum*: Toxicity, Bioaccumulation, and Biotransformation Products
Reprinted from: *Molecules* **2020**, *25*, 1476, doi:10.3390/molecules25071476 . . . 75

Magdalena Pazda, Magda Rybicka, Stefan Stolte, Krzysztof Piotr Bielawski, Piotr Stepnowski, Jolanta Kumirska, Daniel Wolecki and Ewa Mulkiewicz
Identification of Selected Antibiotic Resistance Genes in Two Different Wastewater Treatment Plant Systems in Poland: A Preliminary Study
Reprinted from: *Molecules* **2020**, *25*, 2851, doi:10.3390/molecules25122851 . . . 93

Aleksandra Pieczyńska, Stalin Andres Ochoa-Chavez, Patrycja Wilczewska, Aleksandra Bielicka-Giełdoń and Ewa M. Siedlecka
Insights into Mechanisms of Electrochemical Drug Degradation in Their Mixtures in the Split-Flow Reactor
Reprinted from: *Molecules* **2019**, *24*, 4356, doi:10.3390/molecules24234356 . . . 111

Katarzyna Wychodnik, Grażyna Gałęzowska, Justyna Rogowska, Marta Potrykus, Alina Plenis and Lidia Wolska
Poultry Farms as a Potential Source of Environmental Pollution by Pharmaceuticals
Reprinted from: *Molecules* **2020**, *25*, 1031, doi:10.3390/molecules25051031 . . . 127

Peiyi Li, Yizhao Wu, Yali Wang, Jiangping Qiu and Yinsheng Li
Soil Behaviour of the Veterinary Drugs Lincomycin, Monensin, and Roxarsone and Their Toxicity on Environmental Organisms
Reprinted from: *Molecules* **2020**, *24*, 4465, doi:10.3390/molecules24244465 . . . . **143**

André Pereira, Liliana Silva, Célia Laranjeiro, Celeste Lino and Angelina Pena
Selected Pharmaceuticals in Different Aquatic Compartments: Part I—Source, Fate and Occurrence
Reprinted from: *Molecules* **2020**, *25*, 1026, doi:10.3390/molecules25051026 . . . . **159**

André Pereira, Liliana Silva, Célia Laranjeiro, Celeste Lino and Angelina Pena
Selected Pharmaceuticals in Different Aquatic Compartments: Part II—Toxicity and Environmental Risk Assessment
Reprinted from: *Molecules* **2020**, *25*, 1796, doi:10.3390/molecules25081796 . . . . **193**

Natalia Treder, Tomasz Bączek, Katarzyna Wychodnik, Justyna Rogowska, Lidia Wolska and Alina Plenis
The Influence o f I onic L iquids o n t he E ffectiveness o f A nalytical M ethods U sed i n the Monitoring of Human and Veterinary Pharmaceuticals in Biological and Environmental Samples—Trends and Perspectives
Reprinted from: *Molecules* **2020**, *25*, 286, doi:10.3390/molecules25020286 . . . . **225**

About the Editor

Jolanta Kumirska is Associate Professor at the Faculty of Chemistry of the University of Gdansk, Poland. She is also Head of Team of Chemical Analytics and Diagnostics at the Faculty of Chemistry. She received her PhD degree in 2006. Her research interests include environmental science and analytical chemistry, especially determination of pharmaceutical residues in environmental samples, analysis of the fate of pharmaceuticals in the environment including ecotoxicological studies, and analysis of the mechanisms of the sorption of pharmaceuticals to soils and bottom sediments. Moreover, her research is focused on the development of analytical methods for forensic purposes, structural analysis of oligo- and polysaccharides, and the application of spectroscopic techniques for structural elucidation and analysis of structure–property–activity relationships. She is an author or co-author of more than 80 research works in peer-reviewed journals. She serves as a reviewer for numerous international journals.

Editorial

Special Issue "Pharmaceutical Residues in the Environment"

Jolanta Kumirska

Department of Environmental Analysis, Faculty of Chemistry, University of Gdansk, Wita Stwosza 63, 80-308 Gdansk, Poland; jolanta.kumirska@ug.edu.pl

Academic Editors: Mireia Guardingo and Emity Wang
Received: 23 June 2020; Accepted: 25 June 2020; Published: 26 June 2020

Keywords: pharmaceutical residues; fate in the environment; fate in WWTPs; ecotoxicity; antibiotic resistance; development of methods; environmental risk assessment

---

Pharmaceuticals, due to their pseudo-persistence and biological activity as well as their extensive use in human and veterinary medicine, are a class of environmental contaminants that is of emerging concern. In contrast to some conventional pollutants, they are continuously delivered at low levels, which might give rise to toxicity even without high persistence rates. These chemicals are designed to have a specific physiological mode of action and frequently to resist inactivation before exerting their intended therapeutic effect. These features, among others, make pharmaceuticals responsible for bioaccumulation and toxic effects in aquatic and terrestrial ecosystems. It is extremely important to know how to remove them from the environment and/or how to perform their biological inactivation. Hence, the detection, determination and analysis of the fates of pharmaceuticals and their metabolites in different compartments of the environment are some of the main tasks of modern analytical and environmental chemistry. An important limitation of such studies is the availability of sufficiently sensitive and reliable analytical methods for determining the different pharmaceuticals present in trace amounts in such complex matrices. Although great advances have been made in their detection in aquatic matrices, there are limited analytical methodologies for the trace analysis of target and non-target pharmaceuticals in matrices such as soils, sediments or biota. There are still many gaps in robust data on their fate and behavior in the environment, as well as on their threats to ecological and human health. This Special Issue has included nine current research and three review articles in this field.

Seven research articles deal with the presence of pharmaceuticals in wastewater samples and evaluate their fate, ecotoxicity and/or elimination in wastewater treatment plants (WWTPs) equipped with different purification technologies [1–7].

Giebułtowicz et al. [1] have provided a comprehensive overview of the presence of 26 selected antibiotics in two Polish WWTPs (wastewater and sludge) and have provided crucial information on their removal efficiency and their risk to resistance selection as well as cyanobacteria and eukaryotic species. They established that the removal efficiency of these compounds was more than 50% for both WWTPs. The highest antimicrobial resistance risk was estimated in the influents of WWTPs for azithromycin, ciprofloxacin, clarithromycin, metronidazole and trimethoprim and in the sludge samples for azithromycin, ciprofloxacin, clarithromycin, norfloxacin, trimethoprim, ofloxacin and tetracycline.

Guedes–Alonso et al. [2] have studied the removal efficiencies of 11 pharmaceuticals in three wastewaters treated by conventional or natural purification systems over two years in order to determine the occurrence and removal of pharmaceutical residues in Gran Canaria (Spain). A combination of secondary treatments and reverse osmosis presents favorable removal efficiencies (over 95% for most studied compounds). However, all the target pharmaceuticals were present in the effluent samples.

Zhang et al. [3] have found that constructed wetlands (CWs) could achieve a high removal efficiency of sulfamethoxazole (SMX) (>98%) and that the concentration of SMX in the bottom layer was higher compared with that in the surface layer. A degradation mechanism of SMX was also proposed. Moreover, the relative abundance of *sul* genes exhibited an increase, which tended to be stable throughout the treatment duration.

The effectiveness of CWs for the removal of 15 pharmaceuticals and endocrine disrupting compounds in municipal WWTPs was also investigated by Wolecki et al. [4]. For the first time in such a study, three plants, namely *Cyperus papyrus* (Papyrus), *Lysimachia nemorum* (Yellow pimpernel) and *Euonymus europaeus* (European spindle), were taken into account. The investigation was performed using real municipal WWTP conditions and with the determination of target compounds not only in raw and treated wastewater but also in plant materials (a new ASE-SPE-GC-MS(SIM) method for this purpose was developed and validated in this study). The authors confirmed that the elimination efficiency of the investigated compounds from wastewater was in the range of 35.8% to above 99%. Moreover, *Lysimachia nemorum* was the most effective for the uptake of target compounds among the tested plant species.

Nałęcz–Jawecki et al. [5] have evaluated the biological activity of four antidepressants, fluoxetine, sertraline, paroxetine and mianserin, on the ciliated protozoan *S. ambiguum*. Acute toxicity, bioconcentration and biotransformation studies were performed. The authors observed that sertraline was the most toxic among the studied antidepressants. However, the toxic effects occurred at concentrations at least two orders of magnitude higher than those determined in effluents and freshwaters.

The main aim of the Pazda et al. research article [6] was to compare the occurrence of selected tetracycline- and sulfonamide resistance genes in raw influent and final effluent samples from two Polish WWTPs which were different in terms of size and applied biological wastewater treatment processes (conventional activated sludge (AS)-based in one WWTP and a combined conventional AS-based method with constructed wetlands (CWs) in the other). The genes selected for the study are commonly detected in wastewater samples, represent different resistance mechanisms, and are also located in different genetic elements (especially in mobile genetic elements which significantly influence the spread of antibiotic resistance genes (ARGs)). Furthermore, a method for the isolation of total DNA and the identification of selected ARGs in wastewater samples was developed. All thirteen ARGs coding resistance to tetracyclines, *tet (A, B, C, G, K, L, M, O, Q, X)* and sulfonamides *(sulI, sulII, sulIII)*, were detected in raw influent and final effluent samples from both WWTPs. The results of the comparative quantitative qPCR-based analyses in most cases showed the enrichment of the selected ARGs after the wastewater treatment processes (more than a 10-fold increase in five of the studied resistance genes was observed in the final effluent of a conventional WWTP). The results of this research article allowed the authors to estimate the scale of ARG spread in the environment, depending on the size and type of WWTP system, and highlight the need to implement high-efficiency preventive actions.

Pieczyńska et al. [7] have investigated the degradation of cytostatic drugs (CD), 5-Fuorouracil (5-FU), cyclophosphamide (CP) and ifosfamide (IF) and their mixtures, using an electrochemical filter press cell divided by a Nafion membrane in batch treatment (flow recirculation). The order of the CD degradation rate in single drug solutions and in mixtures was found to be 5-FU < CP < IF. The fundamental reaction mechanism, as well as the effects of natural water constituents on the kinetics and mechanisms of electrochemical oxidation of cytostatic drugs in their mixtures, were studied.

Two research articles describe the occurrence of pharmaceuticals in soil samples and evaluate their mobility and toxicity on environmental organisms [8,9].

Wychodnik et al. [8] established the influence of mass breeding of hens on soil contamination with 26 pharmaceuticals and caffeine (CAF). The results showed that the observed changes in pharmaceutical presence in the analyzed soil samples could be defined as seasonal (in all summer samples, less substances (four pharmaceuticals) were determined in contrast with samples collected in March 2019 (10 pharmaceuticals and CAF)). Moreover, concentration levels of sulfamethazine and sulfanilamide in samples collected in July 2019 were approximately five times higher than those collected in March 2019. The antibiotic resistance of 85 random bacterial strains isolated from soil samples was also determined. The level of bacterial resistance to antibiotics did not differ between the samples from intensive breeding farm surroundings and those from the reference area.

The soil behavior of the veterinary drugs, lincomycin, monensin and roxarsone, and their toxicity on environmental organisms (algae, plants, daphnia, fish, earthworms and quails) have been investigated by Li et al. [9]. Lincomycin and roxarsone were characterized by moderate soil mobility; however, roxarsone's ecotoxicity implied that it is a potential ecological risk. Monensin was the most toxic among the three drugs tested, and its higher affinity for soil made it easier to be accumulated.

Apart from research papers, three interesting review articles [10–12] have been published in this Special Issue. Two of them were written by Pereira et al. [10,11]. The first [10] tackled the source, fate and occurrence of 22 pharmaceuticals, six metabolites and transformation products belonging to seven therapeutic groups, in several aquatic compartments (wastewater influents and effluents, surface waters, groundwaters, seawaters, mineral waters and drinking waters). The second article [11] presents the issues of toxicity and the risk assessment of pharmaceuticals, using the occurrence data obtained in the first paper, highlighting and updating the current knowledge on this subject. Such an approach led to the integration of all of these issues under the scope of the environmental risk assessment, providing new and updated data not only on occurrence and toxicity but also relating to the risk assessment of human pharmaceuticals in the most important water compartments.

The last paper of this Special Issue, written by Treder et al. [12], deals with the use of ionic liquids (ILs) in liquid chromatography, gas chromatography and capillary electrophoresis for the determination of pharmaceuticals in environmental and biological matrices. Based on the large number of reported references, these compounds are very effective for the analysis of different classes of compounds. In addition, they are eco-friendly and therefore very useful in the analytical and preparative fields. However, the limitations that appear during their use show that success in experiments is not easy and this field of research requires further development.

In conclusion, the frequent detection of many pharmaceuticals in the environment has been an increasing concern due to their potential to cause undesirable ecological effects, which may range from endocrine disruption in fish and wildlife to antibiotic resistance in pathogenic bacteria.

Funding: This research received no external funding.

Acknowledgments: The Guest Editor wishes to thank all the authors for their contributions to this Special Issue, all the reviewers for their work in evaluating the submitted articles and the editorial staff of *Molecules*, especially Mireia Guardingo and Emity Wang, the Assistant Editors of this journal, for their kind help in making this Special Issue.

Conflicts of Interest: The authors declare no conflict of interest.

References

1. Giebułtowicz, J.; Nałęcz-Jawecki, G.; Harnisz, M.; Kucharski, D.; Korzeniewska, E.; Płaza, G. Environmental Risk and Risk of Resistance Selection Due to Antimicrobials' Occurrence in Two Polish Wastewater Treatment Plants and Receiving Surface Water. *Molecules* **2020**, *25*, 1470. [CrossRef] [PubMed]
2. Guedes-Alonso, R.; Montesdeoca-Esponda, S.; Pacheco-Juárez, J.; Sosa-Ferrera, Z.; Santana- Rodríguez, J.J. A Survey of the Presence of Pharmaceutical Residues in Wastewaters. Evaluation of Their Removal using Conventional and Natural Treatment Procedures. *Molecules* **2020**, *25*, 1639. [CrossRef] [PubMed]
3. Zhang, S.; Lu, Y.-X.; Zhang, J.-J.; Liu, S.; Song, H.-L.; Yang, X.-L. Constructed Wetland Revealed Efficient Sulfamethoxazole Removal but Enhanced the Spread of Antibiotic Resistance Genes. *Molecules* **2020**, *25*, 834. [CrossRef] [PubMed]
4. Wolecki, D.; Caban, M.; Pazda, M.; Stepnowski, P.; Kumirska, J. Evaluation of the Possibility of Using Hydroponic Cultivations for the Removal of Pharmaceuticals and Endocrine Disrupting Compounds in Municipal Sewage Treatment Plants. *Molecules* **2020**, *25*, 162. [CrossRef] [PubMed]

5. Nałęcz-Jawecki, G.; Wawryniuk, M.; Giebułtowicz, J.; Olkowski, A.; Drobniewska, A. Influence of Selected Antidepressants on the Ciliated Protozoan Spirostomum ambiguum: Toxicity, Bioaccumulation, and Biotransformation Products. *Molecules* **2020**, *25*, 1476. [CrossRef] [PubMed]
6. Pazda, M.; Rybicka-Misiejko, M.; Stolte, S.; Bielawski, K.; Stepnowski, P.; Kumirska, J.; Wolecki, D.; Mulkiewicz, E. Identification of selected antibiotic resistance genes in two different wastewater treatment plant systems in Poland: A preliminary study. *Molecules* **2020**, *25*, 2851. [CrossRef] [PubMed]
7. Pieczyńska, A.; Andres Ochoa-Chavez, S.; Wilczewska, P.; Bielicka-Giełdoń, A.; Siedlecka, E.M. Insights into Mechanisms of Electrochemical Drug Degradation in Their Mixtures in the Split-Flow Reactor. *Molecules* **2019**, *24*, 4356. [CrossRef] [PubMed]
8. Wychodnik, K.; Gałęzowska, G.; Rogowska, J.; Potrykus, M.; Plenis, A.; Wolska, L. Poultry Farms as a Potential Source of Environmental Pollution by Pharmaceuticals. *Molecules* **2020**, *25*, 1031. [CrossRef] [PubMed]
9. Li, P.; Wu, Y.; Wang, Y.; Qiu, J.; Li, Y. Soil Behaviour of the Veterinary Drugs Lincomycin, Monensin, and Roxarsone and Their Toxicity on Environmental Organisms. *Molecules* **2019**, *24*, 4465. [CrossRef] [PubMed]
10. Pereira, A.; Silva, L.; Laranjeiro, C.; Lino, C.; Pena, A. Selected Pharmaceuticals in Different Aquatic Compartments: Part I—Source, Fate and Occurrence. *Molecules* **2020**, *25*, 1026. [CrossRef] [PubMed]
11. Pereira, A.; Silva, L.; Laranjeiro, C.; Lino, C.; Pena, A. Selected Pharmaceuticals in Different Aquatic Compartments: Part II—Toxicity and Environmental Risk Assessment. *Molecules* **2020**, *25*, 1796. [CrossRef] [PubMed]
12. Treder, N.; Bączek, T.; Wychodnik, K.; Rogowska, J.; Wolska, L.; Plenis, A. The Influence of Ionic Liquids on the Effectiveness of Analytical Methods Used in the Monitoring of Human and Veterinary Pharmaceuticals in Biological and Environmental Samples—Trends and Perspectives. *Molecules* **2020**, *25*, 286. [CrossRef] [PubMed]

Article

Environmental Risk and Risk of Resistance Selection Due to Antimicrobials' Occurrence in Two Polish Wastewater Treatment Plants and Receiving Surface Water

Joanna Giebułtowicz [1], Grzegorz Nałęcz-Jawecki [2], Monika Harnisz [3], Dawid Kucharski [1], Ewa Korzeniewska [3] and Grażyna Płaza [4,*]

1. Department of Bioanalysis and Drugs Analysis, Faculty of Pharmacy, Medical University of Warsaw, 1 Banacha, 02-097 Warszawa, Poland; jgiebultowicz@wum.edu.pl (J.G.); dawid.kucharski@wum.edu.pl (D.K.)
2. Department of Environmental Health Sciences, Faculty of Pharmacy, Medical University of Warsaw, 1 Banacha, 02-097 Warszawa, Poland; grzegorz.nalecz-jawecki@wum.edu.pl
3. Department of Environmental Microbiology, Faculty of Environmental Sciences, University of Warmia and Mazury, 5 Oczapowskiego, 10-719 Olsztyn, Poland; monikah@uwm.edu.pl (M.H.); ewa.korzeniewska@uwm.edu.pl (E.K.)
4. Microbiology Unit, Institute for Ecology of Industrial Areas, 6 Kossutha, 40-844 Katowice, Poland

* Correspondence: g.plaza@ietu.pl; Tel.: +48 322546031

Academic Editors: Jolanta Kumirska and Teresa A. P. Rocha-Santos
Received: 21 February 2020; Accepted: 21 March 2020; Published: 24 March 2020

Abstract: In this study, a screening of 26 selected antimicrobials using liquid chromatography coupled to a tandem mass spectrometry method in two Polish wastewater treatment plants and their receiving surface waters was provided. The highest average concentrations of metronidazole (7400 ng/L), ciprofloxacin (4300 ng/L), vancomycin (3200 ng/L), and sulfamethoxazole (3000 ng/L) were observed in influent of WWTP2. Ciprofloxacin and sulfamethoxazole were the most dominant antimicrobials in influent and effluent of both WWTPs. In the sludge samples the highest mean concentrations were found for ciprofloxacin (up to 28 μg/g) and norfloxacin (up to 5.3 μg/g). The removal efficiency of tested antimicrobials was found to be more than 50% for both WWTPs. However, the presence of antimicrobials influenced their concentrations in the receiving waters. The highest antimicrobial resistance risk was estimated in influent of WWTPs for azithromycin, ciprofloxacin, clarithromycin, metronidazole, and trimethoprim and in the sludge samples for the following antimicrobials: azithromycin, ciprofloxacin, clarithromycin, norfloxacin, trimethoprim, ofloxacin, and tetracycline. The high environmental risk for exposure to azithromycin, clarithromycin, and sulfamethoxazole to both cyanobacteria and eukaryotic species in effluents and/or receiving water was noted. Following the obtained results, we suggest extending the watch list of the Water Framework Directive for Union-wide monitoring with sulfamethoxazole.

Keywords: antibiotics; wastewater; sewage sludge; risk assessment; removal efficiency; LC-MS/MS analysis

---

1. Introduction

The fate of contaminants, particularly pharmaceutically active compounds (PhACs) in the environment is receiving considerable attention from researchers. PhACs appear as contaminants in wastewater, soil, surface and ground water, municipal sewage, and in the influents and effluents of wastewater treatment plants [1–3]. There are several sources of PhACs in the environment. The most important is human and veterinary medicine as well as plant agriculture. The main sources of aquatic contamination with human antimicrobials are wastewater treatment plants (WWTPs). The PhACs enter WWTPs along with wastewater from the disposal of unused or expired drugs in toilets. However, human excretion is considered to be the most important source. Generally, WWTPs are not designed to eliminate PhACs during the technological process, and a number of studies have shown the presence of different PhACs in both raw and treated sewage sludge and wastewater [4–8]. There is no data on either the removal efficiency or the concentration of antimicrobials in Polish WWTPs. The concentration of PhACs in the environment depends on the consumption of pharmaceuticals, which is country- and culture-specific, and their pharmacokinetics, and may considerably vary with seasons and physicochemical properties of these compounds, various process operating parameters of WWTPs, and bacterial community structure [9,10]. According to the European Centre for Disease Prevention, in 2018 the corresponding population-weighted mean consumption of antimicrobials (in defined daily dose (DDD) units per 1000 inhabitants per day) in European Union and European Economic Area countries was 18.4 DDD. In Poland the consumption rate was calculated as 23 DDD. Higher values were observed only for France (23.6 DDD), Greece (32.4 DDD), Romania (25.0 DDD) and Spain (24.3 DDD) [11].

Antimicrobials are one of the most extensively investigated PhACs. They belong to contaminants of emerging concern (CEC), which helps assess hazards to human health and ecosystems. They are one of the most popular pharmaceuticals used in veterinary care, farming, and medicine. According to the United States Geological Survey (USGS), CEC includes "any synthetic or naturally occurring chemical or any microorganism that is not commonly monitored in the environment, but has the potential to enter the environment and cause known or suspected adverse ecological and/or human health effects" [12].

WWTPs are not specifically designed for antimicrobial removal, and, consequently, these molecules are released directly into the receiving environment. An important issue is to identify the sources of antimicrobials in water and to assess their concentrations in surface, ground, and potable waters. The presence of antimicrobials in surface and ground waters, and even in drinking water, has been identified worldwide, for example in the UK [13], Italy [14], China [15], Australia [16], and the USA [17]. In our previous study, 20 of the 26 investigated antimicrobials up to a concentration of 1000 ng/L in the river water close to the effluent discharge from the main WWTP in Warsaw (Poland) were detected [18]. Although WWTPs are considered the main source of antimicrobials for surface waters, the current legislation at a European level does not contain an antimicrobial concentration requirement for discharge from WWTPs to receiving water. Antimicrobials have been determined in numerous WWTPs such as those in Germany [8,19], France [7], Croatia [20], Spain [21], China [22], Switzerland [23], Sweden [24], and Norway [25]. Given the number of scientific papers regarding the analysis of the antimicrobials' concentrations in European and global WWTPs, the knowledge of the occurrence of antimicrobials in Polish WWTPs is scare. Moreover, there are only scarce data on risk assessment on resistance selection and on environmental toxicity in WWTPs. To our best knowledge, no such data regarding sludge and sludge-affected soils exists.

The presence of the antimicrobials in the environment may pose an environmental risk. Environmental risk is defined as actual or potential threat of adverse effects on aquatic and/or terrestrial organisms. In the case of antimicrobials, the most endangered are prokaryotes, e.g., nitrification bacteria [26] or cyanobacteria [27]. Antimicrobials can also pose a risk of resistance selection. It is observed as preferential outgrowth of antimicrobial-resistant bacteria and changes in the antibiotic sensitivity of the entire microbial population in antimicrobial concentrations below the minimal inhibitory concentration. As a consequence, the resistant bacteria is able to survive in the presence of an antimicrobial in concentration that is usually sufficient to inhibit or kill microorganisms of the same species [28]. The antimicrobial-resistant genes can be transferred between distantly related bacterial species and to bacteria that colonize the human body and human pathogens [26]. The estimates suggest that 700,000 deaths occur every year because of antimicrobial resistance; moreover, by 2050, there might be 10 million deaths every year [29].

In this context, the aim of the study was to investigate the occurrence, abundance, and removal efficiency of the selected antimicrobials in two Polish wastewater treatment plants. The risk assessment approach based on environmental risk quotients (RQs) was also calculated to assess antimicrobial resistance risks and ecological environmental risk of antimicrobials to cyanobacteria and eukaryotic species. The antimicrobials were selected based on the sales data and occurrence of the antimicrobials in the environment in Europe. Additionally, the satisfactory performance of the analytical method was taken into account.

2. Results

2.1. *Predicted Concentrations of Antimicrobials in WWTPs*

According to sales data for 2018 (published by the Polish National Health Fund (NFZ)), among detected antimicrobials, the highest sale in Poland was noted for clarithromycin (up to 1509 kg/month), sulfamethoxazole (up to 1358 kg/month), ofloxacin (up to 829 kg/month), ciprofloxacin (up to 821 kg/month), and clindamycin (737 kg/month). We compared the predicted load of antimicrobials in the WWTP1 (PLoad) (Poland) with the load calculated based on the measured concentrations of the drugs in the influent (water phase) and in the primary sludge ($Load_{W+S}$) (Table A1, Online Resource). For most of the tested antimicrobials, the $Load_{W+S}$ to PLoad ratio was low (up to 20%) because the high metabolism in the human body results in a lower level of parent compound, e.g., the biotransformation ratio of clindamycin is 85% [30]. Moreover, the measured load of fluoroquinolones (ciprofloxacin and norfloxacin) was very close to that of the predicted load, and unlike other antimicrobials that are primarily present in the water phase, ciprofloxacin and norfloxacin are distributed evenly between water and the primary sludge. For six antimicrobials (erythromycin, metronidazole, oxytetracycline, tetracycline, sulfathiazole, and vancomycin), the $Load_{W+S}$ to PLoad ratio was very high and exceeded 1000% because of the low and very low predicted load value. Note that almost all antimicrobials in Poland are available by prescription and most of them are reimbursed; however, some, such as vancomycin, are primarily used in hospitals, while tetracyclines (oxytetracycline and tetracycline) and sulfathiazole are primarily used in veterinary medicine and are not reported by NFZ.

2.2. *Antimicrobial Concentrations in Influents of WWTPs*

The concentrations of 26 antimicrobials in influents from WWTP1 and WWTP2 and receiving waterbodies are shown in Table 1. The significant differences between the concentrations of antimicrobials in the samples collected in different sampling periods were observed. The differences were due to seasonal variations in antimicrobial use. According to National Health Fund (NFZ) database, in summer the consumption of antimicrobials was low (Table A1). In autumn and winter, the consumption increased significantly, probably due to numerous infections occurring each year in that period [31].

Table 1. Mean and standard deviation (n = 3) of the target antimicrobial concentrations (ng/L) in influent, effluent, and receiving water of two wastewater treatment plants (WWTPs) located in Poland at Silesian (WWTP1) and Warmian-Masurian Voivodship (WWTP2).

	WWTP1 Influent		WWTP1 Effluent		WWTP1 Leachate		River1 Upstream		River1 Downstream		WWTP2 Influent		WWTP2 Effluent		River2 Upstream		River2 Downstream		MDL [1]	MQL [1]
	Mean	SD	Mean	SD	Mean	SD	Mean	SD	Mean	SD	Mean	SD	Mean	SD	Mean	SD	Mean	SD		
AZM	87	71	230	110	320	290	25	15	441	249	360	450	650	680	5.2	2.7	36	21	10	33
CIP	1260	680	184	72	890	410	108	33	95	4	4300	4300	312	73	12	12	182	182	2.4	8.1
CLR	480	190	160	170	102	23	37	9	79	50	560	590	143	40	12.2	8.2	20.5	11.1	0.3	0.9
CLI	134	87	166	60	73	37	78	46	134	3	106	51	290	200	2.3	1.1	25.4	3.5	2.9	9.5
ERY	58	71	21	18	30	42	7	7	10	0	28	20	16	12	<MDL		<MDL		0.7	2.4
LCM	20	15	48	52	24.7	7.8	3	3	9	4	102	46	56	19	<MDL		4.0	1.4	1.4	4.7
MTZ	250	160	69	82	11.1	8.4	9	3	21	11	7400	9600	88	41	<MDL		9.4	3.1	2.4	7.9
NOR	240	130	31	28	210	150	95	93	<MDL		80	110	10	11	<MDL		<MDL		6.3	21
OFX	135	35	26	14	200	91	32	18	4	4	195	21	40	11	8.4	0.5	31	23	1.4	4.6
OTC	0.7	0.9	0.1	0.1	1.1	1.4	<MDL		1	0	<MDL		<MDL		<MDL		<MDL		0.1	0.2
RIF	5.2	3.3	<MDL		<MDL		<MDL		<MDL		5.3	3.4	2.9	2	<MDL		<MDL		2.9	9.6
ROX	18	19	6.6	6.9	1.6	1.6	4	4	5	2	6.2	5.3	6.2	2.1	<MDL		<MDL		0.5	1.6
SD	<MDL		3.4	4.3	2.4	2.5	<MDL		<MDL		<MDL		<MDL		<MDL		<MDL		0.6	1.9
SDM	4.1	3.2	3.4	3.5	<MDL		<MDL		<MDL		8.9	7.2	4.9	3.6	<MDL		<MDL		1.8	6.1
SXT	1300	460	630	220	480	630	644	41	451	95	3000	1900	770	280	<MDL		76.1	4.6	5.9	19
ST	94	46	21	14	136	54	7	6	29	19	180	110	36	16	<MDL		<MDL		2.4	8
TET	190	190	39	55	180	170	<MDL		0	0	210	160	61	48	<MDL		7.2	6.4	0.2	0.7
TBZ	18.4	2.8	22.3	4.5	18	11	12	7	16	0	11	2.8	25.5	3	<MDL		4.3	2.1	2.8	9.4
TMP	254	41	160	190	94	52	38	3	36	8	900	770	220	110	9.1	8.4	24.4	3.9	3	9.9
VAN	350	390	114	60	840	220	27	23	62	8	3200	3600	162	62	<MDL		10.7	3.2	15	50

CFR (270 ng/L), FLRX (3.1 ng/L), LOM (0.9 ng/L), NAL (6.1 ng/L), PEF (12 ng/L), SDD (3.7 ng/L) were not detected. Their MDLs (method detection limits) are provided in parentheses. [1] Presented MDLs and method quantitation limits (MQLs) were calculated for influents. The values for effluents are about two times lower and for surface water about four times lower.

Nineteen and 18 out of 26 analyzed antimicrobials were detected in the wastewater from WWTP1 and WWTP2, respectively. In most cases, the antimicrobial levels were higher in wastewater of WWTP2 than in wastewater of WWTP1. The mean values of antimicrobials' concentration in analyzed samples ranged from <MDL (method detection limit) to 7400 ng/L. In influent from WWTP1, the average concentrations of two antimicrobials, i.e., ciprofloxacin and sulfamethoxazole, were higher than 1000 ng/L and concentrations of 10 antimicrobials was higher than 100 ng/L. While in influent from WWTP2, the concentration of four antimicrobials, i.e., ciprofloxacin, metronidazole, sulfamethoxazole, and vancomycin, exceeded the level of 1000 ng/L and concentrations of 12 antimicrobials exceeded the level of 100 ng/L. The highest average concentrations of antimicrobials in influent from WWTP2 were recorded for metronidazole (7400 ng/L), followed by ciprofloxacin (4300 ng/L), vancomycin (u 3200 ng/L), and sulfamethoxazole (3000 ng/L). In other countries, the concentration of metronidazole was lower than 1000 ng/L, that of ciprofloxacin was up to 3800 ng/L [32] but frequently below 400 ng/L [33], and that of sulfamethoxazole was up to 7900 ng/L [34]. There are not much data on vancomycin occurrence in the environment, and it has not been detected in very high concentrations to date as it is primarily used intravenously to treat severe infections in hospitals [35]. The differences in the concentrations of antimicrobials in the influent between two WWTPs can be due to various consumption of antimicrobials in the sampling period and various sources of antimicrobials in these regions. As an example, near WWTP1 is a hospital (449 beds), whereas near WWTP2 is a hospital (458 beds), poultry plant, and galenic laboratory.

2.3. Antimicrobial Concentrations in Effluent of WWTPs

The concentrations of 26 antimicrobials in effluents from WWTP1 and WWTP2 are shown in Table 1. In effluents from both WWTPs, the highest average concentrations were observed for azithromycin (up to 650 ng/L), sulfamethoxazole (up to 770 ng/L), ciprofloxacin (up to 312 ng/L), and clindamycin (up to 290 ng/L). In effluent from WWTP2, a higher mean concentration of azithromycin but lower of metronidazole was observed. The percentage contribution of the analyzed antimicrobials in influent and effluent of both WWTPs are presented in Figure 1. To summarize, among antimicrobials tested, ciprofloxacin and sulfamethoxazole were the most dominant antimicrobials in influent and effluent from both WWTPs.

Comparing our data from the literature, the results obtained were similar. The concentration of azithromycin was up to 380 ng/L [36] in Switzerland, that of sulfamethoxazole was up to 1300 ng/L in Spain [21], and that of clindamycin was up to 5 ng/L in Australia [32].

2.4. Antimicrobial Concentrations in Receiving Waters

In order to assess the impact of antimicrobials on the receiving water bodies, samples were taken upstream and downstream from the WWTPs' discharge (see Section 3.1). The concentrations of 26 antimicrobials in receiving water are shown in Table 1. The river upstream of WWTP1 was more polluted with analyzed antimicrobials than WWTP2 upstream river (Table 1). The highest concentration of sulfamethoxazole (average value: 644 ng/L) was detected upstream of WWTP1. While in the upstream of WWTP2, the highest concentration of clarithromycin (average value: 12.2 ng/L) was detected. The wastewater discharge from WWTP2 resulted in a significant increase of antimicrobials' concentration in the river. This fact can be explained by the high concentrations of antimicrobials in the case of effluent from WWTP2 (Table 1). In the case of WWTP1, taking into account all antimicrobials, no statistical differences between upstream and downstream were observed. Despite this, the average concentration of azithromycin in WWTP1 downstream was more than 15 times higher compared to the concentration in the upstream. The concentrations of antimicrobials in the receiving waters depend on the concentration of antimicrobials in effluent, distance of sampling, or the different flow rate. In our study, WWTP2 effluent effect on antimicrobials' concentrations in the receiving surface water was observed.

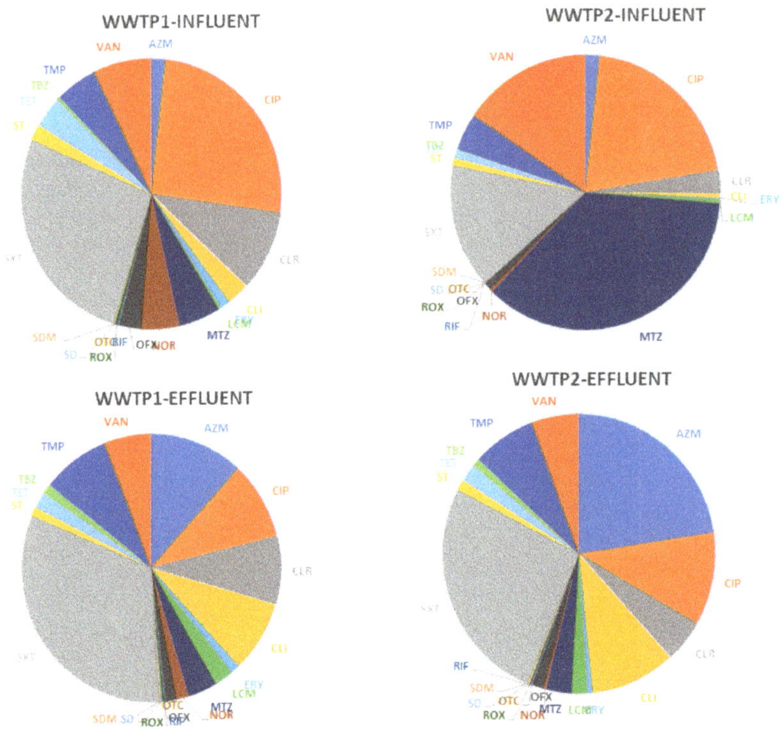

Figure 1. Percentage contribution (%) of tested antimicrobials in influent and effluent of both WWTPs, $n = 3$.

2.5. *Antimicrobials' Removal Efficiency*

The removal efficiency (calculated based on equations described in Section 3.3) of tested antimicrobials was similar for both WWTPs ($p > 0.05$). The average removal efficiency was above 50% for 12 and 10 out of 20 detected antimicrobials for WWTP1 and WWTP2, respectively (Figure 2). Among the antimicrobials from the sulfonamide group, four were detected in effluents and their removal rate ranged from 17% to 80% with an average of 52%. Certain inconsistencies exist in literature about sulfonamide removal, as per the review by Le-Minh et al. [37]. Some researchers have reported an effective removal of sulfonamide [38], although others have mentioned the opposite results [39]. Based on the literature data, this fact might be explained by the differences in operational conditions of each WWTP.

The removal efficiency of trimethoprim was 37% in WWTP1 and 76% in WWTP2 ($p = 0.2$) (the treatment process used in the WWTPs are presented in Section 3.1). Regarding fluoroquinolones, three of them, i.e., norfloxacin, ciprofloxacin, and ofloxacin, were detected in influent of both WWTPs with the average removal being above 80%. The plausible mechanism for the removal of fluoroquinolones from water is sorption to sediment because their concentration was detected as very high in the sludge [40–42] and a slow biodegradation rate was reported [43]. The average removal efficiency for fluoroquinolone antimicrobials estimated by Wang et al. [44] was about 50%. Rodayan et al. [45] reported values of around 60%, while for WWTPs in Switzerland, higher removal efficiencies were observed, reaching up to 87% (for norfloxacin) [46]. In the study of Gao et al. [47], the removal efficiencies of WWTPs for fluoroquinolone antimicrobials ranged from 48% to 72%.

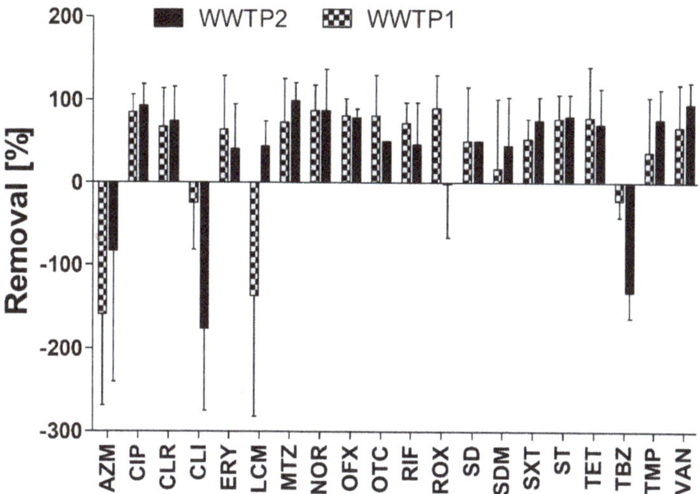

Figure 2. Removal efficiency (%) of selected antimicrobials in WWTPs calculated based on equations described in Section 3.3.

In this study, a negative removal rate for azithromycin, clindamycin, lincomycin, and thiabendazole was observed (Figure 2). Moreover, for these antimicrobials, a high variance in removal rate was noted, which agrees with previous studies in which certain antimicrobials were reported to be more abundant in effluents than influents [6,8,10,25,33,48].

The negative removal rate has been detected in the literature. For example, Chunhui et al. [49] reported that the removal efficiencies for macrolide antimicrobials in WWTPs was –4%. Similarly, Gao et al. [47] reported that the removal rate of macrolide antimicrobials was higher, and reached from –34% to 69%. The negative removal rate is explained in many ways. First, certain metabolites can return to the parent pharmaceutical during primary and secondary treatment because of glucuronide conjugate hydrolytic cleavage. Second, pharmaceuticals that sorb to organic matter and particles, as well as accumulate in sediments and biofilms, might be resuspended in wastewater during storm water events. Typically, deconjugation or resorption from particles is observed for fluoroquinolones (ciprofloxacin, ofloxacin, and norfloxacin), lincomycin, tetracycline, trimethoprim, and sulfamethoxazole. This issue is particularly relevant to pharmaceuticals that are mainly excreted with bile and feces-like macrolides. During wastewater treatment, they are redissolved, and, therefore, their concentrations in water increase [10,33].

HRT (hydraulic retention time) is one of the major factors influencing the antimicrobial removal efficiency, particularly for compounds that are readily biodegradable and have a low K_d (low tendency to absorb to sludge) [50]. In this study, HRT was 12 h in WWTP1 and 24 h in WWTP2. Other factors influencing the removal efficiency are adsorption coefficients and persistence of the antimicrobials. Lower levels in effluents could be also interpreted as the removal of antimicrobials because of biodegradation and/or chemical and physical transformations, e.g., hydrolysis or sorption to solid matter. Biodegradation/biotransformation and sorption are the two primary mechanisms occurring in the WWTPs. The sorption to sewage sludge is the primary removal mechanism for certain antimicrobials in the samples, e.g., the percentage of the daily load of antimicrobials was >40% for azithromycin, ciprofloxacin, ofloxacin, and oxytetracycline in sludge (Table 2), while the lowest value was observed for sulfamethoxazole and lincomycin. The differences between the sorption of antimicrobials depend on their physicochemical properties, such as charge and lipophilicity, the properties of the sludge-like chemical nature, and other factors such as pH and redox potential [51]. Sorption may occur by hydrophobic interactions with lipophilic cell membranes of the microorganisms or the lipid fractions of the suspended solids and electrostatic interactions of positively charged groups of chemicals with negatively charged surfaces of microorganisms or other components of the sludge [33]. Figure 3 shows differences between the concentrations of antimicrobials in wastewater and sludge (separation observed by principal component 1 (PC1) accounted for 45% of total variance). Two groups were designated by the various amounts of oxytetracycline and norfloxacin in the samples. The relative percentage estimated for the antimicrobial was significantly higher in the sludge than in the wastewater. In wastewater, the relative percentage of metronidazole, lincomycin, and sulfamethoxazole was higher compared to that in the sludge. Similarly, a higher contribution of metronidazole and erythromycin in wastewater and ciprofloxacin in sludge was observed by Ostman et al. [24].

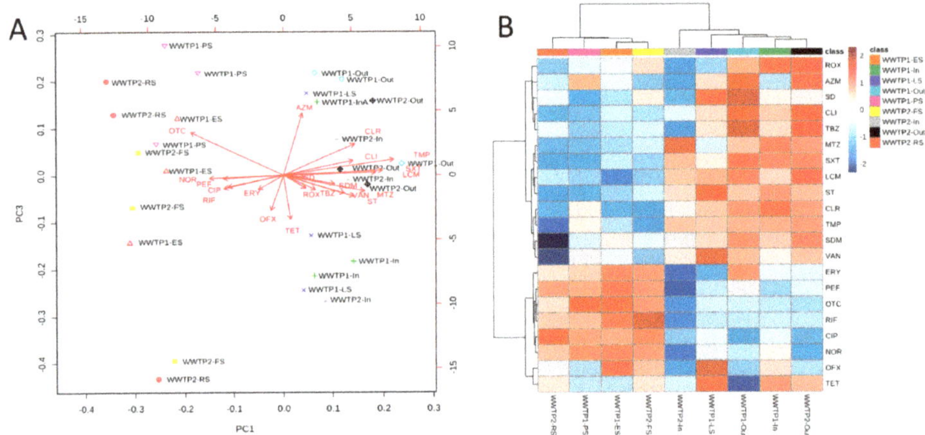

Figure 3. Differences in antimicrobials' concentrations between the wastewater, sludge samples, and leachate collected from WWTP1 and WWTP2. Principal component analysis (PCA) of the all samples (**A**). Variables of a data matrix are represented as arrows. Visualization of the observed relations on the heat map (**B**). ES, excessive sludge; FS, fermented sludge; In, influent; LS, leachate; Out, effluent; PS, primary sludge; RS, residual sludge.

Our results showed that leachate from WWTP1 contained a significant amount of antimicrobials. The mean concentrations of nine antimicrobials were above 100 ng/L (Table 1). The concentrations of azithromycin, ciprofloxacin, norfloxacin, ofloxacin, sulfamethoxazole, and vancomycin were comparable to those in the WWTP1 influent. The results obtained suggest that the antimicrobials are strongly sorbed to solids (Table A1). The antimicrobials bound to solids can be leached to the environment, which should be taken into consideration when conducting the risk assessment. In the Table A1, the K_d values for tested antimicrobials are presented. Generally, K_d is used to estimate the mobility and distribution of the pharmaceuticals in the environment [41,52]. Compounds with low K_d are potentially mobile from soil to the water (through leaching or runoff). According to our study, the most mobile antimicrobials are trimethoprim (log K_d = 5.8), sulfamethoxazole (log K_d = 5.1), lincomycin (log K_d = 5.2), clarithromycin (log K_d = 5.7), clindamycin (log K_d = 5.7), and thiabendazole (log K_d = 5.9).

2.6. Antimicrobials' Concentrations in Sewage Sludge from WWTPs

Fifteen antimicrobials in the primary (WWTP1-PS) and excessive sludge (WWTP1-ES) from WWTP1 as well as in the fermented (WWTP2-FS) and residue sludge (WWTP2-RS) from WWTP2 were detected (Table 2). Results obtained indicated that a part of the antimicrobials were adsorbed to the sludge along the whole technological process.

Table 2. Mean and standard deviation ($n = 3$) of target antimicrobial concentration (ng/g dry weight) in sewage sludge samples in two wastewater treatment plants (WWTPs) located in Poland at Silesian (WWTP1) and Warmian-Masurian Voivodship (WWTP2).

	WWTP1-PS		WWTP1-ES		WWTP2-RS		WWTP2-FS		WWTP1-RS [1]		MDL	MQL
	Mean	SD	Mean	SD	Mean	SD	Mean	SD	Mean	SD		
AZM	1260	690	370	280	82	75	1040	940	100	110	23	79
CIP	12500	6900	6200	2000	6100	5500	28000	22000	7800	4400	490	1670
CLR	289	52	18	14	18	23	150	180	3.0	0.8	2.2	7.4
CLI	40	25	58	35	29	29	47	30	9.1	5.6	1.3	4.5
LCM	4.3	3.2	1.5	2.0	2.5	2.6	19	24	0.13	0.05	0.2	0.6
NOR	5300	3200	3600	2500	1600	1400	3400	2300	<MDL		370	1270
OFX	810	890	640	510	300	190	480	120	36	21	12	41
OTC	590	410	320	230	50	35	190	200	4.0	6.5	0.2	0.7
PEF	250	150	190	110	140	120	256	81	<MDL		29	100
ROX	11.4	7.8	<MDL		<MDL		<MDL		0.22	0.26	12	40
SD	<MDL		<MDL		1.1	0.8	7.7	11	0.14	0.18	1.1	3.7
SDM	12	14	13	14	7.5	11	<MDL		0.29	-	0.6	2.0
SXT	<MDL		81	54	<MDL		<MDL		5.9	9.4	41	140
TET	82	50	40	13	58	29	201	95	25	29	0.7	2.5
TBZ	19.5	5.2	15.6	5.8	43	43	49	22	8.3	6.2	0.6	2.2
TMP	170	110	5.7	4.4	<MDL		<MDL		2.6	1.7	0.5	1.5
VAN	363	55	181	91	36	30	<MDL		370	130	32	110

CFR (350 ng/g), FLRX (12 ng/g), LOM (7.9 ng/g), NAL (3.3 ng/g), MTZ (25 ng/g), RIF (150 ng/g), ST (30 ng/g), SDD (39 ng/g) were not detected. Their MDLs (method detection limits) are provided in parentheses; MQL, method quantitation limit; WWTP-PS, primary sludge; WWTP-ES, excessive sludge; WWTP-FS, fermented sludge, WWTP-RS residual sludge; [1]calculated based to the lowest K_d observed and concentration of antimicrobials in leachate.

In our study the highest concentrations for fluoroquinolones, such as ciprofloxacin (up to 57 μg/g), norfloxacin (up to 9.5 μg/g), ofloxacin (up to 2.0 μg/g), and azithromycin (up to 2.3 μg/g), were determined in the sludge samples (data not presented). In China, the concentration of norfloxacin, ofloxacin, and ciprofloxacin in sludge was determined from all the major provincial cities, ranging ranged from 0.1 to 15.7 μg/g, from 0.3 to 7.9 μg/g, and from 0.1 to 4.7 μg/g, respectively [53]. In Switzerland (up to 3.3 μg/g) and Sweden (up to 4.2 μg/g), the maximum concentration of norfloxacin was lower than that in China and presented in Poland. Furthermore, the maximum level of other fluoroquinolones was higher in Poland than in Switzerland (up to 0.9 μg/g for ciprofloxacin) and Sweden (up to 4.8 μg/g for ciprofloxacin and up to 2.0 μg/g for ofloxacin) [54,55]. The concentration of azithromycin in the activated sludge was up to 0.16 μg/g and 0.056 μg/g in Germany and Switzerland, respectively, and the values were lower compared to those for Poland (up to 0.71 μg/g) [56]. Recently, a study from Germany reported the concentration of azithromycin and ciprofloxacin in sediments to be up to 0.33 μg/g and 0.71 μg/g [41], respectively.

2.7. Antimicrobial Resistance Risk Assessment

The risk factor of antimicrobial resistance selection (named as antimicrobial resistance risk, defined as preferential outgrowth of antimicrobial-resistant bacteria) in wastewater was high for the following antimicrobials: azithromycin, ciprofloxacin, clarithromycin, metronidazole, and trimethoprim (Figure 4A). As expected, this factor was the highest in influent from both WWTPs.

Similar results were obtained by Ostman et al. [24], who observed a high risk of resistance selection in influents in Swedish WWTPs because of the prevalence of ciprofloxacin and metronidazole. No risk for azithromycin was detected in their study.

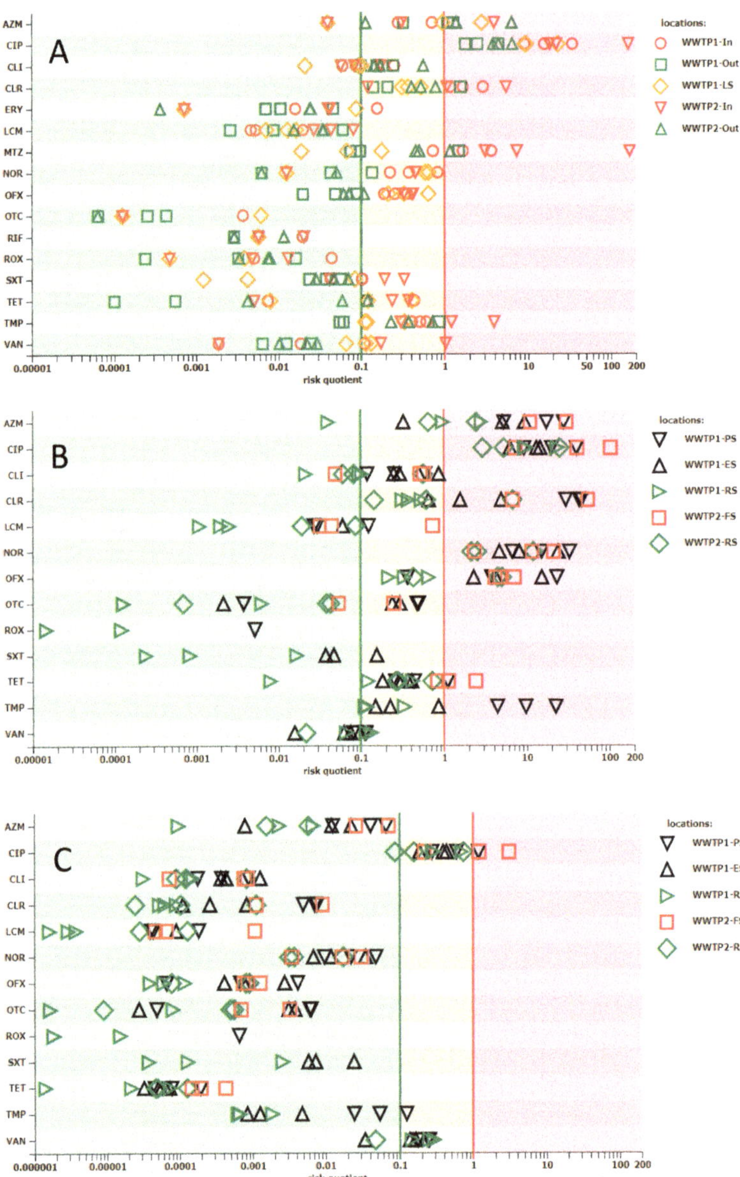

Figure 4. Estimation of risk quotients of resistance selection for the antimicrobials occurring in wastewater (**A**), sludge (**B**), and sludge-amended soil (**C**). Risk quotient below 0.1 indicates minimal risk (green area), between 0.1 and 1 is medium risk (orange area), and over 1 is high risk (red area). ES, excessive sludge; FS, fermented sludge; In, influent; LS, leachate; Out, effluent; PS, primary sludge; RS, residual sludge.

It is interesting, the notable antimicrobial resistance risk was also predicted upstream of both WWTPs. High antimicrobial resistance risk for ciprofloxacin and medium risk for azithromycin, clarithromycin, and clindamycin was observed in the case of WWTP1, while medium antimicrobial resistance risk for ciprofloxacin was noted in the case of WWTP2. Downstream WWTP1, on top of the previous risk, the risk for metronidazole turned medium, whereas the risk for azithromycin rose to high. In the case of WWTP2, the risk for ciprofloxacin rose from medium to high and the medium risk for ofloxacin, clarithromycin, and clindamycin appeared (data not presented).

The antimicrobial resistance risk was high for azithromycin, ciprofloxacin, clarithromycin, norfloxacin, trimethoprim, ofloxacin, and tetracycline in sludge (Figure 4B). Interestingly, residual sludge could pose a risk to the environment, particularly because of the presence of fluoroquinolones and macrolides. Mean PEC_{soil} (predicted environmental concentration in soil) was mainly lower than 1 ng/g and higher concentrations were observed for ciprofloxacin (up to 41 ng/g), norfloxacin (up to 7.8 ng/g), ofloxacin (up to 1.2 ng/g), and azithromycin (up to 2.3 ng/g) (Table 3). When RQ was analyzed, a high risk for soil was noted for ciprofloxacin (Figure 4C). The data on risk assessment on the antimicrobial resistance selection in sludge and sludge-amended soil presented in this paper are the first.

2.8. Environmental Risk Assessment

The high risk was observed for both cyanobacteria (Figure 5A) and eukaryotic species (Figure 5B) due to azithromycin, clarithromycin and sulfamethoxazole in effluents. Furthermore, medium risk was predicted for chronic exposure of cyanobacteria to ciprofloxacin, erythromycin, norfloxacin, and ofloxacin (in effluents). The same level of risk was evaluated for chronic exposure of eukaryotic organisms to ciprofloxacin, erythromycin, and roxithromycin (in effluents). Verlicchi et al. [33] published a review, which reported that six antimicrobials posed a high environmental risk for eukaryotes, i.e., erythromycin, ofloxacin, sulfamethoxazole, clarithromycin, tetracycline, and azithromycin. Harrabi et al. [6] observed a high risk for ofloxacin, ciprofloxacin, azithromycin, sulfamethoxazole, and trimethoprim for eukaryotic species; however, clarithromycin was not evaluated. In their studies, the predicted no effect concentration (PNEC) was, however, calculated differently compared to our study. The authors obtained PNEC values 1000 times lower than the toxicity values found for the most sensitive eukaryotic species that were assayed [6,33]. In our study, other factors, i.e.1000, 100, 50 or10, were used to calculate PNEC. Thus, RQ values obtained for the same concentrations of antimicrobials were lower.

In the case of upstream of WWTP1, there was high risk for cyanobacteria posed by azithromycin, norfloxacin, and sulfamethoxazole, whereas medium risk was observed for ciprofloxacin and clarithromycin. The risk for eukaryote was also observed but the level of the risk was different for the antimicrobials, e.g., for clarithromycin high risk was estimated, and for azithromycin, ciprofloxacin, and sulfamethoxazole medium risk was noted.

None of the antimicrobials posed high risk from upstream of WWTP2. However, medium risk was posed by azithromycin, clarithromycin for cyanobacteria, and medium risk for eukaryote posed by clarithromycin were observed. The medium risk for cyanobacteria posed by ciprofloxacin, ofloxacin, clarithromycin, and sulfamethoxazole in WWTP2 downstream was noted. The risk of toxicity due to azithromycin content increased from medium to high comparing the upstream and downstream. Two more antimicrobials, i.e., azithromycin and ciprofloxacin, were predicted to cause the medium risk for eukaryote.

The RQ calculated for cyanobacteria exposed to sludge was medium for azithromycin (primary sludge up to 0.36, excessive sludge up to 0.12, fermented sludge up to 0.39) and clarithromycin (primary sludge up to 0.13, fermented sludge up to 0.16). No data on risk assessment on antimicrobial presence in sludge was found in literature to compare with our data.

Antimicrobials are designed to act on prokaryotic organisms; thus, environmental bacteria are more likely to be adversely affected compared to other environmental species such as aquatic invertebrates and vertebrates. However, compared to cyanobacteria, certain microalgae and macrophytes are more sensitive to certain antifolate and quinolone antimicrobials [27].

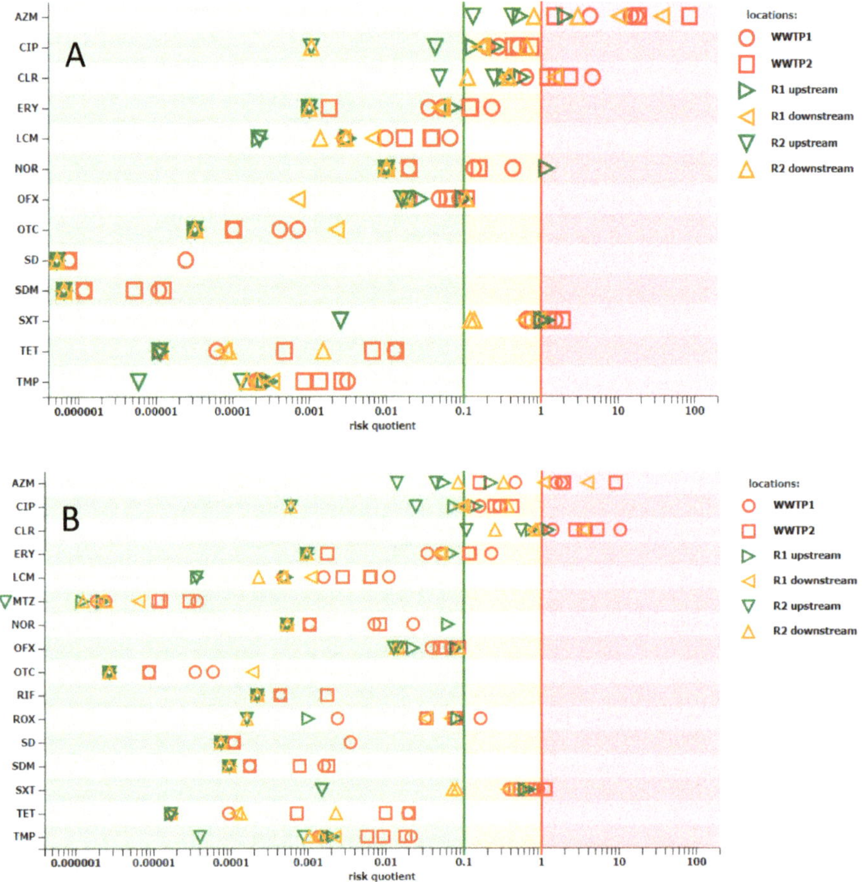

Figure 5. Estimation of risk quotients of chronic exposure of cyanobacteria (**A**) and eukaryotic organisms (**B**) to the antimicrobials occurring in effluents and receiving water. Calculations were based on the measured concentration of antimicrobials in wastewater and PNEC calculated based on [27]. Risk quotient below 0.1 indicates minimal risk (green area), between 0.1 and 1 is medium risk (orange area), and over 1 is high risk (red area).

3. Materials and Methods

3.1. Description of WWTPs and Sample Collection

The raw influent, final effluent, leachate, and sludge samples (i.e., primary sludge and excessive sludge (WWTP1) or fermented sludge and residual sludge (WWTP2)) were collected from two Polish WWTPs during the three sampling campaigns: (1) In June/July 2018, (2) in October 2018, and (3) in December 2018. Simultaneously, the receiving river water samples from upstream and downstream (WWTP1, 0.25 km; WWTP2, 1.8 km) of the outlet were collected. Each sample was collected in triplicate. The rivers' average annual flow is 3.5 m^3/s (WWTP1) and 3.7 m^3/s (WWTP2).

WWTP1 is located in one of the cities from Metropolitan Association of Upper Silesia, one of the largest urban areas in the EU and the center of Poland's industries, particularly coal and metal, with a density of 1600 people per km^2 [57], geographical coordinates: N 50° 5′ 35.881, E 19° 3′ 32.202. WWTP2 is located in the forested and agriculture area of Warmian-Masurian Voivodship in the city with a density of 1960 people per km^2 [57], geographical coordinates: N 53° 48′ 46.700, E 20° 26′ 55.800.

In 2018, WWTP1 had an equivalent population of 189,332 inhabitants, the average flow rate was 26,830 m^3/d, and the plant was operated with a hydraulic retention time (HRT) of ~12 h and a solid retention time (SRT) of 25 days, while WWTP2 served a population equivalent of 250,000 inhabitants, and had a flow rate of 32,130 m^3/d, HRT of 24 h, and SRT of 87 days. Both WWTPs receive domestic, hospital, and industrial wastewaters and treat wastewater with a mechanical-biological system with elevated removal of nutrients. The biological section of WWTP1 included sequential reactors activated sludge chambers (the system of chambers of different oxygen conditions: anaerobic, anoxic, and aerobic), secondary settling tank, and anoxic chamber. In the case of WWTP2, the biological part included a pre-denitrification chamber, phosphorus removal tank, nitrification/denitrification chambers, and secondary settling tanks. Sewage sludge produced in the WWTP1 was subjected to a thickening process, methane fermentation, and dehydration. The leachate produced during sludge fermentation (WWTP1-FS) was returned to the pumping station and again treated. The sewage sludge from WWTP2 was used as a soil improver. The detailed description and technical parameters of both WWTPs are presented by Buta et al. [58].

All samples were collected in triplicate and placed in the sterile glass bottles in volumes of 1–2 liters. Sludge samples were collected after mechanical concentration (primary sludge, WWTP1-PS), after gravity concentration (excessive sludge, WWTP1-ES), after an open fermentation pool (fermented sludge, WWTP2-FS), and after all processes (residue sludge, WWTP2-RS), i.e., sludge for management. Then the samples were transported to the laboratory on the same day and stored at a temperature of 4 °C until analysis.

3.2. *Analysis of Antimicrobials' Concentrations*

3.2.1. Chemicals

This study targeted 26 antimicrobials including: Azithromycin (AZM), cefadroxil (CFR), ciprofloxacin (CIP), clarithromycin (CLR), clindamycin (CLI), erythromycin (ERY), fleroxacin (FLRX), lincomycin (LCM), lomefloxacin (LOM), metronidazole (MTZ), nalidixic acid (NAL), norfloxacin (NOR), ofloxacin (OFX), oxytetracycline (OTC), pefloxacin (PEF), rifampicin (RIF), roxithromycin (ROX), sulfadiazine (SD), sulfadimethoxine (SDM), sulfamethazine (SDD), sulfamethoxazole (SXT), sulfathiazole (ST), tetracycline (TET), thiabendazole (TBZ), trimethoprim (TMP), and vancomycin (VAN). All pharmaceutical standards for target antimicrobials were of high purity grade (>90%). All compounds were purchased from Sigma–Aldrich (Darmstadt, Germany). Only trimethoprim was sourced from The National Drug Research Institute in Warsaw, Poland. Isotopically labeled compounds used as mixture of internal standards (1 µg/mL in methanol), i.e., azithromycin-C13, ciprofloxacin-D8, sulfamethoxazole-D4, clindamycin-D4, erythromycin-C13D3, ofloxacin-D8, tetracycline-D6, and trimethoprim-D9 (Toronto Research Chemicals, Toronto, Canada), were added to each sample before extraction (500 µL). Solvents, such as HPLC-grade methanol, acetonitrile (LiChrosolv), and formic acid (98%), were obtained from Merck (Darmstadt, Germany). Moreover, ultrapure water was obtained from a Millipore water purification system (Milli-Q water). All working solutions were prepared prior to analysis.

3.2.2. Preparation of Water and Sewage Sludge Samples

Aqueous samples were filtered through glass fiber filters (GF/C, Whatman, Pittsburgh, PA, USA) and membrane filters (0.2 µm, Sartorius Goettingen, Germany). To 200 mL volumes of filtrate, 200 mg of ethylenediaminetetraacetic acid was added. Then, solid-phase extraction (SPE) (Oasis HLB cartridges, 3 mL, 400 mg, Waters Corp., Milford, MA, USA) using a Phenomenex vacuum system (Torrance, CA, USA) was performed. The elutions were made with pure methanol (3 × 2 mL). The eluents were evaporated to dryness under a stream of nitrogen (99.999% purity, Multax, Poland) at 40 °C and reconstituted in a methanol-water mixture (10:90, *v*/*v*) (1 mL).

Sludge samples were centrifuged at 5000 rpm for 5 min and the supernatant was discarded. Five g of WWTP2-RS or 10 g of the other samples were placed into 50 mL polypropylene tubes, which were then mixed with 9 mL of 30 mMpotassium phosphate monobasic solution and 1 mL of methanol, and extracted for 20 min using 10 mL of acetonitrile with 1% formic acid and modified QUECHERS salts (4 g $MgSO_4$, 1 g NaCl, 1 g Na_3Citrate, and 0.5 g Na_2Citrate•H_2O). Next, the samples were centrifuged for 5 min at 5000 *g*. The samples were then cleaned by incubating 9 mL of extract with 500 mg octadecyl sorbent and 750 mg of $MgSO_4$. They were then vortexed for 5 min at 1200 rpm and centrifuged for 5 min at 5000 rpm. Eight mL of the extract (organic layer) under a nitrogen stream at 40 °C and reconstituted in 0.5 mL of a mixture of methanol-water (10:90) was evaporated. The isotopically labeled compounds were used as an internal standards mixture to each sample before extraction (50 μL).

3.2.3. Antimicrobial Detection by LC-MS/MS Analysis

Antimicrobial concentrations were analyzed by high-performance liquid chromatography coupled to mass spectrometry with a Hybrid Triple Quadrupole/Linear Ion trap mass spectrometer (QTRAP®4000, AB SCIEX, Framingham, MA, USA). LC analysis was performed using an Agilent 1260 Infinity (Agilent Technologies, Santa Clara, CA, USA) equipped with a degasser, thermostated autosampler, and binary pump, and connected in series to an AB Sciex 4000 QTRAP mass spectrometer equipped with a Turbo Ion Spray source that was operated in both positive mode and negative mode. The curtain gas, ion source gas 1, ion source gas 2, and collision gas (all high purity nitrogen) were set at 35 psi, 60 psi, 40 psi, and "medium" instrument units, respectively, and the ion spray voltage and source temperature were set at 5000 V and 600 °C, respectively. Chromatographic separation was achieved with a Kinetex RP-18 column (100 mm × 4.6 mm, 2.6 μm) supplied by Phenomenex (Torrance, CA, USA). The column was maintained at 40 °C and the flow rate was 0.5 mL/min. The mobile phase consisted of HPLC-grade water with 0.2% formic acid as eluent A and acetonitrile with 0.2% formic acid as eluent B. The gradient (%B) was as follows: 0 min. 10%, 1 min. 10%, 25 min. 90%, and 35 min. 90%. The injection volume was 10 μL. The target compounds were analyzed in multiple reaction monitoring (MRM) mode in positive ionization mode (ESI +), monitoring two transitions between the precursor ion and the most abundant fragment ions for each compound. Internal standards were attributed to analyzed compounds based on similarities between chemical structures of surrogate and analyzed compound (according to the Tanimoto similarity index). The LC-MS method was validated using three quality control levels (low, medium, and high) prepared on effluents. The interday precision higher than 15% was observed for cefadroxil (up to 22%), norfloxacin (up to 22%), lomefloxacin (up to 24%), and azithromycin (up to 35%) in case of wastewater and for cefadroxil (up to 23%), azithromycin (up to 26%), roxithromycin (up to 34%), and clarithromycin (up to 38%) in case of sediments. Significant matrix effect (lower than 85% or higher than 115%) was observed for 12 compounds in the case of wastewater and 21 in the case of sediments. Additionally to internal standard addition to control the matrix effect, each sample was tested without and after fortification with antimicrobials.

The following blanks were used: the HPLC blank (10% methanol) and the method blank (Milli-Q water, calcinated sand) to evaluate the contamination resulting from the complete preparation and analytical procedure. The positive control (water or sand fortified with pharmaceuticals) was also applied. Then, the obtained results were adjusted with recovery and matrix effect. The method detection limit (MDL) and method quantitation limit (MQL) for the entire method (including extraction) were determined as the amount of analyte in matrix spiked with signal-to-noise ratios (S/N) of 3:1 and 10:1, respectively. For the pharmaceuticals already present in samples, MDL and MQL were estimated by determining the S/N of the minimum measured concentrations and extrapolating to S/N values of 3 and 10, respectively.

3.3. Calculations

The following parameters were calculated:

Antimicrobial removal (RE) calculated using Equations (1) and (2) according to Douziech et al. [59]:

$$RE\ [\%] = (1 - \exp(RR)) \times 100\% \quad (1)$$

$$RR = \ln(C_{EWW}/C_{IWW}) \quad (2)$$

where RE is the removal of antimicrobials during treatment [%], RR (response ratio (effect size)) is measured per WWTP and antimicrobial, C_{EWW} is the mean of the effluent wastewater concentration of the antimicrobial ($n = 3$), and C_{IWW} is the mean of the influent wastewater concentration of the antimicrobial ($n = 3$).

PNEC (predicted no effect concentration) for resistance selection in wastewater was calculated according to Bengtsson-Palme and Larsson [28], whereas PNEC for eukaryotic species was calculated as described by Page et al. [27]. Depending on the number of long-term toxicity tests performed, i.e., no test, the test on one, two, or three trophic levels, the lowest obtained NOEC (no observable effect level) or EC_{50} (half maximal effective concentration) was divided by 1000, 100, 50, or 10 [60]. PNEC for cyanobacteria was obtained by dividing the NOEC or EC_{50} by 10. All PNEC values were presented in Table 4.

PNEC for soil ($PNEC_{soil}$) and sludge ($PNEC_{sludge}$) was obtained by multiplying PNEC with K_d (solid/liquid partition coefficient, Table 2). K_d was calculated as the ratio of the concentration of an antimicrobial on a solid phase (sludge) divided by the equilibrium concentration in the contacting liquid phase (wastewater). PNEC values for resistance selection can be found in Table 5.

Risk quotients (RQs) were calculated using Equation (3):

$$RQ_{wastewater} = MEC_{wastewater}/PNEC \quad (3)$$

$$RQ_{sludge} = MEC_{sludge}/PNEC_{sludge} \quad (4)$$

$$RQ_{wastewater} = PEC_{soil}/PNEC_{soil} \quad (5)$$

where PEC_{soil} is predicted environmental concentration in soil (Table 3), MEC is measured environmental concentration, and PNEC is predicted no effect concentration.

The risk ranking criterion was $RQ < 0.1$, minimal risk; $1 > RQ \geq 0.1$, medium risk; and $RQ > 1$, high risk [61].

Predicted daily load of the antimicrobial into WWTP (PLoad) from actual consumption data of the product using Equation (6) is:

$$PLoad = TOTAL \cdot WWTP/(INHAB \cdot 30) \quad (6)$$

where PLoad is the predicted daily load of the antimicrobial into WWTP (mg/d), TOTAL is the total monthly consumption of the pharmaceutical in a country (mg), WWTP is an equivalent number of inhabitants in a WWTP, and INHAB is the number of inhabitants in a country (38,000,000).

Average daily load of the antimicrobial into a WWTP from the measured data using Equation (7) is:

$$Load_{W+S} = Load_W + Load_S = C_W \cdot Flow + C_S \cdot PS \quad (7)$$

where $Load_{W+S}$ is the average total daily load of antimicrobial into WWTP (mg/d), $Load_W$ is the load of a compound in a water phase of an influent, $Load_S$ is the load of a compound in a primary sludge, C_W is the concentration of antimicrobial in the WWTP influent (water phase) (mg/m^3), Flow is the average daily flow rate in the WWTP (m^3/d), C_S is the concentration of antimicrobial in the WWTP primary sludge (mg/m^3), and PS is the average daily volume of the WWTP primary sludge (m^3/d).

3.4. Statistical Analysis

The statistical analysis of the results was performed with the STATISTICA version 13.1 for Windows (TIBCO Software Inc., Palo Alto, CA, USA) and Metaboanalyst 4.0. Student's t-test was used for comparison of samples. Principal components analysis (PCA) was used to determine differences between wastewater and sewage sludge.

4. Conclusions

In this study, 70% of the examined antimicrobials in both wastewater and sewage sludge collected from two Polish WWTPs were detected. To the best of our knowledge, this is the first report on antimicrobial occurrence in Polish WWTPs. The removal efficiency, antimicrobial resistance risk, and ecological threat (RQs) were examined according to the obtained data. The WWTPs removed ~50% of selected antimicrobials with good efficiency, above 50%. The level of antimicrobials in both untreated and treated wastewater, river as well as sewage sludge, poses a risk of resistance selection as shown by RQ calculations. Moreover, influents and river waters posed high and medium risk, particularly for cyanobacteria and eukaryotes due to the presence of ciprofloxacin, macrolides, and sulfamethoxazole. Following the obtained results, the watch list of substances for Union-wide monitoring in the field of water policy (already includes macrolides and ciprofloxacin) should be extended with sulfamethoxazole. Our study also indicates the need for evaluation of antimicrobials' concentrations not only in treated wastewater, but also in sewage sludge because of its usage in the fertilization process, which is environmentally sustainable options for re-use of the WWTP by-products. Several antimicrobials tested were present at levels that have been suggested to promote resistance development in sludge-amended soils (predicted concentrations).

The most important observation made is a possible pressure for the development of antimicrobial resistance in the WWTPs. WWTPs can be considered as potential hot spots for the dissemination of antimicrobial resistance. Leakage of antimicrobials can select for increased resistance among environmental bacteria and influence the virulence of antimicrobial-sensitive bacterial infections directly by reducing the infective dose and transmission [27,29]. Therefore, additional studies on the characterization of wastewater treatment plants' microbial communities and the profiles of antimicrobial-resistant genes are necessary. Our study also highlights the lack of sufficient data to evaluate or predict the risk of resistance development and environmental threat. In fact, data on risk assessment of wastewater and sludge in other European countries is also scarce.

Author Contributions: All authors contributed equally to the development of the manuscript in all aspects: Investigations and conceptualization, J.G., G.N.-J., D.K.; writing—original draft, J.G., G.N.-J., G.P.; writing—review and editing, J.G., G.N-J., E.K., M.H., G.P.; funding acquisition, G.P. and J.G. All authors have read and agreed to the published version of the manuscript.

Funding: This study was done under the project No 2017/26/M/NZ9/00071 funded by National Science Center, Poland. LC-MS/MS analyses were carried out with the use of the CePT infrastructure financed by the European Union, the European Regional Development Fund within the Operational Program "Innovative economy" for 2007–2013.

Acknowledgments: The authors would like to thank Marcin Giebułtowicz for drug sales data analysis and Ryszard Marszałek for his technical support during LC-MS/MS measurement. The authors are also very grateful to Bartosz Kózka and Bartłomiej Sankowski for their contribution in the conduct of the analytical study. We gratefully acknowledge the technical support provided by the WWTPs.

Conflicts of Interest: The authors declare no conflict of interest.

Appendix A

Table A1. The comparison of the predicted (PLoad) and measured load of antimicrobial into WWTP1.

	Sale in 2018 (kg/month)	PLoad (g/d) [1]			$Load_W$ (g/d) [2]			$Load_S$ (g/d) [3]			$Load_W$/PLoad (%)			$Load_S$/PLoad (%)			$Load_{W+S}$ [4]/ PLoad(%)		
	range	S[5]	A	W	S	A	W	S	A	W	S	A	W	S	A	W	S	A	W
AZM[6]	98–352	25.3	26.3	47.2	0.3	4.8	1.9	0.2	2.9	5.4	1	13	4	1	8	11	2	21	15
CIP	601–821	106.5	128.9	112.8	58.7	15.6	27.1	54.2	26.1	60.1	55	12	24	51	20	53	106	32	77
CLR	288–1509	68.9	121.1	145.5	7.9	11.2	19.4	0.04	0.27	0.09	12	9	13	0	0	0	12	9	13
CLI	608–737	109.9	116.4	106.8	1.6	2.4	6.8	0.3	0.2	0.8	1	2	6	0	0	1	1	2	7
NOR	179–214	30.9	35.5	29.9	5.1	11.2	3.0	10.3	18.0	52.7	16	31	10	33	51	176	49	82	186
OFX	609–829	107.3	131.0	113.7	2.6	4.9	3.5	10.2	1.5	2.8	2	4	3	10	1	3	12	5	6
RIF	40–48	7.08	7.19	6.63	0.26	0.08	0.08				4	1	1	0	0	0	4	1	1
ROX	6–28	1.44	2.33	2.69	0.14	0.14	1.19				10	6	44	0	0	0	10	6	44
SXT	606–1358	133.3	196.0	194.1	18.5	44.9	44.0	0.4	0.3	1.2	14	23	23	0	0	1	14	23	24
TMP	122–272	26.7	39.7	38.9	8.1	5.5	6.8	0.02	0.02	0.09	30	14	18	0	0	0	30	14	18

[1] PLoad, predicted load of a compound calculated on the basis of sales data; [2] $Load_W$, load of a compound calculated on the basis of measured concentration in a water phase of an influent; [3] $Load_S$, load of a compound calculated on the basis of measured concentration in a primary sludge; [4] $Load_{W+S}$, total load of a compound (water phase and a primary sludge); [5] S summer; A, autumn; W, winter sampling. [6] The pharmaceuticals used only in hospitals and frequently used in veterinary medicine were not presented

Table 2. Mean values of sorption coefficients (log K_d) calculated for sludge and the percent of the total mass of the compound sorbed in sludge to its total daily mass load to WWTP.

Antimicrobial	% of Total Mass Load Sorbed to Sludge	log K_d Sludge
AZM	73%	7.1 ± 4.3
CIP	55%	7.3 ± 4.2
CLR	18%	5.7 ± 2.5
CLI	9%	5.7 ± 2.6
LCM	7%	5.2 ± 1.9
OFX	43%	7.7 ± 4.8
OTC	62%	9.8 ± 6.9
SDM	32%	6.8 ± 3.4
SXT	2%	5.1 ± 1.7
TET	13%	8.3 ± 5.5
TBZ	23%	5.9 ± 2.6
TMP	19%	4.8 ± 1.6
VAN	24%	6.3 ± 3.2

Table 3. Predicted antimicrobial concentration in sludge-amended soil (single sludge application).

Antimicrobial	WWTP1-PS		WWTP1-ES		WWTP2-RS		WWTP2-FS		WWTP1-RS [1]	
(ng/g)	Mean	SD	Mean	SD	Mean	SD	Mean	SD	Mean	SD
AZM	1.9	1.3	0.55	0.50	0.17	0.14	2.3	1.6	0.14	0.13
CIP	18	13	9.1	3.5	8.9	9.8	41	40	11.4	5.2
CLR	0.42	0.09	0.03	0.02	0.03	0.04	0.33	0.37	<0.01	-
CLI	0.06	0.04	0.08	0.06	0.04	0.05	0.07	0.05	0.01	0.01
LCM	0.01	0.01	0.01	-	0.01	0.01	0.03	0.04	<0.01	-
NOR	7.8	5.8	5.2	4.4	2.4	2.4	5.0	4.1	-	-
OFX	1.2	1.6	0.94	0.91	0.44	0.33	0.70	0.21	0.05	0.02
OTC	0.86	0.74	0.48	0.41	0.07	0.06	0.41	0.37	0.01	0.01
PEF	0.36	0.25	0.28	0.20	0.28	0.22	0.38	0.15	-	-
ROX	0.03	-	<0.01	-	<0.01	-	<0.01	-	<0.01	-
SD	<0.01		<0.01		<0.01	-	0.03	-	<0.01	-
SDM	0.03	0.03	0.03	0.03	0.03	-	<0.01	-	<0.01	-
SXT	<0.03	-	0.12	0.10	<0.03	-	<0.03	-	0.01	0.01
TET	0.12	0.09	0.06	0.02	0.08	0.05	0.29	0.17	0.04	0.03
TBZ	0.03	0.01	0.02	0.01	0.09	0.08	0.07	0.04	0.01	0.01
TMP	0.25	0.19	0.01	0.01	<0.01	-	<0.01	-	<0.01	-
VAN	0.53	0.10	0.27	0.16	0.11	-	<0.01	-	0.55	0.15

[1] WWTP-PS, sludge after mechanical treatment; WWTP-ES, sludge after gravity treatment; WWTP-FS, sludge after an open fermentation pool; WWTP-RS, sludge for management.

Table 4. NOEC/EC_{50} (μg/L) and PNEC (ng/L) used for the risk assessment; nd, no data available. Abbreviations in parentheses indicate the most sensitive taxa: MA, microalgae; MP, macrophytes; IN, invertebrates. The data based on the review of Le Page [27].

Compound	Cyanobacteria NOEC/EC_{50}	Cyanobacteria PNEC	Eukaryote NOEC/EC_{50}	Eukaryote PNEC
AZM	0.19	19	1.8	180(MA)
CIP	5.65	565	10	1000(MP)
CLR	0.84	84	2	40(MA)
ERY	2	200	10.3	206(MA)
LCM	18	1800	548	10976(MA)
MTZ	0	nd	250,000	5000000(IN)
NOR	1.6	160	300	3000(MP)
OFX	5	500	31.2	624(MA)
OTC	3.1	310	183	3660(MA)
ROX	nd	nd	10	100(MA)
SD	3900	390,000	135	2700(MA)
SDM	7800	780,000	100	5290(MA)
SXT	5.9	590	10	1000(MP,IN)
TET	90	9000	300	6000(MP)
TMP	1385	135,800	1000	20000(MP)

Table 5. Predicted no effect concentration values for resistance selection of microbial community in wastewater, sludge, and soil.

Antimicrobial	PNEC Wastewater (μg/L) [1]	PNEC Sludge (μg/kg) [2]	PNEC Soil (μg/kg) [3]
AZT	0.25	1200	47
CIP	0.064	1200	27,000
CLR	0.25	140	66
CLI	1	280	-
ER	2	-	260
LIN	2	340	-
METR	0.125	-	0.07
NFL	0.5	-	300
OFL	0.5	4200	730
OTET	0.5	4,200,000	210
RIF	0.5	-	-
ROX	1	4300	50
STH	16	2000	9.6
TET	1	3000	1100
TRI	0.5	100	3.7
VAN	8	17,000	2.4

[1] According to Bengtsson-Palme and Larsson [28]; [2] calculated based on K_d obtained for sludge; [3] calculated according to [26,40,42,62].

References

1. Fu, W.; Fu, J.; Li, X.; Li, B.; Wang, X. Occurrence and fate of PPCPs in typical drinking water treatment plants in China. *Environ. Geochem Health* **2019**, *41*, 5–15. [CrossRef] [PubMed]
2. Omar, T.F.T.; Aris, A.Z.; Yusoff, F.M.; Mustafa, S. Risk assessment of pharmaceutically active compounds (PhACs) in the Klang River estuary, Malaysia. *Environ. Geochem Health* **2019**, *41*, 211–223. [CrossRef] [PubMed]
3. Zhang, P.; Zhou, H.; Li, K.; Zhao, X.; Liu, Q.; Li, D.; Zhao, G. Occurrence of pharmaceuticals and personal care products, and their associated environmental risks in a large shallow lake in north China. *Environ. Geochem. Health* **2018**, *40*, 1525–1539. [CrossRef] [PubMed]
4. Chen, C.; Li, J.; Chen, P.; Ding, R.; Zhang, P.; Li, X. Occurrence of antibiotics and antibiotic resistances in soils from wastewater irrigation areas in Beijing and Tianjin, China. *Environ. Pollut.* **2014**, *193*, 94–101. [CrossRef]
5. Rodriguez-Mozaz, S.; Chamorro, S.; Marti, E.; Huerta, B.; Gros, M.; Sanchez-Melsio, A.; Borrego, C.M.; Barcelo, D.; Balcazar, J.L. Occurrence of antibiotics and antibiotic resistance genes in hospital and urban wastewaters and their impact on the receiving river. *Water Res.* **2015**, *69*, 234–242. [CrossRef]
6. Harrabi, M.; Varela Della Giustina, S.; Aloulou, F.; Rodriguez-Mozaz, S.; Barceló, D.; Elleuch, B. Analysis of multiclass antibiotic residues in urban wastewater in Tunisia. *Environ. Nanotechnol. Monit. Manag.* **2018**, *10*, 163–170. [CrossRef]
7. Dinh, Q.; Moreau-Guigon, E.; Labadie, P.; Alliot, F.; Teil, M.J.; Blanchard, M.; Eurin, J.; Chevreuil, M. Fate of antibiotics from hospital and domestic sources in a sewage network. *Sci. Total Environ.* **2017**, *575*, 758–766. [CrossRef]
8. Marx, C.; Gunther, N.; Schubert, S.; Oertel, R.; Ahnert, M.; Krebs, P.; Kuehn, V. Mass flow of antibiotics in a wastewater treatment plant focusing on removal variations due to operational parameters. *Sci. Total Environ.* **2015**, *538*, 779–788. [CrossRef]
9. Stadler, L.B.; Delgado Vela, J.; Jain, S.; Dick, G.J.; Love, N.G. Elucidating the impact of microbial community biodiversity on pharmaceutical biotransformation during wastewater treatment. *Microb. Biotechnol.* **2018**, *11*, 995–1007. [CrossRef]
10. Luo, Y.; Guo, W.; Ngo, H.H.; Nghiem, L.D.; Hai, F.I.; Zhang, J.; Liang, S.; Wang, X.C. A review on the occurrence of micropollutants in the aquatic environment and their fate and removal during wastewater treatment. *Sci. Total Environ.* **2014**, *473–474*, 619–641. [CrossRef]

11. ECDC. European Centre for Disease Prevention and Control (2016) An agency of the European Union. Available online: https://www.ecdc.europa.eu/en (accessed on 3 October 2019).
12. Krzeminski, P.; Tomei, M.C.; Karaolia, P.; Langenhoff, A.; Almeida, C.M.R.; Felis, E.; Gritten, F.; Andersen, H.R.; Fernandes, T.; Manaia, C.M.; et al. Performance of secondary wastewater treatment methods for the removal of contaminants of emerging concern implicated in crop uptake and antibiotic resistance spread: A review. *Sci. Total Environ.* **2019**, *648*, 1052–1081. [CrossRef] [PubMed]
13. Mompelat, S.; Le Bot, B.; Thomas, O. Occurrence and fate of pharmaceutical products and by-products, from resource to drinking water. *Environ. Int.* **2009**, *35*, 803–814. [CrossRef]
14. Grenni, P.; Ancona, V.; Barra Caracciolo, A. Ecological effects of antibiotics on natural ecosystems: A review. *Microchem. J.* **2018**, *136*, 25–39. [CrossRef]
15. Zhao, S.; Liu, X.; Cheng, D.; Liu, G.; Liang, B.; Cui, B.; Bai, J. Temporal–spatial variation and partitioning prediction of antibiotics in surface water and sediments from the intertidal zones of the Yellow River Delta, China. *Sci. Total Environ.* **2016**, *569–570*, 1350–1358. [CrossRef]
16. Watkinson, A.J.; Murby, E.J.; Kolpin, D.W.; Costanzo, S.D. The occurrence of antibiotics in an urban watershed: From wastewater to drinking water. *Sci. Total Environ.* **2009**, *407*, 2711–2723. [CrossRef]
17. Loraine, G.A.; Pettigrove, M.E. Seasonal Variations in Concentrations of Pharmaceuticals and Personal Care Products in Drinking Water and Reclaimed Wastewater in Southern California. *Environ. Sci. Technol.* **2006**, *40*, 687–695. [CrossRef]
18. Giebultowicz, J.; Tyski, S.; Wolinowska, R.; Grzybowska, W.; Zareba, T.; Drobniewska, A.; Wroczynski, P.; Nalecz-Jawecki, G. Occurrence of antimicrobial agents, drug-resistant bacteria, and genes in the sewage-impacted Vistula River (Poland). *Environ. Sci. Pollut. Res. Int.* **2018**, *25*, 5788–5807. [CrossRef]
19. Kümmerer, K.; Henninger, A. Promoting resistance by the emission of antibiotics from hospitals and households into effluent. *Clin. Microbiol. Infect.* **2003**, *9*, 1203–1214. [CrossRef]
20. Senta, I.; Terzic, S.; Ahel, M. Occurrence and fate of dissolved and particulate antimicrobials in municipal wastewater treatment. *Water Res.* **2013**, *47*, 705–714. [CrossRef]
21. Radjenovic, J.; Petrovic, M.; Barcelo, D. Fate and distribution of pharmaceuticals in wastewater and sewage sludge of the conventional activated sludge (CAS) and advanced membrane bioreactor (MBR) treatment. *Water Res.* **2009**, *43*, 831–841. [CrossRef]
22. Wang, S.; Cui, Y.; Li, A.; Zhang, W.; Wang, D.; Ma, J. Fate of antibiotics in three distinct sludge treatment wetlands under different operating conditions. *Sci Total Environ.* **2019**, *671*, 443–451. [CrossRef] [PubMed]
23. Gobel, A.; McArdell, C.S.; Joss, A.; Siegrist, H.; Giger, W. Fate of sulfonamides, macrolides, and trimethoprim in different wastewater treatment technologies. *Sci. Total Environ.* **2007**, *372*, 361–371. [CrossRef] [PubMed]
24. Ostman, M.; Lindberg, R.H.; Fick, J.; Bjorn, E.; Tysklind, M. Screening of biocides, metals and antibiotics in Swedish sewage sludge and wastewater. *Water Res.* **2017**, *115*, 318–328. [CrossRef] [PubMed]
25. Plosz, B.G.; Leknes, H.; Liltved, H.; Thomas, K.V. Diurnal variations in the occurrence and the fate of hormones and antibiotics in activated sludge wastewater treatment in Oslo, Norway. *Sci Total Environ.* **2010**, *408*, 1915–1924. [CrossRef]
26. Cycon, M.; Mrozik, A.; Piotrowska-Seget, Z. Antibiotics in the Soil Environment-Degradation and Their Impact on Microbial Activity and Diversity. *Front. Microbiol.* **2019**, *10*, 338. [CrossRef]
27. Le Page, G.; Gunnarsson, L.; Snape, J.; Tyler, C.R. Integrating human and environmental health in antibiotic risk assessment: A critical analysis of protection goals, species sensitivity and antimicrobial resistance. *Environ. Int.* **2017**, *109*, 155–169. [CrossRef]
28. Bengtsson-Palme, J.; Larsson, D.G.J. Concentrations of antibiotics predicted to select for resistant bacteria: Proposed limits for environmental regulation. *Environ. Int.* **2016**, *86*, 140–149. [CrossRef]
29. O'Neill, J. Tackling drug-resistant infections globally:final report and recommendations. *Rev. Antimicrob. Resist.* **2016**. Available online: https://amr-review.org/sites/default/files/160518_Final%20paper_with%20cover.pdf (accessed on 21 February 2020).
30. Wishart, D.S.; Feunang, Y.D.; Guo, A.C.; Lo, E.J.; Marcu, A.; Grant, J.R.; Sajed, T.; Johnson, D.; Li, C.; Sayeeda, Z.; et al. DrugBank 5.0: A major update to the DrugBank database for 2018. *Nucleic Acids Res.* **2018**, *46*, D1074–D1082. [CrossRef]
31. Zachara, J.; Zachara, R.; Zachara, N.; Matuszczak, A.; Kłoda, K. The comparison of use of antibiotics due to acute respiratory infections in the rural population of primary care in 2010 and 2017. *J. Educ. Health Sport* **2019**, *9*, 70–83.

32. Watkinson, A.J.; Murby, E.J.; Costanzo, S.D. Removal of antibiotics in conventional and advanced wastewater treatment: Implications for environmental discharge and wastewater recycling. *Water Res.* **2007**, *41*, 4164–4176. [CrossRef] [PubMed]
33. Verlicchi, P.; Al Aukidy, M.; Zambello, E. Occurrence of pharmaceutical compounds in urban wastewater: Removal, mass load and environmental risk after a secondary treatment–a review. *Sci Total Environ.* **2012**, *429*, 123–155. [CrossRef] [PubMed]
34. Peng, X.; Wang, Z.; Kuang, W.; Tan, J.; Li, K. A preliminary study on the occurrence and behavior of sulfonamides, ofloxacin and chloramphenicol antimicrobials in wastewaters of two sewage treatment plants in Guangzhou, China. *Sci Total Environ.* **2006**, *371*, 314–322. [CrossRef] [PubMed]
35. Cunha, B.A. Vancomycin revisited: A reappraisal of clinical use. *Crit. Care Clin.* **2008**, *24*, 393–420. [CrossRef]
36. Göbel, A.; Thomsen, A.; McArdell, C.S.; Joss, A.; Giger, W. Occurrence and Sorption Behavior of Sulfonamides, Macrolides, and Trimethoprim in Activated Sludge Treatment. *Environ. Sci. Technol.* **2005**, *39*, 3981–3989. [CrossRef]
37. Le-Minh, N.; Khan, S.J.; Drewes, J.E.; Stuetz, R.M. Fate of antibiotics during municipal water recycling treatment processes. *Water Res.* **2010**, *44*, 4295–4323. [CrossRef]
38. Choi, K.J.; Kim, S.G.; Kim, S.H. Removal of tetracycline and sulfonamide classes of antibiotic compound by powdered activated carbon. *Environ. Technol.* **2008**, *29*, 333–342. [CrossRef]
39. Brown, K.D.; Kulis, J.; Thomson, B.; Chapman, T.H.; Mawhinney, D.B. Occurrence of antibiotics in hospital, residential, and dairy effluent, municipal wastewater, and the Rio Grande in New Mexico. *Sci. Total Environ.* **2006**, *366*, 772–783. [CrossRef]
40. Sidhu, H.; D'Angelo, E.; O'Connor, G. Retention-release of ciprofloxacin and azithromycin in biosolids and biosolids-amended soils. *Sci. Total Environ.* **2019**, *650*, 173–183. [CrossRef]
41. Kaeseberg, T.; Zhang, J.; Schubert, S.; Oertel, R.; Siedel, H.; Krebs, P. Sewer sediment-bound antibiotics as a potential environmental risk: Adsorption and desorption affinity of 14 antibiotics and one metabolite. *Environ. Pollut.* **2018**, *239*, 638–647. [CrossRef]
42. Wu, C.; Spongberg, A.L.; Witter, J.D. Sorption and biodegradation of selected antibiotics in biosolids. *J. Environ. Sci Health A* **2009**, *44*, 454–461. [CrossRef] [PubMed]
43. Wang, L.; Qiang, Z.; Li, Y.; Ben, W. An insight into the removal of fluoroquinolones in activated sludge process: Sorption and biodegradation characteristics. *J. Environ. Sci (China)* **2017**, *56*, 263–271. [CrossRef] [PubMed]
44. Wang, W.; Zhang, W.; Liang, H.; Gao, D. Occurrence and fate of typical antibiotics in wastewater treatment plants in Harbin, North-east China. *Front. Environ. Sci. Eng.* **2019**, *13*, 34. [CrossRef]
45. Rodayan, A.; Majewsky, M.; Yargeau, V. Impact of approach used to determine removal levels of drugs of abuse during wastewater treatment. *Sci. Total Environ.* **2014**, *487*, 731–739. [CrossRef]
46. Golet, E.M.; Alder, A.C.; Giger, W. Environmental Exposure and Risk Assessment of Fluoroquinolone Antibacterial Agents in Wastewater and River Water of the Glatt Valley Watershed, Switzerland. *Environ. Sci. Technol.* **2002**, *36*, 3645–3651. [CrossRef]
47. Gao, L.; Shi, Y.; Li, W.; Niu, H.; Liu, J.; Cai, Y. Occurrence of antibiotics in eight sewage treatment plants in Beijing, China. *Chemosphere* **2012**, *86*, 665–671. [CrossRef]
48. Roberts, P.H.; Thomas, K.V. The occurrence of selected pharmaceuticals in wastewater effluent and surface waters of the lower Tyne catchment. *Sci. Total Environ.* **2006**, *356*, 143–153. [CrossRef]
49. Chunhui, Z.; Liangliang, W.; Xiangyu, G.; Xudan, H. Antibiotics in WWTP discharge into the Chaobai River, Beijing. *Arch. Environ. Prot.* **2016**, *42*, 48–57. [CrossRef]
50. Gros, M.; Petrovic, M.; Ginebreda, A.; Barcelo, D. Removal of pharmaceuticals during wastewater treatment and environmental risk assessment using hazard indexes. *Environ. Int.* **2010**, *36*, 15–26. [CrossRef]
51. Kummerer, K. The presence of pharmaceuticals in the environment due to human use–present knowledge and future challenges. *J. Environ. Manag.* **2009**, *90*, 2354–2366. [CrossRef]
52. Osorio, V.; Larrañaga, A.; Aceña, J.; Pérez, S.; Barceló, D. Concentration and risk of pharmaceuticals in freshwater systems are related to the population density and the livestock units in Iberian Rivers. *Sci. Total Environ.* **2016**, *540*, 267–277. [CrossRef] [PubMed]
53. Cheng, M.; Wu, L.; Huang, Y.; Luo, Y.; Christie, P. Total concentrations of heavy metals and occurrence of antibiotics in sewage sludges from cities throughout China. *J. Soils Sediments* **2014**, *14*, 1123–1135. [CrossRef]

54. Lindberg, R.H.; Wennberg, P.; Johansson, M.I.; Tysklind, M.; Andersson, B.A.V. Screening of Human Antibiotic Substances and Determination of Weekly Mass Flows in Five Sewage Treatment Plants in Sweden. *Environ. Sci. Technol.* **2005**, *39*, 3421–3429. [CrossRef] [PubMed]
55. Golet, E.M.; Xifra, I.; Siegrist, H.; Alder, A.C.; Giger, W. Environmental Exposure Assessment of Fluoroquinolone Antibacterial Agents from Sewage to Soil. *Environ. Sci. Technol.* **2003**, *37*, 3243–3249. [CrossRef]
56. Göbel, A.; Thomsen, A.; McArdell, C.S.; Alder, A.C.; Giger, W.; Theiß, N.; Löffler, D.; Ternes, T.A. Extraction and determination of sulfonamides, macrolides, and trimethoprim in sewage sludge. *J. Chromatogr. A* **2005**, *1085*, 179–189. [CrossRef]
57. GUS. *Statistical Yearbook of the Regions—Poland*; Statistics Poland: Warsaw, Poland, 2018.
58. Buta, M.; Hubeny, J.; Zieliński, W.; Korzeniewska, E.; Harnisz, M.; Nowrotek, M.; Płaza, G. The Occurrence of Integrase Genes in Different Stages of Wastewater Treatment. *J. Ecol. Eng.* **2019**, *20*, 39–45. [CrossRef]
59. Douziech, M.; Conesa, I.R.; Benítez-López, A.; Franco, A.; Huijbregts, M.; van Zelm, R. Quantifying variability in removal efficiencies of chemicals in activated sludge wastewater treatment plants—a meta-analytical approach. *Environ. Sci. Processes Impacts* **2018**, *20*, 171–182. [CrossRef]
60. ECHA. *Guidance on information requirements and chemical safety assessment*; European Chemicals Agency: Helsinki, Finland, 2008.
61. Ben, W.; Zhu, B.; Yuan, X.; Zhang, Y.; Yang, M.; Qiang, Z. Occurrence, removal and risk of organic micropollutants in wastewater treatment plants across China: Comparison of wastewater treatment processes. *Water Res.* **2018**, *130*, 38–46. [CrossRef]
62. Rabùlle, M.; Spliid, N.H. Sorption and mobility of metronidazole, olaquindox, oxytetracycline and tylosin in soil. *Chemosphere* **2000**, *40*, 8. [CrossRef]

Sample Availability: Samples of the compounds are not available from the authors.

Article

A Survey of the Presence of Pharmaceutical Residues in Wastewaters. Evaluation of Their Removal Using Conventional and Natural Treatment Procedures

R. Guedes-Alonso *, S. Montesdeoca-Esponda, J. Pacheco-Juárez, Z. Sosa-Ferrera and J. J. Santana-Rodríguez

Instituto Universitario de Estudios Ambientales y Recursos Naturales (i-UNAT), Universidad de Las Palmas de Gran Canaria, 35017 Las Palmas de Gran Canaria, Spain; sarah.montesdeoca@ulpgc.es (S.M.-E.); javier.pacheco@ulpgc.es (J.P.-J.); zoraida.sosa@ulpgc.es (Z.S.-F.); josejuan.santana@ulpgc.es (J.J.S.-R.)
* Correspondence: rayco.guedes@ulpgc.es

Academic Editor: Jolanta Kumirska
Received: 13 March 2020; Accepted: 1 April 2020; Published: 2 April 2020

Abstract: To encourage the reutilization of treated wastewaters as an adaptation strategy to climate change it is necessary to demonstrate their quality. If this is ensured, reclaimed waters could be a valuable resource that produces very little environmental impact and risks to human health. However, wastewaters are one of the main sources of emerging pollutants that are discharged in the environment. For this, it is essential to assess the presence of these pollutants, especially pharmaceutical compounds, in treated wastewaters. Moreover, the different treatment processes must be evaluated in order to know if conventional and natural treatment technologies are efficient in the removal of these types of compounds. This is an important consideration if the treated wastewaters are used in agricultural activities. Owing to the complexity of wastewater matrixes and the low concentrations of pharmaceutical residues in these types of samples, it is necessary to use sensitive analytical methodologies. In this study, the presence of 11 pharmaceutical compounds were assessed in three different wastewater treatment plants (WWTPs) in Gran Canaria (Spain). Two of these WWTPs use conventional purification technologies and they are located in densely populated areas, while the other studied WWTP is based in constructed wetlands which purify the wastewaters of a rural area. The sampling was performed monthly for two years. A solid phase extraction (SPE) coupled to ultra-high performance liquid chromatography tandem mass spectrometry (UHPLC-MS/MS) method was applied for the analysis of the samples, and the 11 pharmaceuticals were detected in all the studied WWTPs. The concentrations were variable and ranged from $ng \cdot L^{-1}$ in some compounds like diclofenac or carbamazepine to $\mu g \cdot L^{-1}$ in common pharmaceutical compounds such as caffeine, naproxen or ibuprofen. In addition, removal efficiencies in both conventional and natural purification systems were evaluated. Similar removal efficiencies were obtained using different purifying treatments, especially for some pharmaceutical families as stimulants or anti-inflammatories. Other compounds like carbamazepine showed a recalcitrant behavior. Secondary treatments presented similar removal efficiencies in both conventional and natural wastewater treatment plants, but conventional treatments showed slightly higher elimination ratios. Regarding tertiary system, the treatment with highest removal efficiencies was reverse osmosis in comparison with microfiltration and electrodialysis reversal.

Keywords: wastewaters; pharmaceutical residues; constructed wetlands; conventional wastewater treatments; solid phase extraction; ultra-high performance liquid chromatography

1. Introduction

The reuse of reclaimed waters for irrigation could represent an alternative to the scarcity of water in arid locations, in addition to saving fertilizers and avoiding the discharge of this water into the environment, which has important ecological impacts [1]. However, the reuse of reclaimed waters is controversial because of the possible presence of emerging contaminants and their entry into the food chain [2]. Wastewaters have been revealed as one of the main pathways of introduction of these types of compounds into the environment. Thus, the increasing demand of water and discovery of the harmful effects of the emerging pollutants over aquatic biota suggest that continuous monitoring of wastewaters is needed [3].

Among these types of pollutants, pharmaceutical compounds are of great concern, not only for legislators, but also the scientific community, due to the variety of physic-chemical properties and the many toxic effects of them. Moreover, this concern is also related to the large consumption of these compounds in modern society, which is estimated as several tons in the European Union alone [4].

The harmful effects of pharmaceuticals over biota are produced when these compounds reach aquatic compartments through wastewaters and organisms are exposed to them [5]. This exposure is due to purification technologies of wastewater treatment plants (WWTPs) which are not designed to eliminate emerging pollutants [6,7]. Moreover, current legislation, does not establish limits for the concentrations of pharmaceuticals in environmental waters. The elimination of pharmaceutical compounds from wastewaters is significantly complex due to the variety of these analytes, their properties, and the possible conjugated compounds formed during metabolization [8]. Nevertheless, in 2015 the European Commission created a watch list which "is a mechanism for obtaining high-quality Union-wide monitoring data on potential water pollutants for the purpose of determining the risk they pose and thus whether Environmental Quality Standards (EQS) should be set for them at EU level" [9]. In this watch list, one anti-inflammatory compound, diclofenac, was added because of the well-known harmful effects of this compound to the aquatic environment [9].

Many studies have focused on determining the occurrence of pharmaceutical compounds in environmental samples. In this sense, some compounds like caffeine and nicotine and their respective metabolites paraxanthine and cotinine, and some broadly used drugs like naproxen or acetaminophen are usually detected at higher concentrations (in the range of $\mu g \cdot L^{-1}$) than other pharmaceuticals. Other common pharmaceuticals like diclofenac, erythromycin or carbamazepine are usually in the range of $ng \cdot L^{-1}$ as previous research has indicated [10–12]. However, many studies evaluate the occurrence of pharmaceuticals during a short period of time; few studies examine these compounds in the long term to evaluate seasonal fluctuations and determine patterns from the presence of these compounds.

In light of the above, it is necessary to carry out monitoring studies of pharmaceuticals to establish not only if the recipient aquatic ecosystems could be contaminated by these kinds of residues, but also to evaluate the efficiency of wastewater treatment facilities in the removal of these compounds. The removal of pharmaceuticals is not always achieved in WWTPs and, as some authors have determined, sometimes this degradation is partial and, in most cases, inefficient [13,14]. For this, it is necessary to evaluate the efficiency of conventional and natural wastewater treatments in the removal of pharmaceutical residues. Conventional wastewater treatments are employed in medium and largely populated areas to purify urban wastewaters, while natural treatments use the purifying power of bacteria, soils, and plants to treat wastewaters. Natural treatments are appropriate for the treatment of domestic wastewaters of small or isolated areas because they require large land areas to be settled. Nonetheless, their advantages are greater than their disadvantages because natural wastewater treatment systems have a low impact on the landscape, use minimal energy and little to no chemicals, and they produce relatively lower amounts of residual solids [15]. In spite of their differences, both systems have shown their capabilities to degrade and remove emerging pollutants from wastewaters [16–18].

In this study, three WWTPs, two based on conventional treatment technologies and one based on natural purification processes, were surveyed for two years in order to determine the occurrence and removal of pharmaceutical residues in Gran Canaria island (Spain). Eleven different compounds (Table 1) belonging to different pharmaceutical families (anti-inflammatories, stimulants, lipid regulators, antihypertensives, anticonvulsants, and antibiotics) were determined in different purifying stages of the studied WWTPs in order to determine the efficiency of conventional and natural wastewater treatments in the removal of them. The choice of these 11 pharmaceuticals was influenced by a previous study that monitored the same WWTPs for six months and showed that these 11 pharmaceuticals were the most detected compounds from a group of 23 different drugs [19]. Due to the expected low concentrations of the compounds, solid phase extraction (SPE) was used as an extraction and preconcentration technique. Then, ultra-high performance liquid chromatography tandem mass spectrometry (UHPLC–MS/MS) was used as the detection and determination technique.

Table 1. Names, identification numbers, molecular weights, and structures of target pharmaceuticals.

Drug Family	Compound	CAS No.	Molecular Weight (g/mole)	Structure
Stimulants	Nicotine	54-11-5	162.230	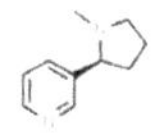
	Caffeine	58-08-2	194.191	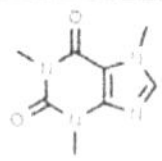
	Paraxanthine	611-59-6	180.164	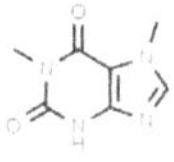
Anti-inflammatories	Naproxen	22204-53-1	230.260	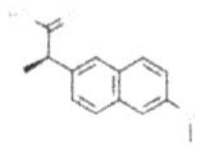
	Ibuprofen	15687-27-1	206.281	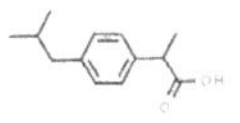
	Diclofenac	15307-86-5	296.149	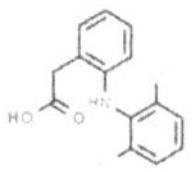
Lipid regulators	Gemfibrozil	25812-30-0	250.333	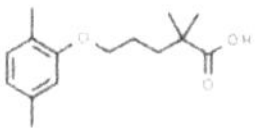
Anti-hypertensives	Atenolol	29122-68-7	266.336	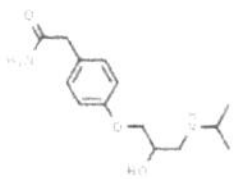
Anti-convulsants	Carbamazepine	298-46-4	236.269	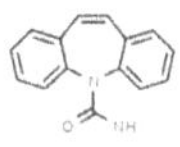
Antibiotics	Trimethoprim	738-70-5	290.318	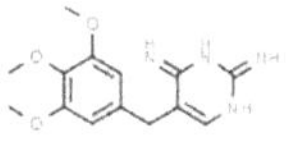
	Erythromycin	114-07-8	733.927	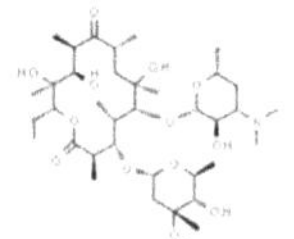

2. Materials and Methods

2.1. Materials

The target pharmaceuticals were purchased from Sigma-Aldrich (Madrid, Spain) and presented purities over 97%. Three internal standards (IS), atenolol-d7 (Toronto Research Chemical Inc, Toronto, Canada), ibuprofen-d3 (Sigma-Aldrich, Madrid, Spain), and sulfamethoxazole-d4 (Dr. Ehrenstorfer GmbH, Ausgburg, Germany) were used to minimize the matrix effect of the studied samples.

Stock solutions were prepared at 1000 mg·L^{-1} by dissolving the compound in high-purity methanol. These solutions were stored in glass-stoppered bottles at –20 °C prior to use. Working solutions were prepared daily from a stock mixture of 10 mg·L^{-1} by diluting proper quantities of the stock mixture in LC–MS grade water.

The mobile phase of UHPLC–MS/MS system was prepared using LC–MS grade methanol and water, both purchased from Panreac (Barcelona, Spain) as well as the modifiers of the mobile phase. The Milli-Q water used in the wash step of SPE was obtained using a water purification system of Millipore (Bedford, MA, USA)

2.2. Sample Collection

Samples were taken in three different wastewater treatment plants from Gran Canaria island (Spain). Two use conventional treatment technologies (C-WWTP1 and C-WWTP2) and receive wastewater from highly-populated areas of the island, located in the northeast and the southeast, respectively. C-WWTP1 has a secondary treatment of activated sludge and a tertiary system based on microfiltration and reverse osmosis. This WWTP treats the water of a population equivalent to 134,000 inhabitants. In C-WWTP2 only the tertiary process based on ultra-filtration was evaluated in order to know the possible risk of the produced water. This WWTP treats the water of a population equivalent to 200,000 inhabitants. The third WWTP surveyed is based in natural treatments (N-WWTP) and is located in a rural zone of the island. It is based in two constructed wetlands (CWs), the first one is a vertical flow wetland and the second one is a planted sub-superficial horizontal flow wetland. This WWTP was designed to treat wastewater equivalent to 500 inhabitants but nowadays it treats a higher volume of wastewater with great results.

In C-WWTP1, the samples were taken in the influent of the plant (point A1), after the biological treatment (point A2), after the microfiltration process (point A3) and lastly, in the final effluent of the plant, after the reverse osmosis process (point A4). In C-WWTP2, the samples were taken before and after the electrodialysis reversal process (points B1 and B2). Finally, in N-WWTP, the samples were taken in the influent (point C1) and after each process of the treatment: Imhoff tank (point C2), vertical flow wetland (point C3), and horizontal sub-superficial flow wetland (point C4) (Figure 1).

The samples were taken monthly for two years from July 2017 to June 2019, at the same time slot and collected in rinsed amber bottles of 1 L. Samples were filtered using 0.65 µm polyvinylidene fluoride (PVDF) membrane filters from Merck Millipore (Cork, Ireland), and acidified to pH below 3 using hydrochloric acid and stored in the dark at 4 °C to inhibit microbial activity.

2.3. Analytical Methodology

To extract and preconcentrate the target pharmaceuticals in wastewater samples, a methodology based on SPE previously optimized [19] was used. Briefly, 250 mL of filtered wastewater at pH 7 were extracted using 500 mg Oasis HLB cartridges (Waters, Barcelona, Spain) in a Varian SPE manifold (Madrid, Spain). After the loading of the samples, the cartridges were washed with 5 mL of Milli-Q water and after that, the retained compounds were eluted with 5 mL of methanol. To achieve a great preconcentration factor, the solvent was evaporated under a gentle nitrogen steam and reconstituted with 1 mL of Milli-Q water with 100 µg·L^{-1} of internal standards. Before analysis the extracts were filtered using Chromafil Xtra PET-20/25 syringe filters with a pore size of 0.20 µm from Machery-Nagel (Düren, Germany).

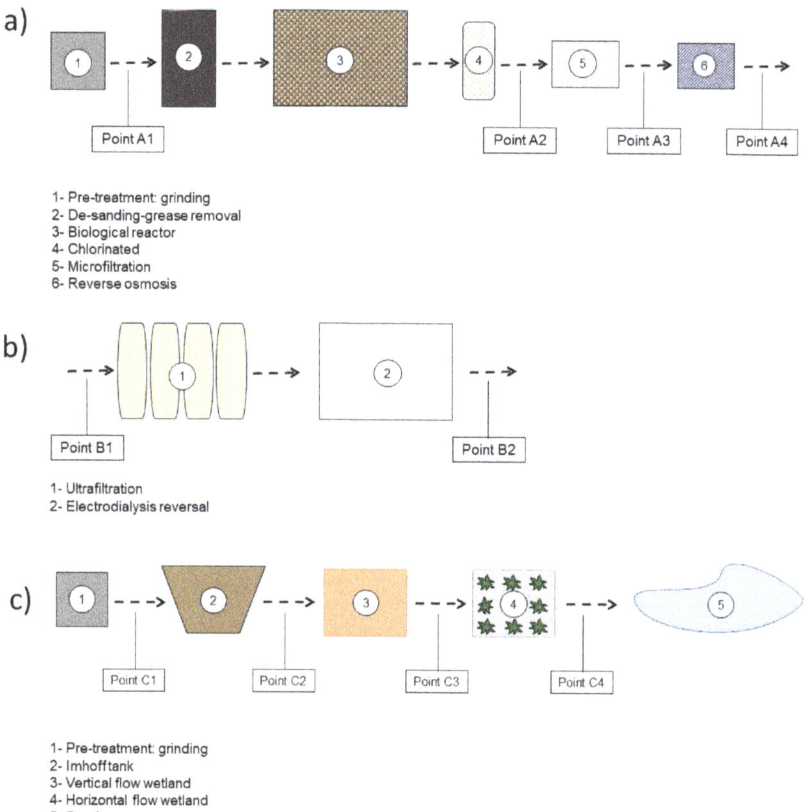

Figure 1. Layout of wastewater treatment plants (WWTPs) surveyed. (**a**) Conventional Wastewater Treatment Plant 1 (C-WWTP1); (**b**) Conventional Wastewater Treatment Plant 2 (C-WWTP2); (**c**) Natural Wastewater Treatment Plant (N-WWTP). Figure adapted from [19].

After the extraction, ultra-high performance liquid chromatography tandem mass spectrometry (UHPLC–MS/MS) was used as a separation and detection technique. The separation of pharmaceuticals was performed using an ACQUITY UPLC BEH Waters C18 column (50 mm × 2.1 mm, 1.7 µm) from Waters Chromatography (Barcelona, Spain). The mobile phase used consisted of LC–MS grade water and methanol both with 0.5% acetic acid and the separation of target compounds was done in gradient mode (Table S1). The chromatographic system used was an AQCUITY UPLC system and consisted of a binary solvent manager pump, an autosampler, a column manager, and a triple quadrupole detector (TQD) controlled by Masslynx software, all of them from Waters Chromatography (Barcelona, Spain). The detection of the target compounds was carried out using electrospray ionization (ESI) in both positive and negative mode and the mass spectrometer parameters had a capillary voltage of 3.00 kV in positive mode and −2.50 kV in negative mode, source temperature of 120 °C and desolvation temperature of 450 °C. The conditions of fragmentation and the ions of the pharmaceuticals under study are presented in Table S2.

This analytical methodology presented great recoveries, between 52.9% and 123.6%, and appropriate detection limits, between 15.3 ng·L^{-1} and 13.3 µg·L^{-1}. The method also showed great linearity, with correlation coefficients (r^2) over 0.99 in all cases and good intra-day and inter-day repeatability, with relative standard deviation (RSD) values below 22% [19].

3. Results and Discussion

In this study, solid phase extraction (SPE) coupled to ultra-high-performance liquid chromatography tandem mass spectrometry (UHPLC-MS/MS) method was applied for analysis of water samples from three WWTPs. After the analysis, a statical study was performed in order to establish the median concentrations of the detected pharmaceuticals. The median concentration was used instead of the average concentration due to the dispersion of the concentrations detected during the two years of study. Moreover, the frequency of detection and the removal efficiency of each purification process were also calculated by dividing the number of samples with detected concentrations of a pharmaceutical by the total number of analyzed samples (positive results in Table 2).

3.1. Occurrence and Concentrations of Target Pharmaceuticals

3.1.1. Conventional Treatment WWTPs

Conventional wastewater purification treatments are based on a primary treatment which is focused on the elimination of fats, sand, and coarse solids. After that, a secondary treatment, usually based on biological processes, is used to degrade organic matter from wastewater as well as to eliminate suspended solids and other organic pollutants. These processes could be aerobic or non-aerobic, and the most used biological process is activated sludge. Finally, if the wastewater will be re-used, tertiary processes must be performed in order to obtain a water with great quality. There are many tertiary treatment technologies and most used are microfiltration, nanofiltration or reverse osmosis [20–23].

C-WWTP1 is based in activated sludge technology and as can be seen in Table 2, the stimulants under study (nicotine, caffeine, and paraxanthine) present the highest concentrations at influent samples, probably because these compounds are used in pharmaceutical formulations but also are excreted by smokers in the case of nicotine and after drinking beverages like coffee in the case of caffeine and its metabolite, paraxanthine. In fact, the median concentrations of these compounds reached values between 45.8 and 95.6 $\mu g \cdot L^{-1}$ during monitoring. This behavior was observed in other studies in which the highest concentrations of pharmaceuticals studied were obtained for these three stimulants as well [24–27]. Moreover, two anti-inflammatory drugs, naproxen and ibuprofen, also presented median concentrations in the range of $\mu g \cdot L^{-1}$ (between 19.9 and 27.3 $\mu g \cdot L^{-1}$) in influent samples and naproxen showed the highest concentration of all the monitored compounds (521.7 $\mu g \cdot L^{-1}$ in an influent sample). Anti-inflammatory compounds show a high rate of consumption in Spain [28] and the high concentrations of ibuprofen can be explained because this compound does not require a medical prescription as some authors have stated [29]. In addition, two pharmaceuticals related to cardiovascular problems and diseases namely, atenolol and gemfibrozil, presented low influent median concentrations of $\mu g \cdot L^{-1}$ (1.13 and 3.35 $\mu g \cdot L^{-1}$, respectively). The median concentrations of the rest of the target compounds were in the range of $ng \cdot L^{-1}$. As the purification process was performed, the median concentration of the compounds decreased. In fact, the sum of the median concentrations of the target compounds ranged from 150.9 $\mu g \cdot L^{-1}$ in the influent, to 0.90 $\mu g \cdot L^{-1}$ after the osmosis treatment, which is indicative of great elimination in the wastewater purification system. Figure 2a shows the changes in the distribution of pharmaceutical median concentrations in the different sampling points. In this sense, it can be observed that the contributed total concentration of stimulants decreased during the purification process, while the contribution of some compounds, like carbamazepine, increased; this is because the concentrations during the purification process remain stable. In fact, carbamazepine's median concentration was similar in points A1 and A2 (0.146 and 0.052 $\mu g \cdot L^{-1}$, respectively), but the contribution to the total concentration changed from 0.1% to 5.7% after secondary and tertiary treatment. The composition of the final effluent obtained in the WWTP is a key factor, because it will be useful to predict the possible effects of effluents used in agriculture or discharged into the environment.

Table 2. Median concentrations, range of detected concentrations, and positive analysis for target pharmaceuticals. (**a**) C-WWTP1, (**b**) C-WWTP2, (**c**) N-WWTP.

(a)

Compound	Point A1		Point A2		Point A3		Point A4	
	Median Concentration (min–max Concentration) ($\mu g \cdot L^{-1}$)	Positive Results (%)	Median Concentration (min–max Concentration) ($ng \cdot L^{-1}$)	Positive Results (%)	Median Concentration (min–max Concentration) ($ng \cdot L^{-1}$)	Positive Results (%)	Median Concentration (min–max Concentration) ($ng \cdot L^{-1}$)	Positive Results (%)
Nicotine	14.179 (0.005–45.82)	89.5	0.151 (0.064–0.449)	88.9	0.136 (0.002–0.586)	86.7	0.065 (0.001–0.251)	87.5
Atenolol	1.134 (0.354–2.899)	84.2	0.218 (0.006–0.643)	83.3	0.287 (0.007–1.949)	86.7	0.061 (0.001–0.088)	81.3
Trimethoprim	0.231 (0.053–0.658)	84.2	0.077 (0.034–0.157)	77.8	0.071 (0.029–0.169)	73.3	0.021 (0.013–0.054)	75.0
Paraxanthine	47.869 (12.46–95.63)	89.5	0.518 (0.101–6.900)	88.9	0.425 (0.036–5.901)	86.7	0.117 (0.031–1.262)	81.3
Caffeine	36.495 (15.47–72.62)	89.5	0.589 (0.095–2.499)	88.9	0.663 (0.120–2.705)	80.0	0.169 (0.051–1.943)	87.5
Erythromycin	0.083 (0.070–0.094)	15.8	0.067 (0.043–0.090)	11.1	0.052 (0.044–0.709)	26.7	0.047 (0.047–0.073)	25.0
Carbamazepine	0.146 (0.045–2.394)	89.5	0.240 (0.137–2.597)	88.9	0.234 (0.078–4.687)	86.7	0.052 (0.013–0.158)	87.5
Naproxen	27.333 (4.910–521.7)	84.2	0.172 (0.072–1.528)	66.7	0.282 (0.057–1.311)	73.3	0.102 (0.011–0.584)	68.8
Ibuprofen	19.894 (0.128–147.5)	84.2	0.208 (0.047–1.585)	77.8	0.214 (0.009–1.184)	80.0	0.139 (0.005–1.751)	56.3
Diclofenac	0.139 (0.006–0.708)	47.4	0.117 (0.007–0.404)	77.8	0.152 (0.034–0.288)	73.3	0.024 (0.006–0.115)	37.5
Gemfibrozil	3.353 (0.053–40.63)	84.2	0.729 (0.012–3.602)	83.3	0.598 (0.033–2.865)	80.0	0.103 (0.004–0.485)	56.3

Table 2. *Cont.*

(b)				
	Point B1		Point B2	
Compound	Median Concentration (min–max Concentration) ($\mu g \cdot L^{-1}$)	Positive Results (%)	Median Concentration (min–max Concentration) ($ng \cdot L^{-1}$)	Positive Results (%)
Nicotine	0.103 (0.038–0.228)	100	0.105 (0.047–0.343)	100
Atenolol	0.286 (0.001–0.521)	100	0.184 (0.167–0.234)	85.7
Trimethoprim	0.055 (0.014–0.130)	100	0.043 (0.014–0.074)	100
Paraxanthine	0.310 (0.037–1.256)	100	0.241 (0.031–2.634)	100
Caffeine	0.378 (0.058–0.817)	100	0.369 (0.050–2.749)	100
Erythromycin	nd *	0.0	nd *	0.0
Carbamazepine	0.209 (0.148–0.312)	85.7	0.193 (0.107–0.327)	85.7
Naproxen	0.198 (0.150–0.237)	57.1	0.116 (0.077–0.555)	57.1
Ibuprofen	0.125 (0.027–0.173)	100	0.074 (0.008–0.969)	85.7
Diclofenac	0.202 (0.070–0.280)	85.7	0.091 (0.027–0.139)	85.7
Gemfibrozil	0.723 (0.392–1.098)	85.7	0.701 (0.306–0.883)	85.7

Table 2. *Cont.*

(c)

Compound	Point C1		Point C2		Point C3		Point C4	
	Median Concentration (min–max Concentration) (μg·L^{-1})	**Positive Results (%)**	**Median Concentration (min–max Concentration) (ng·L^{-1})**	**Positive Results (%)**	**Median Concentration (min–max Concentration) (ng·L^{-1})**	**Positive Results (%)**	**Median Concentration (min–max Concentration) (ng·L^{-1})**	**Positive Results (%)**
Nicotine	9.606 (6.449–63.94)	89.5	7.651 (0.194–37.29)	89.5	3.515 (0.134–12.70)	88.9	0.245 (0.086–1.077)	81.8
Atenolol	2.195 (0.757–10.00)	84.2	1.402 (0.037–2.396)	89.5	0.742 (0.072–1.990)	83.3	0.198 (0.082–0.753)	81.8
Trimethoprim	0.034 (0.017–2.166)	63.2	0.031 (0.013–0.531)	68.4	0.020 (0.014–0.674)	50.0	0.018 (0.015–0.088)	36.4
Paraxanthine	40.050 (9.619–85.77)	89.5	21.243 (6.867–39.84)	89.5	5.084 (0.143–30.56)	88.9	0.596 (0.151–2.480)	81.8
Caffeine	37.888 (10.39–126.4)	89.5	41.497 (11.16–65.22)	89.5	16.771 (0.382–47.58)	88.9	1.633 (0.261–5.595)	81.8
Erythromycin	0.101 (0.071–0.131)	10.5	0.066 (0.025–0.107)	10.5	0.081 (0.045–5.346)	33.3	0.064 (0.063–0.066)	18.2
Carbamazepine	0.299 (0.039–1.306)	89.5	0.306 (0.081–0.969)	89.5	0.432 (0.069–7.797)	88.9	0.604 (0.247–1.009)	81.8
Naproxen	9.653 (0.911–320.1)	84.2	9.383 (2.581–177.0)	84.2	7.713 (0.644–77.47)	83.3	1.397 (0.310–5.179)	81.8
Ibuprofen	16.942 (4.378–121.4)	84.2	21.886 (5.331–133.7)	84.2	14.209 (0.931–43.97)	83.3	9.803 (6.658–18.76)	81.8
Diclofenac	0.035 (0.008–0.055)	21.1	0.028 (0.007–0.718)	47.4	0.139 (0.024–0.591)	61.1	0.525 (0.020–2.813)	81.8
Gemfibrozil	0.222 (0.001–16.11)	73.7	0.647 (0.006–5.632)	78.9	2.904 (0.112–7.403)	83.3	3.972 (1.104–7.670)	81.8

* nd: not detected.

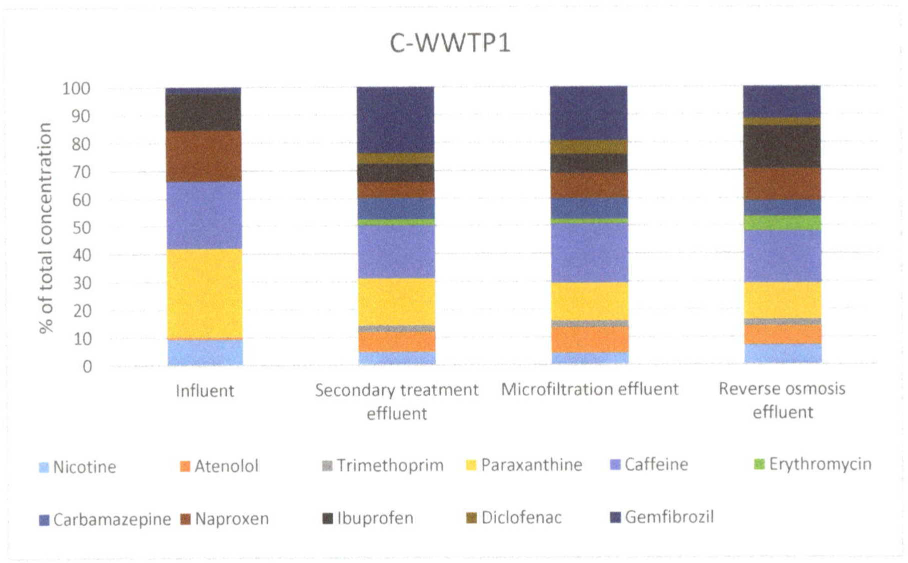

(**a**)

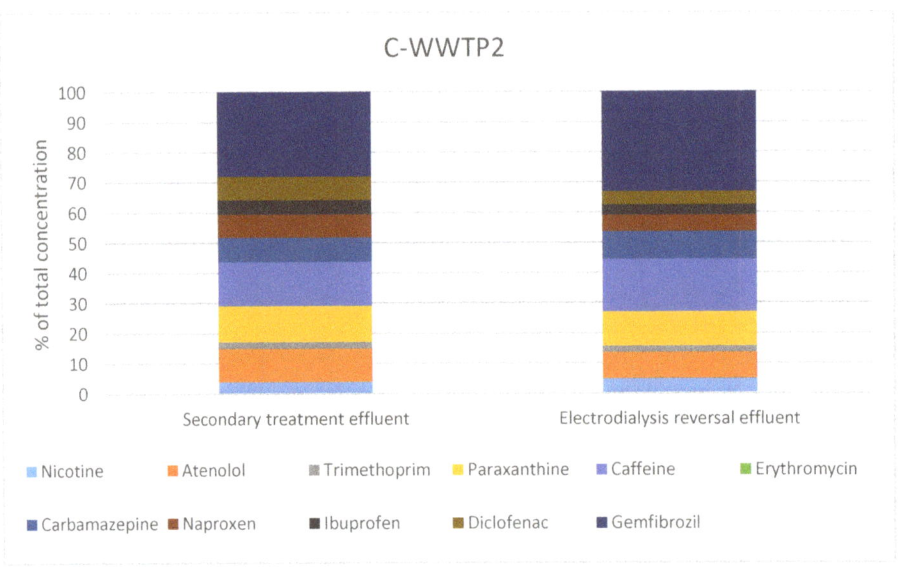

(**b**)

Figure 2. *Cont.*

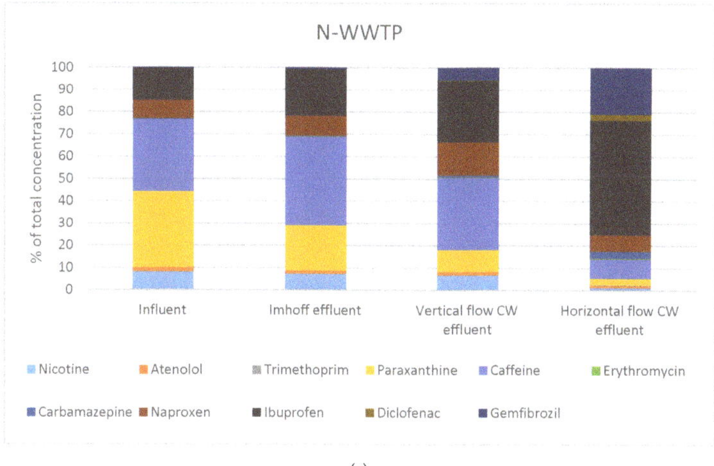

(c)

Figure 2. Distribution of pharmaceuticals from Table 1. (**a**) C-WWTP1, (**b**) C-WWTP2, (**c**) N-WWTP.

C-WWTP2 has an activated sludge treatment too, but in this WWTP the study was conducted in the tertiary process which is based on electrodialysis reversal. This technology has been demonstrated as an effective way to remove some emerging pollutants from drinking water [30] but studies in wastewaters are scarce and for this reason, this tertiary process was evaluated. In this wastewater treatment plant, the samples were taken after the secondary process (point B1) and after the electrodialysis technology (point B2). Before this treatment, an ultrafiltration process was also performed as pretreatment. Ten of the 11 pharmaceuticals under study were detected at concentrations that ranged from 0.055 to 0.723 $\mu g \cdot L^{-1}$ in point B1 and from 0.043 to 0.701 $\mu g \cdot L^{-1}$ in point B2. The highest concentrations were of gemfibrozil, caffeine, and paraxanthine, which ranged from 0.241 to 0.723 $\mu g \cdot L^{-1}$; this coincides with the results of C-WWTP1. This same behavior can be explained because both WWTPs treat the water of big urban areas with similar characteristics. In the same way, the lowest concentrations after secondary treatments (points A2 and B1) coincide in the two conventional WWTPs and correspond to erythromycin, nicotine, and trimethoprim. Figure 2b shows a similar distribution of pharmaceuticals before and after the purification process which means that this technology had a similar impact in the removal of the target compounds.

3.1.2. Natural Treatment WWTP

An alternative to conventional WWTPs is natural WWTPs. From the different types of natural treatment technologies, constructed wetlands have revealed themselves as a great alternative to treat municipal wastewaters from small communities or isolated areas in both vertical flow and horizontal flow layouts [31,32]. In this system, the purification of wastewater is partially done in the vertical flow wetland and the horizontal flow wetland improves the water quality. The highest concentrations of target pharmaceuticals match with C-WWTP1, and were from caffeine, paraxanthine, nicotine, ibuprofen, and naproxen. All of these compounds presented median concentrations that ranged from 9.60 to 40.05 $\mu g \cdot L^{-1}$, while the rest of the target compounds were in the range of $ng \cdot L^{-1}$. The highest concentrations detected in the influent samples of this WWTP were from naproxen (320.07 $\mu g \cdot L^{-1}$) and caffeine (126.40 $\mu g \cdot L^{-1}$). As in the previous WWTPs, in this WWTP the concentrations decreased as the purification process was conducted. This can be observed in the total median concentrations of the WWTP that were 117.02 $\mu g \cdot L^{-1}$ in the influent (point C1), 104.14 $\mu g \cdot L^{-1}$ after Imhoff treatment (point C2), and 51.61 and 19.06 $\mu g \cdot L^{-1}$ after vertical flow (point C3) and horizontal flow (point C4) wetlands, respectively. Figure 2c shows the distribution of pharmaceuticals in the sampling points and it can be observed that the distribution in the influent of this WWTP is similar to C-WWTP1. In the influents of both WWTPs, the majority of compounds found were stimulants and anti-inflammatories. Nevertheless, the pharmaceutical profiles were different in the rest of the sampling points due to the removal efficiency of each WWTP. In the final effluent of the N-WWTP we observed a large contribution of ibuprofen to the total amount of pharmaceuticals; this was not observed for the other WWTPs.

3.2. *Removal of Target Pharmaceuticals*

As previously stated, pharmaceutical compounds have become a concerning group of emerging pollutants to the scientific community and their removal from wastewaters is essential to ensure the environmental quality of recipient ecosystems. Biologically-based WWTPs produce effluents that maintain water quality standards in order to reuse them or dispose into the environment at a reasonable cost, but these WWTPs have limited capability to remove pharmaceuticals [33]. For this reason, it is essential to evaluate the removal efficiency of different technologies, even more so if the purified wastewaters will be re-used in agriculture. To calculate the removal of the different treatments, the following equation was used in the different samplings and WWTPs surveyed.

$$RE\ (\%) = 100 - \left(\frac{[Effl]}{[Inf]} * 100\right)$$

where *RE* is removal efficiency, [*Effl*] is the measured concentration of the pharmaceutical in the effluent of the treatment, and [*Inf*] is the measured concentration of the pharmaceutical in the influent of the treatment.

3.2.1. Conventional Treatment WWTPs

In C-WWTP1, the secondary process provides median removals over 98% for the compounds with the highest concentrations (nicotine, caffeine, paraxanthine, ibuprofen, and naproxen). For the pharmaceuticals used in cardiovascular diseases, namely atenolol and gemfibrozil, the removal efficiency of the biological treatment was also great, with median removals over 72.5%. Diclofenac and trimethoprim showed slightly lower removal efficiencies of 42.9% and 66.5%, respectively. Only carbamazepine showed a negative value of removal. This means that the concentrations after the biological treatment were higher than in the influent. The observed persistence of carbamazepine in treated wastewaters has been the topic of many studies around the world, and for this reason, some authors like Hai et al. have proposed its use as an anthropogenic marker in water [34]. Other authors have attributed the poor removal of carbamazepine to its molecular structure and hydrophilicity [35–37]. In this WWTP, tertiary processes were also evaluated. The microfiltration process was not effective; in fact, the median removals were between −52.9% and 19.1%. Nevertheless, reverse osmosis was an effective method to remove these compounds. The removals of pharmaceuticals in this treatment were high, in all cases over 55%, even for persistent compounds such as carbamazepine. By comparing the concentrations of pharmaceuticals in influent and final effluent, the combination of biological treatment and tertiary processes based on reverse osmosis was effective in the removal of pharmaceutical residues. The median recoveries of the overall purification process were over 90% for all compounds under study, except carbamazepine, which showed an overall removal of 73.7%. This removal could be considered as very satisfactory in comparison with other studies in which the removal of this compound only reached 10–30% [38].

In C-WWTP2, another tertiary process was evaluated; in this case, the combination of an ultrafiltration and electrodialysis reversal process. There is very little literature about the efficiency of this treatment process in wastewaters. In our study, the monitoring was conducted for one year and the results showed that this technology has a moderate efficiency. Only four compounds (atenolol, naproxen, ibuprofen, and diclofenac) showed a median removal between 48% and 58%. The electrodialysis technology was not efficient with the rest of the compounds under study. The target stimulants, nicotine, caffeine, and paraxanthine, showed removal efficiencies between 13.8% and 25.5% while the efficiencies for trimethoprim, carbamazepine, and gemfibrozil were 20.4%, 12.3%, and 5.4%, respectively. These results complement the study of Arola et al. which established that using electrodialysis, diclofenac and ibuprofen were preferentially retained in the diluent [22]. However, the study of Arola et al. was done in a pilot plant and the authors established that the study must be confirmed using real wastewaters, like in this study.

3.2.2. Natural Treatment WWTP

In this WWTP, the Imhoff process showed limited removal efficiency for pharmaceutical residues. In this sense, only two compounds showed 100% elimination during this process, but these two compounds, erythromycin and diclofenac, were detected in less than 20% of the surveyed samples; thus, it is not possible to ensure that this type of purification system is appropriate for the elimination of these two pharmaceuticals. Regarding the anti-inflammatories (naproxen and ibuprofen) and gemfibrozil, they showed negative elimination ratios, which means that the concentrations after the Imhoff process were higher than before. These negative elimination ratios could be explained by daily fluctuations in the concentrations of these compounds because the samples were taken at the same time in each sampling point. Furthermore, some deconjugation processes, in which conjugated compounds are converted into free compounds during the purification process, could explain the negative removal ratios obtained [7,39]. For the rest of the compounds, the eliminations were not high, and the median removals were between 2.9% and 41.8%. Regarding the constructed wetland processes, in most cases both configurations (vertical and horizontal flows) showed similar behaviors. In the vertical flow system, removals over 60% were achieved for the three stimulants, nicotine, caffeine and paraxanthine. Atenolol showed a medium removal of 51.4% and poor elimination efficiencies (between 10.6% and 33.1%) were obtained for trimethoprim, and the three anti-inflammatory compounds. Carbamazepine showed a trend similar to conventional secondary processes and the elimination was negative, which means that the concentration after the purification process was higher, as observed in other studies. In this vertical flow process, the concentrations of gemfibrozil were also higher after treatment which was also observed in a previous study on this WWTP [19]. Finally, the horizontal flow treatment provided slightly higher removal efficiencies than vertical flow treatment. In this sense, five compounds presented median eliminations over 75% (nicotine, caffeine, paraxanthine, atenolol, and naproxen). For trimethoprim and ibuprofen, the removals were low (30.0% and 26.4%, respectively), but higher than those obtained in the vertical flow systems. For diclofenac, a different trend was observed, because the median removal for this compound was −162.5%, which indicates an increase in the concentrations after this treatment which could be explained by the daily fluctuations or by deconjugation processes as in previous treatments. Finally, carbamazepine and gemfibrozil showed negative removals too, but the increase of the concentrations after the treatment was lower than in the vertical flow treatment. Overall, the natural treatment processes performed in this WWTP showed great eliminations for stimulants (over 97.5%), atenolol (90.9%), naproxen (79.4%), and trimethoprim (64.0%). Three compounds (carbamazepine, diclofenac, and gemfibrozil) showed negative removals after the whole purification system and this trend coincided with a previous study performed at this WWTP [19].

3.2.3. Comparison between Conventional and Natural Purification Treatments

To perform comparisons of the treatment technologies, between the two systems of conventional and natural purification, it is necessary to compare the same stages of purification. As can be seen in Figure 3a, both natural and conventional systems provide similar removal efficiencies for target compounds after secondary treatment. The highest elimination rates in both WWTPs were obtained for stimulants, with median removals over 97% and atenolol, which showed a median removal over 90% in both systems. Regarding anti-inflammatory compounds, the trends in the two surveyed WWTPs were different. Conventional treatments showed large removals for naproxen and ibuprofen (between 98.7% and 99.5%), while in the natural WWTP, the removals for these compounds were between 36.1%

and 79.4%. For the third anti-inflammatory compound, diclofenac, C-WWTP1 showed a positive elimination ratio (42.9%) while in the N-WWTP, this efficiency was negative, which implies greater concentration of the free compound in the effluent of the system. In this sense, effluent concentrations of diclofenac were 10 times higher than influent concentrations, but in all cases below 0.5 $\mu g \cdot L^{-1}$. This trend was the same for gemfibrozil which showed a median elimination of 72.5% in C-WWTP1 and more than –300% in the N-WWTP. Finally, for trimethoprim, very similar median removals were achieved (in both wastewater treatment facilities it was over 60%) and this trend was also seen for carbamazepine which reflected the recalcitrant behavior of this compound with median removals near –50% in both WWTPs.

a)

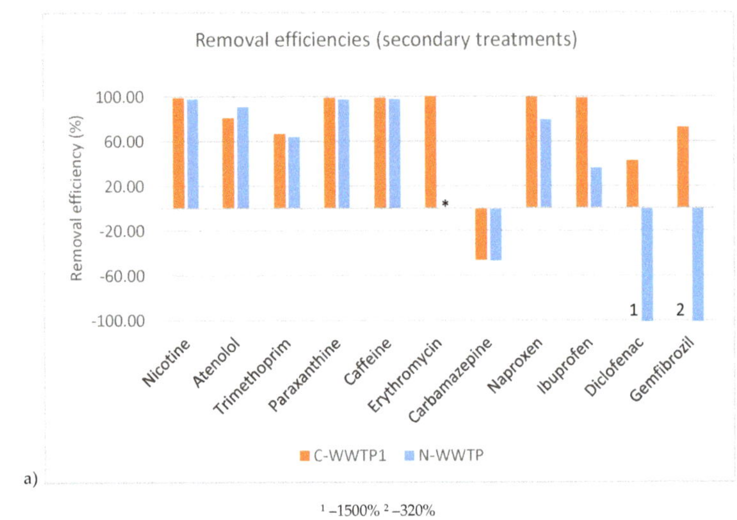

[1] –1500% [2] –320%

b)

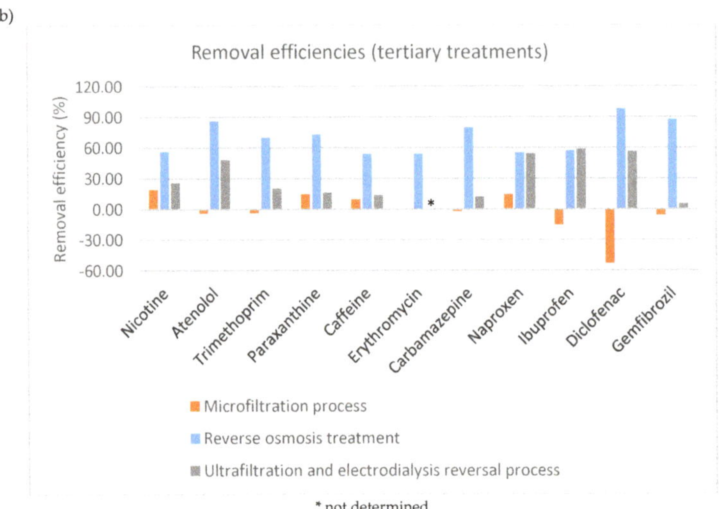

* not determined

Figure 3. Median removal efficiencies for target compounds using (**a**) secondary treatments and (**b**) tertiary treatments.

Regarding tertiary processes, it can be seen in Figure 3b that the reverse osmosis process was the most efficient purification process. The stimulants showed good removals for reverse osmosis process and not-satisfactory removals for microfiltration and electrodialysis reversal technologies. Regarding anti-inflammatories, similar removal efficiencies were obtained for naproxen and ibuprofen using reverse osmosis and electrodialysis reversal (between 54.1% and 58.3%) and for this family of compounds, it was stated that microfiltration technology was not appropriate for the elimination of them. For the rest of the compounds under study, in all cases the removal efficiencies were significantly better using reverse osmosis while electrodialysis reversal showed a better elimination capacity than microfiltration, but poor efficiency (below 30%).

4. Conclusions

In order to provide a safe and valuable resource of water in arid and semi-arid locations by using reclaimed water in agriculture, it is necessary to ensure its quality. One of the reluctances of farmers and legislators is related to the presence of some emerging pollutants, such as pharmaceutical residues, in treated waters. Although the concentration of these compounds in such waters are usually at trace levels, a deep knowledge about their presence and elimination is needed.

Due to the singular characteristics of each family of target pharmaceutical compounds, their removal rates in the studied conventional purification treatments were very variable, but great efficiencies (up to 99.8%) were achieved for some of them. Satisfactory removal values were also provided by the studied natural treatment system, with comparable results for stimulants such as nicotine or caffeine and other drugs like atenolol and naproxen. However, for other compounds, namely ibuprofen, diclofenac, and gemfibrozil, natural treatments were not so effective as conventional ones. In addition, remarkable differences were also observed among tertiary technologies. Reverse osmosis revealed itself as a great option for the elimination of emerging pollutants. In this sense, the combination of secondary treatments and reverse osmosis provided very satisfactory removal efficiencies, over 95% for most compounds under study. Regarding other tertiary technologies, electrodialysis reversal also showed moderate removals for some pharmaceuticals, but in all cases, they were significantly lower than reverse osmosis.

Since all the target pharmaceutical compounds are still present in the studied treated water after using both conventional and natural systems (with concentration levels from $ng \cdot L^{-1}$ to $\mu g \cdot L^{-1}$), further studies are demanded in order to improve the purification systems. Special interest must be paid to some recalcitrant compounds, like carbamazepine, for which very low removal rates were achieved.

The inclusion of emerging compounds, such as pharmaceuticals, in national and European contexts, is also mandatory in order to carry out adequate monitoring programs and to establish reliable control of these pollutants.

Supplementary Materials: The following are available online, Table S1: Gradient used for the chromatographic separation of target pharmaceuticals, title, Table S2: Parent and fragmentation ions and collision conditions for the mass spectrometry detection of target pharmaceuticals.

Author Contributions: Conceptualization, R.G.-A., S.M.-E., Z.S.-F., and J.J.S.-R.; funding acquisition, Z.S.-F. and J.J.S.-R.; sampling, R.G.-A., J.P.-J., and S.M.-E.; methodology, R.G.-A. and J.P.-J.; data analysis, R.G.-A. and J.P.-J.; project administration, R.G.-A., S.M.-E., Z.S.-F., and J.J.S.-R.; supervision, Z.S.-F. and J.J.S.-R.; writing—original draft, R.G.-A. and S.M.-E.; writing—review and editing, R.G.-A., S.M.-E., Z.S.-F., and J.J.S.-R. All authors have read and agreed to the published version of the manuscript.

Funding: This study was supported by Project ADAPTaRES (MAC/3.5b/102) co-funded by the European Union program INTERREG MAC (2014-2020), by means of the European Regional Development Fund.

Acknowledgments: Sarah Montesdeoca-Esponda would like to thank Universidad de Las Palmas de Gran Canaria for her postdoctoral fellowship. The authors would also like to thank Mancomunidad del Sureste de Gran Canaria and Consejo Insular de Aguas de Gran Canaria for their help and support in sampling activities.

Conflicts of Interest: The authors declare no conflict of interest.

References

1. Martínez, S.; Suay, R.; Moreno, J.; Segura, M.L. Reuse of tertiary municipal wastewater effluent for irrigation of *Cucumis melo* L. *Irrig. Sci.* **2013**, *31*, 661–672. [CrossRef]
2. Martínez-Alcalá, I.; Pellicer-Martínez, F.; Fernández-López, C. Pharmaceutical grey water footprint: Accounting, influence of wastewater treatment plants and implications of the reuse. *Water Res.* **2018**, *135*, 278–287. [CrossRef]
3. Barceló, D.; de Alda, M. Contaminación y calidad química del agua: El problema de los contaminantes emergentes. In *Jornadas De Presentación De Resultados: El Estado Ecológico De Las Masas De Agua. Panel Científico-Técnico De Seguimiento De La Política De Aguas*; Fundación Nueva Cultura del Agua PANEL CIENTÍFICO-TÉCNICO DE SEGUIMIENTO DE LA POLÍTICA DE AGUAS Convenio Universidad de Sevilla-Ministerio de Medio Ambiente: Seville, Spain, 2008.
4. Jones, O.A.H.; Voulvoulis, N.; Lester, J.N. Human Pharmaceuticals in the Aquatic Environment a Review. *Environ. Technol.* **2001**, *22*, 1383–1394. [CrossRef] [PubMed]
5. Hernando, M.D.; Mezcua, M.; Fernández-Alba, A.R.; Barceló, D. Environmental risk assessment of pharmaceutical residues in wastewater effluents, surface waters and sediments. *Talanta* **2006**, *69*, 334–342. [CrossRef]
6. Zhou, J.L.; Zhang, Z.L.; Banks, E.; Grover, D.; Jiang, J.Q. Pharmaceutical residues in wastewater treatment works effluents and their impact on receiving river water. *J. Hazard. Mater.* **2009**, *166*, 655–661. [CrossRef]
7. Fatta-Kassinos, D.; Meric, S.; Nikolaou, A. Pharmaceutical residues in environmental waters and wastewater: Current state of knowledge and future research. *Anal. Bioanal. Chem.* **2011**, *399*, 251–275. [CrossRef]
8. Gurke, R.; Rößler, M.; Marx, C.; Diamond, S.; Schubert, S.; Oertel, R.; Fauler, J. Occurrence and removal of frequently prescribed pharmaceuticals and corresponding metabolites in wastewater of a sewage treatment plant. *Sci. Total Environ.* **2015**, *532*, 762–770. [CrossRef] [PubMed]
9. Loos, R.; Marinov, D.; Sanseverino, I.; Napierska, D.; Lettieri, T. *Review of the 1st Watch List under the Water Framework Directive and recommendations for the 2nd Watch List*; European Commission: Luxembourg, 2018.
10. Greenham, R.T.; Miller, K.Y.; Tong, A. Removal efficiencies of top-used pharmaceuticals at sewage treatment plants with various technologies. *J. Environ. Chem. Eng.* **2019**, *7*, 103294. [CrossRef]
11. Botero-Coy, A.M.; Martínez-Pachón, D.; Boix, C.; Rincón, R.J.; Castillo, N.; Arias-Marín, L.P.; Manrique-Losada, L.; Torres-Palma, R.; Moncayo-Lasso, A.; Hernández, F. An investigation into the occurrence and removal of pharmaceuticals in Colombian wastewater. *Sci. Total Environ.* **2018**, *642*, 842–853. [CrossRef]
12. Asimakopoulos, A.G.; Kannan, P.; Higgins, S.; Kannan, K. Determination of 89 drugs and other micropollutants in unfiltered wastewater and freshwater by LC-MS/MS: An alternative sample preparation approach. *Anal. Bioanal. Chem.* **2017**, *409*, 6205–6225. [CrossRef] [PubMed]
13. Kaplan, S. Review: Pharmacological pollution in water. *Crit. Rev. Environ. Sci. Technol.* **2013**, *43*, 1074–1116. [CrossRef]
14. Wang, J.; Wang, S. Removal of pharmaceuticals and personal care products (PPCPs) from wastewater: A review. *J. Environ. Manag.* **2016**, *182*, 620–640. [CrossRef]
15. Crites, R.W.; Middlebrooks, E.J.; Bastian, R.K. *Natural Wastewater Treatment Systems*, 2nd ed.; CRC Press: Boca Raton, FL, USA, 2014; ISBN 9781466583269.
16. Matamoros, V.; Rodríguez, Y.; Albaigés, J. A comparative assessment of intensive and extensive wastewater treatment technologies for removing emerging contaminants in small communities. *Water Res.* **2016**, *88*, 777–785. [CrossRef] [PubMed]
17. Guedes-Alonso, R.; Montesdeoca-Esponda, S.; Herrera-Melián, J.A.; Rodríguez-Rodríguez, R.; Ojeda-González, Z.; Landívar-Andrade, V.; Sosa-Ferrera, Z.; Santana-Rodríguez, J.J. Pharmaceutical and personal care product residues in a macrophyte pond-constructed wetland treating wastewater from a university campus: Presence, removal and ecological risk assessment. *Sci. Total Environ.* **2020**, *703*, 135596. [CrossRef]
18. Durán-Álvarez, J.C.; Sánchez, Y.; Prado, B.; Jiménez, B. The transport of three emerging pollutants through an agricultural soil irrigated with untreated wastewater. *J. Water Reuse Desalin.* **2014**, *4*, 9–17. [CrossRef]

19. Afonso-Olivares, C.; Sosa-Ferrera, Z.; Santana-Rodríguez, J.J. Occurrence and environmental impact of pharmaceutical residues from conventional and natural wastewater treatment plants in Gran Canaria (Spain). *Sci. Total Environ.* **2017**, *599–600*, 934–943. [CrossRef]
20. Rodriguez-Narvaez, O.M.; Peralta-Hernandez, J.M.; Goonetilleke, A.; Bandala, E.R. Treatment technologies for emerging contaminants in water: A review. *Chem. Eng. J.* **2017**, *323*, 361–380. [CrossRef]
21. Oller, I.; Miralles-Cuevas, S.; Aguera, A.; Malato, S. Monitoring and Removal of Organic Micro-contaminants by Combining Membrane Technologies with Advanced Oxidation Processes. *Curr. Org. Chem.* **2018**, *22*, 1103–1119. [CrossRef]
22. Arola, K.; Ward, A.; Mänttäri, M.; Kallioinen, M.; Batstone, D. Transport of pharmaceuticals during electrodialysis treatment of wastewater. *Water Res.* **2019**, *161*, 496–504. [CrossRef] [PubMed]
23. Ibekwe, A.M.; Murinda, S.E. Linking microbial community composition in treated wastewater with water quality in distribution systems and subsequent health effects. *Microorganisms* **2019**, *7*, 660. [CrossRef] [PubMed]
24. Rodriguez-Gil, J.L.; San Sebastián Sauto, J.; González-Alonso, S.; Sánchez Sanchez, P.; Valcarcel, Y.; Catalá, M. Development of cost-effective strategies for environmental monitoring of irrigated areas in Mediterranean regions: Traditional and new approaches in a changing world. *Agric. Ecosyst. Environ.* **2013**, *181*, 41–49. [CrossRef]
25. Sun, Q.; Lv, M.; Hu, A.; Yang, X.; Yu, C.P. Seasonal variation in the occurrence and removal of pharmaceuticals and personal care products in a wastewater treatment plant in Xiamen, China. *J. Hazard. Mater.* **2014**, *277*, 69–75. [CrossRef] [PubMed]
26. Ekpeghere, K.I.; Sim, W.J.; Lee, H.J.; Oh, J.E. Occurrence and distribution of carbamazepine, nicotine, estrogenic compounds, and their transformation products in wastewater from various treatment plants and the aquatic environment. *Sci. Total Environ.* **2018**, *640–641*, 1015–1023. [CrossRef]
27. Rico, A.; Arenas-Sánchez, A.; Alonso-Alonso, C.; López-Heras, I.; Nozal, L.; Rivas-Tabares, D.; Vighi, M. Identification of contaminants of concern in the upper Tagus river basin (central Spain). Part 1: Screening, quantitative analysis and comparison of sampling methods. *Sci. Total Environ.* **2019**, *666*, 1058–1070. [CrossRef] [PubMed]
28. Simó Miñana, J. Use of prescription drugs in Spain and Europe. *Aten. Primaria* **2012**, *44*, 335–347. [CrossRef]
29. Camacho-Muñoz, D.; Martín, J.; Santos, J.L.; Aparicio, I.; Alonso, E. Occurrence, temporal evolution and risk assessment of pharmaceutically active compounds in Doñana Park (Spain). *J. Hazard. Mater.* **2010**, *183*, 602–608. [CrossRef]
30. Gabarrón, S.; Gernjak, W.; Valero, F.; Barceló, A.; Petrovic, M.; Rodríguez-Roda, I. Evaluation of emerging contaminants in a drinking water treatment plant using electrodialysis reversal technology. *J. Hazard. Mater.* **2016**, *309*, 192–201. [CrossRef]
31. Sundaravadivel, M.; Vigneswaran, S. Constructed wetlands for wastewater treatment. *Crit. Rev. Environ. Sci. Technol.* **2001**, *31*, 351–409. [CrossRef]
32. Scholz, M.; Lee, B. Constructed wetlands: A review. *Int. J. Environ. Stud.* **2005**, *62*, 421–447. [CrossRef]
33. Jelic, A.; Gros, M.; Ginebreda, A.; Cespedes-Sánchez, R.; Ventura, F.; Petrovic, M.; Barcelo, D. Occurrence, partition and removal of pharmaceuticals in sewage water and sludge during wastewater treatment. *Water Res.* **2011**, *45*, 1165–1176. [CrossRef]
34. Hai, F.; Yang, S.; Asif, M.; Sencadas, V.; Shawkat, S.; Sanderson-Smith, M.; Gorman, J.; Xu, Z.-Q.; Yamamoto, K. Carbamazepine as a Possible Anthropogenic Marker in Water: Occurrences, Toxicological Effects, Regulations and Removal by Wastewater Treatment Technologies. *Water* **2018**, *10*, 107. [CrossRef]
35. Clara, M.; Strenn, B.; Gans, O.; Martinez, E.; Kreuzinger, N.; Kroiss, H. Removal of selected pharmaceuticals, fragrances and endocrine disrupting compounds in a membrane bioreactor and conventional wastewater treatment plants. *Water Res.* **2005**, *39*, 4797–4807. [CrossRef] [PubMed]
36. Tadkaew, N.; Hai, F.I.; McDonald, J.A.; Khan, S.J.; Nghiem, L.D. Removal of trace organics by MBR treatment: The role of molecular properties. *Water Res.* **2011**, *45*, 2439–2451. [CrossRef]
37. Luo, W.; Hai, F.I.; Kang, J.; Price, W.E.; Guo, W.; Ngo, H.H.; Yamamoto, K.; Nghiem, L.D. Effects of salinity build-up on biomass characteristics and trace organic chemical removal: Implications on the development of high retention membrane bioreactors. *Bioresour. Technol.* **2015**, *177*, 274–281. [CrossRef] [PubMed]

38. Tiwari, B.; Sellamuthu, B.; Ouarda, Y.; Drogui, P.; Tyagi, R.D.; Buelna, G. Review on fate and mechanism of removal of pharmaceutical pollutants from wastewater using biological approach. *Bioresour. Technol.* **2017**, *224*, 1–12. [CrossRef] [PubMed]
39. Jones, O.A.H.; Voulvoulis, N.; Lester, J.N. Human pharmaceuticals in wastewater treatment processes. *Crit. Rev. Environ. Sci. Technol.* **2005**, *35*, 401–427. [CrossRef]

Sample Availability: Samples of the compounds are not available from the authors.

Article

Constructed Wetland Revealed Efficient Sulfamethoxazole Removal but Enhanced the Spread of Antibiotic Resistance Genes

Shuai Zhang [1], Yu-Xiang Lu [2], Jia-Jie Zhang [3], Shuai Liu [4], Hai-Liang Song [2,*] and Xiao-Li Yang [5,*]

1 Jiangsu Key Laboratory of Atmospheric Environment Monitoring and Pollution Control (AEMPC), Collaborative Innovation Center of Atmospheric Environment and Equipment Technology (CIC-AEET), Nanjing University of Information Science &Technology, Nanjing 210044, China; zhangshuai198702@163.com

2 School of Environment, Nanjing Normal University, Jiangsu Engineering Lab of Water and Soil Eco-remediation, Wenyuan Road 1, Nanjing 210023, China; lu_uyx@163.com

3 School of Civil Engineering and Architecture, East China Jiaotong University, Nanchang 330013, China; zhangjiajieedu@163.com

4 College of Environment and Safety Engineering, Qingdao University of Science and Technology, Qingdao 266042, China; liushuai@qust.edu.cn

5 School of Civil Engineering, Southeast University, Nanjing 210096, China

* Correspondence: hlsong@njnu.edu.cn (H.-L.S.); yangxiaoli@seu.edu.cn (X.-L.Y.); Tel.: +86-25-85891243 (H.-L.S.); +86-25-52091223 (X.-L.Y.)

Academic Editor: Derek J. McPhee
Received: 6 January 2020; Accepted: 11 February 2020; Published: 14 February 2020

Abstract: Constructed wetlands (CWs) could achieve high removal efficiency of antibiotics, but probably stimulate the spread of antibiotic resistance genes (ARGs). In this study, four CWs were established to treat synthetic wastewater containing sulfamethoxazole (SMX). SMX elimination efficiencies, SMX degradation mechanisms, dynamic fates of ARGs, and bacterial communities were evaluated during the treatment period (360 day). Throughout the whole study, the concentration of SMX in the effluent gradually increased ($p < 0.05$), but in general, the removal efficiency of SMX remained at a very high level (>98%). In addition, the concentration of SMX in the bottom layer was higher compared with that in the surface layer. The main byproducts of SMX degradation were found to be 4-amino benzene sulfinic acid, 3-amino-5-methylisoxazole, benzenethiol, and 3-hydroxybutan-1-aminium. Temporally speaking, an obvious increase of *sul* genes was observed, along with the increase of SMX concentration in the bottom and middle layers of CWs. Spatially speaking, the concentration of *sul* genes increased from the surface layer to the bottom layer.

Keywords: sulfamethoxazole; antibiotic resistance genes; *sul* genes; bacterial community; constructed wetlands

---

1. Introduction

In recent years, antibiotics have been extensively used as livestock food additives and to fight infections in animal husbandry [1,2]. The overuse of antibiotics results in their continuous release into the environment in China. Additionally, previous investigations have indicated that antibiotics in animal waste could not be entirely removed using traditional lagoon treatment and in wastewater treatment plants (WWTPs) [3]. As a result, antibiotics were widely detected in wastewater, surface water, and groundwater [4,5].

The wide presence of antibiotics in the environment could cause concern because it not only causes serious toxic effects on organisms, but also promotes the spread of antibiotic-resistant genes

(ARGs) [6], even with low concentrations in the environment [7,8]. ARGs could be spread through horizontal gene transfer (HGT) and vertical gene transfer (VGT), and in many cases, could be maintained in microbial populations, even without selection pressure from antibiotics [9–11]. HGT is a major pathway for the transfer of ARGs, including conjugative transposons, integrons, insertion sequences, and plasmids [11,12]. ARGs have often been detected as part of antibiotic resistance super integrons. Therefore, one antibiotic may coselect resistance to other antibiotics when applying multiple antibiotics [13]. Even if antibiotic-resistant bacteria were damaged or killed, ARGs could still be released to the environment and then transformed into other bacteria [14,15]. Previous studies have revealed the high relative abundances of ARGs in wastewater lagoons and municipal wastewater, even after treatment [16–18]. In recent years, ARGs were regarded as fast-growing potential pollution because of the extensive application of antibiotics in the livestock industry [19–21]. Hence, effective treatment processes for antibiotic removal could also prevent the spread of ARGs.

Constructed wetlands (CWs) are designed and constructed exploiting natural processes to treat rural wastewater [22,23]. The advantages of CWs mainly lie in high purification removal, relatively low construction and maintenance costs, reduced energy consumption, convenient operation, and broad application prospects in developing countries or rural areas [24,25]. Recently, CWs were used to remove antibiotics from agricultural and municipal wastewater via natural processes involving plants, soil/sediment, and microorganisms [21,26,27]. CWs have been shown to be more efficient in the removal of antibiotics and ARGs than conventional wastewater treatment systems [16,28]. Liu et al. (2013) [28] found that the total absolute abundances of tetracycline resistance (*tet*) genes and the 16S rRNA were reduced by 50% from swine wastewater using CWs. Huang et al. (2015) [16] reported that the absolute abundances of the ARGs were greatly reduced, with their log units ranging from 0.26 to 3.3. However, previous studies of ARG reduction have always focused on their elimination, rather than the induction of ARGs along with antibiotics removal by CWs. CWs could also be a significant source of ARGs, and may enhance their spread. Therefore, exploring the induction of ARGs in CWs would greatly assist in evaluating their environmental risks.

Sulfamethoxazole (SMX) is a synthetic antibiotic within the sulfonamide antibiotic family, and is largely consumed in the livestock husbandry [29,30]. *Sul* genes (*sulI* and *sulII*) were chosen as the representatives of ARGs for their frequent use [17,31–33]. In this study, four CWs were established to treat a synthetic wastewater containing SMX for 360 day. The objectives of this study were: (1) to investigate the elimination efficiencies and products of SMX, (2) to explore the development of *sul* genes in the reactors, (3) to assess the risks of ARGs in effluent, and (3) the bacterial community during the treatment process.

2. Results and Discussion

2.1. SMX Removal Efficiency

The concentrations of SMX in the effluent, bottom layer, and surface layer of CWs at five sampling points are shown in Figure 1. The concentrations of SMX in the effluent ranged 0.051–0.214, 0.047–0.274, 0.0584–0.342, and 0.098–0.574 μg L^{-1} in CW1, CW2, CW3, and CW4. CW4 fed with 200 μg L^{-1} SMX exhibited significantly higher concentrations of SMX in the effluent than CW1 fed with 20 μg L^{-1} ($p < 0.05$). On D360, the SMX concentrations in the effluent were 0.214, 0.185, 0.342, and 0.574 μg L^{-1} with influent SMX concentration of 20, 50, 100, and 200 μg L^{-1}, respectively. Temporally speaking, the SMX concentrations in effluent gradually increased during the study ($p < 0.05$). In the effluent of CW3, SMX concentrations were 0.0584, 0.142, 0.198, 0.247, and 0.342 on D30, D60, D120, D240, and D360, respectively. Notably, excellent removal efficiencies (>98%) for SMX were obtained using CWs, even when the influent SMX concentration was as high as 200 μg L^{-1}.

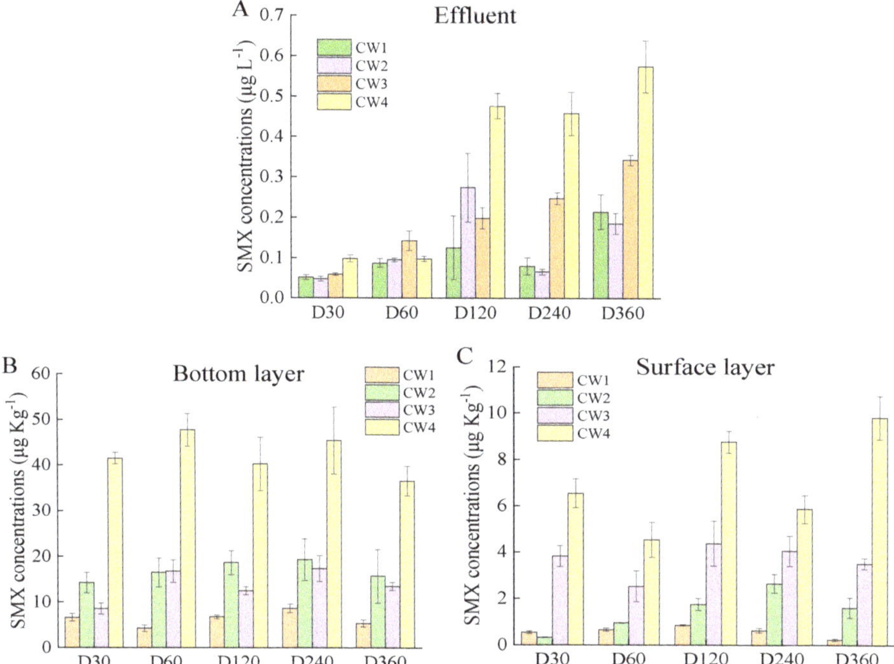

Figure 1. Concentrations of SMX (mean ± SD, $n = 3$) in the effluent water and layers. (**A**) concentration of SMX in the effluent; (**B**) concentration of SMX in the bottom layer; (**C**) concentration of SMX in the surface layer.

SMX has been reported to be easily biodegradable, especially under anaerobic conditions in the CWs [25]. In the present study, it was noteworthy that the removal efficiencies of antibiotics by different CWs were even better than conventional WWTPs [34]. As demonstrated by previous experiments, biodegradation, absorption, hydrolysis, and photodecomposition played significant roles in the removal of antibiotics in the solid and aqueous phases in CWs [34–36].

Previous reports also showed that the adsorption process only accounted for a minor percentage of the total removal in CWs [12].

The SMX concentrations of the bottom layer ranged 4.256–8.620, 14.213–19.281, 8.546–17.322, and 36.564–47.684 μg Kg^{-1} in CW1, CW2, CW3, and CW4. The SMX concentrations of the surface layer ranged 0.239–0.852, 0.327–2.652, 2.526–4.365, and 4.531–9.829 μg Kg^{-1} in CW1, CW2, CW3, and CW4, respectively. The difference of SMX concentration between the bottom and surface layers was affected by the influent SMX concentration. This was probably because the relatively high K_d values prompted SMX to be adsorbed in the substrates of CWs [37]. However, the SMX concentrations in the layers did not significantly increase during the treatment ($p > 0.05$). This result was probably due to the fact that other physicochemical and biological processes occurred jointly or separately after SMX adsorption in substrate [16]. Clearly, the concentration of SMX in the bottom layer was higher compared with that in surface layer. This was not surprising, because that the bottom layer was faced with a higher concentration of SMX due to the vertical up-influent.

2.2. SMX Degradation Products

LC-MS/MS was employed to identify the degradation products in order to clarify the degradation pathway of SMX ($C_{10}H_{11}N_3O_3S$, *m/z* 252.0450). According to the detected mass/charge ratios under positive mode, several potential metabolites were identified, including 4-amino benzene sulfinic acid ($C_6H_7NO_2S$, *m/z* 158.0123), 3-amino-5-methylisoxazole (3A5MI, $C_4H_6N_2O$, *m/z* 99.5315), benzenethiol (C_6H_6S, *m/z* 111.0496), and 3-hydroxybutan-1-aminium ($C_4H_{11}NO$, *m/z* 90.5263) (Figure 2). A previous report indicated that the removal of antibiotics was mainly caused by adsorption rather than degradation in wastewater treatment plants [38]. However, in this study, it was found that the products of SMX may be transformed by bacterial communities. A previous study also found 4-amino benzene sulfinic acid and 3A5MI during the degradation of SMX [39]. During the degradation process, SMX was initially hydrolyzed into 4-amino benzene sulfinic acid and 3A5MI; then, 4-amino benzene sulfinic acid was further transformed into benzenethiol, which was finally degraded to generate methane or carbon dioxide. During the transformation of 3A5MI to 3-hydroxybutan-1-aminium, the isoxazole ring was opened and its nitrogen atom was removed. Since 3-hydroxybutan-1-aminium has a chain structure, it can be easily mineralized into methane by microbes under anoxic conditions [40].

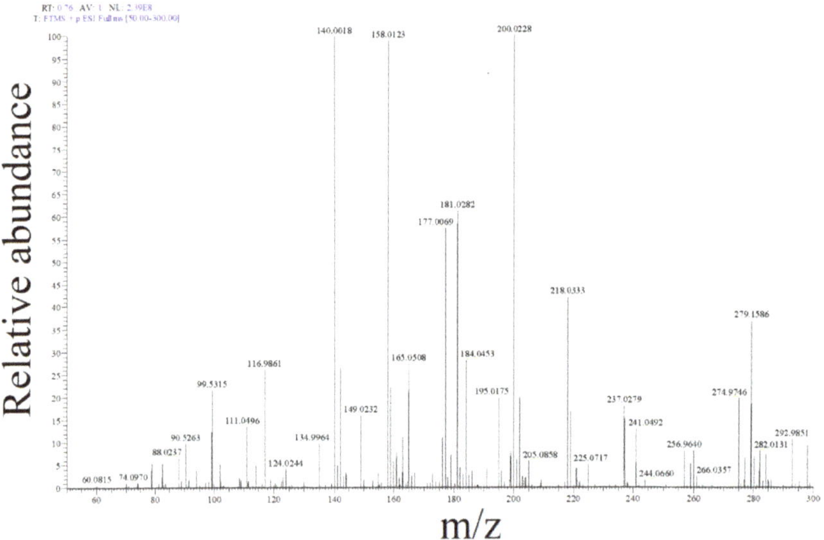

Figure 2. Q-Exactive spectrum degradation products of SMX.

2.3. *Sul Genes in the Effluent and Media*

In order to analyze the changes of corresponding target genes, the relative abundances of *sul* genes were analyzed during the treatment process. The distributions of two *sul* (*sulI* and *sulII*) in the CWs are shown in Figure 3. The relative abundances of *sul* genes showed an obvious increase with the increase of SMX concentrations in the bottom and middle layers. In wastewater treatment installations, the abundances of ARGs were not only determined by their abundances, but were also affected by the concentrations of antibiotics in the influent [6]. Since synthetic wastewater was used in this study, the abundances of *sul* genes in the influent may be ignored. In addition, the *sul* genes were not found in CW0. Therefore, most of the target genes were induced by the antibiotics in the influent. Further, bacteria would also gain the corresponding ARGs via VGT and HGT [10].

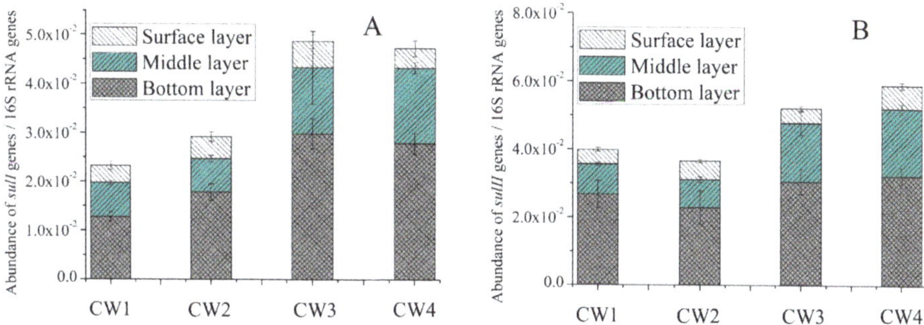

Figure 3. *Sul* genes (*sulI* and *sulII*) normalized to 16S rRNA genes in the layers (bottom layer, middle layer and surface layer) of CW1, CW2, CW3, and CW4, respectively (mean ± SD, $n = 3$). (**A**) *sulI* genes; (**B**) *sulII* genes.

Notably, the concentration of *sul* genes in the bottom layer was higher than that in the middle layer, with the surface layer containing the least (Figure 3). A similar result was obtained in a previous study, which found that the level of ARGs was high in CWs [28]. This observation was mainly influenced by the level of SMX sources and oxygen transport capacity in the bottom layer [28]. Clearly the concentration of SMX in the bottom layer was the highest, followed by the middle and surface layers (Figure 1). The fate of ARGs in the different layers of CWs was mainly attributable to the accumulation of SMX in the substrate. Interestingly, the relative abundances of *sul* genes in the CWs were in the order of *sulII* > *sulI* (Figure 3). This was not surprising, because the different fates of *sul* genes in CWs were mainly caused by their specific mechanisms [6]. The *sulI* gene was generally found to be associated with other ARGs in class 1 integrons, while *sulII* was usually located on small nonconjugative plasmids, or generally located on large transmissible multi-resistance [41]. Therefore, the persistence of *sulII* genes might be attributed to the successive pressure exerted by antimicrobial agents that were transferred via HGT and VGT between pathogens, nonpathogens, and even distantly related organisms [31,41].

ARGs, as a major source of pollution, may be spread in bioreactors [42]. The relative abundances of corresponding *sul* genes in effluent have been reported (Figure 4). The CWs, indicating the rate of spread for *sul* genes, were probably caused by antibiotics. This was comparable to previous reports in which similar abundances of ARGs were observed in the effluent of CWs [21]. In addition, the relative abundance of most *sul* genes was enhanced with higher concentrations of SMX; *sul* genes were not detected in the effluent of CW0. The relative abundance of *sul* genes exhibited an increase, which tended to be stable among the treatment duration. Meanwhile, the relative abundances of *sul* genes in the effluent were in the order of *sulII* > *sulI*. In a word, vertical up-CWs may be a good choice for application as an effective SMX removal method, but the fate of ARGs remains to be further studied in practical applications.

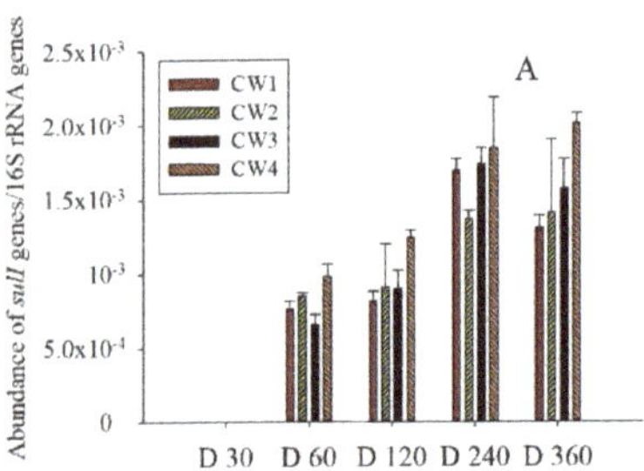

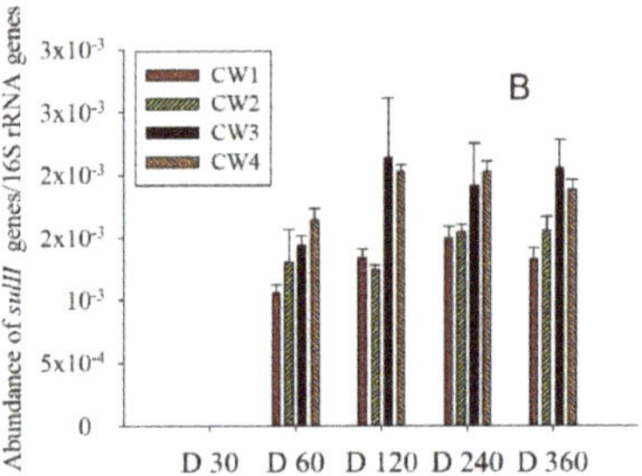

Figure 4. *Sul* genes (*sulI* and *sulII*) normalized to 16S rRNA genes in the effluent of CW1, CW2, CW3, and CW4, respectively (mean ± SD, $n = 3$). (**A**) *sulI* genes; (**B**) *sulII* genes.

2.4. Composition of Bacterial Communities

The microbial community was evaluated in terms of abundance and bacterial structure in response to the different treatments in CWs [43]. There were 30 dominant genera in the phylum level for the bottom layers, occupying > 94.90% of sequences (Figure 5). Twenty genera were identified while 10 remained unknown by taxonomy assignment. *OD1* (49.32%), *Proteobacteria* (44.59%), *Proteobacteria* (31.36%), and *Proteobacteria* (54.53%) had the maximum amounts of dominant phylum and high relative abundances of detection in CW1, CW2, CW3, and CW4, respectively, followed by *Chloroflexi, Bacteroidetes*, and *Acidobacteria*. Previous studies reported that proteobacteria is responsible for the degradation of refractory organic [44]. In addition, *Proteobacteria* (12.79%), *Bacteroidetes, Acidobacteria*, and *Planctomycetes* increased after the SMX treatment process. Because of potential coselection on SMX, the accumulation of antibiotics in substrates may also contribute to the enhancement of some dominant bacteria. Our result was similar to that in a previous report, i.e., that *Proteobacteria* occupied the string majority in antibiotic-correlated bacteria, followed by *Bacteroidetes* and *Actinobacteria* [45]. Some specific bacteria in CWs, like *Nitrospirae*, are well-known to be involved in nitrification and ammonia oxidization [46]. In accordance with our result, *Nitrospirae* was a dominant phylum, with a relative abundance of up to 4.45%.

To assess the variation of dominant bacteria after CW treatment, five samples could be divided into three clusters according to the bacterial composition and relative abundance. Cluster I comprised the original sample, cluster II included the CW1 and CW2 samples, and cluster III comprised all of the remaining samples (CW3 and CW4). Previous study has shown that no significant difference was observed in terms of bacterial abundance, richness, or diversity among different treatments of antibiotics [43]. However, in this experiment, the bacterial communities and their relative abundance were influenced by the SMX content of the influent. Therefore, the phenomenon suggested that the original bacterial structure had to adapt to the variation of different SMX concentration conditions, appearing to change over the 360 days of the experiment period.

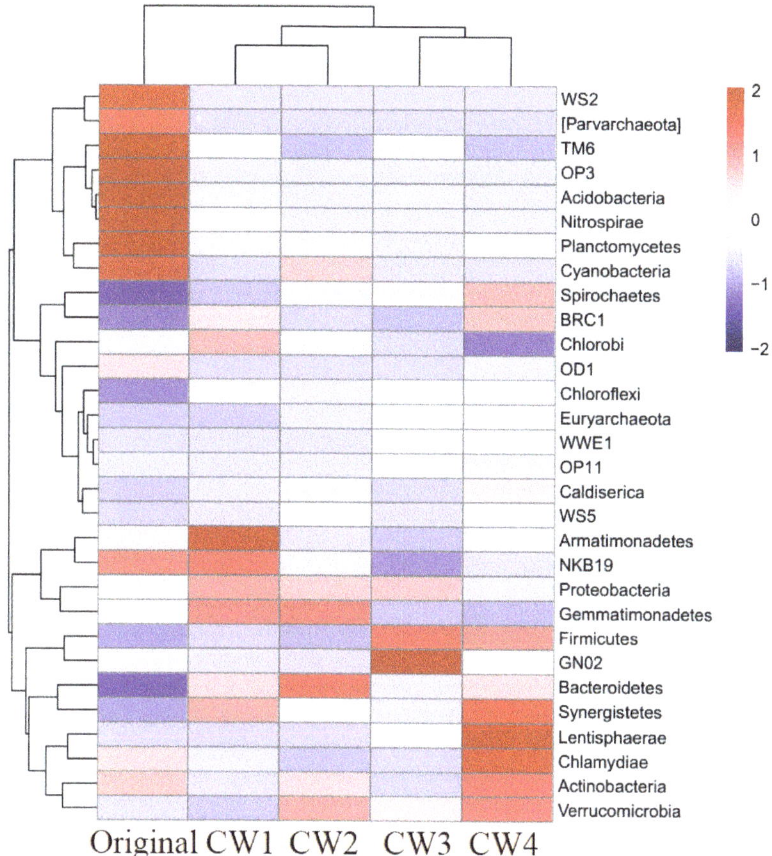

Figure 5. Heat map of bacterial populations in phylum level: Sample names were listed on the x-axis, and genus names were listed on the y-axis.

3. Materials and Methods

3.1. Reactor Configuration

Four CWs were set up with 65 cm in height and 35 cm in diameter with the temperature maintained at 26 ± 3 °C and a relative humidity of 55–65%. The CWs were filled with siliceous gravel and sand (siliceous gravel: sand volume ratio = 1:1; siliceous gravel was 4–7 mm and sand was 5–8 mm in diameter). *Phragmites australis* were transplanted into the top layer. The synthetic wastewater was fed into four CWs from the bottom inlet by peristaltic pump with a hydraulic loading rate of 5 cm d^{-1}. The synthetic wastewater was prepared with tap water containing chemical oxygen demand (500 mg L^{-1}), ammonia nitrogen (40 mg L^{-1}), total nitrogen (150 mg L^{-1}), and total phosphorus (20 mg L^{-1}). Then, 0, 20, 50, 100, and 200 μg L^{-1} of SMX were spiked into synthetic wastewater for CW0, CW1, CW2, CW3, and CW4, respectively.

3.2. SMX Detection

First, 1000 mL of effluent was collected at days 30, 60, 120, 240, and 360, to measure the SMX concentrations. Then, 200 g wetland media was collected from the bottom, middle, and surface layers on day 360. Both water samples and wetland media were taken in triplicate. Water samples were filtered through 0.45 μm fiber filters [16]. Wetland media were extracted by solid-phase extraction (Waters, Millford, MA, USA) [47]. Liquid chromatography-mass spectrometry (LC-MS/MS, Thermo Scientific Q Exactive Hybrid Quadrupole-Orbitrap; Thermo Fisher Scientific, Waltham, MA, USA) was used to analyze the concentrations of SMX in water and wetland media. The mobile phase was composed of pure 30% acetonitrile and 70% water solution (v/v) [39]. Hypersil GOLD C18 column (Acquity UPLC BEH C18; 100 mm × 2.1 mm, 3 μm) was used in this study. The capillary voltage was 3.8 (±) kV, the collision energy was 8 eV, and the capillary temperature was set to 350 °C. SMX products were detected by the LC-MS/MS in positive mode [16].

3.3. DNA Extraction and ARG Analysis

Genomic DNA was extracted from effluents (200 mL) and wetland media (5 g) using a DNA extraction kit (MoBio, Carlsbad, CA, USA) at each sampling point in Section 2.2. The DNA concentrations were tested using ND-1000 NanoDrop spectrophotometer (Thermo Fisher Scientific, Wilmington, DE, USA). The 16S rRNA gene and two *sul* genes (*sulI* and *sulII*) were quantified by a Real-Time PCR System (CFX96, Bio-Rad). The reaction was 25 μL on 96-well plates (Bio-Rad, Shanghai) containing 12.5 μL SYBR Green qPCR mix (Bio-Rad, Shanghai), 0.5 μL of each forward and reverse primers (Bio-Rad, Shanghai), 1 μL DNA templates, and 10.5 μL ddH_2O [48]. PCR protocol and primer sequences of *sul* genes were based on previous studies [49,50].

3.4. High-Throughput Sequencing

Wetland media (5 g) were collected at the bottom layer to analyze microbial community. The V4 region of the bacteria 16S rRNA gene was amplified by PCR (5′-GTGCCAGCMGCCGCGGTAA-3′, 5′-GGACTACHVGGGTWTCTAAT-3′) according to previous reports [51,52]. Each PCR reaction was performed with 50 μL mixture containing 35 ng template DNA, 4 μL PCR Primer Cocktail (5 μM, Qiagen, Valencia, CA, USA), 25 μL PCR Master Mix (Qiagen), and ddH_2O [53]. Illumina MiSeq platform was employed to purify and pool amplicons in equimolar amounts and paired-end sequenced (2 × 250) [51,54]. Data were analyzed with Microsoft Excel 2010, and statistical analyses were conducted by SPSS ver. 19.0.

4. Conclusions

This study clearly demonstrated that excellent SMX removal efficiency among different SMX concentrations was obtained during the treatment in the CWs. The concentration of SMX in the bottom layer was higher compared with that in the surface layer. Good removal efficiencies for SMX were observed using the systems. A degradation mechanism of SMX was proposed. The relative abundances of *sul* genes showed an obvious increase with the increase of SMX content in the bottom and middle layers. The concentration of *sul* genes in the bottom layer was shown to be higher than that in the middle layer; the surface layer presented the lowest concentration. The relative abundance of *sul* genes exhibited an increase, which tended to be stable among the treatment duration. *Proteobacteria* was the dominant phylum in the CWs.

Author Contributions: Methodology, Y.-X.L.; validation, J.-J.Z. and S.L.; investigation, S.Z.; writing—original draft preparation, S.Z.; writing—review and editing, H.-L.S.; funding acquisition, H.-L.S. and X.-L.Y. All authors have read and agreed to the published version of the manuscript.

Funding: This research was funded by National Key Research and Development Program of China (2019YFD1100205), National Major Science and Technology Projects of China (2017ZX07202004), and the National Natural Science Foundation of China (41571476).

Acknowledgments: Hai-Liang Song would like to acknowledge the Qing Lan Project of Jiangsu Province.

Conflicts of Interest: The authors declare no conflict of interest.

References

1. Hou, J.; Wan, W.N.; Mao, D.Q.; Wang, C.; Mu, Q.H.; Qin, S.Y.; Luo, Y. Occurrence and distribution of sulfonamides, tetracyclines, quinolones, macrolides, and nitrofurans in livestock manure and amended soils of Northern China. *Environ. Sci. Pollut. Res.* **2015**, *22*, 4545–4554. [CrossRef] [PubMed]
2. Zhou, L.J.; Ying, G.G.; Zhang, R.Q.; Liu, S.; Lai, H.J.; Chen, Z.F.; Yang, B.; Zhao, J.L. Use patterns, excretion masses and contamination profiles of antibiotics in a typical swine farm, south China. *Environ. Sci. Proc. Imp.* **2013**, *15*, 802–813. [CrossRef] [PubMed]
3. Adegoke, A.A.; Faleye, C.A.; Singh, G.; Stenström, A.T. Antibiotic Resistant Superbugs: Assessment of the Interrelationship of Occurrence in Clinical Settings and Environmental Niches. *Molecules* **2017**, *22*, 29. [CrossRef] [PubMed]
4. Pham, T.D.; Vu, T.N.; Nguyen, H.L.; Le, P.H.P.; Hoang, T.S. Adsorptive removal of antibiotic ciprofloxacin from aqueous solution using protein-modified nanosilica. *Polymers* **2020**, *12*, 57. [CrossRef] [PubMed]
5. Manyi-Loh, C.; Mamphweli, S.; Meyer, E.; Okoh, A. Antibiotic Use in Agriculture and Its Consequential Resistance in Environmental Sources: Potential Public Health Implications. *Molecules* **2018**, *23*, 795. [CrossRef] [PubMed]
6. McKinney, C.W.; Loftin, K.A.; Meyer, M.T.; Davis, J.G.; Pruden, A. Tet and sul antibiotic resistance genes in livestock lagoons of various operation type, configuration, and antibiotic occurrence. *Environ. Sci. Technol.* **2010**, *44*, 6102–6109. [CrossRef] [PubMed]
7. Storteboom, H.; Arabi, M.; Davis, J.G.; Crimi, B.; Pruden, A. Identification of antibiotic-resistance-gene molecular signatures suitable as tracers of pristine river, urban, and agricultural sources. *Environ. Sci. Technol.* **2010**, *44*, 1947–1953. [CrossRef]
8. Ogawara, H. Comparison of Antibiotic Resistance Mechanisms in Antibiotic-Producing and Pathogenic Bacteria. *Molecules* **2019**, *24*, 3430. [CrossRef]
9. Yuan, H.; Miller, J.H.; Abu-Reesh, I.M.; Pruden, A.; He, Z. Effects of electron acceptors on removal of antibiotic resistant Escherichia coli, resistance genes and class 1 integrons under anaerobic conditions. *Sci. Total Environ.* **2016**, *569*, 1587–1594. [CrossRef]
10. Kim, S.; Yun, Z.; Ha, U.H.; Lee, S.; Park, H.; Kwon, E.E.; Cho, Y.; Choung, S.; Oh, J.; Medriano, C.A.; et al. Transfer of antibiotic resistance plasmids in pure and activated sludge cultures in the presence of environmentally representative micro-contaminant concentrations. *Sci. Total Environ.* **2014**, *468–469*, 813–820. [CrossRef]
11. Guo, J.H.; Li, J.; Chen, H.; Bond, P.L.; Yuan, Z.G. Metagenomic analysis reveals wastewater treatment plants as hotspots of antibiotic resistance genes and mobile genetic elements. *Water Res.* **2017**, *123*, 468–478. [CrossRef] [PubMed]
12. Chen, J.; Ying, G.G.; Wei, X.D.; Liu, Y.S.; Liu, S.S.; Hu, L.X.; He, L.Y.; Chen, Z.F.; Chen, F.R.; Yang, Y.Q. Removal of antibiotics and antibiotic resistance genes from domestic sewage by constructed wetlands: Effect of flow configuration and plant species. *Sci. Total Environ.* **2016**, *571*, 974–982. [CrossRef] [PubMed]
13. Beekmann, S.E.; Heilmann, K.P.; Richter, S.S.; García-de-Lomas, J.; Doern, G.V.; Group, G.S. Antimicrobial resistance in Streptococcus pneumoniae, Haemophilus influenzae, Moraxella catarrhalis and group A β-haemolytic streptococci in 2002–2003: Results of the multinational GRASP Surveillance Program. *Int. J. Antimicrob. Agents* **2005**, *25*, 148–156. [CrossRef] [PubMed]
14. Penesyan, A.; Gillings, M.; Paulsen, I. Antibiotic Discovery: Combatting Bacterial Resistance in Cells and in Biofilm Communities. *Molecules* **2015**, *20*, 5286–5298. [CrossRef] [PubMed]
15. Chang, P.H.; Juhrend, B.; Olson, T.M.; Marrs, C.F.; Wigginton, K.R. Degradation of extracellular antibiotic resistance genes with UV254 treatment. *Environ. Sci. Technol.* **2017**, *51*, 6185–6192. [CrossRef] [PubMed]
16. Huang, X.; Liu, C.; Li, K.; Su, J.; Zhu, G.; Liu, L. Performance of vertical up-flow constructed wetlands on swine wastewater containing tetracyclines and tet genes. *Water Res.* **2015**, *70*, 109–117. [CrossRef]
17. Naquin, A.; Shrestha, A.; Sherpa, M.; Nathaniel, R.; Boopathy, R. Presence of antibiotic resistance genes in a sewage treatment plant in Thibodaux, Louisiana, USA. *Bioresour. Technol.* **2015**, *188*, 79–83. [CrossRef]
18. Gao, P.; Munir, M.; Xagoraraki, I. Correlation of tetracycline and sulfonamide antibiotics with corresponding resistance genes and resistant bacteria in a conventional municipal wastewater treatment plant. *Sci. Total Environ.* **2012**, *421–422*, 173–183. [CrossRef]

19. Zhang, S.; Song, H.-L.; Yang, X.-L.; Yang, Y.-L.; Yang, K.-Y.; Wang, X.-Y. Fate of tetracycline and sulfamethoxazole and their corresponding resistance genes in microbial fuel cell coupled constructed wetlands. *RSC Adv.* **2016**, *6*, 95999–96005. [CrossRef]
20. Rosendahl, I.; Siemens, J.; Groeneweg, J.; Linzbach, E.; Laabs, V.; Herrmann, C.; Vereecken, H.; Amelung, W. Dissipation and sequestration of the veterinary antibiotic sulfadiazine and its metabolites under field conditions. *Environ. Sci. Technol.* **2011**, *45*, 5216–5222. [CrossRef]
21. Liu, L.; Liu, C.; Zheng, J.; Huang, X.; Wang, Z.; Liu, Y.; Zhu, G. Elimination of veterinary antibiotics and antibiotic resistance genes from swine wastewater in the vertical flow constructed wetlands. *Chemosphere* **2013**, *91*, 1088–1093. [CrossRef] [PubMed]
22. Hussain, S.A.; Prasher, S.O.; Patel, R.M. Removal of ionophoric antibiotics in free water surface constructed wetlands. *Ecol. Eng.* **2012**, *41*, 13–21. [CrossRef]
23. Almeida, C.M.R.; Santos, F.; Ferreira, A.C.F.; Lourinha, I.; Basto, M.C.P.; Mucha, A.P. Can veterinary antibiotics affect constructed wetlands performance during treatment of livestock wastewater? *Ecol. Eng.* **2017**, *102*, 583–588. [CrossRef]
24. Huang, X.; Zheng, J.; Liu, C.; Liu, L.; Liu, Y.; Fan, H. Removal of antibiotics and resistance genes from swine wastewater using vertical flow constructed wetlands: Effect of hydraulic flow direction and substrate type. *Chem. Eng. J.* **2017**, *308*, 692–699. [CrossRef]
25. Chen, J.; Liu, Y.S.; Su, H.C.; Ying, G.G.; Liu, F.; Liu, S.S.; He, L.Y.; Chen, Z.F.; Yang, Y.Q.; Chen, F.R. Removal of antibiotics and antibiotic resistance genes in rural wastewater by an integrated constructed wetland. *Environ. Sci. Pollut. Res.* **2015**, *22*, 1794–1803. [CrossRef]
26. Fang, H.S.; Zhang, Q.; Nie, X.P.; Chen, B.W.; Xiao, Y.D.; Zhou, Q.B.; Liao, W.; Liang, X.M. Occurrence and elimination of antibiotic resistance genes in a long-term operation integrated surface flow constructed wetland. *Chemosphere* **2017**, *173*, 99–106. [CrossRef]
27. Almeida, C.M.R.; Santos, F.; Ferreira, A.C.F.; Gomes, C.R.; Basto, M.C.P.; Mucha, A.P. Constructed wetlands for the removal of metals from livestock wastewater—Can the presence of veterinary antibiotics affect removals? *Ecotox. Environ. Saf.* **2017**, *137*, 143–148. [CrossRef]
28. Liu, L.; Liu, Y.H.; Wang, Z.; Liu, C.X.; Huang, X.; Zhu, G.F. Behavior of tetracycline and sulfamethazine with corresponding resistance genes from swine wastewater in pilot-scale constructed wetlands. *J. Hazard. Mater.* **2014**, *278*, 304–310. [CrossRef]
29. Cetecioglu, Z.; Ince, B.; Orhon, D.; Ince, O. Anaerobic sulfamethoxazole degradation is driven by homoacetogenesis coupled with hydrogenotrophic methanogenesis. *Water Res.* **2016**, *90*, 79–89. [CrossRef]
30. Soares, S.F.; Fernandes, T.; Trindade, T.; Daniel-da-Silva, A.L. Trimethyl Chitosan/Siloxane-Hybrid Coated Fe3O4 Nanoparticles for the Uptake of Sulfamethoxazole from Water. *Molecules* **2019**, *24*, 1958. [CrossRef]
31. Barkovskii, A.L.; Bridges, C. Persistence and Profiles of Tetracycline Resistance Genes in Swine Farms and Impact of Operational Practices on Their Occurrence in Farms' Vicinities. *Water Air Soil Pollut.* **2011**, *223*, 49–62. [CrossRef]
32. Dan, A.; Yang, Y.; Dai, Y.N.; Chen, C.X.; Wang, S.Y.; Tao, R. Removal and factors influencing removal of sulfonamides and trimethoprim from domestic sewage in constructed wetlands. *Bioresour. Technol.* **2013**, *146*, 363–370. [CrossRef] [PubMed]
33. Yi, X.Z.; Tran, N.H.; Yin, T.R.; He, Y.L.; Gin, K.Y.H. Removal of selected PPCPs, EDCs, and antibiotic resistance genes in landfill leachate by a full-scale constructed wetlands system. *Water Res.* **2017**, *121*, 46–60. [CrossRef] [PubMed]
34. Xu, J.; Xu, Y.; Wang, H.; Guo, C.; Qiu, H.; He, Y.; Zhang, Y.; Li, X.; Meng, W. Occurrence of antibiotics and antibiotic resistance genes in a sewage treatment plant and its effluent-receiving river. *Chemosphere* **2015**, *119*, 1379–1385. [CrossRef]
35. Dordio, A.V.; Carvalho, A.J. Organic xenobiotics removal in constructed wetlands, with emphasis on the importance of the support matrix. *J. Hazard. Mater.* **2013**, *252–253*, 272–292. [CrossRef]
36. Hijosa-Valsero, M.; Fink, G.; Schlusener, M.P.; Sidrach-Cardona, R.; Martin-Villacorta, J.; Ternes, T.; Becares, E. Removal of antibiotics from urban wastewater by constructed wetland optimization. *Chemosphere* **2011**, *83*, 713–719. [CrossRef]
37. Gong, W.; Liu, X.; He, H.; Wang, L.; Dai, G. Quantitatively modeling soil-water distribution coefficients of three antibiotics using soil physicochemical properties. *Chemosphere* **2012**, *89*, 825–831. [CrossRef]

38. Müller, E.; Schüssler, W.; Horn, H.; Lemmer, H. Aerobic biodegradation of the sulfonamide antibiotic sulfamethoxazole by activated sludge applied as co-substrate and sole carbon and nitrogen source. *Chemosphere* **2013**, *92*, 969–978. [CrossRef]
39. Wang, L.; Liu, Y.; Ma, J.; Zhao, F. Rapid degradation of sulphamethoxazole and the further transformation of 3-amino-5-methylisoxazole in a microbial fuel cell. *Water Res.* **2016**, *88*, 322–328. [CrossRef]
40. Wang, L.; Wu, Y.C.; Zheng, Y.; Liu, L.D.; Zhao, F. Efficient degradation of sulfamethoxazole and the response of microbial communities in microbial fuel cells. *RSC Adv.* **2015**, *5*, 56430–56437. [CrossRef]
41. Antunes, P.; Machado, J.; Sousa, J.C.; Peixe, L. Dissemination of sulfonamide resistance genes (sul1, sul2, and sul3) in Portuguese Salmonella enterica strains and relation with integrons. *Antimicrob. Agents Chemother.* **2005**, *49*, 836–839. [CrossRef] [PubMed]
42. Nolvak, H.; Truu, M.; Tiirik, K.; Oopkaup, K.; Sildvee, T.; Kaasik, A.; Mander, U.; Truu, J. Dynamics of antibiotic resistance genes and their relationships with system treatment efficiency in a horizontal subsurface flow constructed wetland. *Sci. Total Environ.* **2013**, *461–462*, 636–644. [CrossRef] [PubMed]
43. Fernandes, J.P.; Almeida, C.M.R.; Pereira, A.C.; Ribeiro, I.L.; Reis, I.; Carvalho, P.; Basto, M.C.P.; Mucha, A.P. Microbial community dynamics associated with veterinary antibiotics removal in constructed wetlands microcosms. *Bioresour. Technol.* **2015**, *182*, 26–33. [CrossRef] [PubMed]
44. Cao, X.; Wang, H.; Zhang, S.; Nishimura, O.; Li, X. Azo dye degradation pathway and bacterial community structure in biofilm electrode reactors. *Chemosphere* **2018**, *208*, 219–225. [CrossRef]
45. Huang, X.; Zheng, J.L.; Liu, C.X.; Liu, L.; Liu, Y.H.; Fan, H.Y.; Zhang, T.F. Performance and bacterial community dynamics of vertical flow constructed wetlands during the treatment of antibiotics-enriched swine wastewater. *Chem. Eng. J.* **2017**, *316*, 727–735. [CrossRef]
46. Arroyo, P.; Saenz de Miera, L.E.; Ansola, G. Influence of environmental variables on the structure and composition of soil bacterial communities in natural and constructed wetlands. *Sci. Total Environ.* **2015**, *506–507*, 380–390. [CrossRef]
47. Srinivasan, P.; Sarmah, A.K. Dissipation of sulfamethoxazole in pasture soils as affected by soil and environmental factors. *Sci. Total Environ.* **2014**, *479*, 284–291. [CrossRef]
48. Wu, D.; Huang, Z.T.; Yang, K.; Graham, D.; Xie, B. Relationships between Antibiotics and Antibiotic Resistance Gene Levels in Municipal Solid Waste Leachates in Shanghai, China. *Environ. Sci. Technol.* **2015**, *49*, 4122–4128. [CrossRef]
49. Rodriguez-Mozaz, S.; Chamorro, S.; Marti, E.; Huerta, B.; Gros, M.; Sanchez-Melsio, A.; Borrego, C.M.; Barcelo, D.; Balcazar, J.L. Occurrence of antibiotics and antibiotic resistance genes in hospital and urban wastewaters and their impact on the receiving river. *Water Res.* **2015**, *69*, 234–242. [CrossRef]
50. Pruden, A.; Pei, R.; Storteboom, H.; Carlson, K.H. Antibiotic Resistance Genes as Emerging Contaminants: Studies in Northern Colorado†. *Environ. Sci. Technol.* **2006**, *40*, 7445–7450. [CrossRef]
51. Wang, W.; Cao, J.; Yang, F.; Wang, X.L.; Zheng, S.S.; Sharshov, K.; Li, L.X. High-throughput sequencing reveals the core gut microbiome of Bar-headed goose (Anser indicus) in different wintering areas in Tibet. *MicrobiologyOpen* **2016**, *5*, 287–295. [CrossRef] [PubMed]
52. Bokulich, N.A.; Subramanian, S.; Faith, J.J.; Gevers, D.; Gordon, J.I.; Knight, R.; Mills, D.A.; Caporaso, J.G. Quality-filtering vastly improves diversity estimates from Illumina amplicon sequencing. *Nat. Methods* **2013**, *10*, 57–59. [CrossRef] [PubMed]
53. Zhang, Y.J.; Zhu, H.; Szewzyk, U.; Geissen, S.U. Enhanced removal of sulfamethoxazole with manganese-adapted aerobic biomass. *Int. Biodeterior. Biodegrad.* **2017**, *116*, 171–174. [CrossRef]
54. Mikkelson, K.M.; Lozupone, C.A.; Sharp, J.O. Altered edaphic parameters couple to shifts in terrestrial bacterial community structure associated with insect-induced tree mortality. *Soil Biol. Biochem.* **2016**, *95*, 19–29. [CrossRef]

Sample Availability: Not available.

Article

Evaluation of the Possibility of Using Hydroponic Cultivations for the Removal of Pharmaceuticals and Endocrine Disrupting Compounds in Municipal Sewage Treatment Plants

Daniel Wolecki, Magda Caban, Magdalena Pazda, Piotr Stepnowski and Jolanta Kumirska *

Department of Environmental Analysis, Faculty of Chemistry, University of Gdansk, Wita Stwosza 63, 80-308 Gdańsk, Poland; daniel.wolecki@phdstud.ug.edu.pl (D.W.); magda.caban@ug.edu.pl (M.C.); magdalena.pazda@phdstud.ug.edu.pl (M.P.); piotr.stepnowski@ug.edu.pl (P.S.)

* Correspondence: jolanta.kumirska@ug.edu.pl; Tel.: +48-58-523-5212

Academic Editor: Teresa Rocha-Santos
Received: 13 November 2019; Accepted: 27 December 2019; Published: 31 December 2019

Abstract: The problem of the presence of pharmaceuticals and endocrine disrupting compounds (EDCs) in the environment is closely related to municipal wastewater and in consequence to municipal wastewater treatment plants (MWWTPs) because wastewater is the main way in which these compounds are transferred to the ecosystem. For this reason, the development of cheap, simple but very effective techniques for the removal of such residues from wastewater is very important. In this study, the analysis of the potential of using three new plants: *Cyperus papyrus* (Papyrus), *Lysimachia nemorum* (Yellow pimpernel), and *Euonymus europaeus* (European spindle) by hydroponic cultivation for the removal of 15 selected pharmaceuticals and endocrine disrupting compounds (EDCs) in an MWWTP is presented. In order to obtain the most reliable data, this study was performed using real WWTP conditions and with the determination of the selected analytes in untreated sewage, treated sewage, and in plant materials. For determining the target compounds in plant materials, an Accelerated Solvent Extraction (ASE)-Solid-Phase Extraction (SPE)-GC-MS(SIM) method was developed and validated. The obtained data proved that the elimination efficiency of the investigated substances from wastewater was in the range of 35.8% for diflunisal to above 99.9% for paracetamol, terbutaline, and flurbiprofen. *Lysimachia nemorum* was the most effective for the uptake of target compounds among the tested plant species. Thus, the application of constructed wetlands for supporting conventional MWWTPs allowed a significant increase in their removal from the wastewater stream.

Keywords: pharmaceuticals; endocrine disrupting compounds; hydroponic cultivation; determining target pollutants in plant materials; municipal wastewater treatment plants

1. Introduction

Pharmaceuticals are used in large quantities around the world. In many cases, they are used not only to prevent disease or for treatment in humans, but also in animals [1,2]. Some of them, for example, natural and synthetic estrogens, are classified also to the group of endocrine disrupting compounds (EDCs). Over the last two decades, the presence of pharmaceuticals and EDCs in the aquatic environment has been confirmed many times [3–6]. They have been found in treated wastewater [7], sewage sludge [8], marine waters [9], and in living organisms [10]. They are extensively introduced into the environment via wastewater and in consequence by wastewater treatment plants (WWTPs) because the classical methods of wastewater treatment in WWTPs (mechanical, biological, chemical) do not completely remove pharmaceuticals and EDCs from the wastewater stream [11–14].

In recent years, some advanced technologies, such as advanced oxidation process (AOP), electrochemical oxidation, activated carbon adsorption, membrane techniques, and membrane bioreactors have been introduced for the removal of pharmaceuticals from wastewater [15–18]. However, these advanced processes are very expensive, and they are often unprofitable in technological systems. Therefore, the choice of cheaper, less complicated but very effective techniques for wastewater treatment with regard to pharmaceutical residues is well-aimed.

One of the options is to combine conventional WWTPs with hydroponic cultures called constructed wetlands (CWs). The basis of this is the use of plants such as macrophytes in the process of biological wastewater treatment [19]. In most cases, CWs are characterized based on water position and flow direction, and they are divided into surface flow (SF) systems, horizontal subsurface flow systems (H-SSF), and vertical subsurface flow systems (V-SSF). In the first case, sewage flows along a pool with growing plants. The root zone and part of the plants are located under the sewage stream. In H-SSF and V-SSF systems, sewage is introduced into a porous medium (generally gravel) in which macrophytes grow [20]. In most cases, constructed wetlands are a separate element of a sewage treatment plant. CWs act as a primary step (raw influent goes to the CW), secondary step (raw influent goes to primary treatment and afterwards to the CW), and tertiary step (raw influent goes to primary treatment, followed by secondary treatment and finally to the CW). This has been clearly and understandably described by Verlicchi et al. [20].

In recent years, constructed wetlands have been reported to be highly efficient in the removal of conventional pollutants from domestic sewage, agricultural sewage, industrial wastewater, mine drainage, leachate, contaminated ground water, and urban runoff [21–26]. Based on the literature, it is known that the mechanisms of pollution removal in CWs can be classified into biotic processes (e.g., microbiological degradation, root, and plant uptake) and abiotic processes like evaporation, photodegradation, oxidation, hydrolysis, retention, or sorption [27]. The efficiency of the removal of a pollutant depends also on such factors as seasonality, weather, and humidity. On the other hand, the use of constructed wetlands to remove pharmaceuticals and EDCs from wastewater is still not fully understood [28]. Most of the research concerning the uptake of pharmaceuticals and EDCs by plants in CWs refers to so-called microcosm, or pilot-scale laboratory systems, as opposed to real systems in WWTPs (Table S1, Supplementary Materials). In addition, only a few scientific papers have examined the actual uptake of emerging environmental pollutants by plants (Table S1, Supplementary Materials). Moreover, in most cases, the removal efficiency of the target compounds from wastewater is determined based only on determining their concentrations in raw and treated wastewater without the examination of their presence in plant tissues [29]. For this reason, it is difficult to assume the influence and mechanism of the removal of EDCs from wastewater by plants, as well as to establish which plants are the best for this purpose.

In this study, in order to obtain the most reliable data, we decided to evaluate the possibility of using hydroponic cultivation for the removal of pharmaceuticals and endocrine disrupting compounds in municipal sewage treatment, both with the application of real WWTP conditions as well as with the determination of the analytes studied in plant materials.

For this reason, the main objectives of this work were as follows: (1) to determine the selected pollutants in raw and treated sewage; (2) to develop an analytical method for determining selected pharmaceuticals and endocrine disrupting compounds in plant materials; (3) to assess the bioaccumulation of the mentioned pollutants in hydroponically cultivated plants; (4) to evaluate which of the tested plant species best meets this purpose.

According to literature data [30–33] and information included in Table S1 in Supplementary Materials, for the isolation of pharmaceuticals and EDCs from plant materials, extraction techniques such as Microwave Assisted Extraction (MAE) [34], QuEChERS (Quick, Easy, Cheap, Effective, Rugged, and Safe) procedures [35,36], Accelerated Solvent Extraction (ASE) [27], Solid-Liquid Extraction (SLE) under gentle mixing [32], as well as Ultrasound-Assisted Extraction (UAE) [33] are used. In this study, we decided to check the usefulness of an ASE-Solid-Phase Extraction (SPE) technique for the isolation of target compounds from plant materials, because in none of the published papers, such investigations have been performed for the same target compounds and with the application of the same plants.

2. Results and Discussion

2.1. Determination of Selected Pharmaceuticals and Endocrine Disrupting Compounds in Treated and Untreated Sewage

The method for the determination of pharmaceuticals and EDCs in untreated and treated sewage was described in Section 3.4. The identification of each analyte was based on retention time, quantitative ion, and conformation ion/s, the quantification analysis on the area peak of quantitative ion (details are presented in Section 3.4). The mass spectra of target compounds with the MS fragments assignation are included in Figure S1 in Supplementary Materials. The determined concentrations of fifteen target compounds in untreated and treated sewage from a municipal wastewater treatment plant (MWWTP) are presented in Table 1.

Table 1. Concentrations of target compounds in untreated and treated sewage samples collected from a municipal wastewater treatment plant (MWWTP), determined using the Solid-Phase Extraction (SPE)-GC-MS(SIM) method (n = 3).

Pharmaceuticals	Concentration in Untreated Sewage	Concentration in Treated Sewage
	(mean ± SD) [ng/L]	
Ibuprofen	2695 ± 916	8 ± 1
Paracetamol	2130 ± 298	<MDL
Terbutaline	154 ± 8	<MDL
Flurbiprofen	51 ± 12	<MDL
Naproxen	3420 ± 342	38 ± 1
Diflunisal	67 ± 12	43 ± 7
Amitriptyline	1676 ± 184	613 ± 38
Imipramine	<MDL	<MDL
Diclofenac	<MDL	<MDL
Clomipramine	169 ± 29	50 ± 5
Nadolol	<MDL	<MDL
Estrone	52 ± 3	19 ± 2
17β-estradiol	110 ± 13	4 ± 0
17α-ethinylestradiol	1622 ± 260	122 ± 18
Estriol	273 ± 27	13 ± 3

Imipramine, diclofenac, and nadolol were not found in both types of samples (concentrations below the method detection limit (MDL) values). The concentrations of the analyzed NSAIDs (ibuprofen, naproxen) in untreated sewage were the highest among the investigated drugs (2695 ± 916 ng/L, 3420 ± 342 ng/L, respectively), which positively correlated with their large consumption and easy availability [6]. Amitriptyline, one of the measured antidepressant drugs, was found in a high concentration in untreated sewage (1676 ± 184 ng/L), and its concentration after wastewater treatment was still significant (613 ± 38 ng/L). Antidepressant drugs were identified in this area in 2014, and also in the Utrata River to which treated sewage is discharged [37].

In this study, for first time in this part of Europe, concentrations of estrogenic hormones in sewage derived from a wastewater treatment plant supported by constructed wetlands were investigated. E1, E2, E3, and synthetic EE2 were found in both types of sewage in concentrations ranging from 52 ± 3 ng/L to 1662 ± 260 ng/L in untreated sewage, and from 4 ± 0 ng/L to 122 ± 18 ng/L in treated sewage. The concentration of EE2 was the highest among the investigated estrogenic hormones, both in untreated and treated sewage (Table 1).

The concentrations of target compounds in untreated and treated sewage determined in this study were in the range presented by other authors for full-scan systems (Table S1, Supplementary Materials).

For example, the concentrations of 86 pharmaceuticals, e.g., diclofenac, ibuprofen, ketoprofen, naproxen, EE2, E2, and E3 in untreated water were in the range from 1 ng/L to 1,000,000 ng/L, and in treated sewage to 901,618 ng/L [32].

However, for the first time, fifteen pharmaceuticals and EDCs were determined in untreated and treated sewage derived from a wastewater treatment plant supported by constructed wetlands working in a temperate climate zone.

2.2. *Evaluation of the Analytical Method for Determining Target Compounds in Plant Materials*

As previously mentioned, understanding the impact of individual plant species on the removal of pharmaceuticals and EDCs from wastewater requires an assessment of the uptake of these compounds by plants, for which an analytical procedure is required. In none of the published papers ([30–33], Table S1, Supplementary Materials), such investigations have been performed for the same target compounds and with the application of the same plants. For this reason, our investigations began from an evaluation of the analytical method for determining 15 target compounds in plant materials.

We decided to check the usefulness of the ASE technique for this purpose. ASE technique has been used only once for the isolation of 18 analytes [27]. In the work, among the tested analytes, were 7 pharmaceuticals (ibuprofen, ketoprofen, naproxen, diclofenac, salicylic acid, caffeine, carbamazepine), which were determined in *Typha angustifolia* and *Phragmites australis* plants (only roots) during an investigation of the behavior of pharmaceuticals and personal care products in mesocosm-scale constructed wetland compartments. A mixture of acetone/hexane (1:1; *v*/*v*), a temperature of 104 °C and two extraction cycles of 13.5 min were applied for ASE. A clean-up step was performed using a florisil column with different elution of the neutral/acidic fractions. In other studies based on other extraction procedures, for the isolation of pharmaceuticals from plant materials, different solvents such as methylene chloride:MeOH (2:1 *v*/*v*) and MTBE (Methyl *tert*-butyl ether):MeOH (1:1 *v*/*v*) [34], hexane:ethyl acetate (1:1, *v*/*v*) [35], 0.1 M HCl:ACN (1:1, *v*/*v*) [36], ACN with 0.5% formic acid (*v*/*v*) [32], and anhydrous sodium sulfate and MeOH [33] were applied.

In this study, the usefulness of three new extraction solvents: MeOH:H_2O (1:1, *v*/*v*), EtOH:H_2O (1:1, *v*/*v*), ACN:H_2O (1:1, *v*/*v*), which were used at two temperatures: 50 °C and 80 °C for the isolation of target compounds by ASE was studied. Moreover, the effect of the acidification of the obtained extracts prior to SPE purification (pH ~ 2) was investigated (Figure 1).

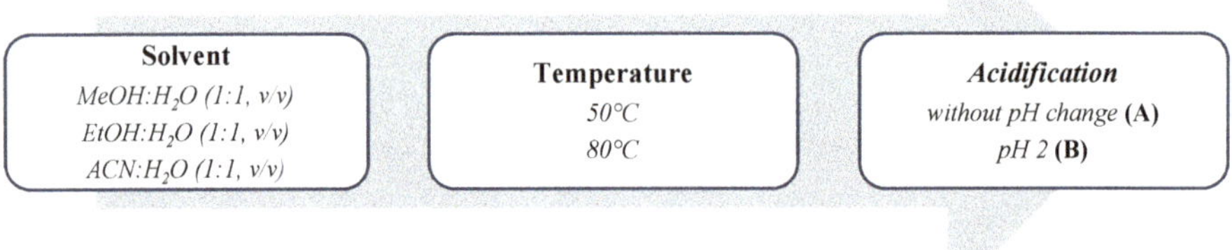

Figure 1. Selection of the optimal Accelerated Solvent Extraction (ASE)-SPE conditions for the isolation of target compounds from plant materials.

The results of the optimization of the ASE procedure are presented in Table 2. As can be seen, among the tested ASE conditions, the best recoveries of target pharmaceuticals (ranging from 49 ± 6 to 111 ± 9%), were obtained during the application of a mixture of MeOH:H_2O (1:1, *v*/*v*) and a temperature of 50 °C (Table 2). Only the recoveries of tri-cyclic antidepressant drugs (amitriptyline and clomipramine) were lower. It should be highlighted that these two compounds were not subjected to the derivatization procedure and the efficiency of ionization of these two compounds in an ion source during the GC-MS(SIM) analysis was high. In effect, the sensitivity of the proposed analytical method for the determination of these analytes was enough despite low recovery results.

Table 2. Recoveries (mean ± SD) of fifteen analytes using different conditions of the ASE procedure (n = 3, conc. 1000 ng/g d.w.).

Type of Solvents/ Temperature/ Pharmaceuticals	EtOH:H_2O (1:1, v/v)		ACN:H_2O (1:1, v/v)		MeOH:H_2O (1:1, v/v)	
	80 °C	50 °C	80 °C	50 °C	80 °C	50 °C
	Value of Recovery [% ± SD (%)]					
Ibuprofen	105 ± 22	154 ± 5	105 ± 16	194 ± 13	65 ± 5	111 ± 9
Paracetamol	26 ± 11	36 ± 7	10 ± 2	25 ± 3	42 ± 5	74 ± 9
Terbutaline	16 ± 5	25 ± 6	19 ± 4	26 ± 3	48 ± 3	51 ± 5
Flurbiprofen	89 ± 15	120 ± 18	80 ± 14	140 ± 8	52 ± 2	89 ± 10
Naproxen	91 ± 9	118 ± 6	80 ± 13	139 ± 7	54 ± 3	89 ± 11
Diflunisal	90 ± 18	115 ± 7	90 ± 11	159 ± 7	55 ± 3	88 ± 12
Amitriptyline	27 ± 8	58 ± 11	52 ± 11	10 ± 3	25 ± 4	26 ± 5
Imipramine	29 ± 8	88 ± 25	74 ± 10	16 ± 2	59 ± 2	55 ± 8
Diclofenac	103 ± 12	10 5± 3	68 ± 8	76 ± 9	98 ± 6	105 ± 5
Clomipramine	16 ± 7	44 ± 19	60 ± 8	6 ± 1	31 ± 6	16 ± 4
Nadolol	73 ± 6	91 ± 12	59 ± 4	97 ± 17	59 ± 3	67 ± 9
Estrone	63 ± 13	80 ± 15	49 ± 6	67 ± 5	38 ± 3	49 ± 6
17β-estradiol	77 ± 10	86 ± 15	71 ± 11	100 ± 5	45 ± 4	61 ± 8
17α-ethinylestradiol	125 ± 16	130 ± 10	235 ± 14	219 ± 12	101 ± 9	108 ± 10
Estriol	65 ± 7	93 ± 14	64 ± 7	88 ± 7	50 ± 4	71 ± 7

The analyzed pharmaceuticals and EDCs contain polar functional groups (pKa are presented in Table S2), therefore the acidification of ASE extracts prior to SPE purification could reduce or stop the dissociation of weak acids and increase the dissociation of weak bases. Oasis HLB cartridges used in the SPE procedure are stable from pH 0–14, and they are useful for the isolation of acidic, basic, and neutral analytes. However, modifications of the pH of the loaded sample could strongly influence the recovery data. For this reason, the effect of the acidification of ASE extracts to pH 2 prior to SPE purification on the recovery of analytes was also evaluated. Such studies were performed for the most effective extraction mixture (MeOH:H_2O 1:1, v/v; Table 2) which was tested at two temperatures: 80 °C and 50 °C. The results are shown in Figure 2.

As we can see (Figure 2), in most cases, the acidification of ASE extracts led to a decrease in recoveries, both at 80 °C and 50 °C. It was observed for weak acids such as naproxen, difunisal, diclofenac), and weak bases (paracetamol, terbutaline, amitriptyline, clomipramine and nadolol; Table S2). In case of acidification of the ASE extracts obtained at 80 °C (Figure 2A) the extraction efficiencies increased for ibuprofen, imipramine, estrone (E1), 17β-estradiol (E2), 17α-ethinylestradiol (EE2), and estriol (E3). However, when the ASE procedure was performed at 50 °C, these values decreased significantly for all mentioned compounds (Figure 2B). The extraction efficiencies of polar drugs: paracetamol (logP 0.46), terbutaline (logP 0.90), and nadolol (logP 0.81), decreased significantly where the ASE extracts obtained at 50 °C were subjected to acidification prior SPE (Figure 2B). Similar results were observed in our previous study concerning the development of a method for determining NSAIDs and natural estrogens in the mussel *Mytilus edulis trossulus* [38]. In summary, based on the obtained results, the application of the mixture of MeOH:H_2O (1:1, v/v) at 50 °C without the acidification of the ASE extract prior to SPE purification was chosen as the most optimal ASE-SPE conditions for the isolation of target compounds from plant material.

In order to compare the recovery data obtained in this work with those presented by other scientists for pharmaceuticals and EDCs isolated from plants used in constructed wetlands, we tried to find the necessary information in all the cited papers [27,33–36]. Unfortunately, the recovery data were presented only in one of them [35], and only for triclosan, methyl-triclosan, and triclocarban. Nuel and coworkers [32] stated that drug detection and quantification methodologies had been fully described in their paper submitted to the journal Science of the Total Environment (Nuel et al. (2018)), but this paper was not published (based on data from Scopus; 31.10.2019).

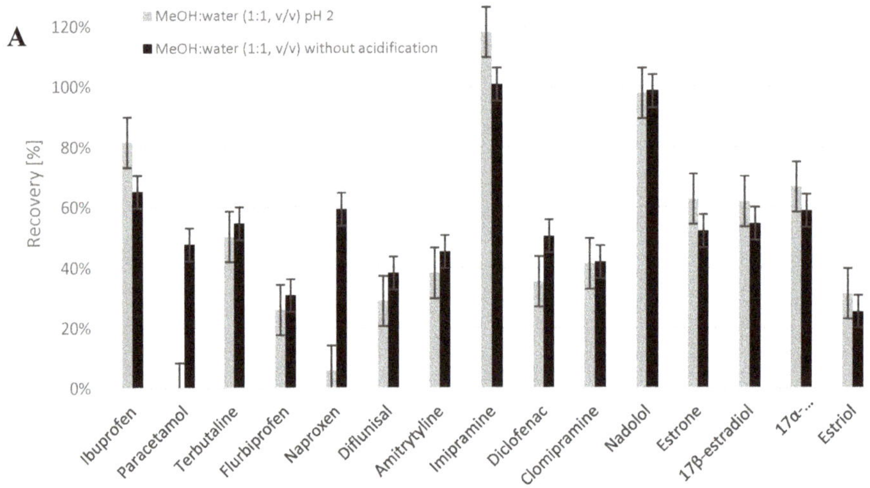

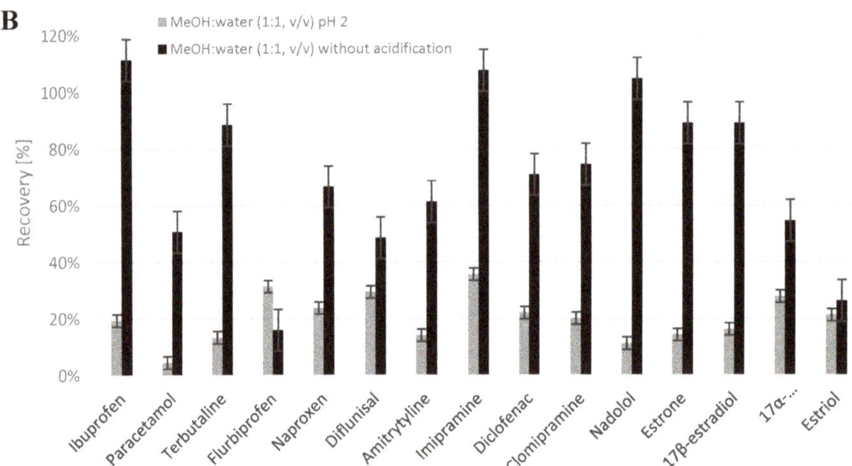

Figure 2. Impact of the acidification of ASE extracts prior to SPE purification on the recoveries of analytes, carried out at 80 °C (**A**) and 50 °C (**B**), respectively.

2.3. Validation Parameters of the Proposed ASE-SPE-GC-MS(SIM) MetHod for Determining Target Compounds in Plant Materials

The ASE-SPE-GC-MS(SIM) methodology for the determination of target pharmaceuticals in plant samples has been validated in accordance with the guidelines described in Section 3.6. The determined validation parameters are presented in Table 3.

Table 3. Selected validation parameters of the ASE-SPE-GC-MS(SIM) method for determining target compounds in plant samples (analytical range from method quantification limit (MQL) to 2500 ng/g). Precision and accuracy were determined for three concentrations from analytical range (78 ng/g d.w., 625 ng/g d.w., and 2500 ng/g d.w.).

Validation Parameters	Calibration Curves	R^2	Measurement Intermediate Precision (RSD%)	Mean Recovery	ME	MQL	MDL
Compound				(%)	%	(ng/g d.w.)	
Ibuprofen	68698 (±635) x + 1208.4 (±648.4)	0.9995	0.44–7.61	80–102	10	0.4	0.1
Paracetamol	74010 (±4738) x − 8672.6 (±4835.8)	0.9760	2.46–20.25	100	19	0.5	0.2
Terbutaline	364393 (±2459) x − 1622.6 (±2510.1)	0.9997	0.39–8.75	99–100	−31	0.8	0.3
Flurbiprofen	67739 (±632) x − 737.2 (621.9)	0.9996	0.46–10.86	95–98	−20	0.4	0.1
Naproxen	67687 (±829) x − 1444.7 (±880.5)	0.9994	3.73–13.65	99–101	43	0.4	0.1
Diflunisal	114510 (±2633) x − 4994.1 (±2590.4)	0.9970	0.05–9.33	97–102	15	0.4	0.1
Amitriptyline	339801 (±3704) x + 30042.5 (3642.7)	0.9994	0.44–16.20	94–102	−9	2	0.7
Imipramine	33490 (±333) x − 752.78 (±339.8)	0.9994	1.08–14.00	97–100	12	0.7	0.2
Diclofenac	18224 (±287) x − 673.8 (±292.7)	0.9990	3.62–21.60	89–101	25	0.4	0.1
Clomipramine	17286 (±128) x − 26.9 (±125.7)	0.9997	0.56–17.09	99–100	−8	2	0.7
Nadolol	550924(±13936) x − 31,586 (±13706.5)	0.9970	1.82–9.86	91–102	28	0.6	0.2
Estrone (E1)	24502(±587) x + 284.4 (±599.1)	0.9970	2.03–14.88	99–100	−2	0.8	0.3
17β-estradiol (E2)	44380 (±1275) x − 276.4 (1302.1)	0.9950	0.18–10.78	80–101	−3	0.6	0.2
17α-ethinylestradiol (EE2)	12263 (±246) x + 560.9 (±251.5)	0.9980	0.16–12.16	80–102	−11	0.4	0.1
Estriol (E3)	9240 (±265) x + 724.7 (±303.3)	0.9970	1.00–15.78	99–100	31	0.5	0.2

The coefficient of determination (R^2) was in the range of 0.9760–0.9997; the intermediate precision measurement in the range of 0.05–21.60%. Mean recoveries were between 80% and 102% (Table 3). The lowest MDL value was recorded for ibuprofen, flurbiprofen, naproxen, diflunisal, diclofenac, and EE2 – 0.4 ng/g d.w.; the highest value of 2 ng/g d.w. was calculated for clomipramine and amitriptyline. Matrix effects were not significant for most of the analyzed compounds, apart from naproxen (ME 43%). The highest suppression of signals was observed for terbutaline (ME −31%) and flurbiprofen (ME −20%). Generally, matrix effect values were similar to those presented in our previous studies for the same target pharmaceuticals isolated from environmental matrices [39].

A comparison of the obtained matrix effect data with those presented by other authors was not possible because matrix effect values were not included in the cited papers [27,32–36]. The method quantification limit (MQL) and MDL values were similar to those we presented (Table S1 in Supplementary Materials).

2.4. Assessment of Bioaccumulation of Selected Pollutants in Hydroponically Cultivated Plants

The determined concentrations of target compounds in three species of hydroponically cultivated plants (in ng/g dry weight) and the elimination efficiency of pharmaceuticals and EDCs in an MWWTP with constructed wetlands are presented in Table 4. The elimination efficiency (EE) was calculated according to Equation (6) described in Section 3.8.

Table 4. Results of determining target compounds in three species of hydroponically cultivated plants from an MWWTP using the developed ASE-SPE-GC-MS(SIM) method (n = 3), and the elimination efficiency of these compounds from wastewater in an MWWTP supported by constructed wetlands.

Pharmaceuticals	*Cyperus papyru* (Papyrus)	*Lysimachia nemorum* (Yellow Pimpernel)	*Euonymus europaeus* (European Spindle)	EE
	(mean ± SD) [ng/g Dry Weight]			%
Ibuprofen	700 ± 28	<MDL	1616 ± 124	99.7
Paracetamol	<MDL	<MDL	<MDL	>99.9
Terbutaline	<MDL	5323 ± 1869	<MDL	>99.9
Flurbiprofen	2107 ± 92	<MDL	<MDL	>99.9
Naproxen	<MDL	1422 ± 216	<MDL	98.9
Diflunisal	1569 ± 321	5569 ± 1298	260 ± 4	35.8
Amitriptyline	404 ± 5	1399 ± 20	435 ± 5	63.4
Imipramine	3533 ± 198	<MDL	<MDL	-[1]
Diclofenac	<MDL	<MDL	<MDL	-[1]
Clomipramine	<MDL	<MDL	<MDL	70.4
Nadolol	<MDL	<MDL	<MDL	-[1]
Estrone (E1)	<MDL	<MDL	<MDL	63.5
17β-estradiol (E2)	<MDL	<MDL	<MDL	96.4
17α-ethinylestradiol (EE2)	4126 ± 821	3136 ± 599	214 ± 3	92.5
Estriol (E3)	<MDL	<MDL	<MDL	95.2

[1] if drug concentrations were below the MDL value in both untreated and treated wastewater, the elimination efficiency (EE) was not calculated.

Among the 15 investigated analytes, three NSAIDs (ibuprofen, flurbiprofen, diflunisal), two antidepressants (amitriptyline, imipramine), and one synthetic hormone (EE2) were found in Papyrus (*C. Papyrus*) (Table 4). In the case of the Yellow pimpernel (*L. nemorum*) species, terbutaline (β_2-agonists), naproxen, and diflunisal as well amitriptyline and EE2 were found. The tissue of the European spindle (*E. europaeus*) plant contained ibuprofen, diflunisal, amitriptyline, and EE2. The highest concentrations were observed for diflunisal and terbutaline determined in Yellow pimpernel (*L. nemorum*) (5569 ± 1298 and 5323 ± 1869 ng/g d.w., respectively). The concentrations of other target compounds were below the MDL value (Table 4). Among three determined estrogenic hormones, only EE2 was found in all the investigated plant species; its concentration was the highest in Papyrus (*C. Papyrus*) (4126 ± 821 ng/g dry weight); E1 and E2 (natural hormones occurring in living organisms) were not detected in any plant tissues. Example chromatograms with marked SIM ions, recorded in this study for real plant samples, are presented in the Supplementary Materials (Figures S3–S5).

Few studies directly point to the transport or uptake of pharmaceuticals by plants in constructed wetlands (Table S1 in Supplementary Materials). For example, 0.2% of the initial amount of diclofenac (1 mg/L) was detected in the roots and leaves of *Typha latifolia* during one week of exposure (laboratory system) [36]. Hijosa-Valsero et al. [27] investigated the possibility of the uptake of 18 selected analytes, including ibuprofen, ketoprofen, naproxen, diclofenac, salicylic acid, caffeine, and carbamazepine by *Typha angustifolia* and *Phragmites australis* plants in constructed wetlands (mesocosm-scale). Ibuprofen, salicylic acid and caffeine were detected in plant tissues, whereas ketoprofen, naproxen, diclofenac and carbamazepine were not found in these species. The main substance detected in root tissues was salicylic acid (123–2560 ng/g). The authors confirmed that *T. angustifolia* is more suitable for the removal of the investigated pollutants than *Phragmites australis* [27]. In another study, performed by Nuel et al. [32], ibuprofen was found in all the investigated plant samples (*Salix alba*, *Iris pseudacorus*, *Juncus effusus*, *Callitriche palustris*, *Carex caryophyllea*) used in constructed wetlands. According to Wang et al. [33] the bioconcentration factors (BCFs) in *Typha angustifolia* of 8 compounds, e.g., caffeine, carbamazepine, ibuprofen, fluoxetine, gemfibrozil ranged between 60 and 2000. Concentrations of the determined compounds in plant tissue were up to several hundred ng/g for caffeine. This confirms that the results obtained in this study are in agreement with the literature data. However, for the first time, such investigations were performed for Papyrus (*Cyperus papyrus*), European spindle (*Euonymus europaeus*) and Yellow pimpernel (*Lysimachia nemorum*) taking into account this selected group of pharmaceuticals and EDCs.

In the study, the obtained elimination efficiency (EE) of the investigated compounds from wastewater was in the range of 35.8% for diflunisal to above 99.9% for paracetamol, terbutaline and flurbiprofen (Table 4). For comparison, the literature EE data for ibuprofen, presented in Table S1, are as follows: 5–88% [40], 96% [41–43], 44–77% [44], 94% [45], 52–85% [46], >99% [47], 51–54% (winter), 85–96% (summer) [48], 42–99% [49]. In our study, this was 99.7%, which confirms that the elimination efficiency of ibuprofen was comparable or higher than that calculated for other studies. Similar the EE data for pharmaceuticals is shown also in papers [50–52]. Thus, the obtained results proved the uptake of the investigated micropollutants by plants allowed an increase in the effectiveness of their removal from wastewater in such a system.

2.5. Assessment of the Usefulness of Hydroponically Cultivated Plants for Removing Target Compounds from Sewage Stream

According to the literature data, due to the considerable size of rhizomes and roots, the most frequently hydroponically cultivated plants are *Typha sp.* and *Phragmites sp.* (Table S1 in Supplementary Materials). In this study, the possibility of using three plants: *Cyperus papyrus* (Papyrus), *Lysimachia nemorum* (Yellow pimpernel), and *Euonymus europaeus* (European spindle) for this purpose was examined for the first time. These plant species are very well adapted to growth in MWWTPs and exhibit the strongest growth during the growing season. In order to assess the usefulness of these species for removing pharmaceuticals and EDCs from the sewage stream, the sum of the masses of all the target compounds taken by the tested species was established and the obtained values are presented in Table 5.

Table 5. The sum of the uptake of selected pharmaceuticals and endocrine disrupting compounds (EDCs) in ng/g dry weight by tested plant species grown in an MWWTP.

Plant Species	*Cyperus papyru*	*Lysimachia nemorum*	*Euonymus europaeus*
$\sum_{\text{Selected pharmaceuticals and ECDs}}$		[ng/g dry weight]	
	12,439	16,849	2525

Therefore, the highest concentration of all target compounds was observed for *the Lysimachia nemorum* (Yellow pimpernel) plant (16,849 ng/g d.w.), followed by *Cyperus papyrus* (Papyrus) (12,439 ng/g d.w.), and the lowest for *Euonymus europaeus* (European spindle) (2525 ng/g d.w.). Taking into account the summary uptake of target pharmaceuticals and EDCs by tested plants, the *Lysimachia nemorum* (Yellow pimpernel) species is the best for this purpose. The differences in the uptake of target compounds by these plants are connected to their morphological structures. *Euonymus europaeus* (European spindle) has a significant number of woody stems and rhizomes, which means that the uptake of target compounds by this plant is much lower than by the green parts of the two other tested species. Thus, the determination of target compounds in untreated wastewater and treated wastewater, as well as in plant tissues, allowed for establishing which hydroponic cultivation system supported the wastewater treatment process the most significantly (Tables 3–5).

3. Materials and Methods

3.1. Chemicals and Materials

Pure standards (>98%) of ibuprofen, paracetamol, flurbiprofen, naproxen, diflunisal, diclofenac sodium salt, nadolol, terbutaline, amitriptyline, imipramine, clomipramine, estrone (E1), 17β-estradiol (E2), 17α-ethinylestradiol (EE2), and estriol (E3), as well as the derivatization reagent *N*,*O*-bis(trimethylsilyl)trifluoroacetamide (BSTFA) containing 1% of trimethylchlorosilane (TMCS) and pyridine (99.8%) were purchased from Sigma-Aldrich (Steinheim, Germany). Solvents: methanol (MEOH), ethanol (EtOH), acetonitrile (ACN) were supplied by POCH (Gliwice, Poland), ethyl acetate (EtOAc) by Sigma-Aldrich, while 37% hydrochloric acid (HCl) for the acidification of extracts was provided by Chempur (Piekary Śląskie, Poland). Solid-phase extraction (SPE) was performed using Oasis HLB cartridges (6 mL, 200 mg, Waters Corporation, Miliford, USA).

Standard stock solutions of target compounds (1 mg × mL^{-1}) were prepared in methanol. All the stock solutions were stored at –20 °C. Working calibration standard solutions were prepared by diluting the standard stock solutions in the appropriate amount of methanol, and they were stored in the dark at –20 °C.

3.2. MWWTP and Constructed Wetlands Characteristic

The research was carried out at the municipal wastewater treatment plant (MWWTP) in Sochaczew (Mazowieckie Voivodeship, Poland). In Sochaczew MWWTP, macrophyte cultures are introduced in combination with an oxygen biological reactor. The efficiency of removing conventional pollution in Sochaczew MWWTP is high, as presented in Table 6.

Table 6. List of sewage test results in the Municipal Wastewater Treatment Plant in Sochaczew (average values from 2017).

Factor	Results for Treated Wastewater	Results for Untreated Wastewater	The Highest Acceptable Concentration	Removal Efficiency
BOD_5	3.4 mg/L O_2	460 mg/L O_2	15 mg/L O_2	99.3%
COD_{Cr}	32.7 mg/L O_2	1144.7 mg/L O_2	125 mg/L O_2	97.1%
N_{total}	8.8 mg/L	96.7 mg/L	15 mg/L	90.5%
P_{total}	0.2 mg/L	11.9 mg/L	2 mg/L	98.8%
Suspensions	8.6 mg/L	567.5 mg/L	35 mg/L	98.5%
pH	7.2	7.9	-	-

BOD_5 biochemical oxygen demand (for 5 days). COD_{Cr} Chemical Oxygen Demand (using Potassium Dichromate).

Sochaczew MWWTP consists of the following elements: 1° mechanical wastewater treatment (grates with a throughput of 515 m^3/h, aerated sand traps with degreasers, aerated at 1.91 m^3/min), 2° biological wastewater treatment (flow reactor with activated sludge with throughput at 6000 m^3/d and secondary settling tank with active capacity at 1142 m^3), sediment trap settling tanks with a diameter of 21 m, excessive sludge dewatering station, and lime treatment. The wastewater treatment plant was designed for 55,925 equivalent number of inhabitants with a maximum daily volume of sewage at 11,636 m^3/d.

Constructed wetlands have been implemented in the second stage of wastewater treatment (biological) as a wastewater polishing treatment system. Wastewater mixed with activated sludge flows around plants. Contact between the plants and wastewater occurs only in the rhyzophytic zone (Figure S2, Supplementary Materials). In this way, the green parts do not have contact with the walls and the benefits are the lack of dieback of the shoots, and faster acclimatization. The plants used in the constructed wetlands are as follows: Papyrus (*Cyperus papyrus*), Reed (*Phragmites australis* (Cav.) Trin. ex Steud, 1841), Spathiphyllum (*Spathiphyllum Adans.*), Grey willow (*Salix cinerea*), Rushes (*Juncus tenageia* Ehrh.), Sweet flag (*Acorus calamus*), Yellow iris (*Iris pseudacorus*), European spindle (*Euonymus europaeus*), Yellow pimpernel (*Lysimachia nemorum*), and Summer lilac (*Buddleja davidii* Franch).

3.3. Sampling Sewage and Plants from Constructed Wetland

The samples of untreated and treated sewage were collected in November 2017 into 2.5 L amber glass bottles. Raw sewage samples were taken before mechanical treatment, and treated sewage samples were collected at the outlet to the Utrata River. In the laboratory, the sewage samples were filtered under pressure using a 0.45 μm nylon filter and frozen at −20 °C until analysis. A list of the monitored pharmaceuticals and EDCs with physical and chemical properties is included in the Supplementary Materials in Table S2.

At the same time, samples of the plants used in the constructed wetlands were collected. Only parts which had no contact with sewage were collected to confirm the migration of pharmaceuticals to green tissue. The plants were washed under tap water, separated and dried for 3 days at room temperature (~23 °C). Finally, the plants were dried at 60 °C for 3 hrs in a heating oven (Pol-Eko Apparatures sp.j., Wodzisław Śląski, Poland). The dried plants were homogenized using a mechanical blender (Kenwood, Havant, UK), and frozen at −20 °C until analysis. The average water content in *Cyperus papyrus, Lysimachia nemorum* and *Euonymus europaeus*, determined based on the weight of the sample before and after desiccation, was 75.4%, 64.7%, and 68.5%, respectively.

3.4. Determination of Selected Pharmaceuticals and Endocrine Disrupting Compounds in Treated and Untreated Sewage

Pharmaceuticals and endocrine disrupting compounds were extracted from treated and untreated sewage using SPE on Oasis HLB cartridges (3 mg, 6 mL) according to a previously optimized and validated procedure [53]. Each cartridge was preconditioned with 3 mL of EtOAc, 3 mL of MeOH, and 3 mL of distilled water adjusted to pH 2 with 1 M HCl. A liquid sample (250 mL of wastewater adjusted to pH 2 with 1 M HCl) was passed through a cartridge at a flow rate of ~5 mL min^{-1} using a vacuum manifold. After loading the sample, the cartridge was flushed with 10 mL of an MeOH:H_2O (1:9, *v*/*v*) mixture and subsequently air-dried under a vacuum for 30 min. The adsorbed analytes were eluted with 6 mL of MeOH and dried completely under nitrogen gas in order to be derivatized. Derivatization was carried out using 100 μL of a mixture of 99% BSTFA/1% TMCS and pyridine (1:1, *v*/*v*) [53]. After shaking for 30 s, the reaction vials were placed in a heating block for 30 min at 60 °C. The standards of pharmaceuticals were derivatized using the same procedure. The samples were analyzed using the GCMS-QP 2010 SE Shimadzu System (Shimadzu, Kyoto, Japan) with an AOC-5000 autosampler. The separation of analytes was done using a Zebron ZB-5MSi fused-silica capillary column (30 m, 0.25 mm I.D., 0.25 μm film thickness, Phenomenex). The injection port temperature was 300 °C and 1 μL samples were injected in the splitless mode (1 min). The carrier gas was helium (100 kPa). The oven temperature programme was: 120 °C for 1 min, from 120 °C to 300 °C at 6 °C/min, and finally, 4 min at 300 °C (total 35 min). The transfer line was held at 300 °C. The scan time was 0.3 s. The parameters used for identifying analytes: retention time, quantitative ions, and conformation ions, and applied time windows for selected analytes are shown in Table 7.

Table 7. Retention parameters (time allowed change ± 0.15 min), time windows, and SIM ions for trimethylsilyl (TMS)-derivatives of target compounds (quantitative ions are marked in bold; confirmation ions with intensity relative to quantitative ions (%)).

Identification Parameters/ Pharmaceuticals	Retention Time (R_t) [min]	Characteristic Ions (*m/z*) (Quantitative and Confirmation Ions)	Time Windows [min]
Ibuprofen	10.39	**160**; 263(18); 278(10)	10.01–13.52
Paracetamol	10.64	**206**; 280(90); 295(60)	
Terbutaline	16.20	**86**; 356(47)	13.52–19.88
Flurbiprofen	17.01	**180**; 165(85); 301(25)	
Naproxen	18.36	**185**; 243(80)	
Diflunisal	18.88	**379**; 380(30)	
Amitriptyline	20.64	**58**; 215(30)	19.88–22.50
Imipramine	21.09	**243**; 58(60); 195(40)	
Diclofenac	21.73	**214**; 242(50); 352(12); 367(25)	
Clomipramine	23.90	**268**; 314(25)	22.50–25.63
Nadolol	24.53	**86**; 73(8); 510(33)	
Estrone (E1)	26.48	**342**; 257(50); 285(36); 327(10)	25.63–27.97
17β-estradiol (E2)	27.11	**416**; 285(90)	
17α-ethinylestradiol (EE2)	28.59	**425**; 440(35)	27.97–29.30
Estriol (E3)	29.77	**504**; 311(81)	29.30–30.39

3.5. Development of the Analytical Method for Determining Target Compounds in Plant Materials

An Accelerated Solvent Extraction (ASE) technique with purification of the obtained extracts by SPE was applied for extracting the pharmaceuticals and EDCs from plant tissues. The ASE extraction was performed using a DIONEX ASE 350 (Dionex Corp., Sunnyvale, CA, USA). Three solvent mixtures (MeOH:H_2O 1:1 *v/v*; EtOH:H_2O 1:1 *v/v*: ACN:H_2O 1:1 *v/v*) were applied at two different temperatures (50 °C and 80 °C) for the ASE extraction of the analytes. Moreover, the obtained extracts, prior to SPE purification, were acidified to pH ~2 as well as being investigated without acidification. In each tested condition, the concentration of the target pharmaceuticals and EDCs in spiked plant samples was 1000 ng/g d.w.

A cellulose filter (19.8 mm, Dionex Corp.) was placed on the bottom of a 33-mL stainless steel extraction cell. The cell was filled with 3 ± 0.01 g of diatomaceous earth (on the cellulose filter). Then, 1 ± 0.01 g of spiked and non-spiked homogenous plant tissues was added. Next, the dead volume of the cell was filled with diatomaceous earth. Each extraction was carried out at constant parameters: heating time 5 min, static time 3 min in 3 cycles, and a 10% purge of 50 s. The extraction pressure was set to 1500 psi. The total extraction time was 18.5 min and the extraction volume was 26 ± 0.1 mL. A total of 5 mL of ASE extract was dissolved in deionized water and subjected to purification using SPE under the conditions presented in Section 3.4. A minimum of three samples for each tested parameter were prepared, and each sample was analyzed a minimum of three times.

3.6. Validation of the Proposed Method for Determining Target Compounds in Plant Species

The validation of the analytical method for determining target compounds in plant species was carried out using the matrix-matched calibration solutions and working calibration standard solutions in accordance with the guidelines of the International Vocabulary of Metrology [54]. The matrix-matched calibration solutions were prepared by spiking samples with eight different concentrations of pharmaceuticals and endocrine disrupting compounds in a range of 19.5–2500 ng/g d.w. Calibration curves were based on the external method, and they were obtained by plotting the peak area for each analyte against its concentration. The linearity, correlation coefficient (R^2), intermediate precision measurement (expressed by RSD, $n = 3$), and mean recovery (MR) were established according to procedures described in our previous paper [55]. Briefly, the mean recovery of the analytical method was determined based on the known concentrations of target compounds in the tested samples (C_{known}) and the concentrations determined by the analysis ($C_{determined}$) using Equation (1):

$$\mathbf{MR} = (C_{determined}/C_{known}) \times 100\% \quad (1)$$

The method quantification limit (MQL) was determined using Equation (2):

$$\mathbf{MQL} = (IQL \times 100\%)/(CF \times AR) \quad (2)$$

whereas the method detection limit (MDL) using Equation (3):

$$\mathbf{MDL} = MQL/3 \quad (3)$$

where IQL is the instrumental quantification limit, CF is the concentration factor, and AR is the absolute recovery of analytes (%).

The matrix effect (ME) and absolute recovery (AR) were determined according to procedures described by Caban et al. [56] using Equation (4) and Equation (5), respectively:

$$\mathbf{ME} = ((B - D/A) - 1) \times 100\% \quad (4)$$

$$\mathbf{AR} = ((C - D)/A) \times 100\% \quad (5)$$

where A is the peak area of the analyte recorded for the standard solution, B is the peak area of the analyte recorded for the sample spiked with the target compound after extraction, C is the peak area of the analyte recorded for the sample spiked with the target compound before extraction, and D is the peak area of the analyte recorded for the non-spiked sample (blank sample).

3.7. Application of the Proposed Method for Determining Target Compounds in Hydroponically Cultivated Plants

The optimized and validated ASE-SPE-GC-MS(SIM) method was used to determine 15 target compounds in 3 species of plants (Papyrus (*Cyperus papyrus*), European spindle (*Euonymus europaeus*), Yellow pimpernel (*Lysimachia nemorum*)) used in the constructed wetlands in Sochaczew MWWTP. These species were chosen due to their best adaptation to growth in the MWWTP. In addition, the mentioned plants exhibit the strongest growth during the growing season, which affects the transport of pollutants to plant tissues.

3.8. Evaluation of the Effectiveness of Removing Pharmaceuticals and Endocrine Disrupting Compounds in MWWTP Sochaczew

Concentrations of pharmaceuticals and EDCs in treated and untreated sewage were used to determine the elimination efficiency (EE%) from the wastewater stream in the MWWTP. The elimination efficiency factor was calculated using Equation (6):

$$\mathbf{EE\%} = (C_{untreated} - C_{treated})/(C_{untreated}) \times 100\% \quad (6)$$

where $C_{untreated}$ is the measured concentration of the pharmaceutical in untreated sewage samples, and $C_{treated}$ is the measured concentration of the pharmaceutical in treated sewage samples. This parameter allowed for establishing the effectiveness of removing pharmaceuticals in a municipal treatment plant supported by constructed wetlands.

4. Conclusions

In this study, the analysis of the possibility of using hydroponic cultivation for the removal of 15 pharmaceuticals and endocrine disrupting compounds in municipal wastewater treatment plants is presented. For the first time, three plants: *Cyperus papyrus* (Papyrus), *Lysimachia nemorum* (Yellow pimpernel), and *Euonymus europaeus* (European spindle) were considered. In order to obtain the most reliable data, the investigation was performed using real MWWTP conditions and with the determination of target compounds not only in raw and treated wastewater, but also in plant materials. The determination of target compounds in raw and treated wastewater samples was performed using a previously proposed method [53]; however, for determining target compounds in plant materials, in this study, a new ASE-SPE-GC-MS(SIM) method was developed and validated. The application of a mixture of MeOH:H_2O (1:1, *v*/*v*) at 50 °C without the acidification of the ASE extract prior to SPE purification was found to be the most optimal ASE-SPE conditions for the isolation of target compounds from plant material. The MQL values of the proposed method were in the range of 0.4 ng/g d.w. to 2 ng/g d.w.; the intermediate precision measurement was in the range of 0.05 to 21.60%; the mean recoveries in the range of 80% to 102%. Among the 15 investigated, 5 analytes were found in Papyrus (*C. Papyrus*); 5 target compounds in Yellow pimpernel (*L. nemorum*), and 4 in the tissue of the European spindle (*E. europaeus*) plant. The highest concentration of all target compounds was observed for *Lysimachia nemorum* (Yellow pimpernel), therefore, taking into account the summary uptake of target pharmaceuticals and ECDs by the tested plants, this species is the best for supporting conventional MWWTPs. The obtained data proved that the elimination efficiency of the investigated compounds from wastewater was in the range of 35.8% to 100%. Thus, the application of constructed wetlands for supporting conventional MWWTPs allowed a significant increase in their removal from the wastewater stream.

Establishing which plants effectively cumulate certain types of compounds is useful for designing effective constructed wetlands. Moreover, in the future, the proposed method for determining pharmaceuticals and EDCs in plant materials could be used for assessing the quality of food of plant origin in the cultivation of which sewage sludge or purified sewage was used.

Supplementary Materials: The following are available online. Table S1 Literature data concerning studies of the usefulness of hydroponically cultivated plants for removing target compounds from the sewage stream; Table S2 Chemical structures and physical and chemical properties of selected non-steroidal anti-inflammatory drugs (NSAIDs), analgesics, β-blockers, β-agonists, antidepressant drugs, and estrogen-based hormones; Figure S1 The mass spectra of target compounds with the MS fragments assignation; Figure S2 Activated sludge chamber with a system of constructed wetlands in the investigated Municipal Wastewater Treatment Plant in Sochaczew (Mazowieckie Voivodeship, Poland); Figure S3 Example chromatogram with marked SIM ions for the determined target compounds in real Papyrus (*Cyperus papyrus*) samples; Figure S4 Example chromatogram with marked SIM ions for the determined target compounds in real Yellow pimpernel (*Lysimachia nemorum*) samples; Figure S5 Example chromatogram with marked SIM ions for the determined target compounds in real European spindle (*Euonymus europaeus*) samples.

Author Contributions: Conceptualization, D.W., M.C. and J.K.; Formal analysis, M.P.; Funding acquisition, D.W. and P.S.; Methodology, D.W. and M.C.; Project administration, J.K.; Resources, M.P.; Supervision, J.K.; Validation, D.W.; Writing—original draft, D.W.; Writing—review & editing, J.K. All authors have read and agreed to the published version of the manuscript.

Funding: Financial support was provided by the Ministry of Science and Higher Education under grant nos. 539-8610-B334-18/19 and DS 531-8616-D593-19-1E.

Acknowledgments: The authors are also grateful to the Municipal Wastewater Treatment Plant (MWWTP) in Sochaczew (Mazowieckie Voivodeship, Poland) for very fruitful cooperation and the possibility to obtain the required samples, and to Agata Miecznikowska for her experimental support.

Conflicts of Interest: The authors declare no conflict of interest.

References

1. Uslu, M.O.; Jasim, S.; Arvai, A.; Bewtra, J.; Biswas, N. A Survey of Occurrence and Risk Assessment of Pharmaceutical Substances in the Great Lakes Basin. *Ozone Sci. Eng.* **2013**, *35*, 249–262. [CrossRef]
2. Zhang, Y.; Geißen, S.U.; Gal, C. Carbamazepine and diclofenac: Removal in wastewater treatment plants and occurrence in water bodies. *Chemosphere* **2008**, *73*, 1151–1161. [CrossRef] [PubMed]
3. Fent, K.; Weston, A.A.; Caminada, D. Ecotoxicology of human pharmaceuticals. *Aquat. Toxicol.* **2006**, *76*, 122–159. [CrossRef]

4. Caban, M.; Lis, E.; Kumirska, J.; Stepnowski, P. Determination of pharmaceutical residues in drinking water in Poland using a new SPE-GC-MS(SIM) method based on Speedisk extraction disks and DIMETRIS derivatization. *Sci. Total Environ.* **2015**, *538*, 402–411. [CrossRef] [PubMed]
5. Wagil, M.; Maszkowska, J.; Białk-Bielińska, A.; Caban, M.; Stepnowski, P.; Kumirska, J. Determination of metronidazole residues in water, sediment and fish tissue samples. *Chemosphere* **2015**, *119*, A28–A34. [CrossRef] [PubMed]
6. Kumirska, J.; Migowska, N.; Caban, M.; Łukaszewicz, P.; Stepnowski, P. Simultaneous determination of non-steroidal anti-inflammatory drugs and oestrogenic hormones in environmental solid samples. *Sci. Total Environ.* **2015**, *508*, 498–505. [CrossRef]
7. Białk-Bielińska, A.; Kumirska, J.; Borecka, M.; Caban, M.; Paszkiewicz, M.; Pazdro, K.; Stepnowski, P. Selected analytical challenges in the determination of pharmaceuticals in drinking/marine waters and soil/sediment samples. *J. Pharm. Biomed. Anal.* **2016**, *121*, 271–296. [CrossRef]
8. Golet, E.M.; Strehler, A.; Alder, A.C.; Giger, W. Determination of Fluoroquinolone Antibacterial Agents in Sewage Sludge and Sludge-Treated Soil Using Accelerated Solvent Extraction Followed by Solid-Phase Extraction. *Anal. Chem.* **2002**, *74*, 5455–5462. [CrossRef]
9. Borecka, M.; Białk-Bielińska, A.; Siedlewicz, G.; Kornowska, K.; Kumirska, J.; Stepnowski, P.; Pazdro, K. A new approach for the estimation of expanded uncertainty of results of an analytical method developed for determining antibiotics in seawater using solid-phase extraction disks and liquid chromatography coupled with tandem mass spectrometry technique. *J. Chromatogr. A* **2013**, *1304*, 138–146. [CrossRef]
10. Caban, M.; Szaniawska, A.; Stepnowski, P. Screening of 17α-ethynylestradiol and non-steroidal anti-inflammatory pharmaceuticals accumulation in Mytilus edulis trossulus (Gould, 1890) collected from the Gulf of Gdańsk. *Oceanol. Hydrobiol. Stud.* **2016**, *45*, 605–614. [CrossRef]
11. Carballa, M.; Omil, F.; Lema, J.M.; Llompart, M.; García-Jares, C.; Rodríguez, I.; Gómez, M.; Ternes, T. Behavior of pharmaceuticals, cosmetics and hormones in a sewage treatment plant. *Water Res.* **2004**, *38*, 2918–2926. [CrossRef] [PubMed]
12. Jiang, J.Q.; Zhou, Z.; Sharma, V.K. Occurrence, transportation, monitoring and treatment of emerging micro-pollutants in waste water—A review from global views. *Microchem. J.* **2013**, *110*, 292–300. [CrossRef]
13. Petrie, B.; McAdam, E.J.; Scrimshaw, M.D.; Lester, J.N.; Cartmell, E. Fate of drugs during wastewater treatment. *Trends Anal. Chem.* **2013**, *49*, 145–159. [CrossRef]
14. Repice, C.; Grande, M.D.; Maggi, R.; Pedrazzani, R. Licit and illicit drugs in a wastewater treatment plant in Verona, Italy. *Sci. Total Environ.* **2013**, *463–464*, 27–34. [CrossRef] [PubMed]
15. Klamerth, N.; Rizzo, L.; Malato, S.; Maldonado, M.I.; Agüera, A.; Fernández-Alba, A.R. Degradation of fifteen emerging contaminants at μg L-1initial concentrations by mild solar photo-Fenton in MWTP effluents. *Water Res.* **2010**, *44*, 545–554. [CrossRef]
16. Molinos-Senante, M.; Reif, R.; Garrido-Baserba, M.; Hernández-Sancho, F.; Omil, F.; Poch, M.; Sala-Garrido, R. Economic valuation of environmental benefits of removing pharmaceutical and personal care products from WWTP effluents by ozonation. *Sci. Total Environ.* **2013**, *461–462*, 409–415. [CrossRef]
17. Trinh, T.; van den Akker, B.; Stuetz, R.M.; Coleman, H.M.; Le-Clech, P.; Khan, S.J. Removal of trace organic chemical contaminants by a membrane bioreactor. *Water Sci. Technol.* **2012**, *66*, 1856–1863. [CrossRef]
18. Naddeo, V.; Belgiorno, V.; Ricco, D.; Kassinos, D. Degradation of diclofenac during sonolysis, ozonation and their simultaneous application. *Ultrason. Sonochem.* **2009**, *16*, 790–794. [CrossRef]
19. Li, W.C. Occurrence, sources, and fate of pharmaceuticals in aquatic environment and soil. *Environ. Pollut.* **2014**, *187*, 193–201. [CrossRef]
20. Verlicchi, P.; Zambello, E. How efficient are constructed wetlands in removing pharmaceuticals from untreated and treated urban wastewaters? A review. *Sci. Total Environ.* **2014**, *470–471*, 1281–1306. [CrossRef]
21. Choudhary, A.K.; Kumar, S.; Sharma, C. Constructed wetlands: An option for pulp and paper mill wastewater treatment. *Electron. J. Environ. Agric. Food Chem.* **2011**, *10*, 3023–3037.
22. Davies, L.C.; Pedro, I.S.; Ferreira, R.A.; Freire, F.G.; Novais, J.M.; Martins-Dias, S. Constructed wetland treatment system in textile industry and sustainable development. *Water Sci. Technol.* **2008**, *58*, 2017. [CrossRef] [PubMed]
23. García, J.; Rousseau, D.P.L.; Morató, J.; Lesage, E.; Matamoros, V.; Bayona, J.M. Contaminant Removal Processes in Subsurface-Flow Constructed Wetlands: A Review. *Crit. Rev. Environ. Sci. Technol.* **2010**, *40*, 561–661. [CrossRef]

24. Stottmeister, U.; Wießner, A.; Kuschk, P.; Kappelmeyer, U.; Kästner, M.; Bederski, O.; Müller, R.A.; Moormann, H. Effects of plants and microorganisms in constructed wetlands for wastewater treatment. *Biotechnol. Adv.* **2003**, *22*, 93–117. [CrossRef]
25. Sundaravadivel, M.; Vigneswaran, S. Constructed Wetlands for Wastewater Treatment. *Crit. Rev. Environ. Sci. Technol.* **2001**, *31*, 351–409. [CrossRef]
26. Vymazal, J. The use constructed wetlands with horizontal sub-surface flow for various types of wastewater. *Ecol. Eng.* **2009**, *35*, 1–17. [CrossRef]
27. Hijosa-Valsero, M.; Reyes-Contreras, C.; Domínguez, C.; Bécares, E.; Bayona, J.M. Behaviour of pharmaceuticals and personal care products in constructed wetland compartments: Influent, effluent, pore water, substrate and plant roots. *Chemosphere* **2016**, *145*, 508–517. [CrossRef]
28. Matamoros, V.; Nguyen, L.X.; Arias, C.A.; Salvadó, V.; Brix, H. Evaluation of aquatic plants for removing polar microcontaminants: A microcosm experiment. *Chemosphere* **2012**, *88*, 1257–1264. [CrossRef]
29. Ranieri, E.; Verlicchi, P.; Young, T.M. Paracetamol removal in subsurface flow constructed wetlands. *J. Hydrol.* **2011**, *404*, 130–135. [CrossRef]
30. Li, X.; Manman, C.; Anderson, B.C. Design and performance of a water quality treatment wetland in a public park in Shanghai, China. *Ecol. Eng.* **2009**, *35*, 18–24. [CrossRef]
31. Vystavna, Y.; Frkova, Z.; Marchand, L.; Vergeles, Y.; Stolberg, F. Removal efficiency of pharmaceuticals in a full scale constructed wetland in East Ukraine. *Ecol. Eng.* **2017**, *108*, 50–58. [CrossRef]
32. Nuel, M.; Laurent, J.; Bois, P.; Heintz, D.; Wanko, A. Seasonal and ageing effect on the behaviour of 86 drugs in a full-scale surface treatment wetland: Removal efficiencies and distribution in plants and sediments. *Sci. Total Environ.* **2018**, *615*, 1099–1109. [CrossRef] [PubMed]
33. Wang, Y.; Yin, T.; Kelly, B.C.; Gin, K.Y.-H. Bioaccumulation behaviour of pharmaceuticals and personal care products in a constructed wetland. *Chemosphere* **2019**, *222*, 275–285. [CrossRef] [PubMed]
34. Park, N.; Vanderford, B.J.; Snyder, S.A.; Sarp, S.; Kim, S.D.; Cho, J. Effective controls of micropollutants included in wastewater effluent using constructed wetlands under anoxic condition. *Ecol. Eng.* **2009**, *35*, 418–423. [CrossRef]
35. Zarate, F.M.; Schulwitz, S.E.; Stevens, K.J.; Venables, B.J. Bioconcentration of triclosan, methyl-triclosan, and triclocarban in the plants and sediments of a constructed wetland. *Chemosphere* **2012**, *88*, 323–329. [CrossRef]
36. Bartha, B.; Huber, C.; Schröder, P. Uptake and metabolism of diclofenac in Typha latifolia—How plants cope with human pharmaceutical pollution. *Plant Sci.* **2014**, *227*, 12–20. [CrossRef] [PubMed]
37. Giebułtowicz, J.; Nałęcz-Jawecki, G. Occurrence of antidepressant residues in the sewage-impacted Vistula and Utrata rivers and in tap water in Warsaw (Poland). *Ecotoxicol. Environ. Saf.* **2014**, *104*, 103–109. [CrossRef]
38. Wolecki, D.; Caban, M.; Pazdro, K.; Mulkiewicz, E.; Stepnowski, P.; Kumirska, J. Simultaneous determination of non-steroidal anti-inflammatory drugs and natural estrogens in the mussels *Mytilus edulis trossulus*. *Talanta* **2019**, *200*, 316–323. [CrossRef]
39. Caban, M.; Lis, H.; Kobylis, P.; Stepnowski, P. The triple-sorbents solid-phase extraction for pharmaceuticals and estrogens determination in wastewater samples. *Microchem. J.* **2019**, *149*, 103965. [CrossRef]
40. Breitholtz, M.; Näslund, M.; Stråe, D.; Borg, H.; Grabic, R.; Fick, J. An evaluation of free water surface wetlands as tertiary sewage water treatment of micro-pollutants. *Ecotoxicol. Environ. Saf.* **2012**, *78*, 63–71. [CrossRef]
41. Dordio, A.; Carvalho, A.J.P.; Teixeira, D.M.; Dias, C.B.; Pinto, A.P. Removal of pharmaceuticals in microcosm constructed wetlands using Typha spp. and LECA. *Bioresour. Technol.* **2010**, *101*, 886–892. [CrossRef] [PubMed]
42. Llorens, E.; Matamoros, V.; Domingo, V.; Bayona, J.M.; García, J. Water quality improvement in a full-scale tertiary constructed wetland: Effects on conventional and specific organic contaminants. *Sci. Total Environ.* **2009**, *407*, 2517–2524. [CrossRef] [PubMed]
43. Matamoros, V.; García, J.; Bayona, J.M. Organic micropollutant removal in a full-scale surface flow constructed wetland fed with secondary effluent. *Water Res.* **2008**, *42*, 653–660. [CrossRef] [PubMed]
44. Matamoros, V.; Salvadó, V. Evaluation of the seasonal performance of a water reclamation pond-constructed wetland system for removing emerging contaminants. *Chemosphere* **2012**, *86*, 111–117. [CrossRef] [PubMed]

45. Verlicchi, P.; Galletti, A.; Petrovic, M.; Barceló, D.; Al Aukidy, M.; Zambello, E. Removal of selected pharmaceuticals from domestic wastewater in an activated sludge system followed by a horizontal subsurface flow bed—Analysis of their respective contributions. *Sci. Total Environ.* **2013**, *454–455*, 411–425. [CrossRef] [PubMed]
46. Ávila, C.; Reyes, C.; Bayona, J.M.; García, J. Emerging organic contaminant removal depending on primary treatment and operational strategy in horizontal subsurface flow constructed wetlands: Influence of redox. *Water Res.* **2013**, *47*, 315–325. [CrossRef] [PubMed]
47. Conkle, J.L.; White, J.R.; Metcalfe, C.D. Reduction of pharmaceutically active compounds by a lagoon wetland wastewater treatment system in Southeast Louisiana. *Chemosphere* **2008**, *73*, 1741–1748. [CrossRef]
48. Hijosa-Valsero, M.; Matamoros, V.; Sidrach-Cardona, R.; Martín-Villacorta, J.; Bécares, E.; Bayona, J.M. Comprehensive assessment of the design configuration of constructed wetlands for the removal of pharmaceuticals and personal care products from urban wastewaters. *Water Res.* **2010**, *44*, 3669–3678. [CrossRef]
49. Hijosa-Valsero, M.; Matamoros, V.; Martín-Villacorta, J.; Bécares, E.; Bayona, J.M. Assessment of full-scale natural systems for the removal of PPCPs from wastewater in small communities. *Water Res.* **2010**, *44*, 1429–1439. [CrossRef]
50. Lee, S.; Kang, S.-I.; Lim, J.-L.; Huh, Y.J.; Kim, K.-S.; Cho, J. Evaluating controllability of pharmaceuticals and metabolites in biologically engineered processes, using corresponding octanol–water distribution coefficient. *Ecol. Eng.* **2011**, *37*, 1595–1600. [CrossRef]
51. Reif, R.; Besancon, A.; Le Corre, K.; Jefferson, B.; Lema, J.M.; Omil, F. Comparison of PPCPs removal on a parallel-operated MBR and AS system and evaluation of effluent post-treatment on vertical flow reed beds. *Water Sci. Technol.* **2011**, *63*, 2411–2417. [CrossRef] [PubMed]
52. Ávila, C.; Pedescoll, A.; Matamoros, V.; Bayona, J.M.; García, J. Capacity of a horizontal subsurface flow constructed wetland system for the removal of emerging pollutants: An injection experiment. *Chemosphere* **2010**, *81*, 1137–1142. [CrossRef] [PubMed]
53. Kumirska, J.; Plenis, A.; Łukaszewicz, P.; Caban, M.; Migowska, N.; Białk-Bielińska, A.; Czerwicka, M.; Stepnowski, P. Chemometric optimization of derivatization reactions prior to gas chromatography-mass spectrometry analysis. *J. Chromatogr. A* **2013**, *1296*, 164–178. [CrossRef] [PubMed]
54. De Bièvre, P. The 2012 International Vocabulary of Metrology: "VIM". *Accredit. Qual. Assur.* **2012**, *17*, 231–232. [CrossRef]
55. Migowska, N.; Caban, M.; Stepnowski, P.; Kumirska, J. Simultaneous analysis of non-steroidal anti-inflammatory drugs and estrogenic hormones in water and wastewater samples using gas chromatography-mass spectrometry and gas chromatography with electron capture detection. *Sci. Total Environ.* **2012**, *441*, 77–88. [CrossRef] [PubMed]
56. Caban, M.; Migowska, N.; Stepnowski, P.; Kwiatkowski, M.; Kumirska, J. Matrix effects and recovery calculations in analyses of pharmaceuticals based on the determination of β-blockers and β-agonists in environmental samples. *J. Chromatogr. A* **2012**, *1258*, 117–127. [CrossRef]

Sample Availability: Samples of the compounds are not available from the authors.

Article

Influence of Selected Antidepressants on the Ciliated Protozoan *Spirostomum ambiguum*: Toxicity, Bioaccumulation, and Biotransformation Products

Grzegorz Nałęcz-Jawecki [1,*], Milena Wawryniuk [1], Joanna Giebułtowicz [2], Adam Olkowski [1] and Agata Drobniewska [1,*]

1 Department of Environmental Health Sciences, Faculty of Pharmacy, Medical University of Warsaw, ul. Banacha 1, 02-097 Warszawa, Poland; mwawryniuk@wum.edu.pl (M.W.); olek_adam@interia.pl (A.O.)
2 Department of Bioanalysis and Drugs Analysis, Faculty of Pharmacy, Medical University of Warsaw, ul. Banacha 1, 02-097 Warszawa, Poland; jgiebultowicz@wum.edu.pl
* Correspondence: gnalecz@wum.edu.pl (G.N.-J.); agata.drobniewska@wum.edu.pl (A.D.); Tel.: +48-22-572-0795 (G.N.-J.); Tel./Fax: +48-22-572-0740 (A.D.)

Academic Editor: Jolanta Kumirska
Received: 21 February 2020; Accepted: 20 March 2020; Published: 25 March 2020

Abstract: The present study aimed to evaluate the effect of the most common antidepressants on aquatic protozoa. *Spirostomum ambiguum* was used as the model protozoan. The biological activity of four antidepressants, namely fluoxetine, sertraline, paroxetine, and mianserin, toward *S. ambiguum* was evaluated. Sertraline was found to be the most toxic drug with EC_{50} values of 0.2 to 0.7 mg/L. The toxicity of the antidepressants depended on the pH of the medium and was the highest in alkaline conditions. Sertraline was also the most bioaccumulating compound tested, followed by mianserin. Slow depuration was observed after transferring the protozoa from the drug solutions to a fresh medium, which indicated possible lysosomotropism of the tested antidepressants in the protozoa. The biotransformation products were identified using a high-resolution mass spectrometer after two days of incubation of the protozoa with the tested antidepressants. Four to six potential biotransformation products were observed in the aqueous phase, while no metabolites were detected in the protozoan cells. Because of the low abundance of metabolites in the medium, their structure was not determined.

Keywords: Spirotox; fluoxetine; sertraline; paroxetine; mianserin; pharmaceuticals in the environment

1. Introduction

Protozoa play an important role in the aquatic food web as primary consumers. They are common in surface waters and activated sludge in waste-water treatment plants (WWTP), where they feed on bacteria and may ingest pollutants directly from water. *Spirostomum ambiguum* is one of the largest ciliated protozoa with long generation time of about 70 h. It tolerates pH changes from 5.5 to 8.0, can be cultured in laboratory, and stored in an inorganic medium for at least eight days [1]. Thus, it is a very convenient organism and has been used in ecotoxicological studies for more than 25 years [1–3].

Antidepressants are one of the major group of pharmaceuticals used worldwide. Sertraline, fluoxetine, and paroxetine, belonging to the most commonly used selective serotonin re-uptake inhibitors (SSRIs), were ranked 14, 31, and 68, respectively, on the top 300 best-selling drugs in 2020 with 38.3, 21.9, and 11.7 million prescriptions, respectively, in the U.S. in 2017 (www.clincalc.com, accessed: 7 February 2020). Mianserin is an atypical, tetracyclic antidepressant used for the treatment of major depressive disorders. Antidepressants, as with other pharmaceutically active compounds (PhACs), are released into freshwaters mainly with waste-water, and they have been detected in effluents, freshwaters, and even drinking waters in many countries [4–9]. Mole and Brooks [10] wrote a comprehensive review of the occurrence of SSRIs in the environment. They found that fluoxetine, citalopram, paroxetine, and sertraline, and their main metabolites norfluoxetine and norsertraline were the most commonly detected antidepressants. Their concentrations in influents and effluents were up to several μg/L. SSRIs have been identified not only in water and sediments but also in fish caught in the effluent waste streams and in molluscs and fish tissues commonly consumed by humans [11]. Most of the SSRIs are slowly degraded under the influence of both biotic and abiotic factors, and due to the continuous discharge are called pseudo-persistent contaminants [11].

High biological activity of antidepressants, especially SSRIs, has been reported for algae [12] and crustaceans and fish [11]. In short-term acute toxicity tests, the EC_{50} values ranged from 0.2 to 10 mg/L [11,12]. However, in chronic toxicity assays, the lowest observed effect concentration (LOEC) values were much lower, as low as 0.0136 mg/L for *Pseudokirchneriella subcapitata* [12]. Moreover, SSRIs are considered to be potentially bioaccumulative [13]. They were detected in fish [14,15] and bivalve [16] tissues. Sertraline was the most bioaccumulating compound in the effluent reach stream [17]. Its bioaccumulation factor (BCF) in two benthic organisms *Hydropsyche* sp. and *Erpobdella octoculata* ranged from 920 to 2100 L/kg and was close to the predicted value. There is a lack of knowledge of the response of protozoa to PhACs, probably due to their small size that entails the use of more sophisticated research techniques.

The analysis of the occurrence of PhACs and their human metabolites in the environment has been restricted to compounds with standards available in the market [8,10,18]. With the development of new analytical mass spectrometers, i.e., Quadrupole time of flight (QTOF) working in all ion MS/MS modes and high-resolution Orbitrap™, it has become possible to detect a huge amount of non-target PhAC metabolites and transformation products [18,19] not only of human origin but also of microbial origin. However, neither PhAC nor their metabolites have yet been reported for their effect on protozoa. Some antidepressant metabolites, e.g., norfluoxetine, have even higher biological activity as parent compounds [20]. Thus, to recognize the complete risk of drugs occurring in the aquatic environment, it is important to know both biotic and abiotic transformation of these compounds.

The present study aimed to evaluate the biological activity of four antidepressants, namely fluoxetine, sertraline, paroxetine and mianserin, on the ciliated protozoan *S. ambiguum*. Our comprehensive study included analysis of acute toxicity, bioconcentration, and biotransformation. Acute toxicity was evaluated with the prolonged, seven-day Spirotox assay. As the tested compounds ionize in water solution, three pH levels from the range of the pH of natural freshwaters (6.0, 6.5 and 7.4) were tested, and the relationship between pH and toxicity was discussed. The bioconcentration of the tested pharmaceuticals was analyzed in the six days uptake followed by six days depuration phases, while the biotransformation products were identified after two days of incubation of the protozoa with the tested antidepressants. In the bioconcentration tests, high-performance liquid chromatography (HPLC) with the mass spectrometer detector QTRAP™ was used, while in the biotransformation tests, UPLC with MS/MS Orbitrap™ was applied with post facto analysis performed by Compound Discoverer Software (Thermo Fisher Scientific, Waltham, MA). To the best of our knowledge, this is the first study on the bioaccumulation and possible biotransformation of PhACs by protozoa.

2. Results

2.1. Toxicity

The protozoan *S. ambiguum* could be stored in an inorganic medium for a long time without losing its viability; thus, the Spirotox test could be prolonged up to seven days. In all tests, the toxic effect percent in the negative control was less than 10%; thus, the results of the tests were valid.

Sertraline was the most toxic antidepressant in all the tested approaches, with EC_{50} in the range of 0.2–0.7 mg/L (Figure 1 and Table A1). Paroxetine and fluoxetine were three-fold, while mianserin was 10-fold less toxic than sertraline. The tested compounds were acutely toxic to *S. ambiguum*, as the

LC_{50} and EC_{50} values were close to each other. This implies that sublethal effects quickly became lethal ones. Moreover, in most cases, the EC_{20} values were less than two times lower than EC_{50} values (Table A1). Only for fluoxetine and mianserin tested at pH lower than seven, the EC_{50} to EC_{20} ratio was higher than two. EC_{20} is a threshold value that indicates the threat to the population of the tested organism. This implies that the EC_{50} value is a good predictive value that can be used to predict the effects of the substances on an entire population. As expected, the toxicity increased with the time of incubation, and the seven-day values were much lower than the one- and the two-day values. The toxicity also depended on the pH of the medium. *S. ambiguum* could be tested in a wide range of pH from 5.5 to 8.0. The toxicity was measured at three pH values 6.0; 6.5 and 7.4 to imitate natural freshwaters. For all tested antidepressants, an increase in toxicity was observed with increasing pH. For SSRIs, the step change can be seen between pH 6.0 and 6.5, while for mianserin, the toxicity increased gradually with the increase in pH, especially after one and two days of incubation. The relationship between toxicity and pH of the medium was previously reported for nitrophenols [21], and to the best of our knowledge, this relationship has not been tested for pharmaceuticals thus far. The toxicity-to-water pH relationship has two consequences. First, the pH of the water should be more strictly defined in the ecotoxicity guidelines to prevent high variability of the results. The present data indicate that the pH shift by only one unit may result in a significant change in toxicity. Second, pH of the water and effluent should be considered in the environmental risk assessment of the ionizable compounds. The tested antidepressants are cationic amphiphilic drugs that ionize in acidic solutions, and the bioavailability of the ionized form of the compound is lower than that of the non-ionized one. For many amphiphilic compounds, the biological activity may be predicted using the pH-dependent water/octanol partition coefficient (log D) instead of log P. Taking into account the whole group of compounds tested there was no correlation between the toxicity of the antidepressants to *S. ambiguum* and lipophilicity expressed by both log P and log D coefficients (Table 1). Thus, their biological activity cannot be explained by the simple non-polar and polar narcosis mechanism of action [22]. The tested drugs inhibit neurotransmitter's (serotonin) re-uptake in vertebrate's tissues. Minguez et al. [23] reported the correlation of SSRI toxicity towards *Daphnia* with the log P coefficient. However, they also observed irreversible cell lysis in the abalone hemocytes, probably due to interactions between the drugs and lysosomal membrane phospholipids [23]. As vacuolization was the first symptom of toxicity of the tested compounds in *S. ambiguum*, we expected that such interactions also occur in protozoa and are the main reason of toxic effects.

Table 1. Physicochemical characteristics of the tested antidepressants.

	Fluoxetine	Paroxetine	Sertraline	Mianserin
	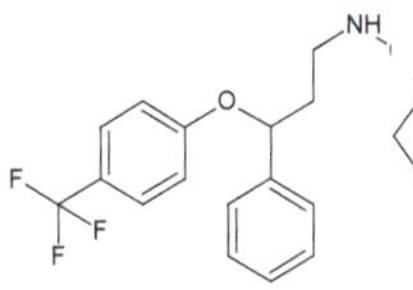	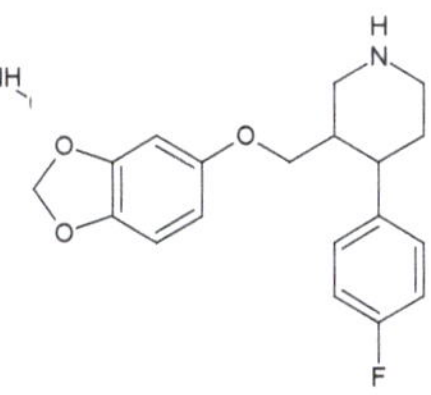	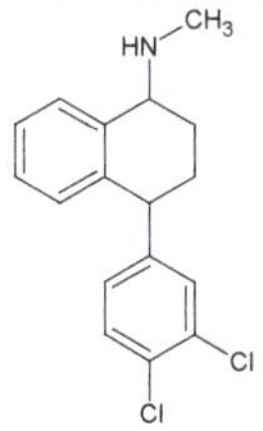	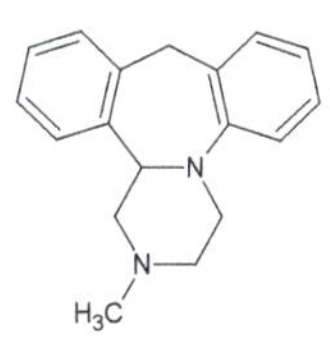
pKa	9.80	9.77	9.85	6.92
Log P	4.09	3.15	5.06	3.83
Log D; pH = 6.0	1.23	0.01	2.17	2.38
Log D; pH = 6.5	1.33	0.11	2.17	2.84
Log D; pH = 7.4	1.81	0.61	2.23	3.50

pKa: acid dissociation constant obtained from www.drugbank.ca; Log P: Octanol-water partition coefficient obtained from www.drugbank.ca; Log D: pH-dependent octanol-water partition coefficient calculated with logD predictor (KLOP algorithm, MarvinSketch 15.2.23. software (https://disco.chemaxon.com/apps).

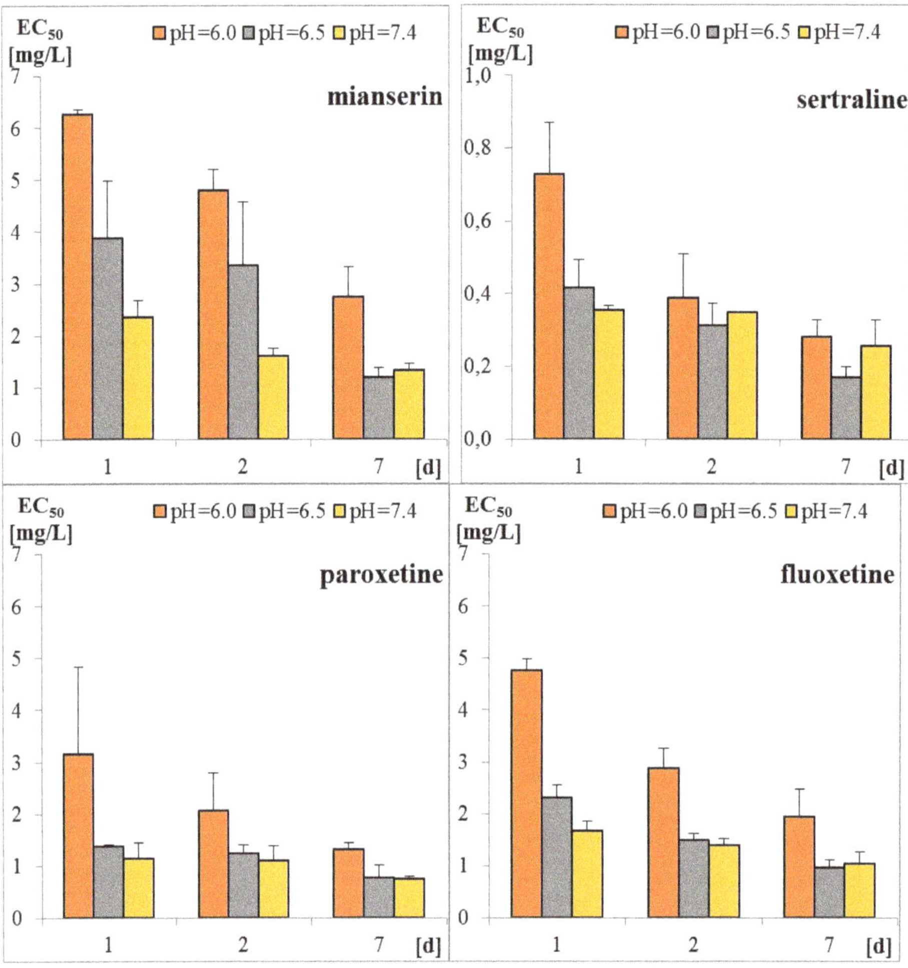

Figure 1. Toxicity of tested antidepressants in Spirotox test after 1, 2, and 7 days incubation (EC_{50} expressed in mg/L).

Antidepressants, especially sertraline, are very potent against parasitic protozoa with IC_{50} of 0.16 mg/L and 0.24 mg/L for *Plasmodium falciparum* and *Trypanosoma brucei rhodosiensis*, respectively, and are considered to be applicable in the treatment of relevant tropical diseases caused by these parasites [24]. Palit and Ali [25] reported high activity of sertraline against another parasite protozoan *Leishmania donovani*. They hypothesized that sertraline induces cell apoptosis by lowering adenosine triphosphate (ATP) levels, resulting in a reduction in oxygen consumption. However, more research is needed to prove this hypothesis and to determine the mode of action of antidepressants towards protozoa.

The protozoan *S. ambiguum* appeared to be comparably sensitive as other organisms used in acute toxicity bioassays. Similar to our results, sertraline was reported to be the most toxic antidepressant to crustaceans with 48-h LC_{50} of 0.12 mg/L for *Ceriodaphnia dubia* [26], 24-h LC_{50} of 0.6 mg/L for *Thamnocephalus platyurus* [27] and 48-h EC_{50} of 0.92 mg/L for *Daphnia magna* [28]. Slightly lower toxicity was reported for fluoxetine, ranging from 0.23 and 0.82 mg/L for *C. dubia* and *D. magna* [12] to 0.85 mg/L for *T. platyurus* [29]. Contrary to the previous two antidepressants, paroxetine was

10-fold less toxic to *D. magna* (6.3 mg/L) [28] than to *C. dubia* (0.58 mg/L) [26]. Very little information is available for mianserin. Wawryniuk et al. [30] reported 24-h LC_{50} of 1.8 mg/L for *T. platyurus*, while Minguez et al. [23] showed 48-h EC_{50} of 7.81 mg/L for *D. magna*. Similar acute toxicity data were reported for fish: 48-h LC_{50} of 0.198 mg/L for fluoxetine towards *Pimephales promelas* [31] and 96-h LC_{50} of 0.38 mg/L for sertraline towards *Oncorhynchus mykiss* [27]. These values are 2–3 orders of magnitude higher than the levels of antidepressants detected in municipal effluents and freshwaters, and therefore, the acute toxicity effect is not expected in the environmental samples.

2.2. Bioaccumulation

To evaluate bioaccumulation of the tested antidepressants in protozoa, *S. ambiguum* was incubated with the antidepressants at three concentrations: low (10 μg/L), medium (25 μg/L), and high (100 μg/L) for six days uptake phase, followed by six days depuration phase.

Whole-body internal concentrations based on the parent compound were measured. The concentrations of the compounds inside the protozoa and in the medium were determined four times in each research phase. The results of the concentration of the tested antidepressants in *S. ambiguum* cells and in the medium are shown in Figure 2 and Table A2, while the BCF values are presented in Figure 3 and Table A3. From the internal concentration data, it can be concluded that uptake and elimination kinetics vary greatly between the tested pharmaceuticals. *S. ambiguum* accumulated significant amounts of sertraline and mianserin, but different bioaccumulation scenarios were observed in each case and for each drug concentration. The concentration of sertraline in the protozoan cells increased gradually during the uptake phase for low and medium drug concentration. For the highest level tested, the highest sertraline concentration was determined after 24 h, followed by a gradual decrease in its concentration. In the depuration phase, the sertraline intracellular concentration remained at a high level, falling by only 40% of the highest concentration (all tested concentrations). Mianserin reached its highest concentration in *S. ambiguum* cells after two days of incubation. After six days, its level dropped to 60–70% and then gradually decreased in the depuration phase. Fluoxetine and paroxetine were not accumulated inside the protozoan cells, and their BCF values during the uptake phase never exceeded 1000 L/kg, while for mianserin and sertraline, the BCF values reached much higher at 4939 and 34,092 L/kg, respectively. The U.S. Environmental Protection Agency has established a BCF ranging from 100 to 1000 L/kg to indicate a medium concern for bioaccumulation [13]; compounds with BCF > 1000 L/kg are considered to be highly bioaccumulating.

The bioaccumulation of SSRIs has been reported in invertebrates and fish by many authors [17,32–34], and the results varied depending on the species. The BCF for sertraline calculated by Grabicova et al. [17] for *E. octoculata* and *Hydropsyche* sp. was higher than 2000 L/kg, while Du et al. [32] found that the BCF value for *Planorbid* sp. was only 990 L/kg. These values were an order of magnitude lower than our results obtained for *S. ambiguum*. The largest spread of results was published for fluoxetine. The value close to our value was obtained by Franzellitti et al. [33] in the marine mussel *Mytilus galloprovincialis*; after seven days of treatment at the concentrations of 30 and 300 ng/L, the BCF ranged from 200 to 800 L/kg. A higher value of 3000 L/kg was reported by Du et al. [32] for *Planorbid* sp. In contrast, Meredith-Williams et al. [34] obtained quite different BCF values of 185,900 L/kg and 1387 L/kg in freshwater shrimp (*Gammarus pulex*) and the water boatman (*Notonecta glauca*), respectively. According to these authors, the 2–3 orders of magnitude higher BCF values for fluoxetine in *G. pulex* resulted from the limited depuration in these animals. Our results (Figures 2 and 3) also indicate low depuration of the tested pharmaceuticals from *S. ambiguum*. In the most cases, after transferring the protozoa to a fresh medium, the intracellular concentration decreased only 2–3 times. The differences in the degree of uptake across the different organisms may be due to differences in the mode of respiration, behavior, and pH of the test system. Moreover, the BCF values are reduced as organism size increases and increase with increasing lipid content [34,35]. However, Rubach et al. [36] found no relationship between lipid content and chlorpyrifos uptake in all 15 species of fish they tested. Lipophilicity is the most often used criterion for predicting the bioaccumulation potential. According to European guidelines on environmental risk assessment of medicinal products for human use [37], all drug substances with log P > 4.5 should be considered to be potentially persistent and should be screened for bioaccumulation; however, OECD uses lower criteria of only log P > 3 [38]. Based on the calculated log P values, Howard and Muir [13] classified sertraline, fluoxetine, and paroxetine as potentially bioaccumulative. However, at neutral pH, the log D values are much lower than log P values (Table 1), and this can explain such low BCF values for fluoxetine (log D: 1.23–1.81) and paroxetine (log D: 0.01–0.61). Grabicova et al. [17] showed that the antidepressive drug citalopram tended to accumulate in organisms, and the extent of accumulation was equivalent to the extent of metabolic transformation and removal from the body.

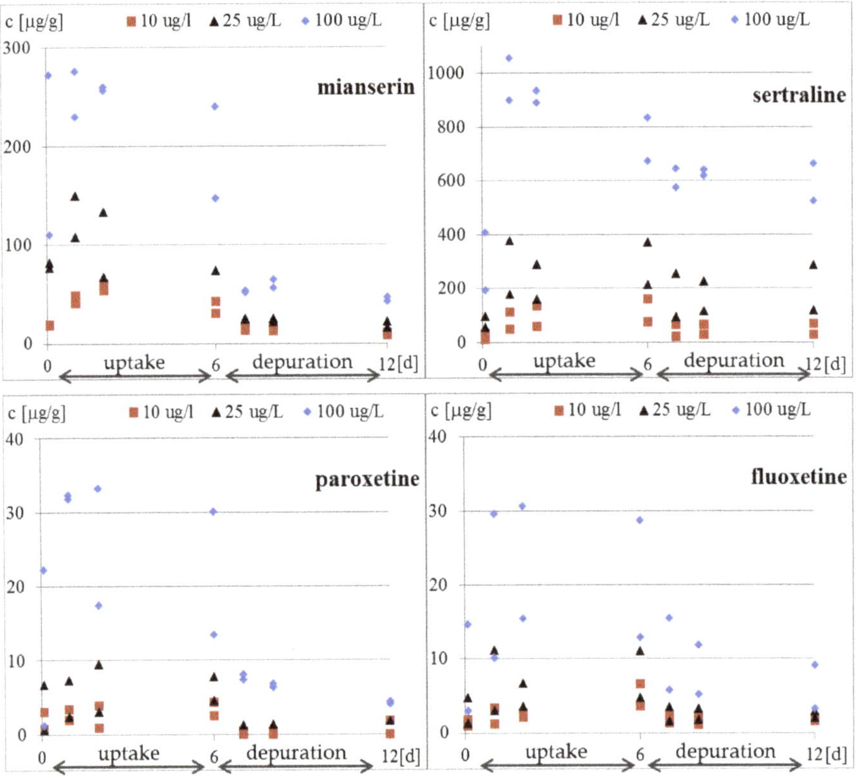

Figure 2. Concentration of tested antidepressants in *S. ambiguum* cells [µg/g].

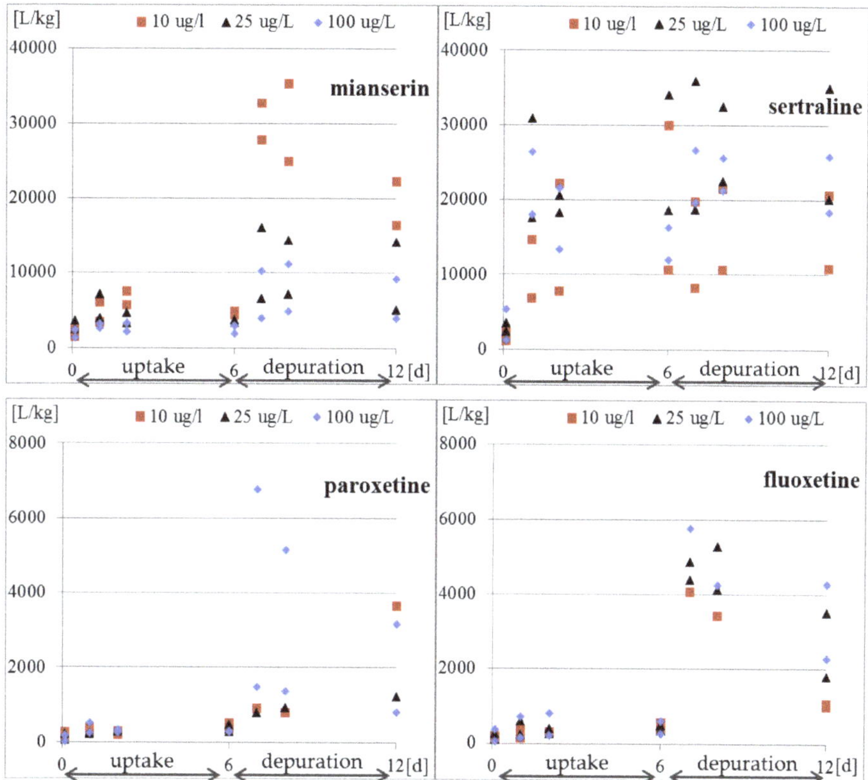

Figure 3. Bioconcentration factor expressed as the ratio of concentration of tested antidepressants in *S. ambiguum* cells to the concentration in water.

After transferring *S. ambiguum* to a clean solution, very slow elimination was observed, and the drugs were detected inside the cells at concentrations up to 11,000 higher than that in the water phase (Figure 3 and Table A3). This indicates that the protozoa were unable to excrete the accumulated antidepressants. The bioaccumulation of drugs in subcellular organelles may eventually result in phospholipidosis and alkalinization of the lysosomes [39]. Two mechanisms are responsible for the accumulation of the basic amphiphilic compounds in cells: binding to phospholipids and lysosomal trapping [40]. The cell membrane and membranes of cellular organelles are permeable to non-ionized compounds [39]. The most acidic pH of protozoa food vacuoles ranges between 3.5 and 4.0. In these conditions, all the tested antidepressants became protonated and cannot pass through the membrane back to the cytosol, which may result in their accumulation within the lysosomes [39]. This phenomenon is called lysosomotropism and has been found in different mammalian cells [39–41]. However, to the best of our knowledge, it has not yet been studied in protozoa. The degree of ion trapping depends on membrane permeability, the pH gradient between the cytosol and lysosome, and physicochemical properties of the compound such as pKa [41]. In our present study, vacuolization of the protozoan cells was observed after six days of incubation with the highest tested concentration of sertraline (100 µg/L) (date not presented). This suggests an effect of the drug on vacuole membrane; however, this hypothesis needs to be confirmed in future research.

2.3. Biotransformation

To evaluate biotransformation, the protozoan *S. ambiguum* was incubated with the antidepressant solution (100 μg/L) in darkness for two days. The Orbitrap™ high-resolution UPLC-MS/MS was used to determine the potential metabolites of the antidepressants in both medium and the protozoan cell. The tentative metabolites of the antidepressant were detected by Compound Discoverer Software (Thermo Fisher Scientific).

The tests were performed twice, and the relative area of the chromatogram peaks are presented in Table 2. The chromatograms of the tested samples were compared to that of the control samples. The peaks observed in two replicates of the samples and not visible in two controls were shown. The predicted transformation products and the difference between the measured and theoretical mass are given. As controls, the antidepressant solutions without the protozoa were incubated under the same conditions. No transformation products were observed in the control samples (data not presented), which confirms the previous findings that these compounds are stable in the aquatic environment [42,43]. Derivatives of only two drugs (fluoxetine and paroxetine) were detected in the protozoa homogenates, whereas four to six transformation products were observed in aquatic media for each antidepressant. The very low levels inside the protozoan cells may be caused by the method of sample preparation. Because of their very low volume, the cell homogenates were analyzed without any enrichment techniques, while the medium was concentrated 100-fold by passing it through Hydrophilic-Lipophilic Balance (HLB) cartridges. The lack of metabolites inside the cells could also be caused by their better solubility in water, high elimination rate from the cells, and lower bioconcentration in the cells than those of the parent compounds.

Table 2. Biotransformation of the tested antidepressants by *S. ambiguum*. Relative abundance of the compounds in the protozoan *S. ambiguum* and in the medium after two days incubation of the protozoans with the parent compounds. The test was performed in duplicate.

[M/z]	Molecular Weight	Delta [ppm]	RDB	RT [min]	Formula	Relative to the Parent Compound	Area of the Chromatographic Peak			
							Medium		*S. ambiguum*	
264.1626	**264.1621**	**2.0**	**10.0**	**10.13**	**C18H20N2**	**Mianserin**	**25,474**	**31,693**	**32,427**	**28,085**
294.1365	294.1363	0.7	11.0	11.16	C18H20N2O2	+O2	48	59	-	-
280.1579	280.1570	3.1	10.0	10.23	C18H20N2O	+O	478	324	-	-
278.1424	278.1414	3.5	11.0	12.34	C18H18N2O	+O; -H2	180	351	-	-
266.1416	266.1414	0.6	10.0	11.12	C17H18N2O	+O; -C H2	12	10	-	-
250.1474	250.1465	3.7	10.0	12.34	C17H18N2	-C H2	25	57	-	-
309.1342	**309.1335**	**2.3**	**8.0**	**10.62**	**C17H18F3NO**	**Fluoxetine**	**20,004**	**33,119**	**9485**	**8925**
527.3015	527.3006	1.7	12.0	11.99	C32H40F3NO2	+C15 H22 O	314	131	-	-
372.1786	372.1781	1.4	6.5	10.56	C19H25F3NO3	+C2 H7 O2	21	11	4	-
365.1603	365.1597	1.7	9.0	13.76	C20H22F3NO2	+C3 H4 O	12	15	-	-
351.1445	351.1441	1.1	9.0	13.26	C19H20F3NO2	+C2 H2 O	74	103	-	-
337.1290	337.1284	1.8	9.0	13.18	C18H18F3NO2	+C O	200	189	-	-
147.1048	147.1043	3.5	5.0	10.62	C10H13N	-C7 H5 F3 O	2360	1179	338	324
329.1423	**329.1422**	**0.3**	**10.0**	**10.33**	**C19H20FNO3**	**Paroxetine**	**12,009**	**405**	**1965**	**2588**
547.3101	547.3092	1.6	14.0	11.77	C34H42FNO4	+C15 H22 O	140	12	-	-
371.1537	371.1527	2.6	11.0	12.83	C21H22FNO4	+C2 H2 O	157	95	-	-
357.1373	357.1371	0.6	11.0	12.72	C20H20FNO4	+CO	201	125	-	-
343.1578	343.1578	0.0	10.0	10.39	C20H22FNO3	+CH2	102	50	52	44
209.1216	209.1210	3.0	5.0	3.51	C12H16FNO	-C7 H4 O2	38	24	-	-
305.0733	**305.0733**	**0.0**	**9.0**	**10.66**	**C17H17Cl2N**	**Sertraline**	**45,092**	**14,502**	**46,409**	**37,880**
523.2410	523.2403	1.3	13.0	12.45	C32H39Cl2NO	+C15 H22 O	134	34	-	-
333.0687	333.0682	1.4	10.0	14.11	C18H17Cl2NO	+CO	85	186	-	-
321.0688	321.0682	2.0	9.0	10.05	C17H17Cl2NO	+O	36	95	-	-
305.0374	305.0369	1.7	10.0	14.29	C16H13Cl2NO	+O; -C H3	46	15	-	-

Five mianserin derivatives were observed in the tested samples, and these were N-demethylation and oxidation products (Table 2). The major mianserin metabolites that are formed in the liver in humans are N-desmethylmianserin, 8-hydroxymianserin and mianserin N-oxide (www.drugbank.ca). Similar products, formed probably by oxidation and oxidative desmethylation, were observed for sertraline, but not fluoxetine (Table 2). Because of the low abundance of these compounds, it was not possible to confirm their structure by fragmentation. Three main sertraline metabolites have been reported in humans: desmethylsertraline, sertraline ketone and sertraline N-carbamoyl glucuronide [44]. In humans, fluoxetine and sertraline are mainly metabolized to N-desmethyl products, which retain their pharmacological activity [18]. N-desmethyl metabolites were also found in aquatic organisms. Silva et al. [18] presented several findings on the occurrence of norfluoxetine and norsertraline in many freshwater fish. These metabolites are more stable than their parent compounds and less polar; thus, their levels in many cases were higher than those of their parent compounds, especially in the liver and brain. However, the authors did not provide the source of these metabolites in aquatic organisms. In organisms collected from the environment, the most probable source of these compounds was the accumulation of metabolites of human origin. Only laboratory tests can prove the occurrence of biotransformation processes in aquatic organisms. Rodriguez et al. [45] detected residual norsertraline in crab cultures incubated with sertraline for two days. Chu et al. [46] found increased concentrations of norfluoxetine in fish incubated with fluoxetine. The mussel *M. galloprovincialis* was exposed to a nominal concentration of fluoxetine (75 ng/L) for 15 days [47]. The authors observed that the concentration of fluoxetine and norfluoxetine increased from 2.53 and 3.06 ng/g dry weight after 3 days up to 9.31 and 11.65 ng/g after 15 days, respectively. These results suggest that fluoxetine accumulated in mussel tissues is likely to be metabolized into norfluoxetine with the increase in the time of exposure.

In humans, paroxetine is metabolized to paroxetine catechol, which is methylated and conjugated into second phase metabolites [42,48]. Cleavage of the paroxetine is also possible, which leads to the formation of the metabolite with a molecular mass of 209 Da [48]. The latter compound was also observed in our studies (Table 2).

Two identical derivatives of SSRIs were observed, which resulted from the addition of CO and C15H22O (Table 2). To the best of our knowledge, such transformation products have not been described either for humans or for aquatic organisms. Their structures were not proposed in the current study because of their very low abundance to perform fragmentation studies. However, this will be the subject of future studies.

2.4. Ecological Implications

The presence of pharmaceuticals in the environment, with a focus on their presence in water, is a potentially major problem with consequences such as toxicity and/or persistence that have not yet been fully understood. Simultaneously, studies involving topical exposure of protozoa to pharmaceuticals in the aquatic environment are very limited [49]. However, protozoa, next to bacteria, constitute the main group of organisms in activated sludge in WWTP, and they are involved in the removal of pollutants from waste-water [50] and in the freshwater self-purification process. Hence, they could have a significant role in removing drugs from the aqueous phase and in their transfer to higher trophic levels. Considering that neuroactive drugs are one of the most ecotoxic pharmaceuticals and that their removal efficiency depends on the condition of conventional activated sludge in WWTP, it is extremely important to know the mechanisms that enable the functioning of protozoa in such conditions and the potential for recovery after contamination. Acute toxicity results obtained in this study were two orders of magnitude higher than the SSRIs concentrations reported in environmental samples. Thus, it can be concluded that the tested antidepressants are unlikely to be toxic to the aquatic protozoa. On the other hand, according to our research and literature review, the SSRIs have been accumulated in biota, and long-term toxic effects cannot be excluded. Thus, future research should be focused on analyzing the transmission of toxic substances, e.g., pharmaceuticals accumulated in vacuoles, and/or their effects on the next generations of organisms and on the next links in the trophic chain.

3. Materials and Methods

3.1. Reagents

Standards of fluoxetine (FLU) and mianserin (MNS) as well as internal standards (IS, nortryptyline and doxepin) were obtained from Sigma-Aldrich (Poznań, Poland), while paroxetine (PAR) and sertraline (SER) were gifts from the National Drug Research Institute, Warsaw, Poland. All the drugs were of high purity grade (>90%). The standard stock solutions of all compounds were prepared in methanol at concentrations of 1 mg/mL and stored at −20 °C. Working solutions were prepared ex tempore by dilution of the stock solutions with the culture medium. IS working solution (500 ng/mL) was prepared ex tempore by dilution of the stock solution with acetonitrile. The solvents, namely HPLC gradient grade methanol, MS grade acetonitrile (LiChrosolv) and formic acid 98%, were provided by Merck (Darmstadt, Germany). Ultrapure water was obtained from a Millipore water purification system (Milli-Q water). The pH-dependent octanol-water partition coefficients (log D) were calculated with logD predictor (https://disco.chemaxon.com/apps).

3.2. Toxicity Assay

Acute toxicity was determined according to the Spirotox assay procedure [1]. Briefly, the assay was performed in 24-well polystyrene microplates. Five 2-fold dilutions were prepared directly in the multi-well plate. Each well contained 1 mL of the test solution and 10 protozoan cells. The microplates were incubated at 25 °C in darkness. Toxic effects (lethality, sublethal responses such as shortening, bending of the cell, immobilization) were noted after 1, 2, and 7 days of incubation. LC_{50}, EC_{50} and EC_{20} values were calculated on the basis of lethal response (L) and all toxic effects (lethal and sublethal) (E), respectively. The toxicity values were expressed in mg/L on the basis of the initial concentrations of the tested compounds. The LC_{50}, EC_{50} and EC_{20} values were determined by graphical interpolation of test response versus toxicant concentration (log scale) [3]. As the diluent and control, Tyrod solution [1] buffered with NaH_2PO_4 and Na_2HPO_4 (50 mM) was used. The toxicity of each compound was tested at pH 6.0; 6.5, and 7.4. All tests were performed in quadruplicate.

3.3. Bioaccumulation Test

The experiments were carried out in 250 mL glass beakers filled with 200 mL of sample or control. As a diluent and control, an inorganic medium (Tyrod solution) was used, with pH adjusted to 7.4. The bioaccumulation experiments were performed for the individual drugs: fluoxetine, sertraline, paroxetine, and mianserin (10, 25 and 100 µg/L). The experimental scheme is presented in Table 3. A total of 1000 protozoan cells were added to each beaker, and the beakers were incubated for 6 days (144 h) at 25 °C in darkness. A total of 100 protozoan cells were subsampled from each test beaker after 2 h and after 1, 2, and 6 days of incubation. Simultaneously, 1 mL of water from each sample was taken for chemical analysis. After 6 days, the protozoa (approximately 500) were transferred to a new glass beaker with the fresh Tyrod solution for testing the depuration of the accumulated drugs. Next, 100 protozoan cells and 1 mL of water were subsampled after 1, 2, and 6 days of incubation. The test was performed in duplicate.

Table 3. Bioaccumulation experiment with protozoan *S. ambiguum*. During the sampling 1 mL of medium and 100 protozoans were collected for chemical analyses.

Preparation:		Accumulation Phase				Depuration Phase		
	0 h	2 h	1 d	2 d	6 d	1 d	2 d	6 d
Medium	+200 mL	−1 mL	−1 mL	−1 mL	−1 mL	−1 mL	−1 mL	−1 mL
S. ambiguum	+1000	−100	−100	−100	−100	−100	−100	−100

Quantitative analyses were performed using Agilent 1260 Infinity (Agilent Technologies, Santa Clara, CA, USA), equipped with a degasser, a thermostated autosampler, a binary pump, and connected in series to a QTRAP®4000 (AB SCIEX, Framingham, MA, USA) equipped with a Turbo Ion Spray source operated in the positive mode. The curtain gas, ion source gas 1, ion source gas 2 and collision gas (all high purity nitrogen) were set at 35 psi, 60 psi, 40 psi and 'medium' instrument units, respectively, and the ion spray voltage and source temperature were set at 5000 V and 600 °C, respectively. Chromatographic separation was achieved with a Kinetex RP-18 column (100 mm, 4.6 mm, particle size 2.6 µm) supplied by Phenomenex (Torrance, CA, USA). The column was maintained at 40 °C at the flow rate of 0.5 mL/min. The mobile phases consisted of HPLC grade water with 0.2% formic acid as eluent A and acetonitrile with 0.2% formic acid as eluent B. The gradient (%B) was as follows: 0 min, 10%; 1 min, 10%; 8 min, 90%; 9 min, 90%. The volume of injection was 10 µL. The target compounds were analyzed in the multiple reaction monitoring (MRM) mode (Table A4) by monitoring two transitions between the precursor ion and the most abundant fragment ions for each compound.

Preparation of *S. ambiguum* samples for HPLC analysis involved mixing 50 µL of sample (100 protozoan cells + medium) with IS (50 µL) and acetonitrile (100 µL). The samples were vortexed (10 min), placed for 10 min in a freezer (at −20 °C) and then centrifuged (5 min at 10,000× *g*). The supernatant (150 µL) was mixed with 375 µL of water and transferred to the autosampler vial. The concentration of pharmaceuticals in organisms was calculated using the measured concentration of the pharmaceutical in the medium and the volume of *S. ambiguum*. An average volume of 100 cells of *S. ambiguum* was 0.50 µL. The preparation of medium samples for HPLC analysis involved centrifugation (10 min at 10,000× *g*), mixing with the IS (9:1) and transferring to vials. No clean up procedure was used.

The validation was performed according to the European Medicines Agency guideline [37]. For *S. ambiguum* extracts, two linearity ranges were selected: 1–100 µg/L and 50–10,000 µg/L of homogenate. For medium samples, the linearity was selected as 0.2–100 µg/L. The coefficients of determination for curves was above 0.99. All validation experiments (accuracy, precision, variation of the relative matrix effect and stability) met the European Medicines Agency (EMEA) acceptance criteria [51].

The concentration of the tested antidepressants in *S ambiguum* was expressed as µg/g assuming the density of the organism as 1 g/mL. The bioconcentration factor was calculated by dividing the substance concentration in organisms to the concentration in the medium and was expressed as L/kg.

3.4. Analysis of Biotransformation of Drugs

The biotransformation of the drugs by the protozoa was analyzed for the four antidepressants: fluoxetine, mianserin, paroxetine, and sertraline. The test beakers were prepared in a manner similar to that for the bioaccumulation experiment. However, only one concentration (100 µg/L) of the drug was tested, and no depuration phase was performed. Concomitant with the tested sample, two control samples were incubated: the abiotic degradation control containing only the same concentration of the tested pharmaceutical (described as "drug control") and the organism control containing only protozoa. After 2 days of incubation in darkness, 500 protozoan cells in 100 µL of medium were transferred to the Eppendorf tube, and 200 µL of acetonitrile was then added. Samples were vortexed (10 min), placed for 10 min in the freezer (at −20 °C) and centrifuged (5 min at 10,000× *g*). The supernatant (150 µL) was mixed with 375 µL of water and transferred to the autosampler vial. Furthermore, 100 mL of medium was sampled at the end of experiment and poured into the preconditioned Oasis HLB (Waters) *spe* cartridge (30 mg). The analytes were eluted with 2 × 3 mL of methanol. The methanol was evaporated under the stream of nitrogen, and the extract was reconstituted with 1 mL of acetonitrile:water (1:9, *v*/*v*). The analysis of transformation products was performed with Ultra High Performance Liquid Chromatography (UHPLC) Dionex Ultimate 3000 with a Q-Exactive hybrid quadrupole-orbitrap mass spectrometer system. Heat electrospray ionization (HESI) was operated in the positive mode. Full MS scans were acquired over *m*/*z* 75–1100 range with the resolution of 70,000 (*m*/*z* 200). Standard mass spectrometric conditions for all experiments were as follows: spray voltage: 3.5 kV; sheath gas pressure: 60 arb; aux gas pressure: 20 arb; sweep gas pressure: 0 arb; heated capillary temperature:

320 °C; loop count: 3; isolation window: *m*/*z* 3.0; and dynamic exclusion: 6.0 s. Chromatographic separation was achieved using a Kinetex RP-18 column (100 mm × 4.6 mm, 2.6 μm) supplied by Phenomenex and equipped with a security guard. The column was maintained at 40 °C at the flow rate of 0.3 mL/min. The mobile phases consisted of HPLC grade water with 0.1% formic acid as eluent A and acetonitrile with 0.1% formic acid as eluent B. The gradient (%B) was as follows: 0 min–10%; 1.5 min–10%; 7.0 min–90%; 12 min–90%. The volume of injection was 10 μL.

All the chromatograms obtained in the biotransformation experiments were integrated with Compound Discoverer Software. The area of the peaks obtained for the sample (protozoa in the drug solution) was divided by the area of the corresponding peaks of the control (protozoa in the medium). Similarly, the area of the peaks obtained for the drug control was divided by the area of the corresponding peaks of the medium. Thus, three values were obtained: tested medium, extract from the protozoan cells, and control medium.

4. Conclusions

We successfully performed a laboratory experiment designed to obtain comprehensive results for acute toxicity, bioconcentration, and biotransformation by determining the biological activity of four antidepressants on the protozoan *S. ambiguum*. The tested compounds were acutely toxic to *S. ambiguum*, and moreover, sublethal effects quickly became lethal ones. Sertraline was the most toxic among the studied antidepressants. However, the toxic effects occur at concentrations at least two orders of magnitude higher than those determined in effluents and freshwaters. Thus, it can be concluded that the tested antidepressants are unlikely to represent a risk to the aquatic protozoa. The results also showed the relationship between pH and toxicity, which has two consequences. First, the pH of the water should be more strictly defined in the aquatic toxicity guidelines to prevent high inter- and intra-laboratory variability of the results. Second, pH of the water and effluent should be considered in the environmental risk assessment, especially for ionizable compounds.

On the basis on the bioconcentration tests, it can be concluded that uptake and elimination kinetics vary greatly between the tested pharmaceuticals. The highest BCF value was obtained for sertraline and mianserin, but different bioaccumulation scenarios can be observed for each pharmaceutical and for each concentration. Our results also indicate that the protozoan cells were unable to excrete the accumulated antidepressants. We suspect that the main reason of the toxic effects and high bioaccumulation ratio were the interactions between the tested drugs and lysosomal membrane phospholipids, which lead to vacuolization. Thus, future research should focus on analyzing the transmission of antidepressants accumulated in vacuoles and/or their effects on the next generations of organisms.

For the first time, the research for the biotransformation products of antidepressants were conducted in the protozoa. However, because of the low abundance of possible biotransformation products, their structure could not be elucidated. This part of the present work revealed a potential for further investigation of pharmaceutical metabolism in protozoa exposed to drugs under natural conditions.

Author Contributions: Conceptualization: A.D., G.N.-J. and M.W.; methodology: J.G., G.N.-J., A.D. and M.W.; formal analysis: J.G., G.N.-J., A.D. and M.W.; investigation: J.G., G.N.-J., A.D., M.W. and A.O.; resources: J.G., G.N.-J., A.D., M.W. and A.O.; data curation: J.G., G.N.-J., A.D. and M.W.; supervision: J.G., G.N.-J. and A.D.; writing—original draft preparation: J.G., G.N.-J., A.O.; A.D. and M.W.; writing—review and editing: J.G., G.N.-J., A.D., M.W. and A.O.; funding acquisition: G.N.-J.; project administration: G.N.-J.; All authors have read and agreed to the published version of the manuscript.

Funding: This research was financially supported by the Medical University of Warsaw (Grant number: FW14/N/2017). LC–MS/MS analysis on QTRAP was carried out using the CePT infrastructure financed by the European Union—the European Regional Development Fund within the Operational Program "Innovative economy" for 2007–2013.

Acknowledgments: The authors wish to thank Ryszard Marszałek for technical assistance during LC–MS/MS analysis.

Conflicts of Interest: The authors declare no conflict of interest.

Appendix A

Table A1. Toxicity of the selected antidepressants in the Spirotox assay [mg/L].

Fluoxetine		pH = 6.0	pH = 6.5	pH = 7.4
1 d	LC_{50}	5.13 ± 0.41	2.78 ± 0.07	1.73 ± 0.17
	EC_{50}	4.75 ± 0.23	2.32 ± 0.24	1.68 ± 0.19
	EC_{20}	2.03 ± 0.82	1.16 ± 0.14	1.21 ± 0.04
2 d	LC_{50}	4.06 ± 0.10	1.78 ± 0.46	1.52 ± 0.16
	EC_{50}	2.88 ± 0.38	1.49 ± 0.14	1.39 ± 0.13
	EC_{20}	1.72 ± 0.51	0.98 ± 0.16	1.06 ± 0.11
7 d	LC_{50}	2.03 ± 0.55	0.99 ± 0.16	1.35 ± 0.07
	EC_{50}	1.94 ± 0.53	0.96 ± 0.16	1.05 ± 0.22
	EC_{20}	0.86 ± 0.15	0.58 ± 0.07	0.70 ± 0.13
Mianserin		**pH = 6.0**	**pH = 6.5**	**pH = 7.4**
1 d	LC_{50}	6.43 ± 0.16	3.95 ± 1.02	2.64 ± 0.50
	EC_{50}	6.27 ± 0.09	3.88 ± 1.11	2.35 ± 0.33
	EC_{20}	3.34 ± 0.76	2.68 ± 0.67	1.58 ± 0.40
2 d	LC_{50}	5.47 ± 0.24	3.51 ± 1.28	2.20 ± 0.44
	EC_{50}	4.80 ± 0.41	3.34 ± 1.23	1.55 ± 0.18
	EC_{20}	2.90 ± 0.25	2.70 ± 0.59	1.16 ± 0.08
7 d	LC_{50}	3.56 ± 0.76	1.34 ± 0.04	1.42 ± 0.07
	EC_{50}	2.75 ± 0.57	1.20 ± 0.18	1.32 ± 0.12
	EC_{20}	2.05 ± 0.23	0.66 ± 0.04	1.01 ± 0.18
Paroxetine		**pH = 6.0**	**pH = 6.5**	**pH = 7.4**
1 d	LC_{50}	4.53 ± 0.55	1.40 ± 0.06	1.17 ± 0.29
	EC_{50}	3.15 ± 1.68	1.37 ± 0.03	1.15 ± 0.29
	EC_{20}	2.46 ± 0.46	1.05 ± 0.05	0.85 ± 0.24
2 d	LC_{50}	2.55 ± 0.75	1.27 ± 0.16	1.15 ± 0.30
	EC_{50}	2.07 ± 0.73	1.24 ± 0.17	1.10 ± 0.29
	EC_{20}	1.71 ± 0.30	0.77 ± 0.45	0.85 ± 0.24
7 d	LC_{50}	1.44 ± 0.21	0.81 ± 0.31	0.77 ± 0.10
	EC_{50}	1.32 ± 0.13	0.76 ± 0.26	0.74 ± 0.06
	EC_{20}	1.12 ± 0.02	0.44 ± 0.16	0.58 ± 0.02
Sertraline		**pH = 6.0**	**pH = 6.5**	**pH = 7.4**
1 d	LC_{50}	0.85 ± 0.09	0.47 ± 0.11	0.41 ± 0.05
	EC_{50}	0.73 ± 0.14	0.42 ± 0.08	0.36 ± 0.01
	EC_{20}	0.46 ± 0.18	0.28 ± 0.04	0.29 ± 0.01
2 d	LC_{50}	0.49 ± 0.10	0.33 ± 0.06	0.35 ± 0.01
	EC_{50}	0.39 ± 0.12	0.31 ± 0.06	0.35 ± 0.02
	EC_{20}	0.19 ± 0.03	0.22 ± 0.08	0.29 ± 0.01
7 d	LC_{50}	0.33 ± 0.08	0.18 ± 0.04	0.27 ± 0.03
	EC_{50}	0.28 ± 0.05	0.17 ± 0.03	0.23 ± 0.04
	EC_{20}	0.20 ± 0.05	0.13 ± 0.03	0.15 ± 0.01

Table A2. Concentration of the tested antidepressants in *S. ambiguum* in µg/g.

Fluoxetine	10 µg/L		25 µg/L		100 µg/L	
Sampling:	**1**	**2**	**1**	**2**	**1**	**2**
0.083 d	1.02	1.80	1.20	4.70	2.95	14.66
1 d	1.27	3.36	2.94	11.04	10.02	29.54
2 d	2.17	2.71	3.57	6.65	15.45	30.51
6 d	3.66	6.65	4.65	10.91	12.82	28.65
7 d	1.32	2.27	1.52	3.37	5.75	15.46
8 d	1.09	2.30	1.73	3.13	5.08	11.77
12 d	1.61	2.05	1.89	2.88	3.15	9.07
Mianserin	**10 µg/L**		**25 µg/L**		**100 µg/L**	
Sampling:	**1**	**2**	**1**	**2**	**1**	**2**
0.083 d	19.88	18.37	81.76	76.13	110.04	272.27
1 d	48.43	40.66	148.83	107.07	229.05	275.21
2 d	54.28	60.35	66.51	132.7	256.15	259.24
6 d	30.32	42.43	72.79	72.59	146.31	239.26
7 d	16.39	13.88	24.21	24.71	51.36	52.78
8 d	17.68	12.49	21.55	24.79	56.04	63.88
12 d	11.29	8.23	21.29	15.04	46.07	42.08
Paroxetine	**10 µg/L**		**25 µg/L**		**100 µg/L**	
Sampling:	**1**	**2**	**1**	**2**	**1**	**2**
0.083 d	0.72	3.01	0.61	6.75	1.27	22.33
1 d	1.89	3.36	2.36	7.33	32.43	31.86
2 d	0.84	3.85	3.03	9.46	17.45	33.26
6 d	2.49	4.34	4.50	7.78	13.46	30.14
7 d	0.45	<LOD	1.19	<LOD	7.44	8.12
8 d	0.40	<LOD	1.40	<LOD	6.90	6.38
12 d	1.82	<LOD	1.85	<LOD	4.11	4.42
Sertraline	**10 µg/L**		**25 µg/L**		**100 µg/L**	
Sampling:	**1**	**2**	**1**	**2**	**1**	**2**
0.083 d	12.01	32.52	54.34	97.41	194.84	409.65
1 d	50.14	112.44	179.26	377.38	900.06	1058.73
2 d	58.24	135.65	161.30	290.20	890.58	935.23
6 d	75.49	159.08	212.90	371.61	673.24	832.82
7 d	22.06	65.37	94.85	255.51	574.47	645.00
8 d	28.55	66.59	116.04	227.17	641.16	619.90
12 d	27.22	68.21	120.04	286.69	525.78	664.49

Table A3. Bioconcentration factor of the tested antidepressants in *S. ambiguum* in L/kg. The test was performed in duplicate.

Fluoxetine	**10 µg/L**		**25 µg/L**		**100 µg/L**	
Sampling:	**1**	**2**	**1**	**2**	**1**	**2**
0.083 d	123	202	81	248	39	359
1 d	157	359	217	585	129	696
2 d	306	283	238	392	191	786
6 d	387	565	328	473	255	600
7 d	4079	11,267	4358	4859	11,122	5738
8 d	3439	9045	5259	4108	10,776	4228
12 d	1000	1065	3500	1779	4256	2257
Mianserin	**10 µg/L**		**25 µg/L**		**100 µg/L**	
Sampling:	**1**	**2**	**1**	**2**	**1**	**2**
0.083 d	2320	1445	3666	2450	1395	2357
1 d	6092	3366	7190	4033	3138	2621
2 d	7497	5721	3276	4666	3318	2137
6 d	4381	4939	3772	3430	1885	2973
7 d	32,774	27,757	16,138	6648	10,271	4008
8 d	35,358	24,976	14,364	7201	11,208	4854
12 d	22,287	16,464	14,191	5169	9214	4004
Paroxetine	**10 µg/L**		**25 µg/L**		**100 µg/L**	
Sampling:	**1**	**2**	**1**	**2**	**1**	**2**
0.083 d	130	267	54	221	18	171
1 d	389	280	214	247	523	252
2 d	189	290	281	324	349	277
6 d	524	357	482	281	293	286
7 d	905		794		1487	6770
8 d	806		934		1379	5148
12 d	3649		1234		823	3179
Sertraline	**10 µg/L**		**25 µg/L**		**100 µg/L**	
Sampling:	**1**	**2**	**1**	**2**	**1**	**2**
0.083 d	1092	2374	2394	3516	1203	5383
1 d	6803	14,603	17,574	30,933	18,037	26,468
2 d	7745	22,238	18,351	20,582	13,352	21,649
6 d	10,573	30,015	18,643	34,092	12,033	16,362
7 d	8261	19,810	18,781	35,987	19,674	26,653
8 d	10,573	21,482	22,532	32,453	21,301	25,616
12 d	10,800	20,669	20,140	34,962	18,358	25,856

Table A4. MS/MS optimized conditions for the analyzed compounds and internal standards.

[M + H]⁺ Ion	*Quantitative Product Ion*	Compound	*DP* [V]	*CE* [V]	*CXP* [V]	*Qualitative Product Ion*	*CE* [V]	*CXP* [V]
280.1	107.0	Doxepin	71	33	6	235.1	25	14
310.3	44.1	Fluoxetine	56	37	6	148.1	13	12
265.1	208.0	Mianserin	96	31	12	118.0	43	8
264.1	233.1	Nortriptyline	71	23	14	91.1	35	6
330.2	192.1	Paroxetine	51	31	14	70.1	49	4
306.0	158.9	Sertraline	51	35	12	275.0	19	16

CE-collision energy; *DP*-declustering potential; *CXP*-collision cell exit potential.

References

1. Nałęcz-Jawecki, G. Spirotox—*Spirostomum ambiguum* acute toxicity test—10 Years of experience. *Environ. Toxicol.* **2004**, *19*, 359–364. [CrossRef] [PubMed]
2. Nałęcz-Jawecki, G.; Demkowicz-Dobrzański, K.; Sawicki, J. Protozoan *Spirostomum ambiguum* as a highly sensitive bioindicator for rapid and easy determination of water quality. *Sci. Total Environ.* **1993**, *134* (Suppl. 2), 1227–1234. [CrossRef]
3. Nałęcz-Jawecki, G. Spirotox test—*Spirostomum ambiguum* acute toxicity test. In *Small-Scale Freshwater Toxicity Investigations. Volume 1—Toxicity Test Methods*; Blaise, C., Férard, J., Eds.; Springer: New York, NY, USA, 2005; pp. 299–322.
4. Barbosa, M.O.; Ribeiro, A.R.; Ratola, N.; Hain, E.; Homem, V.; Pereira, M.F.R.; Blaney, L.; Silva, A.M.T. Spatial and seasonal occurrence of micropollutants in four Portuguese rivers and a case study for fluorescence excitation-emission matrices. *Sci. Total Environ.* **2018**, *644*, 1128–1140. [CrossRef] [PubMed]
5. Burns, E.E.; Carter, L.J.; Kolpin, D.W.; Thomas-Oates, J.; Boxall, A.B.A. Temporal and spatial variation in pharmaceutical concentrations in an urban river system. *Water Res.* **2018**, *137*, 72–85. [CrossRef] [PubMed]
6. Fernández, C.; González-Doncel, M.; Pro, J.; Carbonell, G.; Tarazona, J.V. Occurrence of pharmaceutically active compounds in surface waters of the henares-jarama-tajo river system (madrid, spain) and a potential risk characterization. *Sci. Total Environ.* **2010**, *408*, 543–551. [CrossRef] [PubMed]
7. Giebułtowicz, J.; Nałęcz-Jawecki, G. Occurrence of antidepressant residues in the sewage-impacted Vistula and Utrata rivers and in tap water in Warsaw (Poland). *Ecotoxicol. Environ. Saf.* **2014**, *104*, 103–109. [CrossRef]
8. Metcalfe, C.D.; Chu, S.; Just, C.; Li, H.; Oakes, K.D.; Servos, M.R.; Andrews, D.M. Antidepressants and their metabolites in municipal wastewater, and downstream exposure in an urban watershed. *Environ. Toxicol. Chem.* **2010**, *29*, 79–89. [CrossRef]
9. Paíga, P.; Santos, L.H.; Ramos, S.; Jorge, S.; Silva, J.G.; Delerue-Matos, C. Presence of pharmaceuticals in the Lis river (Portugal): Sources, fate and seasonal variation. *Sci. Total Environ.* **2016**, *573*, 164–177. [CrossRef]
10. Mole, R.; Brooks, B.W. Global scaning of selective serotonin reuptake inhibitors: Occurrence, wastewater treatment and hazards in aquatic systems. *Environ. Pollut.* **2019**, *250*, 1019–1031. [CrossRef]
11. Puckowski, A.; Mioduszewska, K.; Łukaszewicz, P.; Borecka, M.; Caban, M.; Maszkowska, J.; Stepnowski, P. Bioaccumulation and analytics of pharmaceutical residues in the environment: A review. *J. Pharm. Biomed. Anal.* **2016**, *127*, 232–255. [CrossRef]
12. Brooks, B.W.; Turner, P.K.; Stanley, J.K.; Weston, J.J.; Glidewell, E.A.; Foran, C.M.; Slattery, M.; La Point, T.W.; Huggett, D.B. Waterborne and sediment toxicity of fluoxetine to select organisms. *Chemosphere* **2003**, *52*, 135–142. [CrossRef]
13. Howard, P.H.; Muir, D.C.G. Identifying new persistent and bioaccumulative organics among chemicals in commerce II: Pharmaceuticals. *Environ. Sci. Technol.* **2011**, *45*, 6938–6946. [CrossRef] [PubMed]
14. Brooks, B.W.; Chambliss, K.C.; Stanley, J.K.; Ramirez, A.; Banks, K.E.; Johnson, R.D.; Lewis, R.J. Determination of selected antidepressants in fish from effluent-dominated stream. *Environ. Toxicol. Chem.* **2005**, *24*, 464–469. [CrossRef] [PubMed]

15. Schultz, M.M.; Painter, M.M.; Bartell, S.E.; Logue, A.; Furlong, E.T.; Werner, S.L.; Schoenfuss, H.L. Selective uptake and biological consequences of environmentally relevant antidepressant pharmaceutical exposures on male fathead minnows. *Aquat. Toxicol.* **2011**, *104*, 38–47. [CrossRef] [PubMed]
16. Burket, S.R.; White, M.; Ramirez, A.J.; Stanley, J.K.; Banks, K.E.; Waller, W.T.; Chambiliss, C.K.; Brooks, B.W. *Corbicula fluminea* rapidly accumulate pharmaceuticals in an effluent dependent urban steam. *Chemosphere* **2019**, *224*, 873–883. [CrossRef] [PubMed]
17. Grabicova, K.; Grabic, R.; Blaha, M.; Kumar, V.; Cerveny, D.; Fedorova, G.; Randak, T. Presence of pharmaceuticals in benthic fauna living in a small stream affected by effluent from a municipal sewage treatment plant. *Water Res.* **2015**, *72*, 145–153. [CrossRef] [PubMed]
18. Silva, L.J.G.; Pereira, A.M.P.T.; Meisel, L.M.; Lino, C.M.; Pena, A. Reviewing the serotonin reuptake inhibitors (SSRIs) footprint in the aquatic biota: Uptake, bioaccumulation and ecotoxicology. *Environ. Pollut.* **2015**, *197*, 127–143. [CrossRef] [PubMed]
19. Heuett, N.V. Target and Non-Target Techniques for the Quantitation of Drugs of Abuse, Identification of Transformation Products, and Characterization of Contaminants of Emergent Concern by High-Resolution Mass Spectrometry. Ph.D. Thesis, Florida International University, Miami, FL, USA, 2015; p. 2194. [CrossRef]
20. Nałęcz-Jawecki, G. Evaluation of the in vitro biotransformation of fluoxetine with HPLC, mass spectrometry and ecotoxicological tests. *Chemosphere* **2007**, *70*, 29–35. [CrossRef]
21. Nałęcz-Jawecki, G.; Sawicki, J. Influence of pH on the toxicity of nitrophenols to Microtox and Spirotox tests. *Chemosphere* **2003**, *52*, 249–252. [CrossRef]
22. Bearden, A.P.; Schultz, T.W. Structure-activity relationships for Pimephales and Tetrahymena: A mechanism of action approach. *Environ. Toxicol. Chem.* **1997**, *16*, 1311–1317. [CrossRef]
23. Minguez, L.; Di Poi, C.; Farcy, E.; Ballandonne, C.; Benchouala, A.; Bojic, C.; Cossu-Leguille, C.; Costil, K.; Serpentini, A.; Lebel, J.; et al. Comparison of the sensitivity of seven marine and freshwater bioassays as regards antidepressant toxicity assessment. *Ecotoxicology* **2014**, *23*, 1744–1754. [CrossRef] [PubMed]
24. Kaiser, M.; Maser, P.; Tadoori, L.P.; Ioset, J.-R.; Brun, R. Antiprotozoal activity profiling of approved drugs: A starting point toward drug repositioning. *PLoS ONE* **2015**, *10*, e0135556. [CrossRef] [PubMed]
25. Palit, P.; Ali, N. Oral therapy with sertraline, a selective serotonin reuptake inhibitor, shows activity against *Leishmania donovani*. *J. Antimicrob. Chemother.* **2008**, *61*, 1120–1124. [CrossRef] [PubMed]
26. Henry, T.B.; Kwon, J.-W.; Armbrust, K.L.; Black, M.C. Acute and chronic toxicity of five selective serotonin reuptake inhibitors in *Ceriodaphnia dubia*. *Environ. Toxicol. Chem.* **2004**, *23*, 2229–2233. [CrossRef]
27. Minagh, E.; Hernan, R.; O'Rourke, K.; Lyng, F.M.; Davoren, M. Aquatic ecotoxicity of the selective serotonin reuptake inhibitor sertraline hydrochloride in a battery of freshwater test species. *Ecotox. Environ. Saf.* **2009**, *72*, 434–440. [CrossRef]
28. Christensen, A.M.; Faaborg-Andersen, S.; Ingerslev, F.; Baun, A. Mixture and single-substance toxicity of selective serotonin reuptake inhibitors toward algae and crustaceans. *Environ. Toxicol. Chem.* **2007**, *26*, 85–91. [CrossRef]
29. Nałęcz-Jawecki, G.; Szczęsny, Ł.; Solecka, D.; Sawicki, J. Short ingestion tests as alternative proposal for conventional range finding assays with *Thamnocephalus platyurus* and *Brachionus calyciflorus*. *Int. J. Environ. Sci. Technol.* **2011**, *8*, 687–694. [CrossRef]
30. Wawryniuk, M.; Pietrzak, A.; Nałęcz-Jawecki, G. Evaluation of direct and indirect photodegradation of mianserin with high-performance liquid chromatography and short-term bioassays. *Ecotoxicol. Environ. Saf.* **2015**, *115*, 144–151. [CrossRef]
31. Stanley, J.K.; Ramirez, A.J.; Chambliss, C.K.; Brooks, B.W. Enantiospecific sublethal effects of the antidepressant fluoxetine to a model aquatic vertebrate and invertebrate. *Chemosphere* **2007**, *69*, 9–16. [CrossRef]
32. Du, B.; Haddad, S.P.; Scott, W.C.; Chambliss, C.K.; Brooks, B.W. Pharmaceutical bioaccumulation by periphyton and snails in an effluent-dependent stream during an extreme drought. *Chemosphere* **2015**, *119*, 927–934. [CrossRef]
33. Franzellitti, S.; Buratti, S.; Capolupo, M.; Du, B.; Haddad, S.P.; Chambliss, C.K.; Brooks, B.W.; Fabbri, E. An exploratory investigation of various modes of action and potential adverse outcomes of fluoxetine in marine mussels. *Aquat. Toxicol.* **2014**, *151*, 14–26. [CrossRef] [PubMed]
34. Meredith-Williams, M.; Carter, L.J.; Fussell, R.; Raffaelli, D.; Ashauer, R.; Boxall, A.B. Uptake and depuration of pharmaceuticals in aquatic invertebrates. *Environ. Pollut.* **2012**, *165*, 250–258. [CrossRef] [PubMed]

35. Hendriks, A.J.; Linde, A.V.D.; Cornelissen, G.; Sijm, D.T.H.M. The power of size. 1. Rate constants and equilibrium ratios for accumulation of organic substances related to octanol-water partition ratio and species weight. *Environ. Toxicol. Chem.* **2001**, *20*, 1339–1420. [CrossRef]
36. Rubach, M.N.; Baird, D.J.; Van den Brink, P.J. A new method for ranking mode specific sensitivity of freshwater arthropods to insecticides and its relationship to biological traits. *Environ. Toxicol. Chem.* **2010**, *29*, 476–487. [CrossRef]
37. EMA/CHMP. European Medicines Agency/Committee for Medicinal Products for Human use. In *Guideline on Bioanalytical Method Validation;* EMEA/CHMP/EWP/192217/2009; European Medicines Agency: London, UK, 2011. Available online: http://www.ema.europa.eu/docs/en_GB/document_library/ (accessed on 4 December 2017).
38. OECD. Test No. 315: Bioaccumulation in Sediment-dwelling Benthic Oligochaetes. In *OECD Guidelines for the Testing of Chemicals;* Section 3; OECD Publishing: Paris, France, 2008. [CrossRef]
39. Lemieux, B.; Percival, M.D.; Falgueyret, J.-P. Quantitation of the lysosomotropic character of cationic amphiphilic drugs using the fluorescent basic amine Red DND-99. *Anal. Biochem.* **2004**, *327*, 247–251. [CrossRef] [PubMed]
40. Daniel, W.A.; Wójcikowski, J.; Pałucha, A. Intracellular distribution of psychotropic drugs in the grey and white matter of the brain: The role of lysosomal trapping. *Br. J. Pharmacol.* **2001**, *134*, 807–814. [CrossRef]
41. Kuzu, O.F.; Toprak, M.; Noory, M.A.; Robertson, G.P. Effect of lysosomotropic molecules on cellular homeostasis. *Pharmacol. Res.* **2017**, *117*, 177–184. [CrossRef]
42. Kwon, J.-W.; Armbrust, K.L. Hydrolysis and photolysis of paroxetine, a selective serotonin reuptake inhibitor, in aqueous solutions. *Environ. Toxicol. Chem.* **2004**, *23*, 1394–1399. [CrossRef]
43. Kwon, J.-W.; Armbrust, K.L. Laboratory persistence and fate of fluoxetine in aquatic environments. *Environ. Toxicol. Chem.* **2006**, *25*, 2561–2568. [CrossRef]
44. Obach, R.S.; Cox, L.M.; Tremaine, L.M. Sertraline is metabolized by multiple cytochrome P450 enzymes, monoamine oxidases, and glucuronyl transferases in human: An in vitro study. *Drug Metab. Dispos.* **2005**, *33*, 262–270. [CrossRef]
45. Rodriguez, A.P.; Santos, L.H.M.L.M.; Ramalhosa, M.J.; Delerue-Matos, C.; Guimaraes, L. Sertraline accumulation and effects in the estuarine decapod *Cacinus maenas*: Impotrance of the history of exposure to chemical stress. *J. Hazard. Mater.* **2015**, *283*, 250–358. [CrossRef]
46. Chu, S.; Metcalfe, C.D. Analysis of paroxetine, fluoxetine and norfluoxetine in fish tissues using pressurized liquid extraction, mixed mode solid phase extraction cleanup and liquid chromatography-tandem mass spectrometry. *J. Chromatogr. A* **2007**, *1163*, 112–118. [CrossRef] [PubMed]
47. Silva, L.J.G.; Martins, M.C.; Pereira, A.M.P.T.; Meisel, L.M.; Gonzalez-Rey, M.; Bebianno, M.J.; Lino, C.M.; Pena, A. Uptake, accumulation and metabolization of the antidepressant fluoxetine by *Mytilus galloprovincialis*. *Environ. Pollut.* **2016**, *213*, 432–437. [CrossRef] [PubMed]
48. Jornil, J.; Jensen, K.G.; Larsen, F.; Linnet, K. Identification of cytochrome P450 isoforms involved in the metabolism of paroxetine and estimation of their importance for human paroxetine metabolism using a population-based simulator. *Drug Metab. Dispos.* **2009**, *38*, 376–385. [CrossRef] [PubMed]
49. Wawryniuk, M.; Drobniewska, A.; Sikorska, K.; Nałęcz Jawecki, G. Influence of photolabile pharmaceutical on the photodegradation and toxicity of fluoxetine and fluvoxamine. *Environ. Sci. Pollut. Res.* **2018**, *25*, 6890–6898. [CrossRef] [PubMed]
50. Babko, R.; Kuzmina, T.; Suchorab, Z.; Widomski, M.K.; Franus, M. Influence of treated sewage dis charge on the bentos ciliate assembly in the lowland river. *Ecol. Chem. Eng. S* **2016**, *23*, 461–471. [CrossRef]
51. EMEA/CHMP. European Medicines Agency/Committee for Medicinal Products for Human use. In *Guideline on the Environmental Risk Assessment of Medicinal Products for Human Use;* EMEA/CHMP/SWP/4447/00 corr 1; European Medicines Agency: London, UK, 2006. Available online: http://www.ema.europa.eu/docs/en_GB/document_library/ (accessed on 4 December 2017).

Sample Availability: Samples of the compounds are available from the authors.

Article

Identification of Selected Antibiotic Resistance Genes in Two Different Wastewater Treatment Plant Systems in Poland: A Preliminary Study

Magdalena Pazda [1], Magda Rybicka [2], Stefan Stolte [3], Krzysztof Piotr Bielawski [2], Piotr Stepnowski [1,*], Jolanta Kumirska [1], Daniel Wolecki [1] and Ewa Mulkiewicz [1]

1. Department of Environmental Analysis, Faculty of Chemistry, University of Gdansk, Wita Stwosza 63, 80-308 Gdansk, Poland; magdalena.pazda@phdstud.ug.edu.pl (M.P.); jolanta.kumirska@ug.edu.pl (J.K.); daniel.wolecki@phdstud.ug.edu.pl (D.W.); ewa.mulkiewicz@ug.edu.pl (E.M.)
2. Department of Molecular Diagnostics, Intercollegiate Faculty of Biotechnology, University of Gdansk and Medical University of Gdansk, Abrahama 58, 80-307 Gdansk, Poland; magda.rybicka@biotech.ug.edu.pl (M.R.); krzysztof.bielawski@biotech.ug.edu.pl (K.P.B.)
3. Institute of Water Chemistry, Technische Universität Dresden, Bergstraße 66, 01062 Dresden, Germany; stefan.stolte@tu-dresden.de

* Correspondence: piotr.stepnowski@ug.edu.pl; Tel.: +48-58-523-2041

Academic Editor: Teresa A. P. Rocha-Santos
Received: 30 April 2020; Accepted: 18 June 2020; Published: 20 June 2020

Abstract: Antibiotic resistance is a growing problem worldwide. The emergence and rapid spread of antibiotic resistance determinants have led to an increasing concern about the potential environmental and public health endangering. Wastewater treatment plants (WWTPs) play an important role in this phenomenon since antibacterial drugs introduced into wastewater can exert a selection pressure on antibiotic-resistant bacteria (ARB) and antibiotic resistance genes (ARGs). Therefore, WWTPs are perceived as the main sources of antibiotics, ARB and ARG spread in various environmental components. Furthermore, technological processes used in WWTPs and its exploitation conditions may influence the effectiveness of antibiotic resistance determinants' elimination. The main aim of the present study was to compare the occurrence of selected tetracycline and sulfonamide resistance genes in raw influent and final effluent samples from two WWTPs different in terms of size and applied biological wastewater treatment processes (conventional activated sludge (AS)-based and combining a conventional AS-based method with constructed wetlands (CWs)). All 13 selected ARGs were detected in raw influent and final effluent samples from both WWTPs. Significant ARG enrichment, especially for *tet*(*B*, *K*, *L*, *O*) and *sulIII* genes, was observed in conventional WWTP. The obtained data did not show a clear trend in seasonal fluctuations in the abundance of selected resistance genes in wastewaters.

Keywords: antibiotic resistance genes (ARGs); antibiotic-resistant bacteria (ARB); wastewater treatment plants (WWTPs); activated sludge (AS); constructed wetlands (CWs); environmental pollution; spread of resistance; tetracyclines; sulfonamides

1. Introduction

Antibiotic resistance is arguably one of the greatest threats and challenges to public health and contemporary medicine worldwide. As a consequence of the widespread use of antimicrobials for human, veterinary and agricultural purposes, the number of antibiotic-resistant bacteria (ARB) is constantly increasing [1]. According to the published data, in 2015 in the European Union (EU) and European Economic Activity countries, ARB caused 671,689 infections and led to over 33,000 deaths [2]. Furthermore, the cost of hospitalisation of European patients infected by selected multidrug-resistant bacteria was estimated at least EUR 1.5 billion annually [3], while in the United States this cost was even higher and estimated at approximately USD 2.2 billion [4].

Antimicrobial compounds such as pharmaceuticals are classified to a group of contaminants of emerging concern (CECs) [5]. The results of many studies confirmed the presence of trace amounts of numerous antibiotics not only in municipal, hospital or industrial wastewater, but also in surface, ground and drinking water, as well as in soil and bottom sediments [6]. For this reason, and due to the growing awareness that the occurrence of these compounds in the environment may pose a threat to human health and ecological systems, the EU Commission has included three macrolide antibiotics (azithromycin, clarithromycin, erythromycin) (EU Decision, 2015/495 of 20 March 2015) as well as amoxicillin and ciprofloxacin (EU Decision, 2018/840 of 5 June 2018) in the European Union Watch List as substances subjected to monitoring. A comprehensive risk assessment of antimicrobial compounds in the environment for human health is complicated, but the risk is clearly recognized when concentrations sufficient for selection of resistant pathogenic strains are reached. Available data suggest that such minimum selective concentrations may be very low and, even at environmentally relevant antibiotic concentrations, the maintenance and selection of resistant bacteria can occur [7].

Antibiotic resistance in healthcare-associated infections are a common and alarming problem in many countries [8]; e.g., in Poland, where the incidence of these types of infections appear higher than in neighbouring countries [9]. There are several ongoing activities across countries to reduce rising antibiotic resistance problem. The European Surveillance of Antimicrobial Consumption (ESAC) network, and more recently the European Centre for Disease Prevention and Control (ECDC) [10,11] and the World Health Organization (WHO) Europe [12], have researched antibiotic utilisation across Europe including Poland. The available data indicate that Poland has one of the highest rates of total consumption of antibiotics among European countries [13]. Infections in Polish patients demonstrate the relationship between utilisation and resistance levels, i.e., the high sulfamethoxazole (SMX) consumption [14] accompanied by high levels of resistance to it [15]. Concerns with current antibiotic resistance rates in Poland are confirmed by data on severe *Acinetobacter baumannii* infections characterised by the high level of resistance to commonly used antibiotics [16,17], as well as data on MRSA epidemiology both in hospitalised adult patients, as well as new-borns with very low birth weight [18].

It has been confirmed that wastewater treatment plants (WWTPs) are significant reservoirs of antibiotic resistance genes (ARGs) and sources of their spread in the environment [19,20]. Due to the unregulated consumption of antibiotics, excretion by humans and animals or improper medication disposal methods, native forms of these compounds, their metabolites and transformation products end up in hospital and municipal wastewater. According to the published data, tetracyclines are present in WWTPs at concentration levels of 1300 ng/L, measured in raw wastewater [21] and even 1420 ng/L in effluent [22]. Sulfamethoxazole, the most often identified sulfonamide in WWTPs, was detected at concentration levels of 5597 ng/L in influent [23] and up to 6000 ng/L in treated wastewater [24]. The occurrence of SMX and its main derivative N^4-acetylsulfamethoxazole in raw wastewater (1464 and 1763 ng/L, respectively) and effluent (508 and 16 ng/L, respectively) samples collected in WWTP1 were also reported [25]. In the environment, antibiotics are not only chemical pollutants that can exert toxic effects, but they are above all able to cause selection pressure [26]. WWTPs are considered probable hotspots for antibiotic resistance dissemination in the environment. The presence of antibiotic residues, even at low concentrations, combined with the high density and diversity of microorganisms (including pathogenic, commensal, environmental and indigenous) sustained by a nutrient rich environment, might facilitate the ARB proliferation and ARGs horizontal gene transfer (HGT) mediated by mobile genetic elements (MGEs), such as plasmids, transposons, integrons and bacteriophages [27]. Furthermore, WWTPs have a substantial impact on the spread and abundance of ARGs in the environment [28]. The data available in the literature indicate a significant increase in numbers of antibiotic resistance determinants in effluent discharged from WWTPs [1,29,30]. Different tetracycline resistance genes, e.g., *tet*($A, B, C, G, L, M, O, Q, X$), as well as sulfonamide resistance genes, e.g., *sulI*, *sulII*, *sulIII* have been detected in the effluent of WWTPs [31]. The abovementioned data and results for the other ARGs confirm that effluent discharged from WWTPs is one of the major anthropogenic sources of these pollutants in the environment and can pose a real threat of spreading resistance to bacterial pathogens [19,32].

Since conventional activated sludge (AS)-based WWTPs seem to be inefficient in ARG removal, the implementation of additional wastewater cleaning processes is necessary. It has been demonstrated that conventional disinfectants and advanced oxidation processes, or their combinations, are not capable of significantly reducing the amount of ARGs. The membrane bioreactor technology and photocatalytic ozonation seem to be good technological solutions for the future; however, the possibility of their application in full-scale WWTPs, due to high costs, is questionable [33,34]. Constructed wetlands (CWs) have been suggested as a cost-effective, ecological and efficient technology in wastewater treatment. The removal mechanism of contaminants in CWs is complicated and consists of physical, chemical and biological processes among plants, substrates and microorganisms, which can be also affected by CW type, substrate type and plants [35–39]. CWs have been proposed as a promising alternative solution for removing a wide variety of conventional pollutants as well as antibiotics, and even ARB and ARGs [35,40]. The reduction of antibiotics and ARGs in CWs, as demonstrated, could be achieved at relatively similar or even higher rates than in conventional WWTPs [36]. On the other hand, it was also suggested that, although CWs effectively remove antibiotics, they probably stimulate the spread of ARGs [41].

Considering the fact that the abundance of ARGs in final effluent can be influenced by the type of WWTP technological processes and their operating conditions, the main aim of this research was to compare the occurrence and abundance of selected tetracycline and sulfonamide resistance genes before and after the purification process in two different in size and applied wastewater treatment processes (conventional AS-based and combining a conventional biological AS-based method with CWs) WWTPs. To examine possible seasonal fluctuations in gene abundances, the samples were collected over four seasons in 2018. A molecular-based approach and quantitative polymerase chain reaction (qPCR) technique were used to study ARGs in the WWTPs. The genes selected for the study represent different mechanisms of resistance such as: ribosome protection (coded by genes *tet*(*M*, *O*, *Q*)), efflux pump (coded by genes *tet*(*A*, *B*, *C*, *G*, *K*, *L*)), drug modification (coded by gene *tetX*) [42] and target modification (coded by *sul* genes) [43]. Moreover, when choosing genes, their location on different genetic elements (especially on MGEs which significantly influences the ARG dissemination) was taken into account. Tetracycline resistance genes are located on plasmids (*tet*(*A*, *C*, *K*, *O*)), transposons (*tet*(*B*, *M*, *X*) and chromosome (*tetQ*). In addition, some genes are found both on the integrons and chromosome (*tetG*), on plasmids and chromosome (*tetL*) or on transposons and plasmids (*sul* genes) [43,44].

It is well known that the nucleic acid extraction process and quality of isolated DNA are crucial for the subsequent polymerase chain reaction (PCR) gene detection in environmental samples and can influence the research results and its interpretation. Therefore, to select the optimal method for the isolation and purification of total DNA from WWTP samples, 10 commercially available DNA purification kits were tested and assessed in terms of the efficiency of isolation, DNA purity and the suitability of isolated DNA as template for PCR.

To analyze the data from qPCR, a relative quantification method was applied. The method is used in situations when the determination of the absolute copy number of the transcript is not necessary and reporting the relative change in gene expression is sufficient. In that case, the comparative C_T method (also known as $\Delta\Delta C_T$, or $2^{-\Delta\Delta CT}$) is used to calculate relative changes in gene expression determined from qPCR experiments. The data are presented as the fold change in gene expression (normalized to 16S rRNA reference gene) between effluent and influent.

2. Results and Discussion

2.1. Evaluation of Total DNA Extraction Kits

The results of the total amount and quality of DNA for all tested DNA isolation and purification kits are presented in Table 1. The highest amount of DNA was obtained using the FastDNA SPIN Kit for Soil (kit numbered 4); however, the DNA sample isolated by this method did not equal the quality requirements; the determined $A_{260/230}$ ratio indicated a significant content of contaminants, like residues of reagents used in the extraction process. It was found that DNA of a desired purity ($A_{260/230} \geq 1.8$) was obtained only for the tested kits numbered 1, 2, 5, 6, 8, 9, and 10. In addition, the usefulness of all isolated DNA samples as a template for PCR was verified; PCR using the BACT 1369F and PROK 1492R primers (16S rRNA gene) and agarose gel electrophoresis were carried out. The specific 146 bp DNA fragment was identified in all PCR reactions tested (data not presented). The DNA extraction kits selected for the test were also evaluated according to the convenience of their use, time of the process and the cost of a single isolation. Considering all the above criteria, the GeneMATRIX SOIL DNA Purification Kit was selected as the optimal method for extraction and purification of total DNA from WWTP samples, and used in the further part of this study.

Table 1. Results of quantitative and qualitative analysis of DNA extracted from activated sludge collected in WWTP1 using different commercially available DNA isolation and purification kits.

No.	DNA Extraction Kit	µg of Total DNA per 1 g of AS	$A_{260/280}$	$A_{260/230}$
1	Genomic Mini AX Stool (A&A Biotechnology, Gdynia, Poland)	117.5 ± 0.3	1.90 ± 0.02	1.77 ± 0.03
2	Genomic Mini AX Bacteria + (A&A Biotechnology, Gdynia, Poland)	274.6 ± 1.6	1.92 ± 0.01	1.78 ± 0.04
3	Exgene Soil DNA mini (GeneAll Biotechnology, Seoul, Korea)	365.4 ± 4.5	2.00 ± 0.01	1.31 ± 0.62
4	FastDNA SPIN Kit for Soil (MP Biomedicals, Solon, OH, USA)	751.8 ± 3.0	1.91 ± 0.11	0.55 ± 0.01
5	NucleoSpin Soil lysis buffer 1 (Macherey–Nagel, Düren, Germany)	297.9 ± 16.9	1.98 ± 0.01	1.93 ± 0.08
6	NucleoSpin Soil lysis buffer 2 (Macherey–Nagel, Düren, Germany)	306.2 ± 9.8	1.97 ± 0.01	1.97 ± 0.04
7	PowerSoil DNA Isolation Kit (Qiagen, Hilden, Germany)	21.4 ± 1.4	1.76 ± 0.22	1.02 ± 0.31
8	ZymoBIOMICS DNA Minikit (Zymo Research, Irvine, CA, USA)	247.9 ± 0.5	1.89 ± 0.02	1.81 ± 0.02
9	GeneMATRIX SOIL DNA Purification Kit (EURx, Gdańsk, Poland)	274.6 ± 1.6	1.92 ± 0.01	1.78 ± 0.04
10	GeneMATRIX Environmental DNA & RNA Purification Kit (EURx, Gdańsk, Poland)	213.5 ± 0.8	2.18 ± 0.01	2.15 ± 0.01

The above results are the mean values and standard deviations calculated for three replicates of each isolation.

The selection of the appropriate procedure of genetic material isolation and purification at an early stage of the study is an important step for research based on PCR methods. DNA quality and purity significantly affect the course of reaction. The adequate method should be primarily selected for the type of extracted DNA and the specificity of sample matrix. To our knowledge, commercial kits designed especially for DNA isolation from wastewater or AS samples are not available, hence why kits for soil or fecal samples are used in this type of research [45–48]. WWTP samples as a multi-component research material require an individual approach in choosing the optimal method of DNA extraction. This type of sample contains complex polymers, organic matter, and numerous inorganic compounds—potentially enzymatic inhibitors that can be co-extracted with DNA. On the other hand, differences in the cell wall and membrane structures of microorganisms, as well as the susceptibility of different microorganisms to the cell lysis, significantly affects the efficiency of the extraction process. Our research confirmed that only half of the tested protocols achieve a satisfactory efficiency of the isolation process and receive a purified DNA sample without enzymatic reaction inhibitors, suitable as a template for PCR. The DNA extraction methods should be evaluated not only in terms of the quantity and purity of the obtained genetic material, but above all in terms of how faithfully the diversity of the sequences extracted reflects the structure of the microbial population in the studied environmental sample. Zielińska et al. [49] evaluated the usefulness for metagenomic sequencing of 7 out of the 10 DNA isolation and purification kits tested in our study. The high number of good quality reads, which made it possible to classify all of the obtained sequences at phylum level, high values of Shannon and Simpson indexes, which suggest a high level of the species diversity as well as low level of error rate among replicates, were observed for the GeneMATRIX SOIL DNA Purification Kit (EURx) (kit ID C3 in Zielińska et al. [49]). According to Morgan et al. [50], the reproducibility of extraction kit replicates is of high importance when tracking changes in microbial composition over time, between environments, with respect to seasonal and ecological changes. Moreover, results of the rarefication analysis of the obtained data revealed that the sampling of microbial communities is close to being complete for the analyzed kit [49].

2.2. Occurrence and Removal of ARGs in WWTPs

In this study, the role of selected WWTPs in the dissemination of ARGs in the environment have been investigated using a comparative qPCR method. Culture- or molecular-based approaches are followed to study the antibiotic resistance problem in WWTPs, each of them exhibiting some advantages and drawbacks. Culture-based methods are key to understand phenotypic characteristics of isolates and their resistance patterns, but they have limits with environmental bacteria (as the culturable fraction is only 1% of the total). Molecular methods—based on the isolation of the total DNA from the analyzed samples (influent, effluent or activated sludge) and the detection of specific nucleotide sequences coding ARGs using PCR and/or qPCR techniques—are applied to identify specific DNA targets in microorganisms that cannot be grown in the laboratory, or multiply very slowly but significantly contribute to the resistance problem [33].

All 13 selected genes coding resistance to tetracyclines and sulfonamides were detected in raw influent and final effluent samples from both WWTPs. The results of the resistance gene enrichment obtained for WWTP1 are presented in Table 2 and in Figure 1. Generally, an increase of the ARG abundance from the influent to the effluent has been observed. The most significant enrichment (more than 10-fold) was recorded for *tet(B, K, L, O)* and *sulIII* genes. The corresponding data for WWTP2 are shown in Table 3 and in Figure 2. The obtained results indicate a tendency of the enrichment of selected ARGs occurring after the wastewater treatment process; however, these values are usually lower compared to the conventional WWTP1.

Table 2. Results of comparative qPCR analysis for WWTP1.

ARGs	ARG Enrichment			
	Winter	Spring	Summer	Autumn
tetA	2.2	0.9	2.1	1.0
tetB	9.3	5.6	11.1	0.2
tetC	0.8	0.4	2.5	1.0
tetG	3.6	5.5	3.8	1.3
tetK	1.7	15.1	7.7	2.3
tetL	8.2	1.6	12.9	4.0
tetM	7.2	5.3	0.8	0.9
tetO	12.0	1.0	2.9	1.0
tetQ	0.7	4.0	4.2	0.6
tetX	0.6	2.2	2.2	2.5
sulI	0.9	1.2	6.9	0.9
sulII	0.9	1.5	7.7	0.9
sulIII	14.0	2.3	5.4	1.8

The above results are the enrichment factors (R) calculated for three replicates of each reaction.

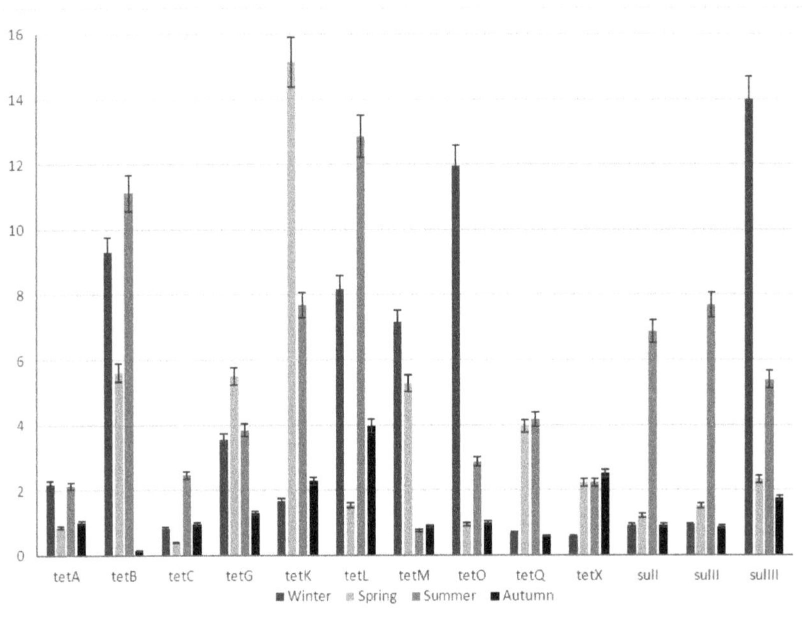

Figure 1. Enrichment of selected ARGs in the WWTP1 samples.

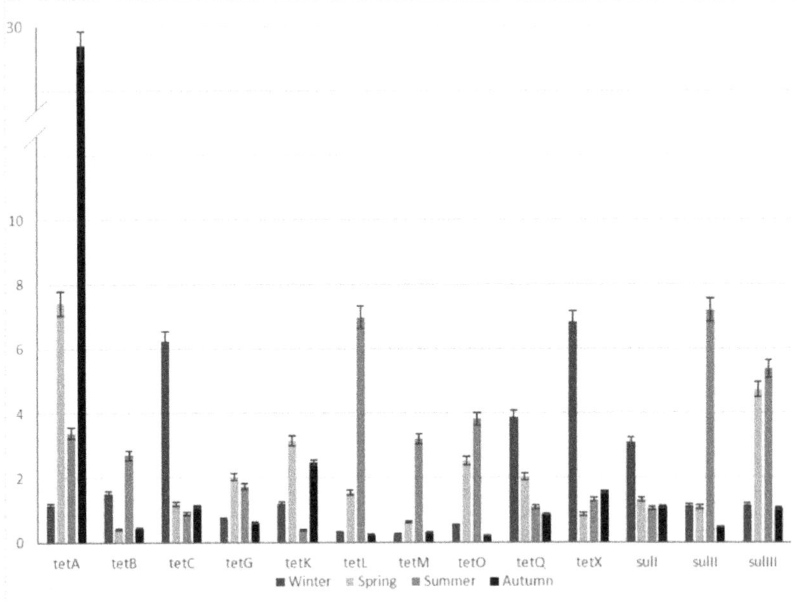

Figure 2. Enrichment of selected ARGs in the WWTP2 samples.

Table 3. Results of comparative qPCR analysis for WWTP2.

ARGs	ARG Enrichment			
	Winter	Spring	Summer	Autumn
tetA	1.2	7.4	3.4	28.8
tetB	1.5	0.4	2.7	0.4
tetC	6.2	1.2	0.9	1.1
tetG	0.7	2.0	1.7	0.6
tetK	1.2	3.1	0.4	2.5
tetL	0.3	1.6	7.0	0.2
tetM	0.3	0.6	3.2	0.3
tetO	0.5	2.5	3.8	0.2
tetQ	3.9	2.0	1.1	0.9
tetX	6.8	0.9	1.3	1.6
sulI	3.1	1.3	1.0	1.1
sulII	1.1	1.1	7.2	0.5
sulIII	1.2	4.7	5.4	1.0

The above results are the enrichment factors (R) calculated for three replicates of each reaction.

The obtained results may arise from the difference in size of compared WWTPs, which means that the average daily wastewater flow in WWTP1 is over 20 times higher than in WWTP2. It can be expected that a higher wastewater flow is associated with a higher load of antibiotic contaminants flowing through WWTP, which may promote selective pressure and ARB proliferation. In addition, even a small inflow of hospital wastewater to WWTP1 may be of key importance for the ARG content, due to the significantly higher concentration of antibiotics, ARB and ARGs compared to domestic wastewater. Therefore, the specially targeted separate treatment of hospital wastewater before discharge to conventional WWTPs is an issue that needs to be addressed by adapting to local circumstances.

An enrichment of ARGs after the conventional AS-based treatment process observed in this study is in agreement with the results reported by other researchers [1,30,51–54]. This could be explained by the selective conditions in WWTPs that may favor bacteria harboring resistance genes, ARGs or/and HGT between bacteria. The genes, for which the highest enrichment after the wastewater treatment process was observed (*tet(B, K, L, O)*, and *sulIII*) in this study, are located on MGEs. These segments of DNA play an important role in adaptation process and are a means to transfer genetic information among and within bacterial species, which may cause their extensive prevalence [55]. On the other hand, some studies reported no change or even a decrease in the relative number of ARGs in the effluent after the conventional treatment process [56,57]. It seems that WWTPs with conventional treatment processes are not efficient in ARG removal. The wastewater treatment processes, which were primarily designed to remove nitrogen, biodegradable organic compounds, ammonia, nitrate, and phosphate are not effective in elimination of microbiological contaminants [33]. The available data from the literature show that biological processes may positively affect the ARB spread and selection as well as the ARG transfer [54,58]. The persistent selective pressure from antibiotic residues at sub-inhibitory concentrations, as well as the high density and diversity of microorganisms (including pathogenic, commensal, environmental and indigenous) sustained by a nutrient-rich environment, create favourable conditions for antibiotic resistance dissemination [27]. Moreover, WWTP sludge was recognized as the main source of tetracycline- and sulfonamide-resistant bacteria and genes discharged into the water environment. A significant enrichment of *tet* (*B, G, H, S, T, X*) and *sul* (*I, II*) genes in relation to 16S rRNA was observed [59]. In the face of the observed ARG enrichment in the wastewater treatment process, Mao et al. [59] emphasize the need for an improved understanding of how to manipulate WWTP operational variables, e.g., by decreasing the nutrient-to-microorganism ratio and thus limiting energy sources for ARB and as a consequence of metabolic deficiencies inhibiting resistance plasmid replication and promoting the loss of antibiotic resistance. According to

Biswal et al. [51], the key factors affecting the prevalence of antibiotic resistance determinants are the operational parameters of WWTPs such as the residence times of hydraulics and solids, which can affect the dynamics of genetic material exchange and the distribution of ARGs in bacteria passing through the system. It was found that the activated sludge process did not affect the removal of ARG-carrying *Escherichia coli*, but increased the abundance of multiple ARGs in the bacterial genome. In turn, it was observed that the physicochemical systems were capable of removing ARB at a high rate, hence the bacterial density (potential ARG donors for HGT) was several orders of magnitude lower, which would significantly reduce the rate of ARG transfers compared to systems with AS. The authors suggested that observed differences in ARG dynamics for the two wastewater treatment types would be the result of the balance between the effectiveness of ARB removal and the HGT rates [51].

Although the quantitative analysis of ARGs in WWTP1 samples has not been studied before, the studies of antibiotic resistance of fecal coliforms and enterococci based on culture methods have been conducted by other authors [25,58,60]. Tetracycline-resistant isolates were detected in all sampling points, including influent and effluent samples, of the WWTP1. It was found that 20% of enterococci and 23% of *E. coli* isolates were resistant to tetracyclines [58]. In turn, resistance to SMX was observed in 11% of *E. coli* isolates. It should be also noted that 75% of SMX-resistant *E. coli* were simultaneously resistant to tetracyclines. The frequency of sulfonamide resistance genes (*sul I–III*) detected in *E. coli* strains of wastewater origin was similar as in other environmental compartments, including clinical ones. The molecular analysis of genes responsible for resistance to sulfonamides among SMX-resistant *E. coli* isolates showed a prevalence of *sulII* (81%) and *sulI* (50%) genes. Moreover, 31% of isolates simultaneously carried both *sul* genes. The *sulIII* gene was rarely detected—it was found in 6% of SMX-resistant isolates from the WWTP effluent [25]. The research concerned not only the analysis of antimicrobial resistance patterns in isolates from wastewater samples, but also on the presence of integrons, which are associated with antibiotic resistance dissemination phenomenon. Class 1 and 2 integrons were detected in 32% and 3% of antibiotic-resistant *E. coli* isolates, respectively, and were related to an increase of resistance to selected antimicrobial agents and multidrug resistance [60]. The positive selection of bacteria with resistance patterns in wastewater processes based on AS was observed. Resistance rates noted for bacteria isolated from treated wastewater was higher than that observed in corresponding influent. The treatment process favored both tetracycline- [58] and SMX-resistant [60] bacteria.

It has been confirmed that the application of CWs allowed a significant increase in the removal of antibiotics and other medicines from the wastewater [34,61–66]. CWs have been also shown to be more efficient in the removal of ARGs than conventional WWTPs [36,38,61–64]. Nolvak et al. [61] reported a decrease in the proportions of different ARGs (including *tet(A, B, M)*, and *sulI*) in effluent (compared to the influent) during the treatment process in horizontal subsurface flow CWs. Moreover, the observed removal efficiency for *sulI* gene was better than in conventional WWTPs [61]. Liu et al. [62] found that the total absolute abundances of *tet* genes and 16S rRNA were reduced by 50% in wastewater using CWs. Significantly reduced absolute abundances of ARGs (*tet(O, M, W, A, X)*, and *intI1*) and 16S rRNA were also reported by Huang et al. [64]. In another study, the removal efficiencies of 16S rRNA, *intI1* and *tet* genes among four different CW treatment systems ranged from 33% to 99% [38]. The observed differences in 16S rRNA in influent and effluent confirmed that studied CWs were capable of bacterial removal from wastewater; however, an increase in the relative abundance of ARGs indicated a risk of the release of relatively more antibiotic-resistant bacteria in proportion to total bacteria into environment [38,62,64]. The results obtained in our study also showed an enrichment of selected ARGs. These values are generally lower compared to the results obtained for the conventional WWTP system which might suggest that the introduction of an additional plant-based purification step increases the efficiency of removing ARGs; however, the amount of ARGs was still higher than in the raw influent. It has been suggested that the ability of the treatment systems to filter out bacteria contributes greatly to the reduction of ARB and, therefore, ARGs from wastewater in CWs (especially those with vertical flow) [38]. Sorption and biological processes occurring in CWs have been proposed as key mechanisms influencing the fate of ARGs during wastewater treatment. Biological processes can both lead to ARG transmission and proliferation as well as cause their degradation [36,62]. When being removed from wastewater, ARGs with their host bacteria were partially deposited in CWs, especially in surface soil which provided optimal conditions for their survival and development. Environmental stressors such as antibiotics and heavy metals accumulated in surface soil could promote horizontal gene transfer between foreign and indigenous bacteria [38]. Other authors suggested [67] that the survival of rhizospheric bacteria of CWs can be also associated with their increased resistance to various environment stressors.

Some authors observed the relationship between ARG reduction efficiency and CW flow type. It was reported that CWs with subsurface flow removed ARGs more effectively [61] compared to CWs with surface flow [68]. The filtration capacity of subsurface flow CWs for bacteria is higher than that of surface flow CWs [38]. In addition, the removal efficiency of horizontal subsurface flow CWs for ARGs was higher (over 50%) than that of vertical wetlands, especially for *sul* genes [37].

The reports available in the literature suggest that, just like conventional WWTPs, CWs could be considered as hotspots for the spread of antibiotic resistance in the environment. Song et al. [69] evaluated the fate of ARGs (*sul* and *tet*) in three lab-scale vertical flow CWs. They found out a positive correlation between abundances of ARGs and the accumulation of SMX and tetracyclines in different layers of CW substrate. Positive correlations were also observed between the abundance of *tet* genes and the antibiotic concentration in the effluent. Although the effluent had lower abundances of ARGs than that in the wetland media, the occurrence of ARGs in effluent might still pose risk for public health. Moreover, the relative abundances of *sul* and *tet* genes showed a significant increase in all samples during the SMX and tetracycline treatment period [69].

2.3. Seasonal ARG Changes

The analysis of obtained results did not show clear trend in seasonal fluctuations in abundance of selected tetracycline and sulfonamide resistance genes in wastewater. However, for conventional WWTP1, enrichment of most studied ARGs was observed in the summer, and to a lesser extent in spring and winter seasons. On the other hand, WWTP2 enrichment was more frequently noted only in summer and spring months. The explanation of ARB and ARG seasonal fate is not clear so far, and the conclusions from numerous studies regarding seasonal fluctuations of antibiotic resistance determinants in WWTP systems are often divergent [70–72]. While some studies showed higher release loads of ARB and ARGs in wastewater samples in spring and summer seasons than in winter months, other authors indicated an increase in numbers of tetracycline, sulfonamide, and vancomycin resistance genes in winter [73]. There are also studies that did not confirm the obvious seasonal fluctuations in the occurrence of ARGs detected in WWTP systems [74,75], which is consistent with the results obtained in this study. The reasons for the quantitative fluctuations of antibiotic resistance determinants in different seasons are complicated and depend on many factors, which include the variability in antibiotic consumption, the microbial composition variation in the wastewater and AS, and the presence of antimicrobial residues in wastewater or the co-selection of heavy metal resistance [73].

3. Materials and Methods

3.1. Characterization of Wastewater Treatment Plants

Two full-scale municipal WWTPs located in northern and central Poland were investigated. Conventional AS-based WWTP1—"Municipal Wastewater Treatment Plant Gdańsk Wschód" collects domestic wastewater from the population of about 570,000 people in the area of Gdańsk, Sopot, Pruszcz Gdański, Żukowo, Kolbudy and in a small part from local industry (5%), as well as hospital wastewater (0.17%). The daily average wastewater flow is 96,000 m^3 with a 24-h retention time. Mechanical treatment units in this WWTP consist of mechanical screens, aerated sand traps with grease removal traps and radial-flow primary sedimentation tanks. Biological treatment units consist of 6 multiphase

MUCT reactors (typical UCT system additionally equipped with a transitional chamber which can optionally serve as a nitrification or denitrification chamber and with deaeration chamber, where the mixture of treated wastewater and AS, recirculated from nitrification chamber to denitrification chamber, is de-oxidized) and 12 radial-flow secondary sedimentation tanks. The plant is also equipped with fermenter, which discharge pre-fermented sludge to sewage before primary sedimentation tanks and thus increases the effectiveness of biological phosphorylation [76]. The effluent from WWTP1 is transported via a pipeline into the Bay of Gdańsk and discharged 2.3 km away from the coastline. WWTP2, "Municipal Wastewater Treatment Plant in Sochaczew", described in detail by Wolecki et al. [66], represents a system which combines the method of biological wastewater treatment with AS and constructed wetlands. CWs have been implemented in the second stage of wastewater treatment (biological). Contact between the plants and wastewater (mixed with AS) occurs only in the rhyzophytic zone. The following plant species were used in CWs culture: European spindle (*Euonymus europaeus*), Grey willow (*Salix cinerea*), Papyrus (*Cyperus papyrus*), Reed (*Phragmites australis*), Rushes (*Juncus tenageia Ehrh.*), Spathiphyllum (*Spathiphyllum Adans.*), Summer lilac (*Buddleja davidii Franch*), Sweet flag (*Acorus calamus*), Yellow iris (*Iris pseudacorus*) and Yellow pimpernel (*Lysimachia nemorum*). CW plants are placed in greenhouse with the total area of 1835.6 m^2, where strict conditions—an optimal air humidity and temperature (35–38 °C)—are maintained for appropriate plant growth. WWTP2 is comparatively smaller than WWTP1, with an average wastewater flow of 4470 m^3 per day. The wastewater collection from the area of Sochaczew city concerns domestic inflow from approximately 37,000 residents. The effluent from WWTP2 is discharged to Utrata River. The average values of the main WWTP1 and WWTP2 technological parameters: biological and chemical oxygen demand, total nitrogen and phosphorus, as well as total suspended solids from the sampling period, are presented in Table 4.

Table 4. Main technological parameters of the studied WWTPs.

Parameter	Unit	WWTP1		WWTP2	
		Influent	**Effluent**	**Influent**	**Effluent**
BOD_5	mgO_2/L	446.0	3.7	449.0	2.6
COD	mgO_2/L	1009.0	35.0	1066.0	32.5
TSS	mg/L	516.0	5.0	501.0	6.1
TP	mg/L	10.8	0.4	10.7	0.2
TN	mg/L	87.0	8.0	87.7	7.9

3.2. Samples Collection and Preparation

Samples of raw influent and final effluent from both studied WWTPs for quantitative analysis of ARGs were collected in January, April, July and October 2018. Additionally, to develop the optimal method of DNA purification, AS samples from the aeration chamber of the biological reactor of WWTP1 were collected in July 2018. Wastewater and AS samples were collected into sterile bottles, transported to the laboratory and stored at 4 °C.

Influent (400 mL) and effluent (2 L) samples were filtered using 1.2-µm glass microfiber filters (VWR, Leuven, France) to remove considerable size contaminants, and next through 0.22-µm mixed cellulose esters membrane filters (Merck Millipore, Cork, Ireland), to retain all biological material. Obtained filters were cut into smaller pieces, transferred to the 15-mL screw cap centrifuge tubes with 6 mL of 1 × PBS and shaken for 20 min (1000 rpm/min) at room temperature. Then the filters were transferred to clean tubes and the procedure was repeated with the new portion of PBS. Both suspensions were combined and centrifuged for 10 min (8000× *g*). All obtained precipitates were stored at −20 °C.

3.3. Selection of the Optimal DNA Isolation and Purification Method

In order to select the optimal method for isolation and purification of total DNA from WWTPs samples, 10 commercially available DNA purification kits (Table 1), dedicated to soil or fecal samples, were tested. The representative research material was AS, which was chosen as the most complex WWTP matrix, due to a significant content of exopolysaccharides and adsorbed on the surface of sludge biopolymers, organic matter and numerous inorganic compounds. AS samples (20 mL) were placed in 50-mL screw cap centrifuge tubes and centrifuged for 10 min (8000× *g*). Obtained precipitate was used for DNA purification procedures performed according to the manufacturer's protocols. All purified DNA samples were subjected to spectrophotometric analyzes using Colibri Microvolume Spectrometer (Titertek Berthold, Pforzheim, Germany) to determine the concentration and purity of DNA ($A_{260/280}$ and $A_{260/230}$ ratios). To ascertain the suitability of isolated DNA as a template for standard PCR, a reaction using the BACT 1369F and PROK 1492R primers (16S rRNA gene) was carried out (primers sequences are listed in Table 5). All reactions were performed in 50 µL volumes in a T100™ Thermal Cycler (Bio-Rad, Hercules, CA, USA) and contained 80 ng of template DNA, 0.4 mM dNTPs (A&A Biotechnology, Gdynia, Poland), 0.5 µM of each primer (Genomed, Warsaw, Poland) and 0.75 units of Marathon DNA polymerase in 1 × Marathon PCR buffer (A&A Biotechnology, Gdynia, Poland). The cycling profile included: 94 °C for 5 min, then 30 cycles of 94 °C for 30 s, 56 °C for 30 s, 72 °C for 30 s, and a final step of 72 °C for 5 min. The presence of PCR product (146 bp in length) was confirmed using agarose gel electrophoresis. 2.0% BASICA LE Agarose gels (Basica Prona™ Agarose, ABO, Gdansk, Poland) were prepared in 1 × TBE buffer, and visualized after staining with ethidium bromide.

Table 5. PCR and qPCR amplification primers.

Target Gene		Primer Sequence (5′-3′)	Amplicon Size (bp)	Source
16S rRNA	F R	CGG TGA ATA CGT TCY CGG GGW TAC CTT GTT ACG ACTT	146	[77]
tetA	F R	GCT ACA TCC TGC TTG CCT TC CAT AGA TCG CCG TGA AGA GG	210	[78]
tetB	F R	TAC GTG AAT TTA TTG CTT CGG ATA CAG CAT CCA AAG CGC AC	206	[42]
tetC	F R	CTT GAG AGC CTT CAA CCC AG ATG GTC GTC ATC TAC CTG CC	418	[78]
tetG	F R	GCT CGG TGG TAT CTC TGC TC AGC AAC AGA ATC GGG AAC AC	468	[78]
tetK	F R	TCG ATA GGA ACA GCA GTA CAG CAG ATC CTA CTC CTT	169	[78]
tetL	F R	TCG TTA GCG TGC TGT CAT TC GTA TCC CAC CAA TGT AGC CG	267	[78]
tetM	F R	AAT AAA TCA TAA ACA GAA AGC TTA TTA TAT AAC AAT AAA TCA TAA TGG CGT GTC TAT GAT GTT CAC	171	This study
tetO	F R	AAC TTA GGC ATT CTG GCT CAC TCC CAC TGT TCC ATA TCG TCA	515	[78]
tetQ	F R	AGA ATC TGC TGT TTG CCA GTG CGG AGT GTC AAT GAT ATT GCA	169	[42]
tetX	F R	CAA TAA TTG GTG GTG GAC CC TTC TTA CCT TGG ACA TCC CG	468	[78]
sulI	F R	GAC GAG ATT GTG CGG TTC TT GAG ACC AAT AGC GGA AGCC	185	[31]
sulII	F R	GAC AGT TAT CAA CCC GCG AC GTC TTG CAC CGA ATG CAT AA	147	[31]
sulIII	F R	ACC ACC GAT AGT TTT TCC GA TGC CTTT TTC TTT TAA AGCC	199	[31]

3.4. Isolation and Purification of Total DNA in Influent and Effluent Samples

Samples were prepared according to the procedure described in Section 3.2 and DNA purification procedures were performed using GeneMATRIX SOIL DNA Purification Kit (EURx, Gdańsk, Poland) according to the manufacturer's protocols. All purified DNA samples were subjected to spectrophotometric analyses using Colibri Microvolume Spectrometer (Titertek Berthold, Pforzheim, Germany) to determine the concentration and purity of DNA ($A_{260/280}$ and $A_{260/230}$ ratios). Isolated DNA samples were stored at −20 °C.

3.5. Quantitative PCR

Quantitative PCR was used to analyze selected genes coding for resistance to tetracyclines and sulfonamides in the collected samples. The 16S rRNA gene was used as the reference gene to account for variability of total amount of bacteria in the samples. Primer sequences adapted from other studies [31,42,77,78] or designed for this study (*tetM*) are listed in Table 5. All qPCR assays were performed in the Roche LightCycler® 480 II (Roche Applied Science, Indianapolis, IN, USA) in 11 μL reaction mixture. Analyzes for each sample were carried out in triplicate. PCR mixtures consisted of 5-μL LightCycler 480 SYBR Green I Master (Roche Applied Sciences, Indianapolis, IN, USA), 2.5 μL each primer of corresponding concentration 5 μM (*tetA, B, C, L, M, O, X, sulI, II*), 2 μM (*tetG, K, sulIII*), 1 μM (*tetQ*) and 1 μL DNA template of 5 ng/μL. In each run, 1 μL microbial DNA-free water as negative control was included. The thermal cycling conditions selected for studied genes are given in Table 6. Specificity of the amplified PCR product was verified by performing melting curve analysis, while amplification efficiency was assessed by monitoring the slope of amplification curves generated for the target genes and the internal controls (16S rRNA). Quantitative analysis of qPCR data was carried out using a comparative C_T method (also known as $\Delta\Delta C_T$, or $2^{-\Delta\Delta CT}$) based on the literature [79]. In the first stage, threshold cycles (C_T) of the tested ARGs and control gene (16S rRNA) amplification reactions in all samples were determined. The C_T value is provided as the result of qPCR analysis for each gene. In turn, for individual samples (raw influent and final effluent), the differences between the C_T values for the tested gene and 16S rRNA were calculated according to Formula (1). The obtained results present the ARG abundance relative to that of 16S rRNA. Then the $\Delta\Delta C_T$ values for each gene were obtained with the use of Formula (2). Computation the normalized value of the relative level of the tested gene in the unknown sample (final effluent) relative to the calibrator sample (raw influent) was carried out based on the Formula (3). The value of parameter R equal to 1 means the relative amount of gene in both samples is comparable, <1 indicates a decrease in the relative amount of the gene in the final wastewater sample, while >1 shows the enrichment of the gene after the wastewater treatment process.

$$\Delta C_{T\,(influent/effluent)} = C_{T\,(ARG)} - C_{T\,(16S\,rRNA)} \tag{1}$$

$$\Delta\Delta CT = \Delta CT_{(effluent)} - \Delta CT_{(influent)} \tag{2}$$

$$R = 2^{-\Delta\Delta CT} \tag{3}$$

Table 6. qPCR conditions used in this study.

Reaction Stage	*tet(B, K, L, M, Q, X), sulIII*		*tet(G, C), sulI, sulII*		*tet(A, O)*	
Pre-incubation	95 °C, 5 min (4.4 °C/s)		95 °C, 5 min (4.4 °C/s)		95 °C, 5 min (4.4 °C/s)	
Amplification	95 °C, 10 s (4.4 °C/s)	45 cycles	95 °C, 10 s (4.4 °C/s)	45 cycles	95 °C, 10 s (4.4 °C/s)	55 cycles
	62 °C, 30 s (2.2 °C/s)		60 °C, 30 s (2.2 °C/s)		60 °C, 30 s (2.2 °C/s)	
	72 °C, 30 s (4.4 °C/s)		72 °C, 30 s (4.4 °C/s)		72 °C, 30 s (4.4 °C/s)	
Melting curve	95 °C, 5 s (4.4 °C/s)		95 °C, 5 s (4.4 °C/s)		95 °C, 5 s (4.4 °C/s)	
	65 °C, 1 min (2.2 °C/s)		65 °C, 1 min (2.2 °C/s)		65 °C, 1 min (2.2 °C/s)	

4. Conclusions

A comparative quantitative analysis of 10 genes coding resistance to tetracycline and three sulfonamide resistance genes in two Polish WWTPs in most cases showed an enrichment of selected ARGs after the wastewater treatment processes. The results have confirmed that WWTPs are hotspots for the spread of antibiotic resistance determinants in the environment. This is a serious problem due to the introduction of final effluent into surface waters, as well as the widespread use of reclaimed wastewater as a fertilizer in agriculture soils. The results of this study highlight the need to implement effective actions to prevent the spread of antibiotic determinants in the environment, such as advanced wastewater treatment processes application, the implementation of permanent microbiological monitoring, taking into account the antibiotic resistance aspect, as well as increased control of drug intake and appropriate management of medical waste. Moreover, the specially targeted separate treatment of hospital wastewater before discharge to conventional WWTPs is an issue that needs to be addressed by adapting to local circumstances.

Author Contributions: Conceptualization, S.S., K.P.B., P.S., J.K. and E.M.; Data curation, M.P.; Funding acquisition, K.P.B. and P.S.; Investigation, M.P. and M.R.; Methodology, M.R. and E.M.; Project administration, M.P.; Resources, M.P., J.K. and D.W.; Supervision, M.R. and E.M.; Visualization, M.P. and E.M.; Writing—original draft, M.P. and E.M.; Writing—review & editing, M.R., S.S., K.P.B. and P.S. All authors have read and agreed to the published version of the manuscript.

Funding: Financial support was provided by the Ministry of Science and Higher Education under grant no. 538-8610-B774-17/18 and DS 531-8616-D593-19-1E.

Acknowledgments: The authors are grateful to the Municipal Wastewater Treatment Plant in Sochaczew (Mazowieckie Voivodeship, Poland) and Municipal Wastewater Treatment Plant Gdańsk Wschód (Pomeranian Voivodeship, Poland) for the possibility to obtain the samples and very fruitful cooperation.

Conflicts of Interest: The authors declare no conflict of interest.

References

1. Alexander, J.; Bollmann, A.; Seitz, W.; Schwartz, T. Microbiological characterization of aquatic microbiomes targeting taxonomical marker genes and antibiotic resistance genes of opportunistic bacteria. *Sci. Total. Environ.* **2015**, *512*, 316–325. [CrossRef]
2. Cassini, A.; Högberg, L.D.; Plachouras, D.; Quattrocchi, A.; Hoxha, A.; Simonsen, G.S.; Colomb-Cotinat, M.; Kretzschmar, M.E.; Devleesschauwer, B.; Cecchini, M.; et al. Faculty Opinions recommendation of Attributable deaths and disability-adjusted life-years caused by infections with antibiotic-resistant bacteria in the EU and the European Economic Area in 2015: A population-level modelling analysis. *Fac. Opin. Post-Publ. Peer Rev. Biomed. Lit.* **2019**, *19*, 56–66. [CrossRef]
3. ECDC/EMEA. *Technical Report. The Bacterial Challenge: Time to React*; ECDC/EMEA: Stockholm, Sweden, 2009.

4. Thorpe, K.E.; Joski, P.; Johnston, K.J. Antibiotic-Resistant Infection Treatment Costs Have Doubled Since 2002, Now Exceeding $2 Billion Annually. *Health Aff.* **2018**, *37*, 662–669. [CrossRef]
5. Jimenez-Tototzintle, M.; Ferreira, I.J.; Duque, S.; Barrocas, P.R.G.; Saggioro, E.M. Removal of contaminants of emerging concern (CECs) and antibiotic resistant bacteria in urban wastewater using UVA/TiO2/H2O2 photocatalysis. *Chemosphere* **2018**, *210*, 449–457. [CrossRef]
6. Gothwal, R.; Shashidhar, T. Antibiotic Pollution in the Environment: A Review. *Clean Soil Air Water* **2014**, *43*, 479–489. [CrossRef]
7. Gullberg, E.; Cao, S.; Berg, O.G.; Ilbäck, C.; Sandegren, L.; Hughes, D.; Andersson, D.I. Selection of Resistant Bacteria at Very Low Antibiotic Concentrations. *PLoS Pathog.* **2011**, *7*, e1002158. [CrossRef] [PubMed]
8. WHO. *Antimicrobial Resistance: Global Report on Surveillance*; WHO: Geneva, Switzerland, 2014.
9. Wałaszek, M.; Różańska, A.; Bulanda, M.; Wojkowska-Mach, J. Alarming results of nosocomial bloodstream infections surveillance in Polish intensive care units. *Przeglad Epidemiol.* **2018**, *72*, 33–44.
10. ECDC. European Surveillance of Antimicrobial Consumption Network (ESACNet). Available online: http://www.ecdc.europa.eu/en/activities/surveillance/ESAC-Net/Pages/index.aspx (accessed on 3 June 2020).
11. ECDC. Quality Indicators for Antibiotic Consumption in the Community in Europe. Available online: http://ecdc.europa.eu/en/healthtopics/antimicrobial-resistance-and-consumption/antimicrobial_resistance/EARS-Net/Pages/EARS-Net.aspx (accessed on 3 June 2020).
12. WHO. Antimicrobial Medicines Consumption (AMC) Network. 2017. Available online: http://www.euro.who.int/en/publications/abstracts/antimicrobial-medicines-consumption-amc-network.-amc-data-20112014-2017 (accessed on 3 June 2020).
13. Wojkowska-Mach, J.; Godman, B.; Glassman, A.; Kurdi, A.; Pilc, A.; Różańska, A.; Skoczyński, S.; Wałaszek, M.; Bochenek, T. Antibiotic consumption and antimicrobial resistance in Poland; findings and implications. *Antimicrob. Resist. Infect. Control.* **2018**, *7*, 136. [CrossRef]
14. ECDC. *Antimicrobial Consumption—Annual Epidemiological Report for 2015*; ECDC: Stockholm, Sweden, 2018.
15. Pomorska-Wesołowska, M.; Różańska, A.; Sobońska, J.; Gryglewska, B.; Szczypta, A.; Dzikowska, M.; Chmielarczyk, A.; Wojkowska-Mach, J. Longevity and gender as the risk factors of methicillin-resistant Staphylococcus aureus infections in southern Poland. *BMC Geriatr.* **2017**, *17*, 51. [CrossRef] [PubMed]
16. Chmielarczyk, A.; Pilarczyk-Zurek, M.; Kaminska, W.; Pobiega, M.; Romaniszyn, D.; Ziółkowski, G.; Wojkowska-Mach, J.; Bulanda, M. Molecular Epidemiology and Drug Resistance of Acinetobacter baumannii Isolated from Hospitals in Southern Poland: ICU as a Risk Factor for XDR Strains. *Microb. Drug Resist.* **2016**, *22*, 328–335. [CrossRef] [PubMed]
17. Chmielarczyk, A.; Pomorska-Wesołowska, M.; Szczypta, A.; Romaniszyn, D.; Pobiega, M.; Wojkowska-Mach, J.; Agnieszka, C.; Monika, P.-W.; Anna, S.; Dorota, R.; et al. Molecular analysis of meticillin-resistant Staphylococcus aureus strains isolated from different types of infections from patients hospitalized in 12 regional, non-teaching hospitals in southern Poland. *J. Hosp. Infect.* **2017**, *95*, 259–267. [CrossRef]
18. Romaniszyn, D.; Różańska, A.; Chmielarczyk, A.; Pobiega, M.; Adamski, P.; Helwich, E.; Lauterbach, R.; Borszewska-Kornacka, M.; Gulczyńska, E.; Kordek, A.; et al. Epidemiology, antibiotic consumption and molecular characterisation of Staphylococcus aureus infections—Data from the polish neonatology surveillance network, 2009–2012. *BMC Infect Dis.* **2015**, *15*, 1–9. [CrossRef] [PubMed]
19. LaPara, T.M.; Burch, T.R.; McNamara, P.J.; Tan, D.T.; Yan, M.; Eichmiller, J.J. Tertiary-Treated Municipal Wastewater is a Significant Point Source of Antibiotic Resistance Genes into Duluth-Superior Harbor. *Environ. Sci. Technol.* **2011**, *45*, 9543–9549. [CrossRef] [PubMed]
20. Pruden, A.; Larsson, D.G.J.; Amézquita, A.; Collignon, P.J.; Brandt, K.K.; Graham, D.W.; Lazorchak, J.M.; Suzuki, S.; Silley, P.; Snape, J.R.; et al. Management Options for Reducing the Release of Antibiotics and Antibiotic Resistance Genes to the Environment. *Environ. Health Perspect.* **2013**, *121*, 878–885. [CrossRef]
21. Gulkowska, A.; Leung, H.; So, M.; Taniyasu, S.; Yamashita, N.; Yeung, L.W.; Richardson, B.J.; Lei, A.; Giesy, J.; Lam, P.K. Removal of antibiotics from wastewater by sewage treatment facilities in Hong Kong and Shenzhen, China. *Water Res.* **2008**, *42*, 395–403. [CrossRef]

22. Minh, T.B.; Leung, H.W.; Loi, I.H.; Chan, W.H.; So, M.K.; Mao, J.; Choi, D.; Lam, J.C.W.; Zheng, G.; Martin, M.; et al. Antibiotics in the Hong Kong metropolitan area: Ubiquitous distribution and fate in Victoria Harbour. *Mar. Pollut. Bull.* **2009**, *58*, 1052–1062. [CrossRef]
23. Peng, X.; Tan, J.; Tang, C.; Yu, Y.; Wang, Z. Multiresidue Determination of Fluoroquinolone, Sulfonamide, Trimethoprim, and Chloramphenicol Antibiotics in Urban Waters in China. *Environ. Toxicol. Chem.* **2008**, *27*, 73. [CrossRef]
24. Batt, A.L.; Bruce, I.B.; Aga, D.S. Evaluating the vulnerability of surface waters to antibiotic contamination from varying wastewater treatment plant discharges. *Environ. Pollut.* **2006**, *142*, 295–302. [CrossRef] [PubMed]
25. Łuczkiewicz, A.; Felis, E.; Ziembińska, A.; Gnida, A.; Kotlarska, E.; Olanczuk-Neyman, K.; Surmacz-Górska, J. Resistance of Escherichia coli and Enterococcus spp. to selected antimicrobial agents present in municipal wastewater. *J. Water Health* **2013**, *11*, 600–612. [CrossRef] [PubMed]
26. Birosova, L.; Mackuľak, T.; Bodík, I.; Ryba, J.; Škubák, J.; Grabic, R. Pilot study of seasonal occurrence and distribution of antibiotics and drug resistant bacteria in wastewater treatment plants in Slovakia. *Sci. Total. Environ.* **2014**, *490*, 440–444. [CrossRef]
27. Rizzo, L.; Manaia, C.M.; Merlin, C.; Schwartz, T.; Dagot, C.; Ploy, M.; Michael, I.; Fatta-Kassinos, D. Urban wastewater treatment plants as hotspots for antibiotic resistant bacteria and genes spread into the environment: A review. *Sci. Total. Environ.* **2013**, *447*, 345–360. [CrossRef] [PubMed]
28. Storteboom, H.; Arabi, M.; Davis, J.G.; Crimi, B.; Pruden, A. Tracking Antibiotic Resistance Genes in the South Platte River Basin Using Molecular Signatures of Urban, Agricultural, And Pristine Sources. *Environ. Sci. Technol.* **2010**, *44*, 7397–7404. [CrossRef] [PubMed]
29. Makowska, N.; Koczura, R.; Mokracka, J. Class 1 integrase, sulfonamide and tetracycline resistance genes in wastewater treatment plant and surface water. *Chemosphere* **2016**, *144*, 1665–1673. [CrossRef]
30. Bengtsson-Palme, J.; Hammarén, R.; Pal, C.; Östman, M.; Björlenius, B.; Flach, C.-F.; Fick, J.; Kristiansson, E.; Tysklind, M.; Larsson, D.G.J. Elucidating selection processes for antibiotic resistance in sewage treatment plants using metagenomics. *Sci. Total. Environ.* **2016**, *572*, 697–712. [CrossRef] [PubMed]
31. Szczepanowski, R.; Linke, B.; Krahn, I.; Gartemann, K.-H.; Gützkow, T.; Eichler, W.; Pühler, A.; Schlüter, A. Detection of 140 clinically relevant antibiotic-resistance genes in the plasmid metagenome of wastewater treatment plant bacteria showing reduced susceptibility to selected antibiotics. *Microbiology* **2009**, *155*, 2306–2319. [CrossRef]
32. Pruden, A.; Arabi, M.; Storteboom, H.N. Correlation Between Upstream Human Activities and Riverine Antibiotic Resistance Genes. *Environ. Sci. Technol.* **2012**, *46*, 11541–11549. [CrossRef]
33. Pazda, M.; Kumirska, J.; Stepnowski, P.; Mulkiewicz, E. Antibiotic resistance genes identified in wastewater treatment plant systems—A review. *Sci. Total Environ.* **2019**, *697*, 1–21. [CrossRef]
34. Sharma, V.K.; Johnson, N.; Cizmas, L.; McDonald, T.J.; Kim, H. A review of the influence of treatment strategies on antibiotic resistant bacteria and antibiotic resistance genes. *Chemosphere* **2016**, *150*, 702–714. [CrossRef]
35. Li, Y.; Zhu, G.; Ng, W.J.; Tan, S.K. A review on removing pharmaceutical contaminants from wastewater by constructed wetlands: Design, performance and mechanism. *Sci. Total. Environ.* **2014**, *468*, 908–932. [CrossRef]
36. Chen, J.; Ying, G.-G.; Wei, X.-D.; Liu, Y.-S.; Liu, S.-S.; Hu, L.-X.; He, L.-Y.; Chen, Z.-F.; Chen, F.-R.; Yang, Y.-Q. Removal of antibiotics and antibiotic resistance genes from domestic sewage by constructed wetlands: Effect of flow configuration and plant species. *Sci. Total. Environ.* **2016**, *571*, 974–982. [CrossRef]
37. Chen, J.; Wei, X.-D.; Liu, Y.-S.; Ying, G.-G.; Liu, S.-S.; He, L.-Y.; Su, H.-C.; Hu, L.-X.; Chen, F.-R.; Yang, Y.-Q. Removal of antibiotics and antibiotic resistance genes from domestic sewage by constructed wetlands: Optimization of wetland substrates and hydraulic loading. *Sci. Total. Environ.* **2016**, *565*, 240–248. [CrossRef] [PubMed]
38. Huang, X.; Zheng, J.; Liu, C.; Liu, L.; Liu, Y.; Fan, H. Removal of antibiotics and resistance genes from swine wastewater using vertical flow constructed wetlands: Effect of hydraulic flow direction and substrate type. *Chem. Eng. J.* **2017**, *308*, 692–699. [CrossRef]

39. Gorito, A.M.; Ribeiro, A.R.; Almeida, C.M.R.; Silva, A.M. A review on the application of constructed wetlands for the removal of priority substances and contaminants of emerging concern listed in recently launched EU legislation. *Environ. Pollut.* **2017**, *227*, 428–443. [CrossRef]
40. Liu, L.; Li, J.; Fan, H.; Huang, X.; Wei, L.; Liu, C. Fate of antibiotics from swine wastewater in constructed wetlands with different flow configurations. *Int. Biodeterior. Biodegrad.* **2019**, *140*, 119–125. [CrossRef]
41. Zhang, S.; Lu, Y.-X.; Zhang, J.-J.; Liu, S.; Song, H.-L.; Yang, X.-L. Constructed Wetland Revealed Efficient Sulfamethoxazole Removal but Enhanced the Spread of Antibiotic Resistance Genes. *Molecules* **2020**, *25*, 834. [CrossRef] [PubMed]
42. Aminov, R.; Garrigues-Jeanjean, N.; Mackie, R.I. Molecular Ecology of Tetracycline Resistance: Development and Validation of Primers for Detection of Tetracycline Resistance Genes Encoding Ribosomal Protection Proteins. *Appl. Environ. Microbiol.* **2001**, *67*, 22–32. [CrossRef]
43. Wang, N.; Yang, X.; Jiao, S.; Zhang, J.; Ye, B.; Gao, S. Sulfonamide-Resistant Bacteria and Their Resistance Genes in Soils Fertilized with Manures from Jiangsu Province, Southeastern China. *PLoS ONE* **2014**, *9*, e112626. [CrossRef]
44. Tuckman, M.; Petersen, P.J.; Howe, A.Y.M.; Orlowski, M.; Mullen, S.; Chan, K.; Bradford, P.A.; Jones, C.H. Occurrence of Tetracycline Resistance Genes among Escherichia coli Isolates from the Phase 3 Clinical Trials for Tigecycline. *Antimicrob. Agents Chemother.* **2007**, *51*, 3205–3211. [CrossRef]
45. Du, J.; Geng, J.; Ren, H.; Ding, L.; Xu, K.; Zhang, Y. Variation of antibiotic resistance genes in municipal wastewater treatment plant with A2O-MBR system. *Environ. Sci. Pollut. Res.* **2014**, *22*, 3715–3726. [CrossRef]
46. Zhang, T.; Shao, M.-F.; Ye, L. 454 Pyrosequencing reveals bacterial diversity of activated sludge from 14 sewage treatment plants. *ISME J.* **2011**, *6*, 1137–1147. [CrossRef]
47. Park, K.Y.; Jang, H.M.; Park, M.-R.; Lee, K.; Kim, D.; Kim, Y.-M. Combination of different substrates to improve anaerobic digestion of sewage sludge in a wastewater treatment plant. *Int. Biodeterior. Biodegrad.* **2016**, *109*, 73–77. [CrossRef]
48. Kim, Y.B.; Jeon, J.H.; Choi, S.; Shin, J.; Lee, Y.; Kim, Y.-M. Use of a filtering process to remove solid waste and antibiotic resistance genes from effluent of a flow-through fish farm. *Sci. Total. Environ.* **2018**, *615*, 289–296. [CrossRef] [PubMed]
49. Zielińska, S.; Radkowski, P.; Blendowska, A.; Ludwig-Gałęzowska, A.; Łoś, J.M.; Łoś, M. The choice of the DNA extraction method may influence the outcome of the soil microbial community structure analysis. *Microbiology* **2017**, *6*, e00453. [CrossRef] [PubMed]
50. Morgan, J.L.; Darling, A.E.; Eisen, J.A. Metagenomic Sequencing of an In Vitro-Simulated Microbial Community. *PLoS ONE* **2010**, *5*, e10209. [CrossRef]
51. Biswal, B.K.; Mazza, A.; Masson, L.; Gehr, R.; Frigon, D. Impact of wastewater treatment processes on antimicrobial resistance genes and their co-occurrence with virulence genes in Escherichia coli. *Water Res.* **2014**, *50*, 245–253. [CrossRef]
52. Rodriguez-Mozaz, S.; Chamorro, S.; Marti, E.; Huerta, B.; Gros, M.; Sànchez-Melsió, A.; Borrego, C.M.; Barceló, J.; Balcázar, J.L. Occurrence of antibiotics and antibiotic resistance genes in hospital and urban wastewaters and their impact on the receiving river. *Water Res.* **2015**, *69*, 234–242. [CrossRef] [PubMed]
53. Amador, P.P.; Fernandes, R.M.; Prudencio, M.C.; Barreto, M.P.; Duarte, I.M. Antibiotic resistance in wastewater: Occurrence and fate of Enterobacteriaceae producers of class A and class C β-lactamases. *J. Environ. Sci. Health A Tox. Hazard. Subst. Environ. Eng.* **2015**, *50*, 26–39. [CrossRef]
54. Auerbach, E.A.; Seyfried, E.E.; McMahon, K.D. Tetracycline resistance genes in activated sludge wastewater treatment plants. *Water Res.* **2007**, *41*, 1143–1151. [CrossRef]
55. Guo, J.; Li, J.; Chen, H.; Bond, P.L.; Yuan, Z. Metagenomic analysis reveals wastewater treatment plants as hotspots of antibiotic resistance genes and mobile genetic elements. *Water Res.* **2017**, *123*, 468–478. [CrossRef]
56. Laht, M.; Karkman, A.; Voolaid, V.; Ritz, C.; Tenson, T.; Virta, M.; Kisand, V. Abundances of Tetracycline, Sulphonamide and Beta-Lactam Antibiotic Resistance Genes in Conventional Wastewater Treatment Plants (WWTPs) with Different Waste Load. *PLoS ONE* **2014**, *9*, e103705. [CrossRef]
57. Hultman, J.; Tamminen, M.; Pärnänen, K.M.M.; Cairns†, J.; Karkman, A.; Virta, M. Host range of antibiotic resistance genes in wastewater treatment plant influent and effluent. *FEMS Microbiol. Ecol.* **2018**, *94*, 1–10. [CrossRef] [PubMed]

58. Łuczkiewicz, A.; Jankowska, K.; Fudala-Książek, S.; Olańczuk-Neyman, K. Antimicrobial resistance of fecal indicators in municipal wastewater treatment plant. *Water Res.* **2010**, *44*, 5089–5097. [CrossRef] [PubMed]
59. Mao, D.; Yu, S.; Rysz, M.; Luo, Y.; Yang, F.; Li, F.; Hou, J.; Mu, Q.; Alvarez, P.J.J. Prevalence and proliferation of antibiotic resistance genes in two municipal wastewater treatment plants. *Water Res.* **2015**, *85*, 458–466. [CrossRef] [PubMed]
60. Kotlarska, E.; Łuczkiewicz, A.; Pisowacka, M.; Burzyński, A. Antibiotic Resistance and Prevalence of Class 1 and 2 Integrons in Escherichia coli Isolated from Two Wastewater Treatment Plants, and Their Receiving Waters (Gulf of Gdansk, Baltic Sea, Poland). *Wastewater Public Health* **2015**, *22*, 67–98. [CrossRef]
61. Nõlvak, H.; Truu, M.; Tiirik, K.; Oopkaup, K.; Sildvee, T.; Kaasik, A.; Mander, Ü.; Truu, J. Dynamics of antibiotic resistance genes and their relationships with system treatment efficiency in a horizontal subsurface flow constructed wetland. *Sci. Total. Environ.* **2013**, *461*, 636–644. [CrossRef]
62. Liu, L.; Liu, C.; Zheng, J.; Huang, X.; Wang, Z.; Liu, Y.; Zhu, G. Elimination of veterinary antibiotics and antibiotic resistance genes from swine wastewater in the vertical flow constructed wetlands. *Chemosphere* **2013**, *91*, 1088–1093. [CrossRef]
63. Liu, L.; Liu, Y.-H.; Wang, Z.; Liu, C.-X.; Huang, X.; Zhu, G.-F. Behavior of tetracycline and sulfamethazine with corresponding resistance genes from swine wastewater in pilot-scale constructed wetlands. *J. Hazard. Mater.* **2014**, *278*, 304–310. [CrossRef]
64. Huang, X.; Liu, C.; Li, K.; Su, J.; Zhu, G.; Liu, L. Performance of vertical up-flow constructed wetlands on swine wastewater containing tetracyclines and tert genes. *Water Res.* **2015**, *70*, 109–117. [CrossRef]
65. Chen, J.; Liu, Y.S.; Su, H.C.; Ying, G.G.; Liu, F.; Liu, S.S.; He, L.Y.; Chen, Z.F.; Yang, Y.Q.; Chen, F.R. Removal of antibiotics and antibiotic resistance henes in rural wastewater by an integrated constructed wetland. *Environ. Sci. Pollut. Res.* **2015**, *22*, 1794–1803. [CrossRef]
66. Wolecki, D.; Caban, M.; Pazda, M.; Stepnowski, P.; Kumirska, J. Evaluation of the Possibility of Using Hydroponic Cultivations for the Removal of Pharmaceuticals and Endocrine Disrupting Compounds in Municipal Sewage Treatment Plants. *Molecules* **2019**, *25*, 162. [CrossRef]
67. Nowrotek, M.; Kotlarska, E.; Łuczkiewicz, A.; Felis, E.; Sochacki, A.; Miksch, K. The treatment of wastewater containing pharmaceuticals in microcosm constructed wetlands: The occurrence of integrons (int1–2) and associated resistance genes (sul1–3, qacEΔ1). *Environ. Sci. Pollut. Res.* **2017**, *24*, 15055–15066. [CrossRef] [PubMed]
68. Anderson, J.C.; Carlson, J.C.; Low, J.E.; Challis, J.K.; Wong, C.S.; Knapp, C.W.; Hanson, M. Performance of a constructed wetland in Grand Marais, Manitoba, Canada: Removal of nutrients, pharmaceuticals, and antibiotic resistance genes from municipal wastewater. *Chem. Central J.* **2013**, *7*, 54. [CrossRef] [PubMed]
69. Song, H.L.; Zhang, S.; Guo, J.; Yang, Y.L.; Zhang, L.M.; Li, H.; Yang, X.L.; Liu, X. Vertical up-flow constructed wetlands exhibited efficient antibioticremoval but induced antibiotic resistance genes in effluent. *Chemosphere* **2018**, *203*, 434–441. [CrossRef] [PubMed]
70. Goossens, H.; Ferech, M.; Vander Stichele, R. Outpatient antibiotic use in Europe and association with resistance: A crossnational database study. *Lancet* **2005**, *365*, 579–587. [CrossRef]
71. Sun, L.; Klein, E.Y.; Laxminarayan, R. Seasonality and Temporal Correlation between Community Antibiotic Use and Resistance in the United States. *Clin. Infect. Dis.* **2012**, *55*, 687–694. [CrossRef] [PubMed]
72. Yuan, Q.-B.; Guo, M.-T.; Yang, J. Monitoring and assessing the impact of wastewater treatment on release of both antibiotic-resistant bacteria and their typical genes in a Chinese municipal wastewater treatment plant. *Environ. Sci. Process. Impacts* **2014**, *16*, 1930–1937. [CrossRef] [PubMed]
73. Yang, Y.; Li, B.; Ju, F.; Zhang, T. Exploring Variation of Antibiotic Resistance Genes in Activated Sludge over a Four-Year Period through a Metagenomic Approach. *Environ. Sci. Technol.* **2013**, *47*, 10197–10205. [CrossRef]
74. Börjesson, S.; Melin, S.; Matussek, A.; Lindgren, P.-E. A seasonal study of the mecA gene and Staphylococcus aureus including methicillin-resistant S. aureus in a municipal wastewater treatment plant. *Water Res.* **2009**, *43*, 925–932. [CrossRef]
75. Börjesson, S.; Mattsson, A.; Lindgren, P.-E. Genes encoding tetracycline resistance in a full-scale municipal wastewater treatment plant investigated during one year. *J. Water Health* **2009**, *8*, 247–256. [CrossRef]
76. Olańczuk-Neyman, K.; Geneja, M.; Quant, B.; Dembińska, M.; Kruczalak, K.; Kulbat, E.; Kulik-Kuziemska, I.; Mikołajski, S.; Gielert, M. Microbiological and Biological Aspects of the Wastewater Treatment Plant "Wschód" in Gdańsk. *Pol. J. Environ. Stud.* **2003**, *12*, 747–757.

77. Suzuki, M.T.; Taylor, L.T.; Delong, E.F.; Burnett, S.L.; Chen, J.; Beuchat, L.R. Quantitative Analysis of Small-Subunit rRNA Genes in Mixed Microbial Populations via 5′-Nuclease Assays. *Appl. Environ. Microbiol.* **2000**, *66*, 4605–4614. [CrossRef] [PubMed]
78. Ng, L.-K.; Martin, I.; Alfa, M.; Mulvey, M. Multiplex PCR for the detection of tetracycline resistant genes. *Mol. Cell. Probes* **2001**, *15*, 209–215. [CrossRef] [PubMed]
79. Schmittgen, T.D.; Livak, K.J. Analyzing real-time PCR data by the comparative CT method. *Nat. Protoc.* **2008**, *3*, 1101–1108. [CrossRef] [PubMed]

Sample Availability: Samples of the compounds are not available from the authors.

Article

Insights into Mechanisms of Electrochemical Drug Degradation in Their Mixtures in the Split-Flow Reactor

Aleksandra Pieczyńska [1], Stalin Andres Ochoa-Chavez [2], Patrycja Wilczewska [1], Aleksandra Bielicka-Giełdoń [1] and Ewa M. Siedlecka [1,*]

1 Faculty of Chemistry, University of Gdansk, Wita Stwosza 63, 80-308 Gdansk, Poland; Aleksandra.pieczynska@ug.edu.pl (A.P.); p.wilczewska@gmail.com (P.W.); a.bielicka-gieldon@ug.edu.pl (A.B.-G.)

2 Centro de Investigación y Control Ambiental, Departamento de Ingeniería Civil y Ambiental, Escurla Politécnica Nacional, Ladrón de Guevara E11-253, Quito P.O. Box 17-01-2759, Ecuador; stalinandres123@outlook.com

* Correspondence: ewa.siedlecka@ug.edu.pl; Tel.: +48-58-523-5228

Academic Editor: Jolanta Kumirska

Received: 21 October 2019; Accepted: 27 November 2019; Published: 28 November 2019

Abstract: The recirculating split-flow batch reactor with a cell divided into anolyte and catholyte compartments for oxidation mixture of cytostatic drugs (CD) was tested. In this study, kinetics and mechanisms of electrochemical oxidization of two mixtures: 5-FU/CP and IF/CP were investigated. The order of the CD degradation rate in single drug solutions and in mixtures was found to be 5-FU < CP < IF. In the 5-FU/CP mixture, k_{app} of 5-FU increased, while k_{app} of CP decreased comparing to the single drug solutions. No effect on the degradation rate was found in the CP/IF mixture. The presence of a second drug in the 5-FU/CP mixture significantly altered mineralization and nitrogen removal efficiency, while these processes were inhibited in IF/CP. The experiments in the different electrolytes showed that •OH and sulphate active species can participate in the drug's degradation. The k_{app} of the drugs was accelerated by the presence of Cl^- ions in the solution. Chlorine active species played the main role in the production of gaseous nitrogen products and increased the mineralisation. Good results were obtained for the degradation and mineralisation processes in mixtures of drugs in municipal wastewater-treated effluent, which is beneficial from the technological and practical point of view.

Keywords: ifosfamide; cyclophosphamide; 5-fluorouracil; cytostatic drug; BDD anode; electrochemical oxidation; intermediates

1. Introduction

5-Fuorouracil (5-FU), cyclophosphamide (CP) and ifosfamide (IF) have been the most commonly used cytostatic drugs worldwide since the 1960s. CP has been classified as the most dangerous, carcinogenic compound, meanwhile the 5-FU and IF have been described as compounds with potential mutagenic and teratogenic effects [1,2]. The drugs are polar, poorly biodegradable and under solar light they are photolytic persistent compounds [2,3]. Hospitals and municipal wastewater treatment plants (WWTPs) act as major point sources of cytostatic drugs discharge to the environment, because they are incompletely removed by the mechanical-biological wastewater treatment. The effluents from municipal or hospital wastewater treatment plants generally contain a mixture of different types of pharmaceuticals with their metabolic products.

The efficient removal of cytostatic drugs from wastewater by powerful transformation treatments such as Advanced Oxidation Processes (AOPs) seems necessary to avoid its continuous discharge into waterbodies. However, the reports of investigations of drugs removal by alternative methods in their mixtures have been limited so far. The degradation of a mixture of 16 cytostatic drugs by hydrolysis, photolysis, biodegradation process, and H_2O_2/UV system was studied by Franquet-Griell et al. [4]. The physical-chemical and biological processes were selected for the study of the organic compound's fate during water and wastewater treatment. The results revealed that IF and CP are very persistent compounds in both the natural environment and during wastewater treatment. Another paper reported how mineralization efficiency of chloramphenicol, ciprofloxacin and dipyrone individually and from equimolar mixtures varied in Fenton and photo-Fenton processes. The authors found that degradation efficiency was suppressed by Cl^- or F^- ions, which were released into the solution upon cleavage of drugs present in the mixture [5]. AOPs including H_2O_2/UV or Fenton process are considered promising alternative techniques for the decomposition of poorly biodegradable micropollutants [6–8]. However, the application of AOPs to remove pharmaceuticals in wastewater and in single drug solutions led to different levels of effectiveness of individual drugs removal, as was previously proved [9–11]. Competition among organic compounds for •OH radical could cause the inhibition or acceleration of the drug oxidation rate in wastewater. Moreover, the inorganic anions e.g., Cl^-, HCO_3^-, can also participate in reactions with •OH radicals, causing a decrease in their concentration available to remove organic matter. The presence of inorganic anions affects the degradation rate of drugs, which depends on their concentration and the type of used AOPs [10,11]. Therefore, the composition of wastewater should be taken into account when assessing the effectiveness of pollution removal by AOPs.

Electrochemical advanced oxidation (EAO) is one of the powerful techniques which are being developed for drugs wastewater remediation [12]. The key factors in the effectiveness of EAO processes are the applied electrode material, the electrolyte composition and the type of cell [13,14]. Among different anode materials, boron-doped diamond (BDD) thin films have attracted great attention. BDD with the weak •OH radicals interaction and the large overpotential for water discharge, enables regulating the yield of generated in situ strong oxidants such as •OH [2]. BDD electrode has been used as an anode for the decomposition of various micropollutants [15] including cytostatic drugs [10]. Another key factor that affects the kinetics and effectiveness of EAO is the type of reactor. A large variety of electrochemical oxidation systems have been tested for the treatment of dyes and drugs wastewater such as: divided or undivided electrode cells, flow cells with parallel electrodes, and flow plants with a three-phase three-dimensional electrode reactor [16–18]. It was shown that flow cells with planar electrodes in a parallel plate configuration can increase the oxidation rate of pollutants. Radjenovic and Petrovic applied a BDD electrode in a plate-and-frame electrolytic cell divided by the cation exchange membrane for the anolyte and catholyte compartment to eliminate X-ray contrast media [19].

The aim of this work was to assess whether the EAO in a split-flow reactor with a divided cell of anolyte and catholyte compartments is able to effectively degrade and mineralize cytostatic drugs in mixtures: 5-FU/CP and IF/CP. The impact of electrolyte composition including actual effluents from WWTP on the rate and efficiency of individual drugs oxidation in mixtures: IF/CP and 5-FU/CP was tested. The mechanism of cytostatic drugs degradation was explored through identification of oxidizing species and formed intermediates. To our best knowledge, until now there have been no studies that bring insight into the mechanisms of cytostatic drug removal in a mixture in the split-flow reactor of a divided electrolytic cell.

2. Results and Discussion

2.1. Comparison of Electrochemical Decomposition in Recirculating Split-Flow Batch Reactor of IF, CF and 5-FU in Single Drug Solution

In the first step of study, cyclic voltammograms for 5-FU, CF and IF (50 mg/L of drug in 42 mM Na_2SO_4 and pH of 6.6) at the Si/BDD electrode were recorded. As we suspected, the drugs did not reveal any peaks in the range of potential values from −1.5V to +3.0V, which indicated that electrochemical oxidation of cytostatic drugs in our experimental conditions was possible only in an indirect way by electrogenerated oxidative entities (Figure S1).

Next, in galvanostatic experiments with the recirculating split-flow batch reactor, the time course of the electrochemical degradation of drugs: IF, CF and 5-FU in single drug solution was monitored by HPLC-UV. The initial concentrations of pharmaceuticals used in experiments were 25 mg·L^{-1} and 50 mg·L^{-1}. They were employed in order to determine the effect of the presence of additional organic matter in the mixture on the kinetics and mineralization of the individual drug. The concentrations of pharmaceuticals were much higher than that found in the waterbodies, due to minimizing the error in the analysis of inorganic ions released and the intermediates of the drugs oxidation process. The operating parameters: volumetric flow rate (13 L·h^{-1}) and current density (150 Am^{-2} ($j_{app} > j_{lim}$)), were previously experimentally selected [20]. The experimental data were well described by pseudo-first-order rate kinetics (R^2 in range 0.95–0.99). In Table 1, the values of apparent rate constants (k_{app}) of 5-FU, CP and IF electrochemical degradation in the single drug solution are listed.

Table 1. The k_{app} for 5-FU, CP and IF degradation in single drug solutions and in 5-FU/CP and IF/CP mixtures in Na_2SO_4 electrolyte.

Single Drug Solution	**C = 50 mg L^{-1}**		**C = 25 mg L^{-1}**			
	$k_{app} \times 10^{-3}$ min^{-1}	**R^2**	**$k_{app} \times 10^{-3}$ min^{-1}**	**R^2**		
5-FU	3.8	0.99	8.9	0.98		
CP	10.4	0.99	13.5	0.99		
IF	19.1	0.99	27.7	0.99		
Mixture	**C = 25 mg L^{-1} 1:1**		**C = 25 mg L^{-1} 1:1 + 10 mg L^{-1} Phosphates**		**C = 25 mg L^{-1} 1:1 + 100 mg L^{-1} Chlorides**	
	$k_{app} \times 10^{-3}$ min^{-1}	**R^2**	**$k_{app} \times 10^{-3}$ min^{-1}**	**R^2**	**$k_{app} \times 10^{-3}$ min^{-1}**	**R^2**
5-FU/CP						
5-FU	11.0	0.96	10.5	0.97	13.6	0.98
CP	11.0	0.99	10.8	0.96	67.4	0.99
CP/IF						
CP	13.9	0.98	13.0	0.99	47.4	0.95
IF	27.9	0.97	27.7	0.98	46.6	0.99

Independently of drug concentration, the degradation rate coefficient (k_{app}) value increased in the following order k_{app}(5-FU) < k_{app}(CP) < k_{app}(IF). The k_{app} values were higher for the concentration of 25 mg·L^{-1} (8.9×10^{-3}–27.7×10^{-3} min^{-1}) than k_{app} values for the concentration of 50 mg·L^{-1} (3.8–19.1 × 10^{-3} min^{-1}) (Table 1). For the lower concentration, the efficiency of the removal of drugs was complete in 300, 150 and 120 min for 5-FU, CP and IF, respectively.

Figure 1 presented the efficiency of the removal of TOC, COD, drug, and TN after 6 h of electrochemical processes. As we can see, TOC conversion reached 13, 14 and 60% respectively for 5-FU, CP and IF indicating that 5-FU and CP were more persistent compounds than IF in the mineralization process. Moreover, the total amount of nitrogen (TN) was still present in the solution in organic or mineral forms for 5-FU and CP, while the TN removal for IF reached 34%. The relatively high value of TN removal for IF degradation indicated that the nitrogen-containing drug was converted to organic and mineral intermediates which existed in the solution, and to gaseous nitrogen species. The release of NOx during mineralization of *N*-containing pollutants by the semi-quantitative method using differential electrochemistry mass spectroscopy was ascertained by Garcia-Segura et al. [21]. The process was found to low extent for 5-FU decomposition and it was not observed for CP degradation. It is worth noting that IF is an isomer of CP, but due to the results it can be suspected that its electrochemical degradation performed in a different way than CP.

The COD/TOC removal ratio characterised the organic matter oxidation progress. The analysis of the parameter shows that, (i) intermediates generated in the electrodegradation of CP were accumulated in solution and (ii) they were higher oxidized compounds than intermediates remaining in the solution and produced in the IF and 5-FU degradation processes. The relatively low TOC conversion (14%) and high COD removal (89%) (TOC/COD_{rem} = 16%) implied that CP was degraded to small molecular carboxylic acids. It is known that direct electron transfer from an organic compound to a BDD electrode is a mechanism of the oxalic acid oxidation [22] and, therefore, the small molecular acids appear to be an inhibitor for their own oxidation and also for other compounds (e.g., HCOOH) that undergo oxidation at high anodic potentials [23,24]. Therefore, the phenomena can inhibit the oxidation of small molecular acids at the BDD anode. The CP degradation to small molecular carboxylic acids was in contrast to IF degradation, where TOC and COD removal were 60 and 62 %, respectively (TOC/COD_{rem} ratio of 96%). In this case, the generated intermediates were mineralized to CO_2 to a higher extent than in the CP degradation, and the small molecular carboxylic acids were not accumulated in the solution.

The mineralization of IF and CF in the recirculating split-flow batch reactor in the anolytic compartment of the cell was additionally analysed based on the percentage of quantity of mineral products such as Cl^-, NH_4^+, NO_3^- and PO_4^{3-} released from the organic matter (Figure 2). The amount of inorganic products was expressed as a the percentage of the total expected amount of Cl, N or P released from the initial concentration of each drug. The N-NO_3^- and P-PO_4^{3-} concentrations were increasing linearly with the time progress, and at the end of the process attained 24 (for IF)-38 (for CP)% and 15 (for IF)-17 (for CP)% of N and P amount initially presented in the drugs solution, respectively. The concentration of N-NH_4^+ was very low (below 1 mg·L^{-1}) as time passed. A similar trend of Cl^- ions releasing was found in our previous study [10]. After 3 h of electrolysis, the amount of Cl^- ions reached 50% of total theoretical amount and after that decreased to below detection levels (0.1 mg·L^{-1}) for both studied drugs. The results suggested that the organic intermediates existing in the solution after electrolysis still contained in their structures the N and P heteroatoms, while the chloric atoms can be transformed to different forms. Released Cl^- ions can be oxidized to Cl_2 at the anode and, depending on the pH, form $HClO/ClO^-$. These chlorine species can undergo electrochemical oxidation to chlorates [25]. Unfortunately, the presence of a high concentration of sulfates in solution (6 g·L^{-1}) makes impossible the analysis of a probable trace amount of ClO_3^-/ClO_4^- in the effluent and these data cannot be shown. Moreover, the other volatile chlorine species (ClO_2, Cl_2O) can be formed. The ClO2 and Cl2O reactions with organic matter do not form significant levels of THMs. They are formed from Cl^- in the presence of •OH radicals, mainly in the acid conditions [26]. It is worth noticing that in our experimental conditions in the anolyte compartment, where the oxidation of drugs took place, the pH was about 3.

The 5-FU released to the solution 25% of F^- ions and 38% of N-NO_3^-, respectively, while the ammonium was in a low concentration (below 1 mg N·L^{-1}). The finding suggested that the remaining amount of fluoride and nitrogen was in the organic by-products. The F^- ions adsorption onto the BDD electrode surface may also be considered.

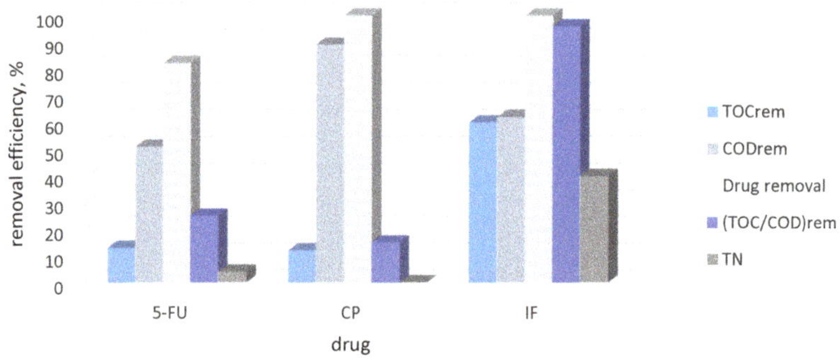

Figure 1. The removal efficiency of TOC, COD, drug concentration, TN and (TOC/COD) ratio after 6 h of electrolysis in single drug solutions in a concentration of 50 mg·L^{-1}.

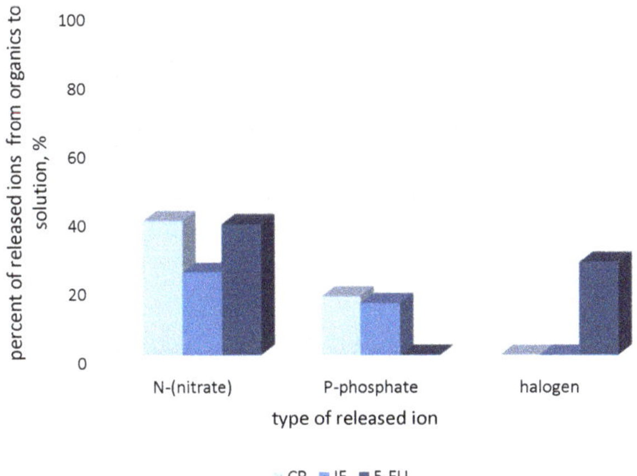

Figure 2. Inorganic ions released in single-drug solution after 6 h of electrochemical oxidation.

2.2. *Decomposition of Cytostatic Drugs in Their Mixtures*

The electrochemical oxidation of mixtures of drugs: CP/5-FU and CP/IF, was investigated. Each compound in the mixture had a 25 mg·L^{-1} concentration. Comparative treatment of drugs in the mixtures (25 mg·L^{-1}/25 mg·L^{-1}) and in the single compound solutions at two different concentrations (25 and 50 mg·L^{-1}) were made in order to clarify the effect of additional organic matter on kinetics and efficiency removal of an individual drug. Moreover, the electrochemical oxidation of mixtures of cytostatic drugs in different electrolytes including effluent from WWTP was examined. The electrochemical experiments were performed in the same operating conditions as described in Section 2.1.

2.2.1. Decomposition of CP and 5-FU Mixture

The decomposition rate of the mixture of drugs was well described by pseudo-first-order rate kinetics. In the CP/5-FU mixture, CP decomposition rate expressed as k_{app} was 11 min^{-1}. This value was lower than that found in the single-drug solution, with a concentration of 25 mg·L^{-1} (k_{app} 13.5 min^{-1}), and it was similar to the value obtained for CP in the single-drug solution with a concentration of 50 mg·L^{-1}. The results indicated that the decomposition of CP was inhibited by concomitantly 5-FU degradation, as an additional organic matter in the solution. On the other hand, the removal rate of 5-FU in the presence of CP was faster than its degradation rate in the single-drug solution regardless of the concentration used (Table 1). The value of k_{app} of 5-FU was likely accelerated by active chlorine species produced from Cl^- ions released during CP degradation. In experimental conditions, where the pH of anolyte was strong acid, the electrochemical generation of HClO and Cl_2•, more selective oxidizing entities than •OH radicals was preferable [27,28].

In order to confirm this hypothesis, the effect of Cl^- ions on the drugs mixture decomposition performance was studied. Cl^- ions in concentrations of 10 mg·L^{-1} (similar amount of Cl^- ions completely released form 50 mg·L^{-1} CP during degradation) and 100 mg·L^{-1} (Table 1) accelerated the rate of both drugs' oxidation and confirmed the speculation that active chlorine species participated in the degradation of 5-FU even in the presence of a low concentration of Cl^- ions released from CP oxidation. During CP decomposition, PO_4^{3-} ions are also released, therefore, the effects of these ions at concentrations of 5 mgP·L^{-1} (similar to maximal concentration of P released during 50 mg·L^{-1} CP degradation) and 10 mgP·L^{-1} (Table 1) were tested. These ions seems to have an insignificant effect on the electrochemical oxidation of 5-FU and CP.

In the next step, the mineralization (TOC_{rem}) and total nitrogen conversion (TN_{rem}) to gaseous intermediates were analyzed after 6 h of electrochemical oxidation. TOC and TN results presented in Figure 3A showed that mineralization and nitrogen removal were higher in the mixture compared to the single-drug solutions at a concentration of 50 mg·L^{-1}.

The significant increase in efficiency of mineralization and the formation of a higher amount of nitrogen gaseous products in the mixture than in the single-drug solutions were likely the result of two phenomenon: (i) the generation of active chlorine species by Cl^- ions being released from CP oxidation and its participation in the decomposition of 5-FU and its intermediates; and (ii) change of the parent compounds' degradation pathways in the presence of a second organic compound and the generation of less-persistent intermediates for the oxidation process.

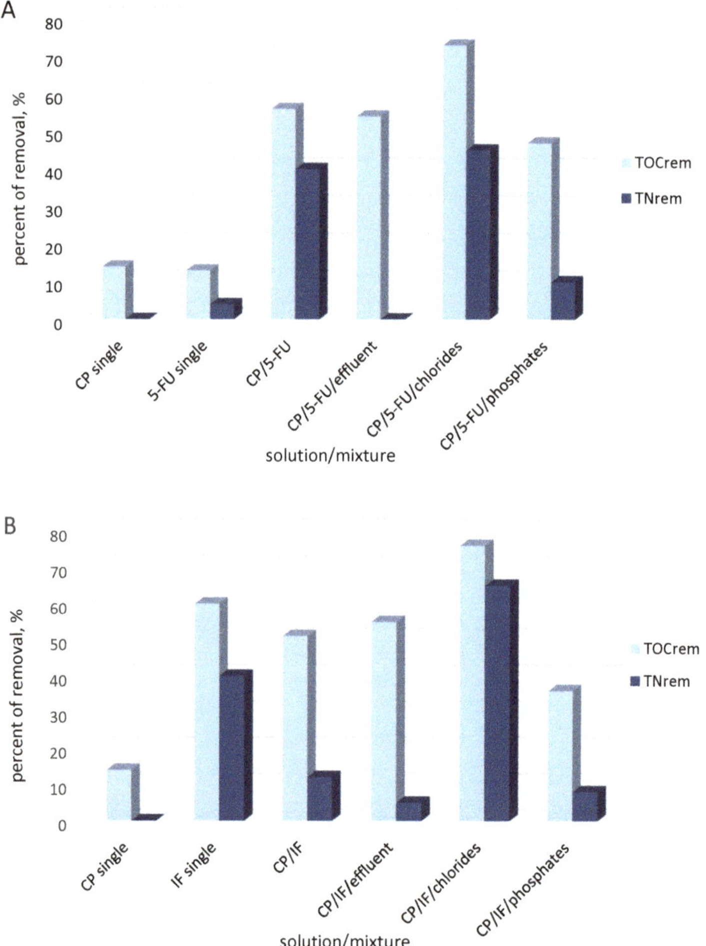

Figure 3. TOC removal and TN conversion efficiency in electrochemical oxidation of single-drug solutions and mixtures of drugs: 5-FU/CP (**A**) and CP/IF (**B**) in different matrices.

The matrix effect was studied in the mixture of 5-FU/CP with the effluent from WWTP. The decomposition rate of drugs was faster in the effluent than in the electrolyte Na_2SO_4. The COD was completely removed in the mixture of drugs in the WWT effluent, which is a beneficial from the practical point of view. The TOC conversion of CP/5-FU in the effluent from WWTP was similar to the

TOC conversion reached in the mixture of drugs in the Na_2SO_4 electrolyte, while the generation of nitrogen gaseous products in the effluent was totally inhibited. The further experiments of CP/5-FU oxidation in the chlorides solution (100 mg·L^{-1}) showed the increase of evolution of gaseous nitrogen products and suggested that chlorine active species participated in their generation. In contrast, the phosphates (10 mg·L^{-1}) negatively affected this process as well as drugs mineralisation, but to a larger extent inhibited the TN conversion.

We inferred that the faster TOC and TN removal in the presence of Cl^- ions in Na_2SO_4 was the consequence of the synergy between •OH, SO_4^-• radicals and active chlorine species in organic and nitrogen compounds oxidation. The •OH/SO_4^-• radicals reacted with the compounds to form the derivatives, which can be more easily attacked by Cl_2• and HClO. Even the low concentration of Cl^- as released form drug degradation is enough to improve the treatment process. The other research group reported that the oxidation potential for the formation of SO_4^-• radicals, and their lifetime, show that chlorine active species are more effective in the oxidation of the organic matter in the bulk of the solution [28,29]. The presence of other inorganic ions such as PO_4^{3-} inhibited this process.

Moreover, depending on the concentration of chlorine formed in direct electrochemical oxidation of Cl^- at anode, the ammonium ions transformed to chloramines (Equation (1)) and then to N_2 and N_2O (Equations (2) and (3)) in the bulk of the solution were decreasing nitrogen content [30,31]:

$$NH_4^+ + HOCl \leftrightarrow NH_2Cl + H_2O + H^+ \tag{1}$$

$$4NH_2Cl + 3Cl_2 + H_2O \rightarrow N_2O + N_2 + 10\,HCl \tag{2}$$

$$2NH_4^+ + 3HOCl \rightarrow N_2 + 3H_2O + 5H^+ + 3Cl^- \tag{3}$$

The conversion of amines to nitrates could be a minor pathway (Equation (4))

$$NH_4^+ + 4HOCl \rightarrow NO_3^- + H_2O + 6H^+ + 4Cl^- \tag{4}$$

Comparing the results acquired in the single drug solutions of CP with results obtained in the mixture of CP/5-FU, it was supposed that reaction 4 was possible for electrochemical oxidation of CP in a single-drug solution (Section 2.1). However, the direct oxidation of organic nitrogen by •OH radicals, without previously releasing from CP ammonium ions, could be a more likely pathway of nitrates production [21].

The nitrogen evolution is attained when the chlorine breaking point with the molar ratio Cl_2:NH_4^+ 1:1 was reached. After 6 h of electrolysis of CP solution, most of the organic nitrogen was converted to N-NO_3^- by •OH/SO_4^-•, and the value of TN removal was low. The reactions 2 and 3 were favored in the mixture of CP/5-FU, due to the proper ratio of the concentration of NH_4^+ and Cl^- released from drugs.In the case of the 5-FU single-drug solution, HOCl was absent and •OH/SO_4^-• likely participated in the oxidation of both organic nitrogen to N-NO_3^- and organic matter to intermediates and CO_2.

The results indicated that chlorine active species in the mixture of drugs elevated the 5-FU/CP mineralization and ammonium conversion to nitrogen gaseous products. Thus, the reaction of drugs with active chlorine species can be deemed to be a meaningful transformation mechanism under our experimental conditions.

2.2.2. IF/CP Mixture Decomposition

IF and CF are isomers. However, the quality and quantities of products found during their electrolysis in the bath reactor [10] and their photocatalytic degradation in the presence of Pt–TiO_2 under solar light was different [11]. Therefore, the examination of the electrochemical decomposition of the CP/IF mixture in the split-flow reactor seems to be interesting.

In the recirculating split-flow batch reactor, the removal rate of IF was twice faster than the CP removal rate. Moreover, the results (Table 1) clearly demonstrated that the degradation rate of the mixture of CP and IF (25 + 25 mg·L^{-1}) and the single-drug solutions with a concentration of 25 mg·L^{-1} were similar. It was unexpected that the increase of total organic matter (second drug presence in solution) did not produce any effect on the degradation rate of both drugs. This different behavior in the drug removal process can be related with the presence of different oxidative entities generated during electrochemical process and their participation in cytostatic drugs oxidation.

The PO_4^{3-} ions slightly inhibited the decomposition of CP, while they did not have any impact on the IF removal. It is possible while the kinetics of oxidants scavenging is slow (e.g., k(SO_4^{-}•+ HPO_4^{-2}) = 1.2 × 10^6 M^{-1}·s^{-1}). The presence of Cl$^-$ ions in the CP/IF mixture mostly increased oxidation of both compounds. The value of k_{app} for CP and IF significantly increased and reached similar values.

The removal efficiency of TOC and TN are presented in Figure 3B. As was found in the mixture of CP/IF, the mineralisation and TN conversion to gaseous products were inhibited comparing to single-drug solutions. After 6 h of electrolysis of the drugs mixture, the nitrogen in the drugs was converted to nitrogen gaseous products to a lower degree than in the IF single-drug solution. In the case of mineralization and TN conversion, the effect of additional organic matter (CP) was observed.

The chlorides significantly elevated CP/IF mixture mineralization and TN conversion to nitrogen gaseous products, while phosphates and WWTP effluents (PO_4^{3-} 0.2 mg·L^{-1}, COD 62 mg·L^{-1}, Cl$^-$ 92 mg·L^{-1}) slightly inhibited the mineralisation process and significantly suppressed the nitrogen removal. This fact implied that the nitrogen removal was the most sensitive process for the radicals scavenging by e.g., phosphates or carbonates. A possible explanation is that the mineralization in the Na_2SO_4 electrolyte was related to the formation of •OH radicals, which occur near the electrode surface and in the lower degree sulfate active species which exist in the bulk solution. In the presence of Cl$^-$, the indirect oxidation of organic matter and ammonium ions happened by chlorine active species acting in the bulk solution. The scavenging ions such as phosphates especially in the low concentration can inhibit at first the oxidation species in the bulk solution and next the •OH radicals due to the diffusion process. The negative influence on the drugs mineralization of effluents from WWTP was also associated with natural organic matter (COD 62 mg·L^{-1}) presence.

Based on the amount of TOC removed, the mineralization current efficiency (MCE) values for each drug and the mixtures were calculated. The order of MCE values was found to be: IF (2.9%) > IF/CP (2.5%) ≈ 5-FU/CP (1.6%) > 5-FU (0.6%) > CF (0.5%). In the case of IF and both mixtures CP/5-FU and IF/CP, the higher the MCE, the less energy was wasted on the side reaction.

2.3. *Active Species Participated in Cytostatic Drugs Removal*

Some authors reported that the Na_2SO_4 electrolyte used for oxidation of organic matter at the BDD anode can be the source of •OH and SO_4^{-}• radicals. Compared to •OH, SO_4^{-}• is similarly oxidative (E^0 (SO_4•$^-$/SO_4^{2-}) = 2.5–3.1 V) but more selective toward electron-rich organic contaminants. Therefore, SO_4•$^-$ is more likely to share in electron transfer reactions from the aromatic ring [32], while •OH radicals mainly participate in hydrogen abstraction or addition reactions [33].

In order to check the hypothesis that •OH/SO_4^{-}• took part in electrochemical oxidation of cytostatic drugs in Na_2SO_4 electrolyte, the degradation rates of the mixtures of drugs: 5-FU/CP and CP/IF have been compared. Two different electrolytes were applied: $NaNO_3$ (42 mM) and Na_2SO_4 (42 mM). In the experiments, t-butanol (*t*-but) was used as a scavenger of •OH radicals, because it reacts about 1000 times faster with •OH (6 × 10^8 M^{-1}·s^{-1}) compared to SO_4^{-}• (9.1 × 10^5 M^{-1}·s^{-1}) [34].

Figure 4 shows the k_{app} of electrochemical oxidation of the mixtures of drugs (5-FU/CP and IF/CP) in different types of electrolytes.

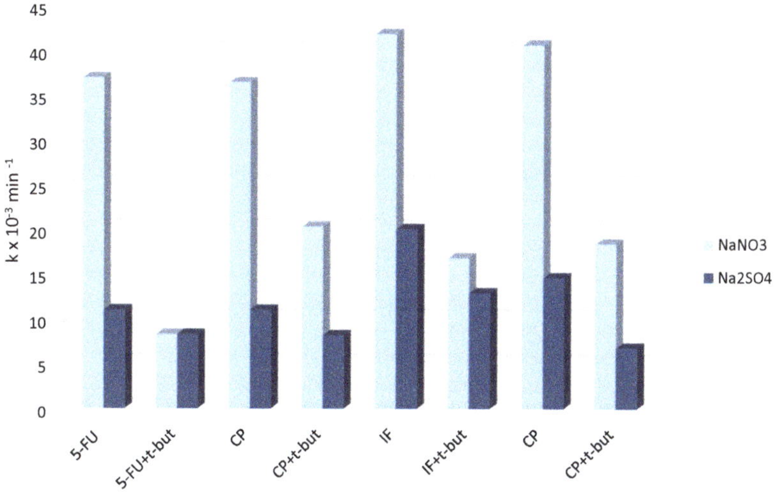

Figure 4. The k_{app} of drugs electrochemical oxidation in their mixtures (5-FU/CP and of IF/CP) in different types of electrolyte with and without t-butanol.

The degradation rate mixtures of drugs in the $NaNO_3$ electrolyte were significantly higher than in Na_2SO_4. The values of k_{app} were similarly demonstrating that the role of •OH radical in drugs degradation is powerful and the oxidation process is non-selective in this electrolyte. The oxidation process with *t*-but addition to the mixture of 5-FU/CP showed that the degradation rate was more inhibited for 5-FU than for CP. As was expected, the scavenging effect of •OH radicals by t-but in CP/IF mixture resulted in the same suppression of both drugs' degradation rates. The order of reaction rate of drugs with •OH radicals can be estimated as follows: 5-FU < CP = IF.

In Na_2SO_4 electrolyte, the value of k_{app} for IF degradation was higher than values of k_{app} for CP and 5-FU oxidation, (k_{app}5-FU = k_{app}CP < k_{app}IF). Moreover, the degradation of drugs in this electrolyte was less suppressed by OH• scavenger (*t*-but) than in $NaNO_3$. The results suggested that in Na_2SO_4, •OH radicals participated in the drug degradation concomitantly with SO_4^-• radicals. Long-life SO_4^-• occurred in the bulk solution as a product of SO_4^{2-} ions with the •OH radicals reaction. Consequently, •OH radicals were partially consumed. As it was mentioned previously, SO_4^-• radicals react more selectively with organic matter and oxidize drugs less rapidly as •OH radicals. SO_4^-• are also less sensitive on the scavenging by *t*-butanol.

In the presence of Cl^- ions released in CP and IF degradation, SO_4^-• radicals could react with Cl^- to produce secondary radicals of Cl• and then Cl_2^-• according to the following reactions (Equations (5) and (7)).

$$SO_4\bullet^- + Cl^- \rightarrow SO_4^{2-} + Cl\bullet, k = 3.2 \times 10^8 M^{-1} \cdot s^{-1} \quad (5)$$

$$Cl^- + \bullet OH \rightarrow HOCl, k = 6.1 \times 10^9 M^{-1} \cdot s^{-1} \quad (6)$$

$$Cl\bullet + Cl\bullet \rightarrow Cl_2^-\bullet, k= 6.5 \times 10^9 M^{-1} \cdot s^{-1} \quad (7)$$

The chlorine active species (HClO, Cl•, Cl_2•$^-$) can also be formed by direct electrolysis of chloride ions [35] released during drugs degradation. The behaviour of cytostatic drugs in mixtures in the Na_2SO_4 electrolyte confirms its different reactivity with SO_4^-• and/or chlorine active species in contrast to •OH radicals.

To improve the speculation that IF, CP and 5-FU react with chlorine active species, the influence of Cl^- ions on the drugs degradation in $NaNO_3$ and Na_2SO_4 electrolytes was compared (Figure 5). In

$NaNO_3$, chlorides inhibited the degradation of studied drugs, due to the scavenging of OH• radicals and the formation of weaker and more selective oxidants such as Cl• and Cl_2•. In sulphate, chlorides can promote the SO_4•$^-$ induced oxidation of cytostatic drugs. In Na_2SO_4, this can happen because the reaction rate of CP degradation was less inhibited in nitrates and more accelerated in sulphates than the degradation of 5-FU (Figure 5), while 5-FU removal was significantly suppressed by chlorides in the $NaNO_3$ electrolyte and slightly accelerated by these ions in Na_2SO_4. The conclusion from the results was that the chlorine active species significantly participated in CP degradation, while favourable radicals for 5-FU oxidation were •OH and SO_4^-• rather than chlorine active species. Generated BDD electrode chlorine active species play the main role in NH_4^+ released from 5-FU oxidation to gaseous products (see Section 2.2.1). The CP/IF mixture's oxidation in the presence of chlorides in different electrolytes confirmed that IF and CP were oxidized by chlorine active species, but their reactivity was likely different.

In order to study the influence of the type of radicals on the mineralization process, the TOC values in effluents of electrolysis in $NaNO_3$ and Na_2SO_4 electrolytes were compared. After 3 h of oxidation, TOC removal in $NaNO_3$ was 50.8 and 48.0% for CP/IF and 5-FU/CP mixtures, respectively, while TOC removal in these mixtures in Na_2SO_4 reached a similar TOC efficiency after 6 h of electrolysis. Due to the high $NaNO_3$ concentration (42 mM), TN removal efficiency by •OH radical was not possible to examine in the conditions employed. Based on the results, the •OH/SO_4^-• and chlorine active species were effective in cytostatic drugs mineralisation. However, the time needed to obtain the same mineralization efficiency by •OH/SO_4^-• and chlorine active species in Na_2SO_4 electrolyte as that reached by •OH radicals in $NaNO_3$ electrolyte required a twice longer time of electrolysis. In waterbodies, the sulphates and chlorides naturally exist, while nitrates in higher concentration are pollutants and responsible for the eutrophication. Therefore, knowledge about the mechanism of cytostatic drugs degradation in the presence of sulphates, chlorides and phosphates is necessary.

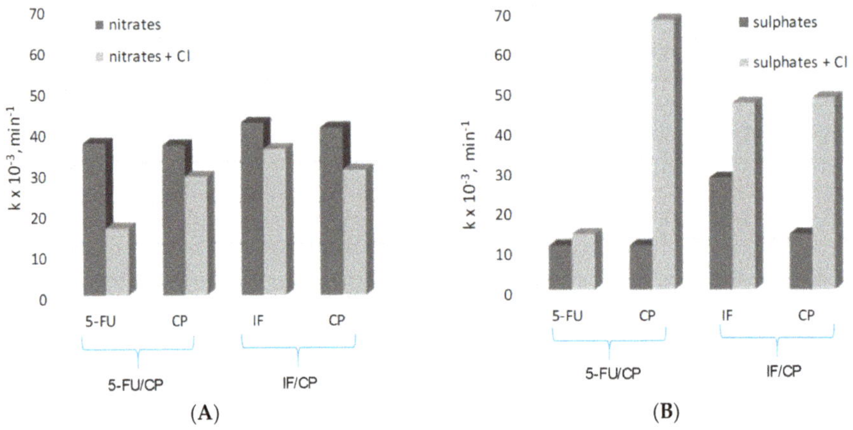

Figure 5. The influence of chlorides on the degradation drugs in 5-FU/CP and CP/IF mixtures in $NaNO_3$ (**A**) and Na_2SO_4 (**B**) electrolytes.

2.4. Identification of CP and IF Intermediates in Mixture in Na_2SO_4

The intermediates were recorded by LC/MS technique after 2h of CP/IF mixture electrolysis in the Na_2SO_4 electrolyte. The intermediates in the CP/IF mixture were found in a higher number but a smaller amount than in the single-drug solutions with a mass number in the range from $[M + H]^+ = 165$ to $[M + H]^+ = 311$. The intermediates, which were identified both in single-drug solutions and in the mixture were with the molecular weights $[M + H]^+ = 165, 199, 259, 277$ and 293. Similar intermediates were reported by other research groups which tested the CP and/or IF degradation by photocatalytic or electrochemical AOPs [10,15,36,37]. Independent of the type of CP and IF solution (single or mixture), the intermediates with mass numbers $[M + H]^+$ = 249, 259, 277 and 293 were recognized. The aldophosphamide ($[M + H]^+$ = 277) and carboxyphosphamide (($[M + H]^+$ = 293) were shown to be the human metabolites of CP and IF. The fact is that carboxyphosphamide had little or no anticancer activity [38]. The structures of possible intermediates identified in the CP/IF mixture are presented in Table 2.

Table 2. The structures of possible intermediates identified in the CP/IF mixture comparing to single-drug solutions.

Drug			
Ifosfamide	**Cyclophosphamide**	**Structure of Intermediate**	**Molecular ion (Fragmentation Ions)**
	✓		$[M + H]^+$ = 293 (275, 227)
✓	✓		$[M + H]^+$ = 277
✓	✓		$[M + H]^+$ = 259 (140)
✓	✓		$[M + H]^+$ = 249 (164)
✓			$[M + H]^+$ = 199 (171; 79)
✓	✓	?	$[M + H]^+$ = 165

3. Materials and Methods

3.1. Chemicals

The CP, IF and 5-FU standards were purchased from Sigma-Aldrich (Steinheim, Germany); Acetonitrile (ACN) and the sodium sulphate (supported electrolyte) were obtained from P.P.H. Stanlab (Lublin, Poland). Sodium chloride and monopotassium phosphate were purchased from POCH S.A. (Gliwice, Poland).

3.2. Voltammetric Measurements

Cyclic voltammetry analysis was accomplished with the BDD (Adamant Technologies, B/C radio about 500 ppm with a surface of 1 cm^2) as the working electrode, Ag/AgCl as the reference electrode, and platinum wire as the counter electrode in a solution of 42 mM Na_2SO_4 containing 50 $mg \cdot L^{-1}$ of the drug. The experiment was carried out in a 10 mL electrochemical cell, at room temperature. Cyclic voltammograms were performed with a PGSTAT 30 Autolab potentiostat/galvanostat.

3.3. Electrochemical Systems

Degradation experiments were performed in a recirculating split-flow batch reactor divided into anolyte and catholyte compartments. The cell contains two parallel electrodes, boron-doped diamond (polycrystalline BDD film on monocrystalline *p*-type Si wafer, Adamant Technologies, B/C radio about 500 ppm) used as an anode, and stainless steel (SS) as a cathode. Both were flat (11.7 mm diameter × 2 mm) with an inert-electrode distance of 3.5 cm. Two peristaltic pumps were used to supply the cytostatic drug solution/mixture at constant volumetric flow rate. The total volume of both the anolyte and catholyte was 100 mL. In the divided cell, the anodic and cathodic compartments were separated into two equal spaces by a cation exchange membrane Nafion® 424 to allow the passage of protons. The current was passed through the cell with a potentiostat/galvanostat. Experiments were performed at room temperature 25 °C ± 3 °C. The pH was monitored with the pH-meter. The reactor and glassware used were protected from light. The electrochemical system applied in the study is presented in Figure 6. All experiments were performed in duplicate. The optimal experimental conditions were previously investigated [20]. Electrochemical degradation of the mixtures (IF/CP and CP/5-FU) in the supported electrolyte solution (42 mM Na_2SO_4 or 42mM $NaNO_3$) was carried out for the initial drug concentration of 25 mg·L^{-1}. A comparison between the oxidative processes occurring in the mixtures of these medicaments and single-drug solutions was conducted. The initial concentrations of each drug in a single-drug solution were 25 and 50 mg·L^{-1}. Subsequently, the degradation experiment was performed in the same operating conditions with the mixture of the drugs, adding 5 and 10 mg·L^{-1} of PO_4^{3-} or 10 and 100 mg·L^{-1} of Cl^-, or with a mixture of drugs added to the actual effluent from the wastewater treatment plan, as seen in Table 3. The following operating conditions were used in the assays: current density of 15 mA·cm^{-2} and flow rate of 13 L·h^{-1}. For all kinetics experiments, the drug concentration was monitored by means of the HPLC-UV analysis. The mineralisation process was estimated based on the total organic carbon (TOC) removal and total nitrogen (TN) removal.

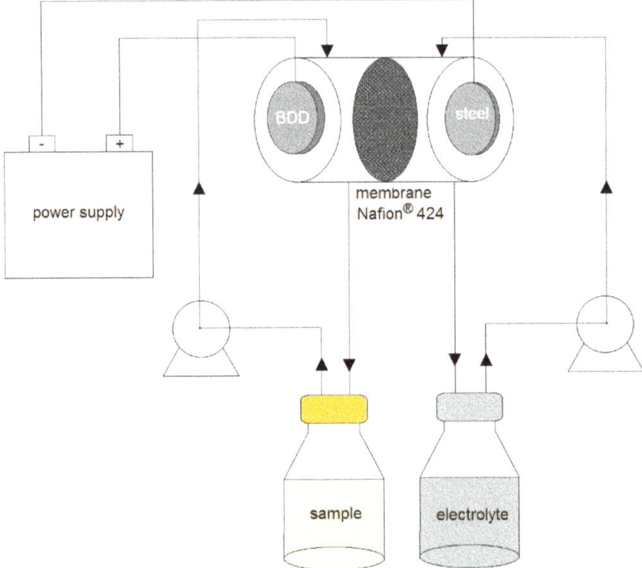

Figure 6. The electrochemical set-up.

The mineralization current efficiency (MCE) according to El-Ghenymy et al. [39] for mixtures of drugs was estimated from Equation (8):

$$\text{MCE\%} = \frac{n\text{FV}(\Delta\text{TOC})}{4.32 \times 10^7\ \text{mIt}} \times 100\% \quad (8)$$

where n is the number of electrons consumed per one molecule of drug assuming the total mineralization (based on the reactions Equation (9) for 5-FU and Equation (10) for IF, CF and molar ratio drugs in the mixture), F is the Faraday constant (96,487 C mol^{-1}), V (dm^3) is the solution volume, (ΔTOC) (mg L^{-1}) is the difference between TOC before and after electrochemical process, 4.32×10^7 is a factor to homogenize units, m is the number of carbon atoms of drugs, I (A) is the applied current, and t (h) is electrolysis time.

$$C_4H_3N_2O_2F + 6H_2O \rightarrow 4CO_2 + 2NH_4^+ + F^- + 7H^+ + 8e^- \quad (9)$$

$$C_7H_{15}N_2O_2Cl_2P + 8H_2O \rightarrow 7CO_2 + 2NH_4^+ + 2Cl^- + PO_4^{3-} + 23H^+ + 20e^- \quad (10)$$

Table 3. Characteristics of effluents from municipal wastewater treatment plant (MWWTP) with mechanical and biological stages in Gdańsk in Poland.

Parameter	Units	Value
pH		6.8
COD	mgO_2/L	62
N-NH_4^+	mgN/L	0.225
NO_3^-	mgN/L	4.463
Cl^-	mg/L	62
PO_4^{3-}	mgP/L	0.159
SO_4^{2-}	mg/L	89
Conductivity	µS/cm	435

3.4. Analytical Methods

The concentration of selected drugs was measured by HPLC (Perkin Elmer, Series 200, Shimadzu Europa GmbH, Dulsburg, Germany). The set-up was equipped with a UV detector (SPD-MZOA Shimadzu Dulsburg, Germany) and a C-18 column (Phenomenex 150 × 4.6 nm, 2.6 µm). A detailed description of the analytical methods of CP, IF and 5-FU determination has been reported in our previous works [10].

The values of chemical oxygen demand were measured using standard cuvette tests (HACH) and an Odyssey spectrophotometer. The concentration of chloride was measured by argentometric method, and fluoride (444-49), phosphate (Cadmium reduction), ammonium (Nessler method) and nitrate (21061-69) were measured using cuvette tests (HACH). Additionally, the reaction's intermediate products were identified with the use of the LC/MS system [37]. The chromatographic parameters were identical with those used during the HPLC-UV analysis. The MS analysis was conducted using positive and negative mode electrospray ionization (ESI) over a mass scan range of 50–350 m/z (target mass 250 m/z) under the conditions described in our previous work [37]. TOC and total amount of nitrogen compounds (TN) were measured using TOC/TN analyzer (Shimadzu, TOC-L CSH Dulsburg, Germany).

4. Conclusions

In the present work, the electrochemical oxidation of three cytostatic drugs and their mixtures in a recirculating split-flow batch reactor equipped with a BDD anode was investigated. The oxidation of cytostatic drugs was performed in an anodic compartment separated from the cathodic compartment by a cation-exchanged membrane. The values of k_{app} for the single-drug solution and for mixtures of cytostatic drugs in Na_2SO_4 as electrolyte were in the following order: k_{app} 5-FU < k_{app} CP < k_{app}

IF. The degradation of drugs in their mixtures showed that the PO_4^{3-} has no significant effect on the drugs' degradation, while Cl^- mainly accelerated this process. Based on the TOC removal and TN conversion to gaseous products, it was found that the degradation pathway of IF was different than that of CP. The organic intermediates found after 6 h of electrolysis were also different for these drugs.

Comparing the results in $NaNO_3$ and Na_2SO_4 electrolytes, it was suggested that the different oxidising species with different levels of reactiveness participated in cytostatic drugs degradation. In Na_2SO_4, 5-FU was mainly oxidized by OH• and SO_4^-• radicals, while chlorine active species simultaneously with OH• and SO_4^-• entities participated in the CP and IF degradation and nitrogen conversion to gaseous products. Given that Cl^- is abundant in natural waters, the involvement of chlorine active species in the electro-oxidation fate of drugs should not be neglected. This was confirmed by the high mineralization and efficiency of drugs removal in the effluent from WWTP, and this is beneficial from the practical point of view. This study is helpful in understanding the fundamental reaction mechanism, as well as the effects of natural water constituents on the kinetics and mechanisms of electrochemical oxidation of cytostatic drugs in their mixtures. Further studies should address the effects of composition of contaminated water (by natural organic matter, carbonates and other anions and cations) on the transformation of cytostatic drugs and the examination of effluents toxicity obtained from AOPs.

Supplementary Materials: The following are available online. Figure S1: Cyclic voltammograms of IF, CP and 5-FU (25 mg/L) in 42 mM Na_2SO_4 (pH = 6.6), BDD as a working electrode, counter electrode (CE)—Pt, scan rate = 100 mV·s^{-1}, T = 20 ± 2 °C.

Author Contributions: E.M.S. conceptualized and designed the experiments; S.A.O.-C., A.P., A.B.-G. and P.W. performed the experiments and analyzed the data; E.M.S wrote the paper; and polished the paper; E.M.S. and A.P. acquired funding for the research. All authors read and approved the final manuscript.

Funding: This research was supported by the Polish Ministry of Research and Higher Education, Poland under the Grant DS530-8626-D596-18 and DS530-8626-D596-19.

Acknowledgments: Thank you very much for the help in lab Katarzyna Bachlińska and Agnieszka Fiszka Borzyszkowska.

Conflicts of Interest: The authors declare no conflicts of interest.

References

1. Huitema, A.D.R.; Tibben, M.M.; Kerbusch, T.; Jantien Kettenes-Van Den Bosch, J.; Rodenhuis, S.; Beijnen, J.H. Simple and selective determination of the cyclophosphamide metabolite phosphoramide mustard in human plasma using high-performance liquid chromatography. *J. Chromatogr. B Biomed. Sci. Appl.* **2000**, *745*, 345–355. [CrossRef]
2. Zhang, J.; Chang, V.W.C.; Giannis, A.; Wang, J.Y. Removal of cytostatic drugs from aquatic environment: A review. *Sci. Total Environ.* **2013**, *445–446*, 281–298. [CrossRef] [PubMed]
3. Pieczyńska, A.; Fiszka Borzyszkowska, A.; Ofiarska, A.; Siedlecka, E.M. Removal of cytostatic drugs by AOPs: A review of applied processes in the context of green technology. *Crit. Rev. Environ. Sci. Technol.* **2017**, *47*, 1282–1335. [CrossRef]
4. Franquet-Griell, H.; Medina, A.; Sans, C.; Lacorte, S. Biological and photochemical degradation of cytostatic drugs under laboratory conditions. *J. Hazard. Mater.* **2017**, *323*, 319–328. [CrossRef] [PubMed]
5. Giri, A.S.; Golder, A.K. Decomposition of drug mixture in Fenton and photo-Fenton processes: Comparison to singly treatment, evolution of inorganic ions and toxicity assay. *Chemosphere* **2015**, *127*, 254–261. [CrossRef]
6. Siedlecka, E.M.; Stepnowski, P. Decomposition rates of methyl tert-butyl ether and its by-products by the Fenton system in saline wastewaters. *Sep. Purif. Technol.* **2006**, *52*, 317–324. [CrossRef]
7. Lai, W.W.-P.; Lin, H.H.-H.; Lin, A.Y.-C. TiO_2 photocatalytic degradation and transformation of oxazaphosphorine drugs in an aqueous environment. *J. Hazard. Mater.* **2015**, *287*, 133–141. [CrossRef]
8. Yu-Chen Lin, A.; Han-Fang Hsueh, J.; Andy Hong, P.K. Removal of antineoplastic drugs cyclophosphamide, ifosfamide, and 5-fluorouracil and a vasodilator drug pentoxifylline from wastewaters by ozonation. *Environ. Sci. Pollut. Res.* **2015**, *22*, 508–515.

9. Siedlecka, E.M.; Stepnowski, P. Effect of chlorides and sulfates on the performance of a Fe^{3+}/H_2O_2 Fenton-like system in the degradation of MTBE and its by-products. *Water Environ. Res.* **2007**, *79*, 2318–2324. [CrossRef]
10. Fabiańska, A.; Ofiarska, A.; Fiszka-Borzyszkowska, A.; Stepnowski, P.; Siedlecka, E.M. Electrodegradation of ifosfamide and cyclophosphamide at BDD electrode: Decomposition pathway and its kinetics. *Chem. Eng. J.* **2015**, *276*, 274–282. [CrossRef]
11. Ofiarska, A.; Pieczyńska, A.; Fiszka-Borzyszkowska, A.; Stepnowski, P.; Siedlecka, E.M. Pt-TiO_2-assisted photocatalytic degradation of the cytostatic drugs ifosfamide and cyclophosphamide under artificial sunlight. *Chem. Eng. J.* **2016**, *285*, 417–427. [CrossRef]
12. Feng, L.; van Hullebusch, E.D.; Rodrigo, M.A.; Esposito, G.; Oturan, M.A. Removal of residual anti-inflammatory and analgesic pharmaceuticals from aqueous systems by electrochemical advanced oxidation processes. A review. *Chem. Eng. J.* **2013**, *228*, 944–964. [CrossRef]
13. Siedlecka, E.M.; Fabiańska, A.; Stolte, S.; Nienstedt, A.; Ossowski, T.; Stepnowski, P.; Thöming, J. Electrocatalytic oxidation of 1-butyl-3-methylimidazolium chloride: Effect of the electrode material. *Int. J. Electrochem. Sci.* **2013**, *8*, 5560–5574.
14. Pieczyńska, A.; Ofiarska, A.; Fiszka-Borzyszkowska, A.; Białk-Bielińska, A.; Stepnowski, P.; Stolte, S.; Siedlecka, E.M. A comparative study of electrochemical degradation of imidazolium and pyridinium ionic liquids: A reaction pathway and ecotoxicity evaluation. *Sep. Purif. Technol.* **2015**, *156*, 522–534. [CrossRef]
15. Coria, G.; Nava, J.L.; Carreño, G. Electrooxidation of diclofenac in synthetic pharmaceutical wastewater using an electrochemical reactor equipped with a boron doped diamond electrode. *Mex. J. Chem. Soc.* **2014**, *58*, 303–308. [CrossRef]
16. Brocenschi, R.F.; Rocha-Filho, R.C.; Bocchi, N.; Biaggio, S.R. Electrochemical degradation of estrone using a boron-doped diamond anode in a filter-press reactor. *Electrochim. Acta* **2016**, *197*, 186–193. [CrossRef]
17. Kobayashi, T.; Hirose, J.; Sano, K.; Hiro, N.; Ijiri, Y.; Takiuchi, H.; Tamai, H.; Takenaka, H.; Tanaka, K.; Nakano, T. Evaluation of an electrolysis apparatus for inactivating antineoplastic in clinical wastewater. *Chemosphere* **2008**, *72*, 659–665. [CrossRef]
18. Queiroza, N.M.P.; Sirésb, I.; Zantac, C.L.P.S.; Tonholoc, J.; Brillas, E. Removal of the drug procaine from acidic aqueous solutions using aflowreactor with a boron-doped diamond anode. *Sep. Purif. Technol.* **2019**, *216*, 65–73. [CrossRef]
19. Radjenovic, J.; Petrovic, M. Sulfate-mediated electrooxidation of X-ray contrast media on boron-doped diamond anode. *Water Res.* **2016**, *94*, 128–135. [CrossRef]
20. Ochoa-Chavez, A.S.; Pieczyńska, A.; Fiszka Borzyszkowska, A.; Espinoza-Montero, P.J.; Siedlecka, E.M. Electrochemical degradation of 5-FU using a flow reactor with BDD electrode: Chomparison of two electrochemical systems. *Chemosphere* **2018**, *201*, 816–825. [CrossRef]
21. Garcia-Segura, S.; Mostafa, E.; Baltruschat, H. Could NOx be released during mineralization of pollutants containing nitrogen by hydroxyl radical? As certaining the release of N-volatile species. *Appl. Catal. B Environ.* **2017**, *207*, 376–384. [CrossRef]
22. Weiss, E.; Groenen-Serrano, K.; Savall, A.; Comninellis, C. A kinetic study of the electrochemical oxidation of maleic acid on boron doped diamond. *J. App. Electrochem.* **2007**, *37*, 41–47. [CrossRef]
23. Kapałka, A.; Fóti, G.; Comninellis, C. Investigation of the Anodic Oxidation of Acetic Acid on Boron-Doped Diamond Electrodes. *J. Electrochem. Soc.* **2008**, *155*, 27–32. [CrossRef]
24. Da Costa, T.F.; Santos, J.E.L.; da Silva, D.R.; Martinez-Huitle, C.A. BDD-Electrolysis of Oxalic Acid in Diluted Acidic Solutions. *J. Braz. Chem. Soc.* **2019**, *30*, 1541–1547. [CrossRef]
25. Sánchez-Carretero, A.; Sáez, C.; Cañizares, P.; Rodrigo, M.A. Electrochemical production of perchlorates using conductive diamond electrolysis. *Chem. Eng. J.* **2011**, *166*, 710–714. [CrossRef]
26. Mostafa, E.; Reinsberg, P.; Garcia-Segura, S.; Baltruschat, H. Chlorine species evolution during electrochlorination on boron-doped diamond anodes: In-situ electrogeneration of Cl2, Cl2O and ClO2. *Electrochimica Acta* **2018**, *281*, 831–840. [CrossRef]
27. Wu, Y.; Yang, Y.; Liu, Y.; Zhang, L.; Feng, L. Modelling study on the effects of chloride on the degradation of bezafibrate and carbamazepine in sulfate radical-based advanced oxidation processes: Conversion of reactive radicals. *Chem. Eng. J.* **2019**, *258*, 1332–1341. [CrossRef]
28. Cañizares, P.; Sáez, C.; Sánchez-Carretero, A.; Rodrigo, M.A. Synthesis of novel oxidants by electrochemical technology. *J. Appl. Electrochem.* **2009**, *39*, 2143–2149. [CrossRef]

29. Burgos-Castillo, R.C.; Sirés, I.; Sillanpää, M.; Brillas, E. Application of electrochemical advanced oxidation to bisphenol A degradation in water. Effect of sulfate and chloride ions. *Chemosphere* **2018**, *194*, 812–820. [CrossRef]
30. Pérez, G.; Saiz, J.; Ibañez, R.; Urtiaga, A.; Ortiz, I. Assessment of the formation of inorganic oxidation by-products during the electrocatalytic treatment of ammonium from landfill leachates. *Water Res.* **2012**, *46*, 2579–2590. [CrossRef]
31. Garcia-Segura, S.; Mostafa, E.; Baltruschat, H. Electrogeneration of inorganic chloramines on boron-doped diamond anodes during electrochemical oxidation of ammonium chloride, urea and synthetic urine matrix. *Water Res.* **2019**, *160*, 107–117. [CrossRef] [PubMed]
32. Mahdi Ahmed, M.; Barbati, S.; Doumenq, P.; Chiron, S. Sulfate radical anion oxidation of diclofenac and sulfamethoxazole for water decontamination. *Chem. Eng. J.* **2012**, *197*, 440–447. [CrossRef]
33. Liang, C.; Wang, Z.S.; Bruell, C.J. Influence of pH on persulfate oxidation of TCE at ambient temperatures. *Chemosphere* **2007**, *66*, 106–113. [CrossRef] [PubMed]
34. Hazime, R.; Nguyen, Q.H.; Ferronato, C.; Slavador, A.; Jaber, F.; Chovelon, J.-M. Comparative study of imazalil degradation in three systems: UV/TiO_2, UV/$K_2S_2O_8$ and UV/TiO_2/K_2SO_8. *App. Catal. B* **2014**, *144*, 286–291. [CrossRef]
35. Cho, K.; Qu, Y.; Kwon, D.; Zhang, H.; Cid, C.A.; Aryanfar, A.; Hoffmann, M.R. Effects of anodic potential and chloride ion on overall reactivity in electrochemical reactors designed for solar-powered wastewater treatment. *Environ. Sci. Technol.* **2014**, *48*, 2377–2384. [CrossRef]
36. Siedlecka, E.M.; Ofiarska, A.; Fiszka Borzyszkowska, A.; Białk-Bielińska, A.; Stepnowski, P.; Pieczyńska, A. Cytostati drug removal using electrochemical oxidation with BDD electrode: Degradation pathway and toxicity. *Water Res.* **2018**, *144*, 235–245. [CrossRef]
37. Česen, M.; Kosjek, T.; Busetti, F.; Kompare, B.; Heath, E. Human metabolites and transformation products of cyclophosphamide and ifosfamide: Analysis, occurrence and formation during abiotic treatments. *Environ. Sci. Pollut. Res.* **2016**, *23*, 11209–11223. [CrossRef]
38. Struck, R.F.; Kirk, M.C.; Mellett, L.B.; el Dareer, S.; Hill, D.L. Urinary metabolites of the antitumor agent cyclophosphamide. *Mol Pharmacol.* **1971**, *7*, 519–529.
39. El-Ghenymy, A.; Rodríguez, R.M.; Arias, C.; Centellas, F.; Garrido, J.A.; Cabot, P.L.; Brillas, E. Electro-Fenton and photoelectro-Fenton degradation of the antimicrobial sulfamethazine using a boron-doped diamond anode and an airdiffusion cathode. *J. Electroanal. Chem.* **2013**, *701*, 7–13. [CrossRef]

Sample Availability: Samples of the compounds are not available from the authors.

Article

Poultry Farms as a Potential Source of Environmental Pollution by Pharmaceuticals

Katarzyna Wychodnik [1], Grażyna Gałęzowska [1], Justyna Rogowska [1], Marta Potrykus [1], Alina Plenis [2] and Lidia Wolska [1,*]

1 Department of Environmental Toxicology, Faculty of Health Sciences with Institute of Maritime and Tropical Medicine, Medical University of Gdansk, Debowa 23A St., 80-204 Gdańsk, Poland; k.wychodnik@gumed.edu.pl (K.W.); grazyna.galezowska@gumed.edu.pl (G.G.); justyna.rogowska@gumed.edu.pl (J.R.); marta.potrykus@gumed.edu.pl (M.P.)

2 Department of Pharmaceutical Chemistry, Faculty of Pharmacy, Medical University of Gdansk, Hallera 107 St., 80-416 Gdansk, Poland; aplenis@gumed.edu.pl

* Correspondence: lidia.wolska@gumed.edu.pl; Tel.: +48-58-349-19-39

Academic Editor: Jolanta Kumirska
Received: 7 February 2020; Accepted: 23 February 2020; Published: 25 February 2020

Abstract: Industrial poultry breeding is associated with the need to increase productivity while maintaining low meat prices. Little is known about its impact on the environment of soil pollution by pharmaceuticals. Breeders routinely use veterinary pharmaceuticals for therapeutic and preventive purposes. The aim of this work was to determine the influence of mass breeding of hens on the soil contamination with 26 pharmaceuticals and caffeine. During two seasons—winter and summer 2019—15 soil samples were collected. Liquid extraction was used to isolate analytes from samples. Extracts were analyzed using ultra-high performance liquid chromatography coupled with tandem mass spectrometry detection (UPLC-MS/MS). The results showed the seasonal changes in pharmaceutical presence in analyzed soil samples. Ten pharmaceuticals (metoclopramide, sulphanilamide, salicic acid, metoprolol, sulphamethazine, nimesulide, carbamazepine, trimethoprim, propranolol, and paracetamol) and caffeine were determined in soil samples collected in March, and five pharmaceuticals (metoclopramide, sulphanilamide, sulphamethazine, carbamazepine, sulfanilamid) in soil samples collected in July. The highest concentrations were observed for sulphanilamide, in a range from 746.57 ± 15.61 ng/g d.w to 3518.22 ± 146.05 ng/g d.w. The level of bacterial resistance to antibiotics did not differ between samples coming from intensive breeding farm surroundings and the reference area, based on antibiotic resistance of 85 random bacterial isolates.

Keywords: pharmaceuticals; soil; poultry farms; ultra-high performance liquid chromatography; antibiotics, antibiotic resistance

---

1. Introduction

Pharmaceuticals are a group of compounds designed to elicit specific biological effects at relatively low concentrations. These compounds can be used to treat both humans and animals. According to Regulation (EU) 2019/6 of the European Parliament and of the Council of 11 December, 2018, on veterinary medicinal products, and repealing Directive 2001/82/EC (OJ L 4, 7.1.2019, p. 43–167), "veterinary medicinal product" means any substance or combination of substances that has properties for treating or preventing disease in animals; or its purpose is to be used in, or administered to animals, with a view to restoring, correcting, or modifying physiological functions by exerting a pharmacological, immunological, or metabolic action. The definition also includes pharmaceutical products used on animals to make a medical diagnosis and in euthanasia of animals [1]. More than 2000 veterinary pharmaceutical products are manufactured from 400 active chemical ingredients to treat various species of animals [2]. Veterinary pharmaceuticals prevent and treat disease and increase the efficiency of food production [3]. These pharmaceuticals belong to several pharmacological categories: antiparasitics (ectoparasiticides, endectocides, and endoparasiticides, including antiprotozoals and anthelmintics), antimicrobials, hormones, antifungals, anti-inflammatory (steroidal and non-steroidal drugs), anaesthetics, tranquilizers, sedatives, bronchodilators, antacids, diuretics, and emetics [4]. The largest group are antimicrobial agents. Antimicrobials are compounds that can kill or inhibit the growth of microorganisms (bacteria, archaea, protozoa, microalgae, and fungi) [5]. Some of these drugs also show growth-promoting effects and are commonly misused for this reason [6].

Pharmaceuticals, massively used in veterinary medicine, can end up in the environment via several routes. Consumed drugs, including antimicrobial agents, are continuously discharged into the natural ecosystems via excretion (urine and feces) after a short time of residence in animal organisms. That main sources of veterinary pharmaceuticals in the environment are intensive livestock activities and farming practices [7]. The agglomeration of large numbers of birds in a small area causes the need for pharmaceuticals, including anti-microbial agents, to prevent and treat microbial infections, as well as to increase feed efficiency [8]. Moreover, to prevent and control contagious poultry diseases vaccines are used. This results in lower bird mortality and, thus, increases the profitability of breeding.

The production of poultry results in hatchery wastes, manure (bird excrement), litter (bedding materials such as sawdust, wood shavings, straw, and peanut or rice hulls), and on-farm mortalities [9]. Intensive livestock farming practices that are used to breed thousands of poultry, often in small areas, face problems with safe and proper disposal of tons of animal excreta produced every day [2]. The practice of using manure for soil fertilization purposes is the major contributor of veterinary pharmaceutical contamination in the environment [10]. These may cause a raise in antibiotic resistance of microbial strains isolated from chickens. Braykov et al. [11] showed that the number of resistant strains isolated from poultry bred in intensive breeding farms is much larger than the number of resistant strains isolated from henhouses. Poultry farms are the biggest emitters of dust, microorganisms, and organic compounds (including pharmaceuticals) in manure, litter, dust, and air [12]. Odor emissions, consisting of a large number of compounds, including ammonia, volatile organic compounds (VOCs), and hydrogen sulfide, adversely affect the life of people living in the vicinity of poultry farms [13]. Dust is one of the components present in poultry production; it originates from poultry residues, molds, and feathers, and is biologically active as it contains microorganisms, some of which may be pathogens [14]. Consequently, fertilization with antibiotics containing animal manure, dust, and sewage sludge seems to be the likely pathway for the release of antibiotics into soil [2]. Moreover, spiking animal manure with antibiotics and applying it on soil showed changes in microbial soil composition and rising antibiotic resistance of soil microbial community [15]. Additionally, antibiotics can migrate from soil to groundwater, or into plants; thus, can cause negative consequences for human health [16].

Commercial livestock production has increased rapidly in the past few years, which has promoted the construction of large production units. The world leader in intensive poultry breeding is the United States of America. In Europe, Poland is the main shareholder in terms of poultry production. In addition, it should be noted that gross domestic production of poultry meat in Poland almost doubled between 2012 and 2017 (from 1,712,000 to 3,110,000 tons carcass weight) [17]. Poland is the only European Union (EU) country in which such significant growth has been observed. The number of farms involved in industrial animal husbandry and the scale of the impact on the environment is different in individual regions of Poland. The results of the audit, carried out by the Supreme Audit Office in 2011–2013, showed the lack of proper supervision by administrative authorities over the functioning of animal farms, ineffective cooperation between the Veterinary Inspection and the Inspection of Environmental Protection in the field of control, and a lack of legislation aimed at solving the problem of odors caused by farms [18]. Veterinary Inspection bodies, and to some extent, the State Sanitary Inspection, are obliged to constantly monitor the presence of prohibited substances and antibiotic resistance in food products of animal origin, as well as to supervise the marketing and use of pharmaceuticals in the process of animal breeding [19]. The audit carried out by the Supreme Audit

Office in 2015–2016 regarding the use of antibiotics in animal production showed that the scale of use of antibiotics in animal production is not exactly known in Poland. This is because the Ministry of Agriculture and Rural Development only has data provided by pharmaceutical wholesalers on the number of veterinary antibiotics sold on the domestic market. At the same time, it was found that the lack of the above-mentioned data is due to the lack of a nationwide system for collecting, monitoring, and comparing data on the use of antibiotics at farm level. The problem was also related to deficiencies in the documentation regarding animal treatment [19]. In 2015–2017, the Supreme Audit Office carried out supervision over the marketing and use of products containing anabolic, hormonal, narcotic, and psychotropic substances in the treatment of animals in Poland. As a result, it was found that both the Veterinary Inspection and Pharmaceutical Inspection activities are insufficient to limit the risk of using these substances contrary to their intended use. The audit found use of drugs not in accordance with their intended use, discrepancies as to the drugs used, or their dose [20].

In effect, an unidentified number of pharmaceuticals can migrate into the environment, including soils. Therefore, it is necessary to study the pharmaceutical content in soils around poultry farms.

The aim of the study was to determine whether mass breeding of hens influences the contamination level of the surrounding soil with pharmaceuticals. The choice of analyzed compounds resulted from the information obtained from the breeder and the review of literature on the routine use of pharmaceuticals in the process of intensive breeding. The first stage of research was to develop a procedure for isolating and determining the selected pharmaceuticals. In the next step, the method was validated. In the last stage, the content of pharmaceuticals in soil samples taken around the poultry farmhouse was determined. Additionally, the prevalence of antibiotic resistant bacterial strains in the soil samples taken around the poultry house was verified.

2. Materials and Methods

2.1. Study Area

2.1.1. Weather Conditions

The farm is located in one of the communes in the northeastern part of Poland. Geographic location translates into specific climate conditions. The warmest and the sunniest month is July (average temperature 17 °C). It is coldest in January (average temperature −1.7 °C). The highest rainfall is recorded in July (85 mm) and the lowest in March (33 mm). Winds from the west and southwest sectors dominate in this area. The average wind speed is 3 m/s, with the highest winds occurring in the winter and the lowest in the summer [21].

In March, on the day of sampling, it was sunny and rainless and the humidity was 55%. The air temperature was around 6–7 °C. The wind was negligible (2–4 m/s), southwest, and spread along the shed wall. In July, on the day of sampling, the weather was mostly sunny, with temporary cloud cover, rainless, and humidity was 47%–58%. The air temperature was between 17 and 21 °C. West wind gusts up to 8.3 m/s.

2.1.2. Poultry House

The whole farm consists of five poultry houses and two outbuildings. Four of the poultry houses are large with an area of 1700 m^2, and one is smaller, at a distance around 80 m from them. The impact of the smaller poultry house, on the soil, was the object of this study (Figure 1). The farm is engaged in broiler breeding. The chicks are bought from an external company, which provides the necessary vaccines and injections of a lincomycin and spectinomycin mixture. The breeding cycle lasts 42 days (6 weeks) with the initial temperature of 31 °C, falling down to 18 °C during the breeding, and constant air circulation (supply, exhaust, heating). Food given to the chickens is supplied in the form of a mixture of soybean meal, wheat, triticale, corn, soybean oil, and premix as a supplement (ready mix of minerals and vitamins, contains 2.5 kg of salt per ton of additives). However, during breeding, the necessary amount of antibiotics (such as amoxicillin, doxycycline, and enrofloxacin) are given ad hoc in water troughs (the line is not cleaned after supply). Notwithstanding, the antibiotic withdrawal period is respected, under threat of punitive measures by the meat purchasing company.

According to the breeder, proceedings involving distant discharge of polluted air and its high frequency are to maintain hygiene in the poultry house itself, and eliminate the controversial odor around the farm. The poultry house is cleaned once, before the start of every cycle (one cycle = 42 breeding days) by an external company. As part of cleaning, the litter is replaced, and the surface is swept and wet-cleaned each time with the use of cleaning agents, fired and whitewashed with lime, with the addition of ammonium sulfate for disinfection. Before settling the shed, disinfectant is sprayed in mist form. The composition of cleaning and disinfecting agents is unknown, and at the same time considered irrelevant in these tests. After replacement, the litter is moved outside the building at point 7A_M and then taken outside the farm.

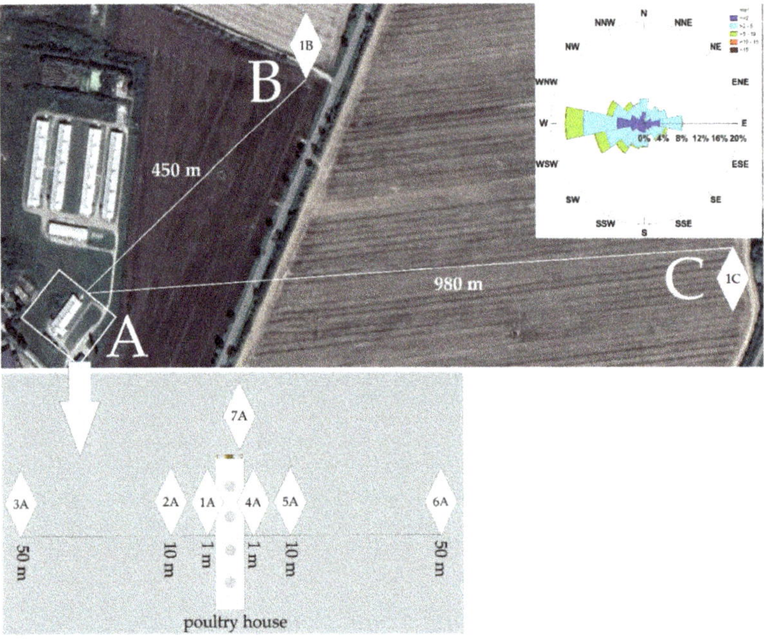

Figure 1. Soil sampling point distribution. 1–7A, 1B, and 1C soil samples point, collected in area A (around poultry house), area B (450 m away), and area C (980 m away).

2.2. Sample Handling

Soil samples were collected on the farm area (samples 1–7A). Six sampling locations, three on each of the two sides of the poultry house, were selected by measuring 1, 10, and 50 m from the longer wall. One sample was collected near the litter storage place. Aside from the samples in the area of direct impact of the poultry house, reference samples were taken (Figure 1). The reference sample 1B was taken at a distance of 450 m in a straight line to the North East. The references sample 1C was taken at a distance of about 1000 m in a straight line to the East. Samples were taken twice (in March, samples suffix M, and in July samples suffix J) to check for potential seasonal variation, in both cases, about 5 weeks after the poultry house was settled.

Using a shovel, about 1 L of soil was collected from each point at a depth of about 20 cm. In each case, the collected soil material was transported to the laboratory and stored in glass containers at 4 °C.

Before sample preparation, each sample passed through a 1.5 mm-mesh sieve (to remove roots and stones) and was freeze-dried using the Scanvac CoolSafe 110-4 PRO, LyoAlfa (Telstar, Denmark).

For microorganisms' determination in soil samples, the samples were collected into sterile plastic containers. The samples were stored at 10 °C until processing, which was done within 3 days after samples collection.

2.3. Analytical Procedure

2.3.1. Solvents and Standards

Methanol, acetonitrile (LiChrosolv®Hypergrade) were purchased from Merck (Darmstadt, Germany). Isopropanol was purchased from Sigma-Aldrich (Saint Louis, MO, USA). Formic acid 98%–100% pure (LiChrosolv®Hypergrade) to acidify the mobile phases and extraction solutions was purchased from Merck (Darmstadt, Germany). Ultra-pure water was produced using Hydrolab system (Hydrolab, Straszyn, Poland).

All pharmaceutical standards were of high purity (for HPLC analysis). Most of the native standards were purchased from Sigma-Aldrich (Saint Louis, MO, USA): ampicillin (AMP), caffeine (CAF), carbamazepine (CBZ), ciprofloxacin (CIP), diclofenac (DIC), enrofloxacin (ENR), ibuprofen (IB), metoclopramide (MTC), metoprolol (MET), nimesulide (NIM), paracetamol (PAR), propranolol (PROP), salicylic acid (SA), acetylsalicylic acid (ASA), sulfacetamide (SFC), spectinomycin (SPEC), streptomycin (STREP), sulfacarbamide (SCA), sulfadiazine (SDA), sulfaguanidine (SGA), sulfamerazine (SMA), sulfamethazine (SMZ), sulfamethoxazole (SMX), sulfanilamide (SNA), sulfathiazole (STA), tetracycline (TET) and trimethoprim (TMP).

2.3.2. Sample Pretreatment

Soil samples were prepared by weighing 8 g of dry soil (from each sample, separately) in 20 mL vials. Samples were extracted using two-stage extraction [22]. In the first step, a mixture of acetonitrile and water (1:1 *v*/*v*) with the addition of 0.1% formic acid was used. The second stage of extraction consisted of using a mixture of acetonitrile, 2-propanol, and water (3:3:4 *v*/*v*/*v*), also with the addition of 0.1% formic acid. For both the 1st and 2nd stage of extraction, 2 mL of extraction solvent was added for every 1 g of soil—16 mL of solvent for every step. The soil and solvent were mixed using a vortex mixer for 30 s, and then subjected to ultrasound using an IS-5,5 ultrasonic cleaner (Intersonic, Olsztyn, Poland) for 15 min. The decanted supernatant, collected after each of the steps, was filtered through a syringe filter (hydrophilic PTFE 0.2 μm membrane, Merck Millipore, (Tullagreen, Carrigtohill, Co. Cork, Ireland) into 40 mL glass vials. The filtered extracts from both steps were combined (in total, 32 mL) and evaporated to dryness under a stream of nitrogen. The residue was dissolved in 2 mL mixture of 25% methanol in water with the addition of 0.1% formic acid.

In order to eliminate the possible effect of the matrix, the same procedure of extraction was carried out with the addition of pharmaceutical standards (corresponding to a current-like concentration of compounds in soil). The mixture of the standards was added to 8 g of soil from each sample point. The rest of the procedure was carried out in accordance with the above-described method.

2.3.3. UPLC-MS/MS Analysis

The prepared samples were analyzed by ultra-performance liquid chromatography. The Nexera X2UPLC-MS/MS (Shimadzu Corp., Kyoto, Japan) contained two LC-30AD pumps, SiL-30AC autosampler, CTO-20AC column oven, CBM-20A communication bus module, and mass spectrometer LC-MS8050 with electro spray ionization with positive and negative ion mode (Table 1). Chromatographic separation was performed using an analytical column Kinetex 2.6 μm, Phenyl-Hexyl 100 Å 4.6 mm, 150 mm (Phenomenex, Torrance, CA, USA). The gradient program consisted of the following: 7-min sequence of linear gradient flows of solvent B (acetonitrile:methanol 1:1 *v*/*v*) balanced with solvent A (water with 0.1% formic acid) at a flow rate of 0.6 mL/min: 50–80% B over 1 min,

80–100% B over 2 min, isocratic 100% B for 3 min, and finally, 100–50% B over 1 min. The injection volume was 1 μL and column temperature was 40 °C.

The main parameters used to identify analytes were their retention times and multiple reaction monitoring (MRM) ratio, which were obtained at 0.25 μg/mL working standard solutions for most standards, and 2.5 μg/mL for AMP, IB, SPEC, STREP, SFC, SNA and TET. Chromatographic data processing was carried out using LabSolution® software (Shimadzu Corp., Kyoto, Japan).

Table 1. The analysis parameters for the monitored ion transitions and triple quadruple mass spectrometer (MS/MS) operation (CE- collision energy, Q1 and Q3Pre Bias—quadruple pre-rod bias voltage).

Parameters for the Monitored Ion Transition						
Name	**Short**	**Polarity**	**Parent Ion > Fragment Ion**	**Q1 Pre Bias [V]**	**CE [V]**	**Q3 Pre Bias [V]**
acetylsalicylic acid	ASA	−	179.10 > 136.95	13	11	24
ampicillin	AMP	+	349.9 > 106.05	20	25	20
caffeine	CAF	+	194.90 > 138.00	20	20	20
carbamazepine	CBZ	+	237.05 > 194.00	20	20	20
ciprofloxacin	CIP	+	322.00 > 314.10	20	21	20
diclofenac	DIC	−	294.05 > 249.90	15	12	16
enrofloxacin	ENR	+	360.10 > 316.10	20	20	20
ibuprofen	IB	−	205.00 > 161.20	14	9	16
metoclopramide	NIM	−	307.05 > 228.95	16	17	20
metoprolol	MTC	+	300.05 > 227.00	20	20	20
nimesulide	MET	+	268.15 > 116.00	20	20	20
paracetamol	PAR	+	152.00 > 110.10	17	18	21
propranolol HCl	PROP	+	260.10 > 116.10	20	19	20
salicylic acid	SA	−	137.30 > 93.15	10	17	20
spectinomycin	SPEC	+	332.80 > 98.10	16	29	14
streptomycin	STREP	+	582.00 > 263.20	28	33	11
sulfacetamide Na	SFC	+	214.95 > 92.05	11	23	21
sulfacarbamide	SCA	+	216.00 > 92.05	30	25	30
sulfadiazine	SDA	+	251.00 > 92.05	18	26	19
sulfaguanidine	SGA	+	214.95 > 92.05	15	26	21
sulfanilamide	SNA	+	172.9 > 86.15	19	16	13
sulfamerazine	SMA	+	264.95 > 92.05	13	29	14
sulfamethazine	SMZ	+	278.50 > 186.50	14	19	27
sulfamethoxazole	SMX	+	253.95 > 92.00	20	30	20
sulfathiazole	STA	+	255.90 > 156.00	20	15	20
tetracycline	TET	+	445.00 > 409.95	20	20	20
trimethoprim	TMP	+	291.05 > 230.10	20	25	20
MS/MS Operation Parameters						
			Interface Temperature (°C)		300	
			Desolvation LineTemperature (°C)		250	
			Nebulizing Gas Flow (L/min)		3	
			Heating Gas Flow (L/min)		9	
			Heating Block (°C)		350	
			Drying Gas Flow (L/min)		10	

2.3.4. Quality Assurance/Quality Control (QA/QC)

Ensuring the quality of the measurements results was carried out by:

- calibration of the LC-MS/MS system,
- determination of the selected validation parameters based on calibration curves.

The calibration step involves the preparation of calibration curves and calculation of response factor (RF) for internal standards. Eleven working standard solutions have been prepared. Limit of determination (LOD) has been calculated from the parameters of a calibration curve constructed on the basis of the three lowest concentrations of working standards solutions. The limit of determination has been calculated according to the formula:

$$LOD = \frac{3.3 \times s}{a} \quad (1)$$

where: a—slope of the calibration curve; s—the standard deviation.

For the calculation of the validation parameters, residual standard deviation (sxy) and standard deviation of the intercept (sa) were taken. Based on LOD(sxy) and LOD(sa) the mean values (LOD) were calculated (see Equation (1)). The coefficients of variation (CV) and uncertainty (U) were calculated based on standard deviations and the average area of the chromatographic peaks obtained during separation of standard mixtures.

To check the matrix effect, two samples were prepared using real soil samples with addition of standards, and using the procedure described in Section 2.3.2. Sample Pretreatment and additionally using solide phase extraction (SPE) cleaning. Based on peak areas, the recovery (%) for standard solutions was calculated.

2.4. Antibiotic Resistance in Microorganisms Isolated from Soil

The soil samples (1–6A; 1B and 1C, Figure 1) were plated on different growth media for evaluation of microbial growth (Luria Bertani Agar—LA, MacConkey Agar, Chapman Agar, Medium with cetrimide, Merck, Darmstadt, Germany). Briefly, 10 g of the sample was mixed with 90 mL of 0.85% NaCl and shaken for 0.5 h, and then serially diluted to 10^{-4}. In addition, 100 µL of each dilution was plated on LA, while solely the dilution 10^{-1} was spread on MacConkey Agar, Chapman Agar and Medium with cetrimide. The plates were incubated at 30 °C for 24–72 h to obtain sufficient growth of the colonies. From each sample, several differently-looking colonies were picked and grown to pure cultures on LA, and then stored frozen with 20% glycerol at −40 °C.

The isolated bacteria were then subjected to antibiotic resistance disc diffusion test with different antibiotics (TMP 5 µg, SMZ 20 µg, TET 10 µg, SPEC 25 µg, and CIP 10 µg). TET, SPEC, and CIP discs were purchased from Oxoid, UK, while trimethoprim and sulfametazine discs were prepared in the laboratory. Briefly, the cotton discs were cut from MN85/220 paper filters (Macherey-Nagel, Germany) and autoclaved. The appropriate amount of the antibiotic stock solution was poured onto the paper disk with automatic pipette and let dry. For the antibiotic resistance disc test, the bacteria were cultured in LB (Luria Bertani boullion, Btl, Łódź, Poland) with 150 rpm shaking at 30 °C, overnight. Then, the OD_{600} measurements of the bacterial cultures were taken with Spectroquant Prove 600 (Merck, Darmstadt, Germany) spectrophotometer and diluted in LB to achieve OD_{600} equal to 0.1. The dilution accurateness was verified with repeated spectrophotometer measurements. In addition, 100 µL of each bacterial suspension was spread on a plate with Muiller-Hinton medium (Graso, Owidz, Poland). Discs with antibiotics were then put on the surface of the plate and the plates were incubated for 24 h at 30 °C. Afterwards, the diameter of the growth inhibition zone around each antibiotic disc was measured.

3. Results

3.1. Quality Assurance/Quality Control (QA/QC)

Based on analysis of working standard solutions in optimal conditions, calibration curves were prepared. Regression coefficients and selected validation parameters are presented in Table 2. The obtained results are satisfactory, showing that the proposed method is suitable for analysis of the selected pharmaceuticals. The recovery for standards were calculated based on peak areas. Recoveries ranged from 65% to 121%, for all pharmaceuticals (mean values for individual compounds: ASA-65%, AMP-81%, CAF-111%, CMZ-98%, CIP-102%, DIC-67%, ENR-111%, IB-92%, MTC-99%, MET-121%, NIM-102%, PAR-81%, PROP-115%, SA- 78%, SFC-75%, SPEC-71%, STREP-74%, SFC-82%, SDA-89% SGA-94% SMA-98%, SMZ-100%, SMX-106%, SNA-98%, STA-82%, TET-67%, TMP-79%). The recoveries of analytes were dependent on the matrix type. This suggests that the use of standard addition or analysis with labeled standards are necessary in order to improve the correction results.

Table 2. Summary of validation data for quantification of selected pharmaceuticals and caffeine in environmental samples by UHPLC-MS/MS method (n = 6). LOD—Limit of dectection, LOQ—Limit of quantification, MDL—Method Detection Limit, MQL—Minimum Quantification Limit.

Pharmaceuticals	Short	Coefficient of Calibration Curve (y = ax + b)		Linearity		Limits				Repeatability	
		a	b	Regression Coefficient R^2	min–max (ng/mL)	LOD (ng/mL)	LOQ (3 × LOD) (ng/mL)	MDL (ng/g)	MQL (ng/g)	CV (%)	U (k = 2)
acetylsalicylic acid	ASA	107058	−1301	0.9931	176.57–5000	58.86	176.57	14.72	44.15	2.5–2.8	0.95
ampicillin	AMP	663156	2850	0.9996	14.25–5000	4.75	14.25	1.19	3.56	1.1–3.9	0.78
caffeine	CAF	29004046	128181	0.999	3.9–500	0.75	2.25	0.19	0.56	0.79–3.9	0.98
carbamazepine	CMZ	17335651	9152	0.9991	8.8–500	2.93	8.80	0.73	2.20	1.0–11	0.65
ciprofloxacin	CIP	2687453	1143	0.9982	9.9–5000	3.3	9.9	0.8	2.5	1.9–11.0	1.4
diclofenac	DIC	6826340	130246	0.9984	8.86–1000	2.89	8.68	0.72	2.17	0.71–9.6	0.64
enrofloxacin	ENR	12634935	−64163	0.9974	22.4–1000	7.5	22.4	1.9	5.6	1.4–5.1	1.4
ibuprofen	IB	10369	16715	0.9912	384.37–5000	316.8	950.4	79.2	237.6	2.2–3.8	1.1
metoclopramide	MTC	14227180430	1348134	0.998	0.39×10^{-3}	0.13×10^{-3}	0.39×10^{-3}	0.03×10^{-3}	0.09×10^{-3}	0.24–8.4	0.39
metoprolol	MET	8726542	10467	0.9991	3.3–1000	1.1	3.3	0.3	0.8	0.41–4.4	1.5
nimesulide	NIM	134064871	3435909	0.9934	12.62–500	4.21	12.62	1.05	3.16	1.6–7.5	0.60
paracetamol	PAR	1696703	1073	0.9993	14.01–5000	4.67	14.01	1.17	3.50	0.76–9.9	0.87
propranolol HCl	PROP	12583700	38410	0.9993	2.8–1000	0.92	2.8	0.2	0.7	0.26–4.6	1.2
salicylic acid	SA	585092	90750	0.9964	49.06–5000	15.80	47.43	3.95	11.85	2.4–8.9	0.89
sulfacetamide Na	SFC	1590160	1393	0.9989	8.7–5000	2.9	8.7	0.7	2.2	1.2–7.6	1.4
spectinomycin	SPEC	55623	−8960	0.998	173.2–5000	57.8	173.2	14.5	43.4	0.64–7.0	1.3
streptomycin	STREP	20759	−23309	0.9983	533.4–5000	177.8	533.4	44.5	133.4	0.60–6.3	1.4
sulfacarbamide	SFC	41286	−8452	0.9966	2.07–5000	0.69	2.07	0.17	0.52	2.1–3.3	0.40
sulfadiazine	SDA	5002325	−12328	0.9995	19.39–1000	6.46	19.39	1.62	4.85	0.36–5.7	0.89
sulfaguanidine	SGA	1264343	−1292	0.999	10.03–5000	3.34	10.03	0.84	2.51	0.80–12	0.87
sulfamerazine	SMA	7308819	−14598	0.9998	8.45–1000	2.82	8.45	0.71	2.12	0.84–6.1	0.66
sulfamethazine	SMZ	583452	2984	0.9863	0.08–5000	0.03	0.08	0.01	0.02	1.0–8.8	0.75
sulfamethoxazole	SMX	8078106	2984	0.9995	0.38–1000	0.13	0.38	0.03	0.10	1.7–8.2	0.89
sulfanilamide	SNA	53617	−25963	0.9973	1500.6–5000	502.9	1508.6	125.7	377.2	0.39–2.7	1.4
sulfathiazole	STA	4356784	615	0.9996	6.5–1000	2.2	6.5	0.6	1.7	0.64–6.0	1.1
tetracycline	TET	2063937	−33350	0.9990	57.2–1000	19.1	57.2	4.8	14.3	0.50–10	1.6
trimethoprim	TMP	26514921	74017	0.9994	2.9–500	0.98	2.9	0.2	0.7	1.7–3.7	1.1

3.2. Analysis of Real Samples

Using the developed and validated method described above, a chromatographic analysis of extracts of real soil samples was performed. Each of the 15 samples was analyzed for the presence of 27 substances. Of these, 10 pharmaceuticals (MTC, SNA, SA, MET, SMZ, NIM, CBZ, TMP, PROP, PAR) and CAF were determined in soil samples collected in March (Figure 2a) and five pharmaceuticals (MTC, SA, SMZ, CBZ, SNA) in soil samples collected in July (Figure 2b). The largest number of various pharmaceuticals (MTC, TMP, SA, SMZ, PROP, CBZ) was determined at 7A_M (the sample collected in March, location where the dirty litter was thrown away). None of the 27 pharmaceuticals sought was determined in sample 6A_J. In reference samples (1B_M, 1B_J and 1C_J) two of the pharmaceuticals (PAR, SNA) and CAF were determined, which were not found in the farm area. Moreover, none of the pharmaceuticals determined on the farm area were determined in reference samples. Therefore, farm area samples (area A) and reference location samples (area B and C) will be described separately, below.

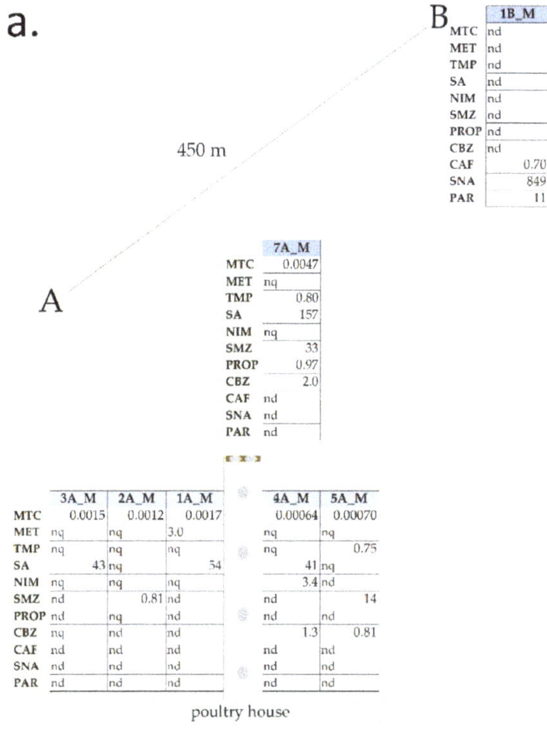

Figure 2. *Cont.*

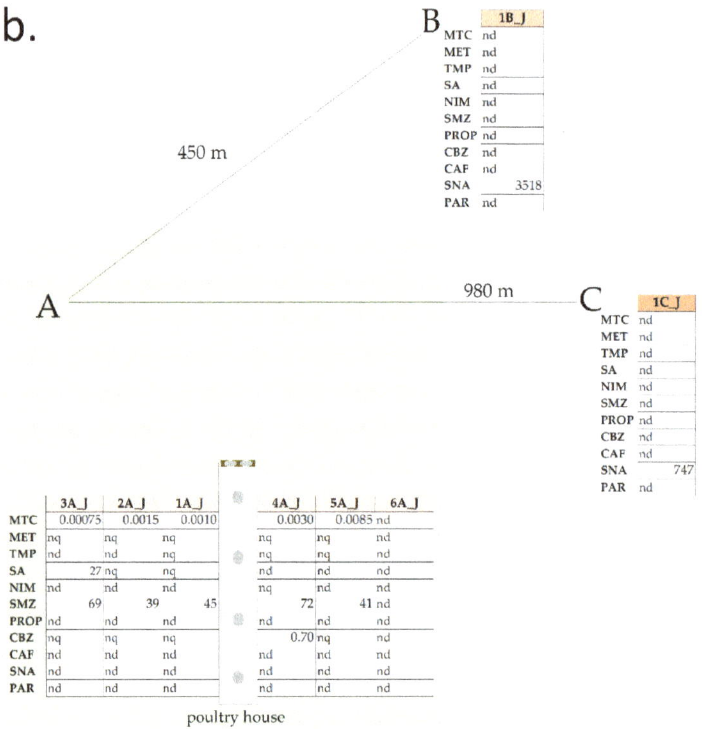

Figure 2. Concentration of detected pharmaceuticals and caffeine in soil samples (ng/g d.w); (**a**) 1–7A_M, 1B_M, soil samples, collected in area A (around poultry house) and area B (450 m away) in March (M); (**b**) 1–7A_J, 1B_J, 1C_J soil samples, collected in area A (around poultry house) area B (450 m away) and area C (980 m away) in July (J). MTC—matoclopramide, MET—metoprolol, TMP—trimpetoprim, SA—salicic acid, NIM—nimesulide, SMZ—supfamethazine, PROP—propranolol, CBZ—carbamazepine, CAF—caffeine, SNA—sulfanilamide, PAR—paracetamol.

3.2.1. Area A

MTC was determined in each of the tested samples around the farm, regardless of the sampling season (except point 6A_J). However, the concentrations in the samples were significantly lower—ranging from $4.69 \times 10^{-3} \pm 0.16$ ng/g d.w in 7A_M to $7.08 \times 10^{-4} \pm 0.263$ ng/g d.w in 5A_M. Similarly, SMZ was present in eight samples, and was determined in relatively high concentration levels. The range was from 68.70 ± 1.74 ng/g d.w. in 3A_J, to 0.81 ± 0.19 ng/g d.w in 2A_M. SA was detected in a range from 157.30 ± 4.65 ng/g d.w in 7A_M, to 27.19 ± 6.01 ng/g d.w in 3A_J. CBZ was found in four samples, three of which were collected in March. The highest concentration of CBZ was determined in sample 7A_M and amounted to 1.96 ± 0.17 ng/g d.w. In turn, the lowest concentration of CBZ was determined in the July 4A_J sample, with concentration equal to 0.70 ± 0.26 ng/g d.w. TMP was detected only in two samples from March on a similar level (7A_M 0.80 ± 0.03 ng/g d.w, 5A_M 0.75 ± 0.02 ng/g d.w.). The other four determined pharmaceuticals were found in individual soil samples. NIM was determined in the 4A_M sample in a concentration level of 3.43 ± 0.03 ng/g d.w., MET in the 1A_M sample at a concentration level of 2.99 ± 0.04 ng/g d.w., and PROP in sample 7A_M at a concentration level equal to 0.97 ± 0.005 ng/g d.w.

3.2.2. Areas B and C

As was mentioned, three other pharmaceuticals were found in the reference samples: CAF, PAR, and SNA. CAF (0.70 ± 0.13 ng/g d.w.) and PAR (11.31 ± 0.29 ng/g d.w.) were determined only in sample 1B_M. In turn, SNA was determined in all three reference samples with the highest concentration (3518.22 ± 146.05 ng/g d.w.) in sample 1B_J. In sample 1B_M, which is the winter counterpart of 1B_J, the concentration of SNA was 848.83 ± 1.19 ng/g d.w. In sample 1C_J, SNA was determined in the concentration level of 746.57 ± 3.90 ng/g d.w.

3.3. *Analysis of Microbial Resistance in Soil Samples*

The bacterial communities able to grow from soil samples collected around the poultry breeding building were shown to be diverse (Supplementary Materials Figure S1). From each collecting time point, a few differently-looking bacterial colonies were picked and further analyzed for antibiotic resistance. Microbial resistance of selected soil bacterial isolates was performed for five different antibiotics, two of which (TMP and SMZ) were determined in soil samples in this study. Three other antibiotics (TET, SPEC and CIP) were analyzed since they prove to be the ones that are commonly used in chicken treatment during their growth in intensive breeding farms [23,24]. The average inhibition zones for bacterial strains isolated from different sampling points situated in zones A, B and, C around the poultry breeding farm are presented in Table 3. The least susceptibility of the tested strains was observed for sulfametazine and trimethoprim, where only six and seven susceptible isolates were found in chicken farm soil, respectively. In the soil sample collected in zone B, this number was even lower, as only two strains were susceptible to TMP and none was susceptible to SMZ. In zone C, none of the strains was susceptible to TMP and SMZ; however, here the number of tested strains was only four. On the other hand, the highest susceptibility was shown for CIP (all strains from each type of soil were susceptible), and an intermediate one was shown for TET and SPEC. Taking into account the results obtained for 85 random bacterial isolates, coming from different areas around the poultry breeding building, we found no significant difference between the bacterial resistance tested for different areas (Table 3).

Table 3. Antibiotic resistance of microbial strains isolated from soil collected in three different zones around the poultry farm building. A (0–50 m around the poultry farm building), B (approx. 500 m from poultry farm building), and C (approx. 1000 m from the poultry farm building). Inhibition zones around different antibiotics: trimethoprim (TMP) 5 μg, sulfametazine (SMZ) 20 μg, tetracycline (TET) 10 μg, spectinomycin (SPEC) 25 μg, and ciprofloxacin (CIP) 10 μg, are presented as average diameter (mm) of inhibition zone with standard deviation for different soil sampling zones.

Soil Samples Localization		**A**	**B**	**C**
N. of Strains Tested		**66**	**15**	**4**
Antibiotics	CIP	29.0 ± 7.5	28.8 ±5.2	28.0 ± 4.0
	SPEC	6.8 ± 8.0	6.2 ± 8.6	0.0
	SMZ	2.5 ± 7.1	0.0	0.0
	TET	13.1 ± 11.2	12.6 ± 10.1	9.5 ± 7.0
	TMP	2.7 ± 8.2	3.7 ± 9.8	0.0

4. Discussion

There is no doubt that, especially in large-scale farms, pharmaceuticals are routinely used for both therapeutic and prophylactic purposes [8,22].

Literature reports on the presence of pharmaceuticals in soil that are potentially related to the poultry industry mainly relate to contaminants resulting from direct soil exposure to chicken droppings. Wei et al. [25] found the presence of sulfonamide and fluoroquinolone drugs in soil in areas loaded with poultry manure. Significant concentrations of sulfonamides were determined at the level of 682–1784 μg/kg soil. Jing et al. [26] presented the detection of antibiotics (tetracyclines and sulfonamides) in soils potentially exposed to chicken manure in northeast China. The concentration levels of pharmaceuticals were 0.29–1590.16 ug/kg soil, with the maximum concentration observed for chlorotetracycline.

Basic information on pharmaceuticals administrated to chickens was obtained from the breeder. According to the interview regarding the drugs used during breeding, lincomycin, SPEC, amoxycilin, doxycyclin, and ENR were administered to chickens. From the above, ENR and SPEC were selected for testing, but none of them was determined.

SMZ, together with TMP, is administered to chickens as a broad-spectrum antibiotic against most gram-negative organisms [23]. Farmers reported administrating them with water (in water line) [11]. Nevertheless, although both substances (TMP and SMZ) are given to chickens in the same quantitative ratio, their concentration level in the soil sample is significantly different. TMP is at the limit of quantification (LOQ) and has only been measured in two samples (7A_M and 5A_M), both in March. In turn, SMZ was determined in three samples collected in March, and in almost all samples collected around the poultry house in July, at higher concentration levels than in March. The difference between March and July values for SMZ can be explained by higher SMZ supply to chickens caused by greater probability of infection in July, because of the high temperature outside. Moreover, given the proportionally lower content of TMP compared to SMZ, TMP may be present in the same samples in which SMZ was determined, but below limit of quantification.

MTC is administered as an antiemetic and intestinal peristalsis drug, and must be given via subcutaneous or intramuscular injections every 12 h. MTC is given to inhibit, among others, the process of defecation—which is also intended to maintain greater hygiene in the poultry house during breeding [27]. The highest MTC concentrations were observed at 7A_M where used litter is temporarily stored. Taking all of this into consideration, it can be concluded that poultry litter is a serious source of pharmaceuticals.

The presence of SA in the environmental samples can be explained by salicylate administration to humans and/or animals, including acetylsalicylic acid (ASA) [28]. The use of salicylates has been systematically increasing for over 100 years. One route of metabolization for this group of drugs is rapid deacetylation to SA in a reaction catalized by a nonspecific enzyme. In effect, only about 68% of the dose reaches the systemic circulation as ASA, while the serum half time duration of ASA is approximately 20 min. Finally, SA and its metabolites are renally excreted [29]. So far, SA was detected in the effluent and river streams [30], and found—in low nanogram per liter concentrations—in groundwater from several areas in Ontario, Canada [31], waters from three watersheds in Nova Scotia, Canada [32], and spring water in Mexico [33]. In our research, ASA was not detected in any collected sample, but SA was determined. Moreover, SA was determined at a higher concentration, as compared to other pharmaceuticals, around the poultry house.

In addition to MTC, SMZ, TMP, and SA described earlier, PROP, CBZ were also determined. Their presence, especially in the 7A_M sample, may be associated with drug use in the process of chicken breeding.

Apart from pharmaceuticals presence in the samples, sample distribution around the farm building was considered. The obtained results show that as the distance from the poultry house increases, the concentration of pharmaceuticals increases as well. This may be due to the way of removing air from the poultry house, as fans on the roof expel air and thus spread contaminated air around the farm building.

The sample containing the largest load of pharmaceuticals was 7A_M (Figure 2). This sample was taken at the location of temporary litter storage removed from the poultry house after the end of the breeding cycle. The 7A_M sample was taken in the fifth week of a new breeding cycle, after litter removal. The presence of litter in this place before taking the sample may be the reason why 6 out of 11 substances were determined in this sample. Moreover, they were detected in the highest concentration levels compared to other samples. It can be assumed that the presence of specific pharmaceuticals in this sample may indicate their use in this breeding cycle.

Qualitative pharmaceuticals composition between the samples from farm area and reference area is diverse. The set of compounds detected in area A differs from the one detected in areas B and C. The reference sampling distance was far enough not to overlap with the poultry house impact.

The enhanced antibiotic resistance of bacterial strains was recently shown in *Escherichia coli* isolates from chicken manure originating from intensive breeding farms [24]. In regards to the presence of antibiotic resistant strains in the soil around intensive breeding farms, the information is scarce. In our study, we observed no difference between the average bacterial resistance of randomly selected soil isolates from intensive farm surroundings and agricultural soil when challenging them with five different antibiotics. However, Zhang et al. [34] observed that spiking poultry manure with antibiotics changes the average resistance in the microbiome of a soil treated with antibiotics containing manure. Indeed, the risk for soil contamination, and enhancing the number of antibiotic resistant microorganisms, may be lower in the surroundings of the intensive breeding farm than in the fields where the poultry manure was spread, as no significant difference was observed between the strains isolated in the vicinity of poultry farm buildings (area A, B, or C).

5. Conclusions

Soil, due to its specific features, such as sorption capacity, distribution coefficient (Kd), and hydrophobicity, is a pharmaceutical receiving matrix more durable than water [35]. Manure is one of the sources of various substances entering soil, including pharmaceuticals. Nevertheless, there is a high probability that the distribution of concentration around the poultry house (tendency-dependent changes depending on the distance) is associated with the transfer of pharmaceuticals with dust exhausted through the fans present on the building roof.

The results showed that the observed changes in pharmaceutical presence in analyzed soil samples can be defined as seasonal (Figure 2a,b). In all summer samples, less substances (five pharmaceuticals) were determined in contrast with samples collected in March (10 pharmaceuticals and CAF). This may indicate that low temperature is a parameter responsible for prolonged half-life time of determined contaminants [35]. However, the MTC concentration in both samples collected in March and July are, relatively, at a similar level. Moreover, concentration levels of SMZ and SNA in samples collected in July was approximately five times higher than in March.

Although advances in instrumentation and the opportunities that liquid chromatography methodologies bring (especially advanced ones, e.g., those coupled with tandem mass spectrometry (LC-MS/MS), facilitating the detection of even trace amounts of pharmaceuticals, in the case of environmental samples, the analyst still faces problems. The matrix effect, which is often difficult to compensate for, and with which we undoubtedly deal with in soil samples, may be the source of numerous discrepancies in qualitative assessment. The developed procedure allows separation and identification of 26 pharmaceuticals and CAF in a swift way (LC-MS/MS analysis—7 min), which proves to be an efficient way of real soil sample analysis. There is no doubt that this type of research needs to be further developed to provide a foundation for risk assessment of drugs entering the environment. It may influence human and animal health, which have been ignored in environmental monitoring programs.

Supplementary Materials: Supplementary Materials are available online.

Author Contributions: K.W., G.G., J.R., M.P.—writing—original draft of the manuscript; K.W., G.G.—performed the analysis, characterized samples (with UPLC-MS/MS); J.R. draft correction; A.P., L.W.—supervision, content support; L.W.—aided in interpreting the results. All authors have read and agreed to the published version of the manuscript.

Funding: This work was supported by the Department of Environmental Toxicology Medical University of Gdansk, and POWR 03.02.00-IP.08.00-DOK.

Acknowledgments: We thank dr Goran Gržnić from Department of Environmental Toxicology Medical University of Gdansk, for language critical review of our manuscript.

Conflicts of Interest: The authors declare no conflict of interest.

References

1. European Parliament and the Council of the European Union. *Regulation (EU) 2019/6 of the European Parliament and of the Council of 11 December 2018 on Veterinary Medicinal Products and Repealing DIRECTIVE 2001/82/EC*; European Parliament and the Council of the European Union: Luxembourg, 2018; Volume L4, pp. 43–167.
2. Tasho, R.P.; Cho, J.Y. Veterinary antibiotics in animal waste, its distribution in soil and uptake by plants: A review. *Sci. Total Environ.* **2016**, *563–564*, 366–376. [CrossRef]
3. Caracciolo, A.B.; Topp, E.; Grenni, P. Pharmaceuticals in the environment: Biodegradation and effects on natural microbial communities. A review. *J. Pharm. Biomed. Anal.* **2015**, *106*, 25–36. [CrossRef] [PubMed]
4. Bártíková, H.; Podlipná, R.; Skálová, L. Veterinary drugs in the environment and their toxicity to plants. *Chemosphere* **2016**, *144*, 2290–2301. [CrossRef] [PubMed]
5. Grenni, P.; Ancona, V.; Caracciolo, A.B. Ecological effects of antibiotics on natural ecosystems: A review. *Microchem. J.* **2018**, *136*, 25–39. [CrossRef]
6. Dahshan, H.; Abd-Elall, A.M.M.; Megahed, A.M.; Abd-El-Kader, M.A.; Nabawy, E.E. Veterinary antibiotic resistance, residues, and ecological risks in environmental samples obtained from poultry farms, Egypt. *Environ. Monit. Assess.* **2015**, *187*, 2. [CrossRef] [PubMed]
7. Kaczala, F.; Blum, S.E. The Occurrence of Veterunary Pharmaceuticals in the Environment: A Review. *Curr. Anal. Chem.* **2016**, *12*, 169–182. [CrossRef] [PubMed]
8. Furtula, V.; Farrell, E.G.; Diarrassouba, F.; Rempel, H.; Pritchard, J.; Diarra, M.S. Veterinary pharmaceuticals and antibiotic resistance of Escherichia coli isolates in poultry litter from commercial farms and controlled feeding trials. *Poult. Sci.* **2010**, *89*, 180–188. [CrossRef] [PubMed]
9. Farrell, D. *Poultry Development Review*; Food and Agriculture Organization of the United Nations: Rome, Italy, 2013.
10. Carvalho, I.T.; Santos, L. Antibiotics in the aquatic environments: A review of the European scenario. *Environ. Int.* **2016**, *94*, 736–757. [CrossRef]
11. Braykov, N.P.; Eisenberg, J.N.S.; Grossman, M.; Zhang, L.; Vasco, K.; Cevallos, W.; Muñoz, D.; Acevedo, A.; Moser, K.A.; Marrs, C.F.; et al. Antibiotic Resistance in Animal and Environmental Samples Associated with Small-Scale Poultry Farming in Northwestern Ecuador. *mSphere* **2016**, *1*, 1–15. [CrossRef]
12. Skóra, J.; Matusiak, K.; Wojewódzki, P.; Nowak, A.; Sulyok, M.; Ligocka, A.; Okrasa, M.; Hermann, J.; Gutarowska, B. Evaluation of microbiological and chemical contaminants in poultry farms. *Int. J. Environ. Res. Public Health* **2016**, *13*, 192. [CrossRef]
13. Maheshwari, S. Environmental Impacts of Poultry Production. *Poultry Fish. Wildl. Sci.* **2013**, *1*, 1–2. [CrossRef]
14. Viegas, S.; Faísca, V.M.; Dias, H.; Clérigo, A.; Carolino, E.; Viegas, C. Occupational exposure to poultry dust and effects on the respiratory system in workers. *J. Toxicol. Environ. Heal.—Part A Curr. Issues* **2013**, *76*, 230–239. [CrossRef] [PubMed]
15. Zhang, Y.; Price, G.W.; Jamieson, R.; Burton, D.; Khosravi, K. Sorption and desorption of selected non-steroidal anti-inflammatory drugs in an agricultural loam-textured soil. *Chemosphere* **2017**, *174*, 628–637. [CrossRef] [PubMed]
16. Hu, X.; Zhou, Q.; Luo, Y. Occurrence and source analysis of typical veterinary antibiotics in manure, soil, vegetables and groundwater from organic vegetable bases, northern China. *Environ. Pollut.* **2010**, *158*, 2992–2998. [CrossRef]
17. AVEC. *Annual Report of the Assiciation of Poultry Processors and Poultry Trade in the EU Countries*; AVEC: Brussels, Belgium, 2018.
18. Supreme Audit Office. *Supervision of the Functioning of Animal Farm*; Supreme Audit Office: Warsaw, Poland, 2014.
19. Supreme Audit Office. *Antibiotics Use in Animal Production in Lubuskie Voivodeship*; Supreme Audit Office: Warsaw, Poland, 2017.

20. Supreme Audit Office. *Supervision of the Marketi ng and Use of Products Containing Anabolic, Hormonal, Incuring and Psychotropical Substances in Animal Treatment, Including Accompanying Products*; Supreme Audit Office: Warsaw, Poland, 2018.
21. Kasperowicz, A.; Wacław, B.; Wasilewski, T. Environment protection program for the Nidzica district for the years 2018–2021, with a perspective for 2022–2025. 2017. Available online: http://www.bip.powiatnidzicki.pl/system/obj/8640_8.3.1.Program_Ochrony_Srodowiska_Powiatu_Nidzickiego.pdf (accessed on 24 February 2020).
22. Golovko, O.; Koba, O.; Kodesova, R.; Fedorova, G.; Kumar, V.; Grabic, R. Development of fast and robust multiresidual LC-MS/MS method for determination of pharmaceuticals in soils. *Environ. Sci. Pollut. Res.* **2016**, *23*, 14068–14077. [CrossRef]
23. Baert, K.; de Baere, S.; Croubels, S.; de Backer, P. Pharmacokinetics and Oral Bioavailability of Sulfadiazineand Trimethoprim in Broiler Chickens. *J. Vet. Pharmacol. Ther.* **2003**, *27*, 301–309.
24. Roth, N.; Käsbohrer, A.; Mayrhofer, S.; Zitz, U.; Hofacre, C.; Domig, K.J. The application of antibiotics in broiler production and the resulting antibiotic resistance in Escherichia coli: A global overview. *Poult. Sci.* **2019**, *98*, 1791–1804. [CrossRef]
25. Wei, R.; Gec, F.; Zhangb, L.; Houb, X.; Caoa, Y.; Gongb, L.; Chenb, M.; Wangb, R.; Bao, E. Occurrence of 13 veterinary drugs in animal manure-amended soils in Eastern China. *Chemosphere* **2016**, *144*, 2377–2383. [CrossRef]
26. An, J.; Chen, H.; Wei, S.; Gu, J. Antibiotic contamination in animal manure, soil, and sewage sludge in Shenyang, northeast China. *Environ. Earth Sci.* **2015**, *74*, 5077–5086. [CrossRef]
27. Metoclopramide for Chickens and Ducks. Available online: http://www.poultrydvm.com/drugs/metoclopramide (accessed on 2 February 2020).
28. Celiz, M.D.; Tso, J.; Aga, D.S. Pharmaceutical metabolites in the environment: Analytical challenges and ecological risks. *Environ. Toxicol. Chem.* **2009**, *28*, 2473–2484. [CrossRef]
29. Zaugg, S.; Zhang, X.; Sweedler, J.; Thormann, W. Determination of salicylate, gentisic acid and salicyluric acid in human urine by capillary electrophoresis with laser-induced fluorescence detection. *J. Chromatogr. B Biomed. Sci. Appl.* **2001**, *752*, 17–31. [CrossRef]
30. Ternes, T.A. Occurrence of drugs in German sewage treatment plants and rivers. *Water Res.* **1998**, *32*, 3245–3260. [CrossRef]
31. Carrara, C.; Ptacek, C.J.; Robertson, W.D.; Blowes, D.W.; Moncur, M.C.; Sverko, E.; Backus, S. Fate of pharmaceutical and trace organic compounds in three septic system plumes, Ontario, Canada. *Environ. Sci. Technol.* **2008**, *42*, 2805–2811. [CrossRef] [PubMed]
32. Comeau, F.; Surette, C.; Brun, G.L.; Losier, R. The occurrence of acidic drugs and caffeine in sewage effluents and receiving waters from three coastal watersheds in Atlantic Canada. *Sci. Total Environ.* **2008**, *396*, 132–146. [CrossRef]
33. Gibson, R.; Becerril-Bravo, E.; Silva-Castro, V.; Jiménez, B. Determination of acidic pharmaceuticals and potential endocrine disrupting compounds in wastewaters and spring waters by selective elution and analysis by gas chromatography-mass spectrometry. *J. Chromatogr. A* **2007**, *1169*, 31–39. [CrossRef]
34. Zhang, Y.J.; Hu, H.W.; Gou, M.; Wang, J.T.; Chen, D.; He, J.Z. Temporal succession of soil antibiotic resistance genes following application of swine, cattle and poultry manures spiked with or without antibiotics. *Environ. Pollut.* **2017**, *231*, 1621–1632. [CrossRef]
35. Awad, Y.M.; Kim, S.-C.; El-Azeem, S.A.M.A.; Kim, K.-H.; Kim, K.-R.; Kim, K.; Jeon, C.; Lee, S.S.; Ok, Y.S. Veterinary antibiotics contamination in water, sediment, and soil near a swine manure composting facility. *Environ. Earth Sci.* **2014**, *71*, 1433–1440. [CrossRef]

Sample Availability: Samples of the compounds are not available from the authors.

Article

Soil Behaviour of the Veterinary Drugs Lincomycin, Monensin, and Roxarsone and Their Toxicity on Environmental Organisms

Peiyi Li, Yizhao Wu, Yali Wang, Jiangping Qiu and Yinsheng Li *

School of Agriculture and Biology, Shanghai Jiao Tong University, Shanghai 200240, China; lipeiyi@sjtu.edu.cn (P.L.); wyzuihao@sjtu.edu.cn (Y.W.); wangyali.98@163.com (Y.W.); jpq@sjtu.edu.cn (J.Q.)
* Correspondence: yinshengli@sjtu.edu.cn

Academic Editor: Jolanta Kumirska
Received: 12 October 2019; Accepted: 3 December 2019; Published: 5 December 2019

Abstract: Lincomycin, monensin, and roxarsone are commonly used veterinary drugs. This study investigated their behaviours in different soils and their toxic effects on environmental organisms. Sorption and mobility analyses were performed to detect the migration capacity of drugs in soils. Toxic effects were evaluated by inhibition or acute toxicity tests on six organism species: algae, plants, daphnia, fish, earthworms and quails. The log K_d values (Freundlich model) of drugs were: lincomycin in laterite soil was 1.82; monensin in laterite soil was 2.76; and roxarsone in black soil was 1.29. The R_f value of lincomycin, roxarsone, monensin were 0.4995, 0.4493 and 0.8348 in laterite soil, and 0.5258, 0.5835 and 0.8033 in black soil, respectively. The EC_{50} for *Scenedesmus obliquus*, *Arabidopsis thaliana*, *Daphnia magna* and LC_{50}/LD_{50} for *Eisenia fetida*, *Danio rerio*, and *Coturnix coturnix* were: 13.15 mg/L,32.18 mg/kg dry soil,292.6 mg/L,452.7 mg/L,5.74 g/kg dry soil and 103.9 mg/kg (roxarsone); 1.085 mg/L, <25 mg/kg dry soil, 21.1 mg/L, 4.76 mg/L, 0.346 g/kg dry soil and 672.8 mg/kg (monensin); 0.813 mg/L, 35.40 mg/kg dry soil, >400 mg/L, >2800 mg/L, >15 g/kg dry soil, >2000 mg/kg (lincomycin). These results showed that the environmental effects of veterinary drug residues should not be neglected, due to their mobility in environmental media and potential toxic effects on environmental organisms.

Keywords: lincomycin; monensin; roxarsone; migration; residual; toxicity

---

1. Introduction

The environmental effects of drug residues have received increasing attention as a new type of pollutant. In addition to their therapeutic purposes, veterinary drugs are also incorporated into animal feed as additives to improve animal growth rate and feed efficiency [1]. In the United States, 112,000 tons antibiotics were used for cattle and pig livestock every year for non-therapeutic purposes [2]. Many studies have indicated that only a small fraction of the veterinary drugs consumed by livestock and pets are metabolized; the majority are released into the environment in their original forms, and can potentially enter the food chain to pose human health risks [3–5]. Drug residues can exist widely in the environment, inducing antibiotics, which can lead to increased drug resistances and reduced effectiveness in human and veterinary medicine [6]. One study found that more than 500 different types of veterinary drugs and more than 100 metabolites in the aquatic environment of 71 countries covering all continents, and the concentrations of some drugs, such as diclofenac, were measured at higher than the safe dose level in some countries [7]. Previous research on the occurrence of 13 veterinary drugs in 23 vegetable fields in eastern China where animal manure was used found that animal feces, especially those from poultry farms, was an important source of veterinary drug accumulation in soil [8]. It has also been shown that people can excrete carbamazepine and its metabolites after consuming food from crops irrigated by waste water, suggesting that human beings can ingest drugs from residue pollution [9]. Therefore, drugs can accumulate in soil, be absorbed by plants, and may have harmful effects on organisms.

Lincomycin is a commonly used lincosamide antibiotic for veterinary purposes, especially in China. It is persistent in the environment due to its pyranose ring, amide, pyrrolidine ring and other structures [10], and exhibits inhibitory effects and toxicity in organisms [11]. Lincomycin has numerous detrimental effects on the haematological and biochemical properties of blood, and can interfere with liver and kidney functions [12]. The EC_{50} value of lincomycin for *Artemia* (brine shrimp) was 283.1 mg/L [13], and the IC_{50} values of lincomycin for *Cylindrotheca closterium* and *Navicula ramosissima* were 14.16 mg/L and 11.08 mg/L, respectively [14]. The toxic effect of lincomycin, tylosin and ciprofloxacin mixture were found synergistic against *C. closterium* and additive for *N. ramosissima* [14]. Recently, the occurrence of antibiotic resistance has heightened concerns over lincomycin usage [15].

Monensin, the most widely used coccidiostat in the U.S. [16], is a class of polyether ion-carrier antibiotic used for the livestock and poultry industry. Monensin in fresh chicken manure can pose an environmental risk under certain conditions, and the use of compost was shown to be a method to degrade monensin [17]. The half-life of monensin ranges from 4–15 d in high-intensity management (i.e., soil amending, watering, and turning) and from 8–30 d in low-intensity management [18]. Because of its widespread use and high persistence, monensin was detected in multiple environments: 0.3 ± 4.5 mg/L in manure; 0.0004 μg/kg in soil; 0.01 ± 0.05 μg/L in surface water; 0.04 ± 0.39 μg/L in underground water; and 1.5 ± 31.5 μg/kg in sediment [16]. When the monensin concentration was 0.05 μmol/L, the adsorption coefficient of various soils was in the range of 0.915–78.6 L/kg [19]. Based on the K_d value, monensin is more mobile than tetracycline and has similar mobility to sulfamethazine. Its toxicity is highly species dependent [19]. It was reported that 50 mg/kg monensin could inhibit the reproduction and survival of earthworms [20]. Due to its high usage, high toxicity characteristics, and unevaluated potential environmental impacts, monensin has been classified as a high-priority environmental pollutant requiring further assessment [21].

Roxarsone is commonly added to the feed for farmed broiler chickens, and nearly all roxarsone is excreted unchanged in the manure [22]. Roxarsone can significantly induce CYP1A2 activity in the pig liver microsome, and the induction effect of it was stronger than enrofloxacin [23]. Zhang [24] found that the average elimination half-life of roxarsone in soil was 26.6–44.9 d and its adsorption in different soil depths was consistent with the Fetter linear adsorption model. Makris et al. [25] found that roxarsone had a higher adsorption capacity than inorganic arsenic (IV) in soil, which may be caused by the organic properties of roxarsone. After entering the soil and water system, roxarsone can be transformed into inorganic arsenic (III) and (V) with stronger migration capacity and greater toxicity [26], and can affect the growth and development of various plants. However, further studies are required to more thoroughly assess the potential environmental effects of roxarsone.

This study focused on the migration of lincomycin, monensin, and roxarsone in several soil environments, and their toxic effects on representative environmental organisms: *Scenedesmus obliquus* (algae), *Arabidopsis thaliana* (plant), *Eisenia fetida* (earthworm), *Danio rerio* (zebrafish), *Daphnia magna* (crustacean) and *Coturnix coturnix* (quail). The aim was to evaluate the environmental risks of the three drugs and provide a foundation for an impact assessment of their environmental residues for pollution management and prevention.

2. Results & Discussion

2.1. Adsorption-Desorption Test

The adsorption rates are shown in Table 1, and the parameters of the Freundlich model fitted are shown in Table 2. The log K_d value of lincomycin and monensin in laterite soil were 1.82 and 2.76. The log K_d value of roxarsone in black soil was 1.29. The corresponding 1/n values of the three equations were less than 2. According to the K_d value, the adsorption of laterite to monensin was greater than to lincomycin, which could be due to the higher molecular weight and lower water solubility of monensin. The 1/n value of lincomycin was 0.0935 in laterite soil. Therefore, the isotherm adsorption line of lincomycin belonged to the "L" type, meaning that at a certain concentration range, the adsorption capacity of soil to lincomycin decreases with increasing drug concentration. This result was consistent with the observed trend of the lincomycin adsorption rate. Another study showed that when the initial drug concentration was 1 mg/L, the adsorption rate of ethanamizuril in podzol soil was approx. 70% and decreased with increasing drug concentration, indicating that podzol soil displayed a stronger adsorption to lincomycin and monensin than ethanamizuril [27]. The lower organic matter contents and higher pH of the laterite and podzol soil are properties that may have contributed the smaller adsorption of roxarsone (pKa_1 = 3.49). Rutherford et al. [28] found that when 2 < pH < 8, the adsorption of roxarsone in soils decreased as pH increased. When the pH of the soil solution increased, the negative charge of the soil surface increased, causing an increased repulsive force between roxarsone and soil colloid, and a lower soil adsorption capacity of roxarsone. The zero adsorption of roxarsone was observed in field soil (pH 8.4, organic matter 6.84%) and wasteland soil (pH 8.2, organic matter 4.73%) [29].

Table 1. Adsorption rates of drugs in different soils (%).

Drug	Soil	Initial Drug Concentration (mg/L)				
		1	2	3	4	5
Lincomycin	Laterite soil	66.67	30.56	35.19	19.44	0
	Podzol soil	97.22	54.17	54.63	0	0
	Black soil	100.0	100.0	86.11	16.67	1.111
Monensin	Laterite soil	56.16	93.68	55.26	90.92	93.66
	Podzol soil	100.0	92.60	87.50	89.99	88.25
	Black soil	100.0	93.21	94.12	95.19	98.78
Roxarsone	Laterite soil	0	0	0	0	0
	Podzol soil	0	0.8080	0	0	0
	Black soil	18.80	38.18	97.41	51.95	42.00

Table 2. Parameters of the Freundlich model.

Drug	Soil	Lg K_d	1/n	R^2
Lincomycin	Podzol soil	1.5672	0.1116	0.2043
	Laterite soil	1.8235	0.0935	0.5282
	Black soil	1.7091	−1.2543	0.7329
Monensin	Podzol soil	1.5194	−0.1842	0.0355
	Laterite soil	2.7601	0.6117	0.8910
	Black soil	1.8502	−0.3915	0.2411
Roxarsone	Black soil	1.2855	1.9393	0.8759

2.2. Soil Mobility Test

Soil mobility test results are shown in Table 3. In laterite soil, the ranking of mobility capacity of drugs was monensin > lincomycin > roxarsone, and in black soil, the ranking was monensin > roxarsone > lincomycin. According to the classification in GB/T 31270-2014 (Test guidelines on environmental safety assessment for chemical pesticides) [30], lincomycin and roxarsone were moderately mobile in laterite and black soils, while monensin was highly mobile. Generally, adsorption ability has a negative correlation with water solubility of drugs, and a positive correlation with soil organic content; migration ability has the opposite correlations [31]. A previous study showed that the R_f value of roxarsone was 0.66 in soil (pH 6.5, organic matter 19.08 g/kg), indicating strong mobility [32]. However, in the present study, the R_f value was 0.4993 in laterite soil (pH 6.710, organic matter 3.92 g/kg), and 0.5835 in black soil (pH 5.257, organic matter 263.69 g/kg). These differences may have resulted from other physicochemical properties [33]. Florasulam was shown to be primarily distributed in a 9–18 cm soil layer in soil (organic matter 23.57 g/kg), and its R_f value was 0.918 [34]. According to the R_f values in the presents study, the migration ability of the three drugs tested could be weaker than florasulam in a similar soil environment.

Table 3. Content distribution of drugs using soil thin layer chromatography.

Drug	Soil	Drug Content in Each Moving Distance (%)						R_f
		0–3 cm	3–6 cm	6–9 cm	9–12 cm	12–15 cm	15–18 cm	
Lincomycin	Laterite soil	9.250	27.74	19.99	10.54	12.09	20.40	0.4995
	Black soil	16.11	15.63	15.01	14.66	17.07	21.51	0.5258
Monensin	Laterite soil	8.154	0	0	4.154	0	87.69	0.8348
	Black soil	0	0	0	0	68.02	31.98	0.8033
Roxarsone	Laterite soil	17.51	20.93	17.96	20.98	13.31	9.310	0.4493
	Black soil	12.45	12.99	10.85	13.10	26.86	23.74	0.5835

2.3. Algae Growth Inhibition Test

The absorbance of algae culture remained unchanged or even decreased with time, indicating that the concentrations of the algae decreased and their growth was restrained. Figure 1A–C show that the inhibition rate of algae (*S. obliquus*) growth by the three drugs was positively correlated with drug concentration. The pH of the treatment groups and controls were 6.9–7.4 at the end of the algae growth inhibition test. As the culture time increased, the percentage inhibition of algae growth rate increased. Based on the 96-h inhibition rate, the calculated EC_{50} (96h) were 0.813 mg/L (lincomycin, 95% confidence interval [0.791, 1.489] mg/L), 1.085 mg/L (monensin, [0.554, 1.193] mg/L) and 13.15 mg/L (roxarsone, [9.70, 17.83] mg/L). These results suggested that lincomycin, monensin and roxarsone showed low, medium, and low toxicity levels to algae, respectively. The EC_{50} (96h) of lincomycin and monensin suggested that these two drugs demonstrated higher toxicity to algae than ofloxacin (6.20), sulfamethoxazole (7.20) and sulfamethazine (9.89) [35], which was consistent with the results of Peng et al. [36].

Peng et al. also found that lincomycin and ofloxacin presented a high ecological risk to algae; the risk quotient (RQ) of lincomycin was 1.93, followed by ofloxacin (1.33), tetracycline (0.41) and erythromycin (0.20). According to another study, lincomycin can inhibit the synthesis of the D1 protein in the algal photosynthesis system, and may lead to algal death [37].

2.4. Plant Sensitivity Test

The effects of drugs on plant growth are shown in Tables 4 and 5. The seedling rates of blank controls and methanol controls (for monensin) were all above 50%, suggesting that experiments were valid.

Table 4. Number of plants emerged in drug groups (n = 10 seeds).

Drug									
Monensin	Concentration (mg/L)		25	50	75	100	125	0	Methanol control
	Plants Emerged	7 d	0.3 ± 0.6 [1]	0	0	0	0	8	6
		14 d	0.3 ± 0.6	0.3 ± 0.6	1.0 ± 1.7	0.3 ± 0.6	0	10	5
Lincomycin	Concentration (mg/L)		6	12	18	24	30	0	Methanol control
	Plants Emerged	7 d	9.0 ± 1.0	8.0 ± 2.0	6.7 ± 2.1	6.7 ± 1.5	7.3 ± 0.6	8	/
		14 d	7.7 ± 2.1	8.0 ± 2.6	5.3 ± 3.5	6.3 ± 0.6	7 ± 1.0	10	/
Roxarsone	Concentration (mg/L)		5	25	45	65	80	0	Methanol control
	Plants Emerged	7 d	9.3 ± 0.6	8.0 ± 1.7	9.7 ± 0.6	9.0 ± 1.0	6.3 ± 0.6	8	/
		14 d	6.0 ± 3.0 *	4.3 ± 1.5 *	3.3 ± 2.3 *	2.7 ± 2.1 *	1.7 ± 0.6 *	10	/

[1] data represents mean ± standard deviations, n = 3, the same below. * represents significant difference at $p < 0.05$ compared with the control group.

Table 5. Effect of lincomycin and roxsarsone on *A. thaliana* height and biomass (14 d).

Drug							
Lincomycin	Concentration (mg/L)	6	12	18	24	30	0
	Total Gross Dry Weight (g)	0.0033 ± 0.0008	0.0046 ± 0.0012	0.0040 ± 0.0030	0.0031 ± 0.0011	0.0074 ± 0.0007	0.0045
	Max. Plant Height (cm)	2.5 ± 0.5	2.7 ± 1.0	3.0 ± 2.8	2.9 ± 1.4	4.2 ± 1.3	3.6
Roxarsone	Concentration (mg/L)	5	25	45	65	80	0
	Total Gross Dry Weight (g)	0.0038 ± 0.0045	0.0010 ± 0.0005	0.0011 ± 0.0012	0.0019 ± 0.0005	0.0005 ± 0.0004	0.0045
	Max. Plant Height (cm)	1.7 ± 0.3	0. 8± 0.3	1.0 ± 0.5	1.8 ± 1.0	0.4 ± 0.1	3.6

(A)

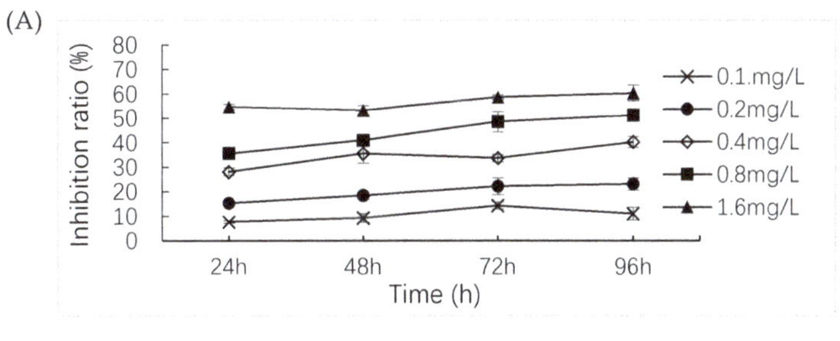

(B)

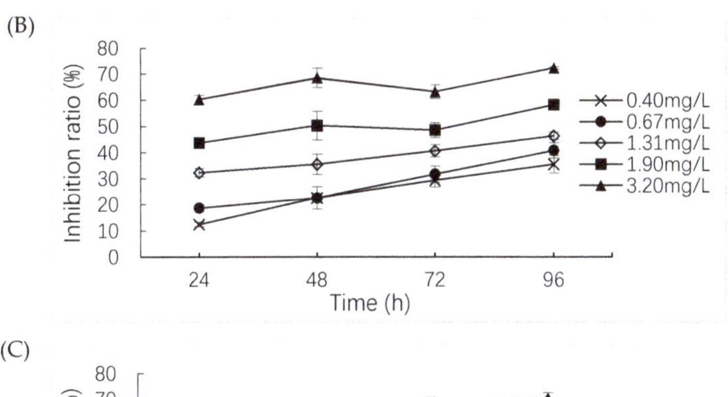

(C)

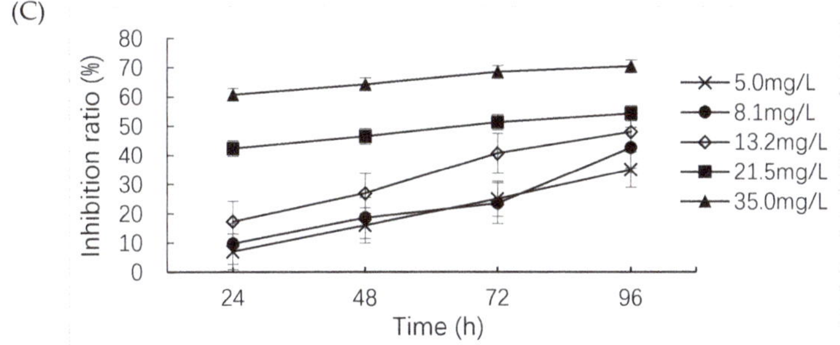

Figure 1. Effect of drugs on the growth of *S. obliquus.* (**A**) lincomycin, (**B**) monensin, (**C**) roxarsone.

2.4.1. Lincomycin

The EC_{50} (14 d) of lincomycin on *A. thaliana* seedling rate was 35.40 mg/kg dry soil (95% confidence interval [19.217, 65.211] mg/kg dry soil), however, there was no significant difference in plant emergence on 14 d (Table 4). With increased lincomycin concentration, the number of seedlings, biomass and plant height were not significantly affected (Tables 4 and 5).

2.4.2. Monensin

The seedling rates of the blank control group and methanol solvent control group on 7 d and 14 d were both higher than 50% (Table 4). However, in the experimental groups, only 1 seedling appeared in the 25 mg/L monensin group. On 14 d, only a few seeds sprouted, and the plant heights were all less than 0.5 cm. Therefore, the EC_{50} (14 d) of monensin on *A. thaliana* seedling rate was <25 mg/kg dry soil (the minimum test concentration). The results indicated that monensin may have strong toxicity on plant growth. A study by Hoagland [38] also confirmed the high phytotoxicity of monensin: it caused herbicidal injury to 1–2 week-old seedlings of seven weed and two crop species when applied at 10^{-4} M as a foliar spray in the greenhouse, and all nine species died within 24–72 h after treatment.

2.4.3. Roxarsone

The EC_{50} (14 d) of roxarsone on *A. thaliana* seedling rate was 32.18 mg/kg dry soil (95% confidence interval [11.319, 49.795] mg/kg dry soil). On 14 d, the number of plants that emerged in the blank control group was significantly higher than in each treatment group (Table 4). In the treatment groups, only the 5 mg/kg and the 80 mg/kg roxarsone groups showed significant differences between each other in seedling rate on 14 d. This indicated that roxarsone could reduce the growth of *A. thaliana*.

2.5. *Daphnia Activity Inhibition Test*

In the control group, the daphnia (*D. magna*) body structures were clearly observed, not immobilized, and did not display unusual behaviours. However, in the high drug concentration groups, the body structure of daphnia appeared blurred, and daphnia immobilization was observed. Table 6 shows the inhibition rate of monensin and roxarsone on daphnia activity. The EC_{50} (48 h) values were 21.1 mg/L (monensin, 95% confidence interval [17.9, 24.7] mg/L) and 292.6 mg/L (roxarsone, [283.7, 301.8] mg/L), which suggested that both drugs had low toxicity on daphnia. Acute activity inhibition did not occur in the lincomycin group at all concentrations tested after 48 h. Therefore, the EC_{50} of lincomycin for daphnia was > 400 mg/L (the maximum test concentration).

Table 6. Activity inhibition numbers of daphnia in monensin and roxarsone groups at 48 h (n = 8).

Monensin	**Concentration (mg/L)**	9.5	14.7	20.0	22.7	35.0	0
	Daphnia inhibited	0	1.0 ± 0	2.3 ± 0.6	4.0 ± 1.0	8.0 ± 0	0
Roxarsone	**Concentration (mg/L)**	250.0	268.9	289.3	311.1	334.7	0
	Daphnia inhibited	0	1.3 ± 0.6	3.3 ± 0.6	5.7 ± 0.6	8.0 ± 0	0

2.6. *Acute Toxicity Tests of Zebrafish*

The potassium dichromate toxicity test (reference poison) for zebrafish (*D. rerio*) is shown in Table 7. The LC_{50} of potassium dichromate for zebrafish was 356.6 mg/L (95% confidence interval [333.0, 381.8] mg/L). An LC_{50} value between 200–400 mg/L demonstrated that the experimental method and the fish quality met the appropriate standards [39].

Table 7. Effect of potassium dichromate on zebrafish mortality (n = 8).

Drug Concentration (mg/L)	0	250	300	350	400	450
Mortality	0	0	2	4	6	7

Table 8 shows the toxicity of monensin and roxarsone on zebrafish. There was no significant difference between the methanol control group and the blank control group ($p > 0.05$). The LC_{50} (96 h) values were 4.76 mg/L (monensin, 95% confidence interval [4.670, 4.853] mg/L) and 452.7 mg/L (roxarsone, [447.9, 457.5] mg/L), suggesting that monensin and roxarsone had medium and low toxicity, respectively, for zebrafish. The zebrafish in the lincomycin group did not show any abnormal activity at all concentrations tested after 96 h. Therefore, the LC_{50} (96 h) of lincomycin was >2800 mg/L (the maximum test concentration). Previous research showed that roxarsone could cause rapid DNA destruction of *Carassius auratus*, and low doses of roxarsone could cause more sustainable and increased damage than high doses [40]. The experimental LC_{50} (96 h) values of oxytetracycline and norfloxacin for zebrafish were 1.262×10^{-3} mol/L and 2.026×10^{-3} mol/L, respectively [41], suggesting their toxicity may be similar to roxarsone and lower than monensin.

Table 8. Effect of monensin and roxarsone on zebrafish mortality (n = 8).

Monensin	**Concentration (mg/L)**	4	4.34	4.7	5.1	5.53	0
	Mortality	0	1	4	6	8	0
Roxarsone	**Concentration (mg/L)**	400	423	447	473	500	0
	Mortality	0	1	3	6	8	0

2.7. Acute Toxicity Tests of Earthworm

Table 9 shows the toxicity of monensin and roxarsone on earthworms (*E. fetida*). The mortality rate of the control group was less than 10%, suggesting the test was valid. The appearance of earthworms in the low concentration groups showed no difference compared to the control group. However, in the 20 g/kg roxarsone group, the earthworms atrophied, the girdles were swollen, and some earthworms died.

Table 9. Effect of monensin and roxarsone on earthworm mortality (n = 10).

Monensin	**Dose (g/kg Dry Soil)**		CK [1]	200	258	332	427	550
	Average	**7 d**	0	0	0	1	3	5
	Mortality	**14 d**	0	0	2	4	7	10
Roxarsone	**Dose (g/kg Dry Soil)**		CK	4	5.03	6.33	7.95	10
	Average	**7 d**	0	0	0	2	5	10
	Mortality	**14 d**	0	0	3	6	10	10

[1] CK represents control check.

There was no significant difference between the methanol control group for monensin and its blank control group ($p > 0.05$). The LC_{50} (14 d) values were 346.0 mg/kg dry soil (monensin, 95% confidence interval [309, 387] mg/kg dry soil) and 5.74 g/kg dry soil (roxarsone, [5.14, 6.41] g/kg dry soil), indicating that both drugs had low toxicity for earthworms [30]. No earthworm mortality or abnormal activity was discovered at all concentrations of lincomycin tested after 14 d. Therefore, the LC_{50} (14 d) of lincomycin was >15 g/kg dry soil (the maximum test concentration). The LC_{50} of monensin for earthworms was higher than the reported 75.883 mg/kg observed in an artificial soil test [20].

2.8. Acute Toxicity Test on French Giant Quail

Table 10 shows the toxicity of monensin and roxarsone on French giant quails. The mortality rate of the control group was less than 10%, suggesting the test was valid. At high concentrations of roxarsone and monensin, inappetence, listlessness and other symptoms were observed in quails. There was no significant difference between the carboxymethylcellulose sodium control and the blank control ($p > 0.05$), and no significant difference between males and females ($p > 0.05$). The LD_{50} (7 d) values were 672.8 mg/kg (monensin, 95% confidence interval [451.0, 1003.8] mg/kg) and 103.9 mg/kg (roxarsone, [80.5, 134.2] mg/kg), indicating that monensin and roxarsone displayed low and medium toxicity, respectively. All quails exposed to lincomycin survived after 7 d, and their appetite, activity, and excretion were no different from quails in the control group. Therefore, the LD_{50} (7 d) of lincomycin was > 2000 mg/kg (the maximum concentration). Another study on the effect of roxarsone on laying hens suggested that roxarsone could cause an increase in aspartate aminotransferase (AST), lactate dehydrogenase (LDH), and creatine kinase (CK) activity, a decrease in the liver weight, and histological evidence of liver damage [42]. Therefore, it is possible that roxarsone causes quail mortality via oxidative stress.

Table 10. Effect of monensin and roxarsone on quail mortality (5 male and 5 female).

Monensin	Drug Dose (mg/kg)		0	80	400	800	1100	1500
	Mortality	Male	0	0	1	3	3	5
		Female	0	0	2	2	3	4
Roxarsone	Drug Dose (mg/kg)		0	30	52.7	100	162.3	284.8
	Mortality	Male	0	0	1	2	4	5
		Female	0	0	1	2	3	5

2.9. Summary of Results

Acute toxicity test data and toxicity category of the three drugs to six species are summarised in Table 11. Lincomycin showed low toxicity for all tested organisms, and medium toxicity for *S. obliquus*. Monensin showed medium toxicity for *S. obliquus* and *D. rerio*, and low toxicity for *E. fetida*, *D. magna*, and *C. coturnix*, and no detectable toxicity for *A. thaliana*. Roxarsone showed medium toxicity for *C. coturnix* and low toxicity for all other tested organisms.

Table 11. Summary of acute toxicity data and category of drugs.

Species	Roxarsone		Monensin		Lincomycin	
	Data	Category	Data	Category	Data	Category
S. obliquus EC_{50} (96 h) mg/L	13.15	low	1.085	medium	0.813	medium
A. thaliana EC_{50} (14 d) mg/kg dry soil	32.18	low	<25	/	35.40	low
E. fetida LC_{50} (14 d) g/kg dry soil	5.74	low	0.346	low	>15	low
D. rerio LC_{50} (96 h) mg/L	452.7	low	4.76	medium	>2800	low
D. magna EC_{50} (48 h) mg/L	292.6	low	21.1	low	>400	low
C. coturnix LD_{50} (7 d) mg/kg	103.9	medium	672.8	low	>2000	low

As an important component of aquatic ecosystems, algal photosynthesis accounts for a large proportion (up to 50%) of global primary productivity [43]. Due to its toxicity on algae, the clinical dosage and release of lincomycin should be carefully regulated. Monensin showed a significantly higher toxicity risk on some species compared with other commonly used antibiotics, such as tetracyclines and quinolones. This result was consistent with previous research findings showing that ionophores exhibited higher toxicity than other antibiotics [19]. More attention should be given to the use of monensin, its residue and accumulation in the environment. Roxarsone-contaminated soil and its accumulation in rice could present serious problems for human health [44]. Despite the low ecotoxicity of roxarsone found in this study, its use, and the use of other arsenic-containing drugs, requires strict control to avoid arsenic entering the food chain. Possible methods to reduce the concentration of drugs in the environment include the treatment of animal manure before field application, the use of alternative bio-agents for disease treatment, and a well targeted legalized use of antibiotics [45].

3. Materials and Methods

3.1. Chemicals, Test Soils and Organisms

Roxarsone (purity >98%) was purchased from Shanghai Titan Technology (Shanghai, China), LTD. Sodium morenate (purity >90%) and lincomycin hydrochloride (purity >95%) were purchased from Sangon Biotech LTD (Shanghai, China). The podzol soil was collected from the Shanghai suburban district. The black soil was taken from a forest in Changbai Mountain, Jilin Province, China. The laterite soil was collected from idle farmland in Chuzhou, Anhui Province, China. All test soils were collected from the surface layer (0–10 cm) that had not been farmed for more than 20 years, and were air-dried, grinded and sieved to a 2-mm size. The basic properties of the test soils are shown in Table 12.

Table 12. Basic properties of the test soils.

Test Soil	pH	Organic Matter (g/kg)	Electrical Conductance (μS/cm)	Total P (g/kg)	NH_4^+-N (mg/kg)	Total K (g/kg)
Podzol soil	6.876	26.00	180.3	0.958	12.50	2.53
Laterite soil	6.710	3.62	58.5	0.070	7.13	11.40
Black soil	5.257	263.69	605.5	1.980	19.74	6.97

Scenedesmus obliquus (green algae) were provided by the Chinese Academy of Sciences Institute of Hydrobiology (Wuhan, Hubei Province, China), and were grown in BG-11 cultures with continuous passage for three times and were tested in the logarithmic growth phase. *Arabidopsis thaliana*, wild type, were provided by School of Life Science and Biotechnology, Shanghai Jiaotong University (Shanghai, China). All seeds were treated at 4 °C for more than a week before sowing. *Daphnia magna* was provided by Guangdong Laboratory Animals Monitoring Institute (Guangzhou, Guangdong Province, China). The daphnia was domesticated in laboratory conditions for 7 d. Healthy infant daphnia were selected for experiments. *Danio rerio* was purchased from a pet house in Minhang District, Shanghai, China. The fish were domesticated in laboratory conditions for 7 d and 2 cm-long healthy fish were selected. *Eisenia fetida* were purchased from Wangjun Earthworm Farm in Jiangsu Province, China, and domesticated for 14 d. Healthy adult earthworms weighing 300–600 mg with an obvious girdle band were selected. *Coturnix coturnix* (French giant quails) were purchased from Shanghai Fengxian Quail Farm in Fengxian District, Shanghai, China. They were aged approx. 30 d and weighed approx. 100 g. The quails were domesticated for 7 d under laboratory conditions, and those with no disease were selected. All experimental methods performed were in accordance with National Standards of China [30,39,46] and OECD [47].

3.2. Adsorption-Desorption and Soil Mobility

3.2.1. Adsorption-Desorption Test

The sorption of drugs by soils was investigated using the oscillation balance method following GB/T 21851-2008 [46]. The optimal soil/solution ratios and adsorption-desorption equilibrium time used in the formal test were obtained by a pre-experiment (Table 13).

Table 13. Optimal soil/solution ratios and equilibrium time in sorption tests.

Soil	Lincomycin		Monensin		Roxarsone	
	Ratio	Time (h)	Ratio	Time (h)	Ratio	Time (h)
Podzol soil	50:1	24	100:1	48	50:1	24
Laterite soil	50:1	24	20:1	24	50:1	48
Black soil	50:1	24	50:1	24	50:1	48

Soil samples (0.5–5 g) were weighed and placed in 50 mL centrifuge tubes covered with tin foil. Appropriate amounts of aqueous solution of different drug concentrations (0, 1, 2, 3, 4, and 5 mg/L) were added to the tubes to reach the optimum soil/solution ratios. The tubes were shaken at 25 ± 2 °C at 180 r/min. Three replicates were used for each concentration group.

After the adsorption reached equilibrium, drugs in solution and soils of all groups were extracted and detected using methods described in 3.2.3 and 3.2.4. Adsorption rates were calculated and the Freundlich model ($\lg C_s = \lg K_d + 1/n \lg C_e$) was applied to fit sorption isotherms.

3.2.2. Soil Mobility Test

Mobility was investigated by soil thin-layer chromatography following GB/T 31270-2014 [30]. 10 μg of each drug was spotted 2.0 cm from the lower end of the coated soil sheet plates (containing 10 g of soil, 0.5 mm soil thickness, air-dried). Blank controls and three replicates were used. After the solvent had evaporated, the glass plate was placed at an angle of 30° in a chromatographic bath, and was then unfolded by distilled water at room temperature until the water surface reached the leading edge of the glass plate. The soil layer was divided into 6 × 3 cm segments. Each soil segment was scraped into a 10 mL centrifuge tube to detect drug content (3.2.3 and 3.2.4). The R_f values of drugs were calculated.

3.2.3. Soil Drug Extraction

Drugs were extracted from soils according to [48]. Briefly, 10 mL of methanol was added to the soil in centrifuge tubes. The tubes were then vortexed for 1 min, sonicated at 25 °C for 10 min, and centrifuged at 8000 r/min for 5 min. The supernatant was passed through a 0.45 μm filter to obtain the pre-treated sample solution. Extraction accuracy was checked by a recovery test (Table 14). Addition concentrations were 1, 2, 4 mg/L, and five replicates for each concentration were tested in each soil.

Table 14. Extracted drug recovery in test soils (%).

Drug	Podzol Soil		Laterite Soil		Black Soil	
	Recovery	RSD	Recovery	RSD	Recovery	RSD
Lincomycin	73.2–97.4	0.801–3.10	71.7–108	1.21–4.83	78.1–98.6	0.344–2.12
Monensin	84.6–91.8	1.98–4.62	81.0–111	0.620–4.62	90.3–106	1.72–5.10
Roxarsone	89.9–92.6	1.15–3.18	91.2–99.3	1.34–1.69	79.4–96.7	1.36–3.87

Drug extraction from water: an appropriate volume of testing water was passed through a 0.22 μm filter to obtain the pre-treated sample solution.

3.2.4. Drug Detection

Pre-treated samples from 3.2.3 were analysed using an ultraviolet spectrophotometer (Agilent, Beijing, China) and the method's accuracy was tested.

Lincomycin

Lincomycin and $PdCl_2$ can form a coloured complex with a maximum absorption peak at 380 nm [49] under acidic conditions. 1.6 mL of 0.02 mol/L $PdCl_2$ solution was added to a pre-treated sample solution in a 50 mL volumetric flask, diluted with 1 mol/L hydrochloric acid solution, and mixed in the dark for 30 min. The absorbance was measured at 380 nm. The linear regression equation was: y (absorbance) = 0.0024 × (drug concentration) − 0.0003, $R^2 = 0.9986$.

Monensin

A 3% vanillin solution was used as a derivatization reagent to react with monensin, and the product has a maximum absorption peak at 525 nm [50]. 2 mL of 3% vanillin solution and methanol were added to the pre-treated sample solution in a 50 mL volumetric flask, shaken, diluted, and then heated in a 45 °C water bath for 20 min. After the solution cooled to room temperature, its absorbance was measured at 525 nm. The linear regression equation was: $y = 0.0329x + 0.0008$, $R^2 = 0.9973$.

Roxarsone

Roxarsone has ultraviolet-absorbing groups such as a benzene ring, and its solution has a maximum absorption peak at 264 nm [51]. The pre-treated sample solution was made up to 50 mL with distilled water and the absorbance was measured at 525 nm. The linear regression equation was: $y = 0.0227x$, $R^2 = 0.\ 9981$, average RSD% = 9.77.

3.3. Ecotoxicology Tests

3.3.1. Algae Growth Inhibition Test

S. obliquus was cultured in 100 mL Erlenmeyer flasks filled with 100 mL culture. The BG-11 culture was prepared according to Deng. et al. [52], with the initial pH adjusted to 7.1–7.5 with 1 M NaOH and HCl solutions. Flasks were placed in an artificial climate chamber with 12 h light and 12 h darkness at 29 °C. Light intensity was 5000 lux.

The original algae concentration was measured by a blood cell counting plate, and sequentially diluted 1:2, 1:10, 1:50, and 1:100. The absorbance of the five gradient concentrations of the algal liquid was measured at OD 660 nm by spectrophotometry, and the equation obtained was y (absorbance) = 687.19 × 10^4 × (algae concentration), R^2 = 0.9982.

Before inoculation, the absorbance was measured and the initial concentration of algae in the conical flask was diluted to 10^5/mL according to the absorbance. The absorbance was measured again after inoculation. According to the results of the pre-experiment, the concentration of algae growth inhibition rate exceeding 50% was selected as the highest concentration of the formal test. Five concentrations and blank controls were used for each drug with three replicates (the solvent control for monensin was 0.1 mL/L dimethyl sulfoxide). The following drug concentrations were used: roxarsone: 5.0, 8.1, 13.2, 21.5, 35.0 mg/L; monensin: 0.40, 0.67, 1.31, 1.90, 3.20; lincomycin: 0.10, 0.20, 0.40, 0.80, 1.60. Monensin was solubilized with dimethyl sulfoxide at 0.1 mL/L. The absorbance of the algae solution was measured after shaking for 0 h, 24 h, 48 h, 72 h and 96 h. The pH was measured after shaking for 96 h in accordance with GB/T 21805-2008 [46].

3.3.2. Plant Sensitivity Test

The plant sensitivity test was conducted by soil culture using circular pots with a diameter of 15 cm. A 600 g sample of test soil was used at a thickness of approx. 10 cm, in accordance with GB/T 31270-2014 [30]. Lincomycin and roxarsone were dissolved in water. Monensin was solubilized with methanol at 2 mL/kg dry soil.

According to the results of the pre-experiment, five concentrations with three replicates and blank controls were used for each drug in the formal tests (the solvent control for monensin was methanol, 2 mL/kg dry soil). The following drug concentrations were used: lincomycin: 6, 12, 18, 24, 30 mg/kg dry soil; roxarsone: 5, 25, 45, 65, 80 mg/kg dry soil; and monensin: 25, 50, 75, 100, 125 mg/kg dry soil.

After the soil and the test solution were thoroughly mixed (methanol volatilized completely), 10 seeds of *A. thaliana* were evenly planted on the surface of podzol soil with conditions of 16 h light and 8 h darkness, 25 ± 2 °C (day) and 22 ± 2 °C (night), 75% humidity and 15,000 lux light intensity. The soil moisture was approx. 60% during the test. Drug effects were evaluated 14 d after emergence of the seedlings reached 50% in the control group. As the test endpoint, the seedling emergence rate of each group was calculated on 14 d after sowing. The biomass and plant height were measured on 14 d.

3.3.3. Growth Inhibition Test of Daphnia Activity

Eight *D. magna* organisms were cultured in a 100 mL Erlenmeyer flask per treatment, with 50 mL of experimental water (tap water, naturally placed for 2 d). Young daphnids (aged less than 24 h, not first brood progeny) were selected, according to GB/T 21830-2008 [46]. Specimens were incubated at 20 °C in complete darkness. Daphnids were not fed during the experiment and were moved with droppers. Monensin was solubilized with methanol to 0.1 mL/L.

According to the results of the pre-experiment, five concentrations with three replicates and blank controls were used for each drug (the solvent control for monensin was 0.1 mL/L methanol). The following drug concentrations were used: roxarsone: 250.0, 268.9, 289.3, 311.1, 334.7 mg/L; monensin: 9.5, 14.7, 20.0, 22.7, 35.0 mg/L; and lincomycin: 80, 120, 180, 270, 400 mg/L. After 24 h and 48 h, the number of immobilised daphnids (used as endpoint) were counted, and their behaviour and appearance in each conical flask were examined.

3.3.4. Acute Toxicity Test of Zebrafish

Acute toxicity test of zebrafish was performed according to GB/T 27861-2011 [39]. Each aquarium (0.2 m × 0.2 m × 0.2 m) was filled with 3 L of experimental water (tap water, naturally placed for 2 d) and 8 fish.

Potassium dichromate was used for the reference poison test, and concentrations used were 0, 250, 300, 350, 400, and 450 mg/L. Mortality rate was calculated after 24 h.

According to the pre-experiment, the lowest total lethal concentration and the highest total survival concentration of the drugs were obtained. In the formal experiment, five drug concentrations were used in that range according to the equal ratio series: roxarsone: 316, 354, 397, 445 and 500 mg/L; monensin: 4.00, 4.34, 4.70, 5.10, 5.53 mg/L. Monensin was solubilized with methanol at 0.1 mL/L. Blank solvent controls (0.1 mL/L methanol for monensin) were used. No replicates were used for each concentration group or controls.

The experimental conditions were 25 ± 2 °C, no aeration, with feeding during the experiment. Dead fish were removed to prevent water pollution. The experiment lasted 96 h, and the number of deaths and symptoms of poisoning were observed every 24 h. Temperature, oxygen concentration and pH were monitored daily. All procedures were approved by the Animal Ethical and Welfare Committee of Shanghai Jiaotong University Animal Department (Approval ID 20180301005).

3.3.5. Acute Toxicity Test of Earthworms

A 500 g sample of podzol soil, 125 mL of distilled water and drug solution were added to each experimental container. Monensin was ventilated for 24 h after mixing to volatilize solvent methanol. Ten earthworms dried by filter paper were added, with their intestines cleaned for 24 h before the experiment.

The lowest total lethal concentration and the highest total survival concentration of the drugs were obtained in pre-experiments. In the formal experiments, five concentrations were used in that range according to the equal ratio series: roxarsone: 1.25, 2.50, 4.00, 10.00 and 20.00 g/kg dry soil; and monensin: 30, 60, 200, 350, and 550 mg/kg dry soil. Three replicates for each concentration, blank controls and solvent control (0.1 mL/L methanol for monensin) were used.

The experimental conditions were 20 ± 2 °C, 75% humidity, and approx. 60% soil moisture. The number of deaths and poisoning symptoms were recorded on 7 d and 14 d, and mortality rate was calculated.

3.3.6. Acute Toxicity Test of Quails

The experiment was performed by a one-off oral toxicity test with oral needles and catheters, in accordance with GB/T 31270-2014 [30]. The dose was 1 mL/quail, where lincomycin and roxarsone of low concentration were dissolved in water, and roxarsone of high concentration and monensin were dissolved in 0.05% sodium carboxymethylcellulose solution.

According to the pre-experiment, the lowest total lethal concentration and the highest total survival concentration of the drugs were obtained. In the formal experiment, five concentrations were used in that range according to the equal ratio series: roxarsone: 30.0, 52.7, 100.0, 162.3 and 284.8 mg/kg; monensin: 80, 400, 800, 1100, and 1500 mg/kg. Blank controls and solvent controls (0.05% sodium carboxymethyl cellulose solution) were used. There were no replicates for each concentration group or controls.

Ten quails (5 male and 5 female) were placed in each cage, and the experimental period was 7 d. The death and symptoms of quails were observed every 24 h. All procedures were approved by the Animal Ethical and Welfare Committee of Shanghai Jiaotong University Animal Department (Approval ID 20180102008).

3.4. Statistical Analysis

The data were calculated using Excel 2016 (Microsoft Inc., Redmond, WA, USA). Univariate analysis of variance was performed using IBM SPSS Statistics 19.0 (International Business Machines Corporation, Armonk, NY, USA), and Probit regression model was established to calculate EC_{50}, LC_{50}, LD_{50} and 95% confidence intervals. The Duncan method was used to analyse significant differences at $p < 0.05$ of each group.

4. Conclusions

This study evaluated the adsorption and migration capacity of lincomycin, monensin and roxarsone in different soil environments and their toxic effects on diverse environmental organisms. Moderate soil mobility and high water solubility exhibited by lincomycin could assist drug transfer and accumulation in various environments, and may cause certain problems after enrichment. Roxarsone was moderately mobile and its ecotoxicity implied that it is a potential ecological risk. Monensin was the most toxic among the three drugs tested, and its higher affinity to soil made it easier to be accumulated. The potential environmental impacts identified by the drugs tested as a result of their mobility, persistence and ecotoxicity will help to inform veterinary drug management and drug residue pollution concerns.

Author Contributions: Investigation, P.L. and Y.W. (Yizhao Wu); formal analysis, P.L., Y.W. (Yizhao Wu) and Y.W. (Yali Wang); writing-original draft, P.L. and Y.W. (Yizhao Wu); writing-review & editing, P.L., Y.W. (Yali Wang) and Y.L.; supervision, J.Q. and Y.L.; project administration, Y.L.; funding acquisition, Y.L.

Funding: This work was supported by the National key R&D program of China (2017YFD0501405).

Conflicts of Interest: The authors declare no conflict of interest. The funders had no role in the design of the study; in the collection, analyses, or interpretation of data; in the writing of the manuscript, or in the decision to publish the results.

References

1. Sarmah, A.K.; Meyer, M.T.; Boxall, A.B. A global perspective on the use, sales, exposure pathways, occurrence, fate and effects of veterinary antibiotics (VAs) in the environment. *Chemosphere* **2006**, *65*, 725–759. [CrossRef] [PubMed]
2. Mellon, M.; Benbrook, C.; Benbrook, L.K. *Hogging it: Estimates of Antimicrobial Abuse in Livestock*; Union of Concerned Scientists: Cambridge, MA, USA, 2001.
3. Kong, W.D.; Zhu, Y.G.; Liang, Y.C. Uptake of oxytetracycline and its phytotoxicity to alfalfa (*Medicago sativa* L.). *Environ. Pollut.* **2007**, *147*, 187–193. [CrossRef] [PubMed]
4. Pan, M.; Chu, L.M. Fate of antibiotics in soil and their uptake by edible crops. *Sci. Total. Environ.* **2017**, *599–600*, 500–512. [CrossRef] [PubMed]
5. Hurtado, C.; Dominguez, C.; Perez-Babace, L. Estimate of uptake and translocation of emerging organic contaminants from irrigation water concentration in lettuce grown under controlled conditions. *J. Hazard. Mater.* **2016**, *305*, 139–148. [CrossRef] [PubMed]
6. Lee, H.C.; Chen, C.M.; Wei, J.T. Analysis of veterinary drug residue monitoring results for commercial livestock products in Taiwan between 2011 and 2015. *J. Food Drug Anal.* **2018**, *26*, 565–571. [CrossRef]
7. Aus, D.B.T.; Weber, F.-A.; Bergmann, A. Pharmaceuticals in the environment-Global occurrences and perspectives. *Environ. Toxicol. Chem.* **2016**, *35*, 823–835.
8. Wei, R.; Ge, F.; Zhang, L. Occurrence of 13 veterinary drugs in animal manure-amended soils in Eastern China. *Chemosphere* **2016**, *144*, 2377–2383. [CrossRef]
9. Paltiel, O.; Fedorova, G.; Tadmor, G. Human exposure to wastewater-derived pharmaceuticals in fresh produce: A randomized controlled trial focusing on carbamazepine. *Environ. Sci. Technol.* **2016**, *50*, 4476–4482. [CrossRef]
10. Chen, W.R.; Ding, Y.; Johnston, C.T. Reaction of lincosamide antibiotics with manganese oxide in aqueous solution. *Environ. Sci. Technol.* **2010**, *44*, 4486–4492. [CrossRef]

11. Wang, M.; Zhang, B.; Wang, J. Degradation of lincomycin in aqueous solution with hydrothermal treatment: Kinetics, pathway, and toxicity evaluation. *Chem. Eng. J.* **2018**, *343*, 138–145. [CrossRef]
12. Jwad, S.M.; Abbas, B.; Jaffat, H.S. Study of the protective effect of vitamin C plus E on lincomycin-induced hepatotoxicity and nephrotoxicity. *Res. J. Pharm. Technol.* **2015**, *8*, 177. [CrossRef]
13. Migliore, L.; Civitareale, C.; Brambilla, G. Toxicity of several important agricultural antibiotics to *Artemia*. *Water Res.* **1997**, *31*, 1801–1806. [CrossRef]
14. Hagenbuch, I.M.; Pinckney, J.L. Toxic effect of the combined antibiotics ciprofloxacin, lincomycin, and tylosin on two species of marine diatoms. *Water Res.* **2012**, *46*, 5028–5036. [CrossRef] [PubMed]
15. Domínguez, C.; Flores, C.; Caixach, J. Evaluation of antibiotic mobility in soil associated with swine-slurry soil amendment under cropping conditions. *Environ. Sci. Pollut. Res.* **2014**, *21*, 12336–12344. [CrossRef] [PubMed]
16. Arikan, O.A.; Mulbry, W.; Rice, C. The fate and effect of monensin during anaerobic digestion of dairy manure under mesophilic conditions. *PLoS ONE.* **2018**, *13*, e0192080. [CrossRef] [PubMed]
17. Zizek, S.; Hrzenjak, R.; Kalcher, G.T. Does monensin in chicken manure from poultry farms pose a threat to soil invertebrates? *Chemosphere* **2011**, *83*, 517–523. [CrossRef]
18. Storteboom, H.N.; Kim, S.C.; Doesken, K.C. Response of antibiotics and resistance genes to high-intensity and low-intensity manure management. *J. Environ. Qual.* **2007**, *36*, 1695–1703. [CrossRef]
19. Sassman, S.A.; Lee, L.S. Sorption and degradation in soils of veterinary ionophore antibiotics: Monensin and lasalocid. *Environ. Toxicol. Chem.* **2010**, *26*, 1614–1621. [CrossRef]
20. Wang, Y. Study on the Ecotoxicological Effects of Monensin on Earthworms. Master's Thesis, Hainan University, Hainan, China, 2010.
21. Capleton, A.C.; Courage, C.; Rumsby, P. Prioritising veterinary medicines according to their potential indirect human exposure and toxicity profile. *Toxicol. Lett.* **2006**, *163*, 213–223. [CrossRef]
22. Garbarino, J.R.; Bednar, A.J.; Rutherford, D.W. Environmental fate of roxarsone in poultry litter. Part, I. Degradation of roxarsone during composting. *Environ. Sci. Technol.* **2003**, *37*, 1509–1514. [CrossRef]
23. Li, Y.; Jiang, M.; Thunders, M. Effect of enrofloxacin and roxarsone on CYP450s in pig. *Res. Vet. Sci.* **2018**, *117*, 97–98. [CrossRef]
24. Zhang, Y.M. Ecotoxicological Study on the Residue of Roxarsone. Master's Thesis, Yangzhou University, Yangzhou, China, 2007.
25. Makris, K.C.; Salazar, J.; Quazi, S. Controlling the fate of roxarsone and inorganic arsenic in poultry litter. *J. Environ. Qual.* **2008**, *37*, 963–971. [CrossRef]
26. Brown, B.L.; Slaughter, A.D.; Schreiber, M.E. Controls on roxarsone transport in agricultural watersheds. *Appl. Geochem.* **2005**, *20*, 123–133. [CrossRef]
27. Zhang, W. A study on Ecological Toxicity and Environmental Behavior of Ethanamizuril. Master's Thesis, Shanghai Jiaotong University, Shanghai, China, 2018.
28. Rutherford, D.W.; Bednar, A.J.; Garbarino, J.R. Environmental fate of roxarsone in poultry liner. Part II. Mobility of arsenic in soils amended with chicken manure. *Environ. Sci. Technol.* **2003**, *37*, 1515–1520. [CrossRef]
29. Jiang, C.A.; Zhai, X.F.; Wang, Y. Adsorption of ascorbic acid and roxarsone additives in different soils. *J. South. China Agric. Univ.* **2013**, *34*, 272–276.
30. National Standards of People's Republic of China. In *GB/T31270-2014 (Test Guidelines on Environmental Safety Assessment for Chemical Pesticides)*; Standardization Administration of the People's Republic of China: Beijing, China, 2014.
31. Petrovic, A.M.; Larsson-Kovach, I. Effect of maturing turfgrass soils on the leaching of the herbicide mecoprop. *Chemosphere* **1996**, *33*, 585–593. [CrossRef]
32. Fiolka, M.J. Activity and immunodetection of lysozyme in earthworm *Dendrobaena veneta* (Annelida). *J. Invertebr. Pathol.* **2012**, *109*, 83–90. [CrossRef] [PubMed]
33. Singh, R.P.; Srivastava, G. Adsorption and movement of carbofuran in four different soils varying in physical and chemical properties. *Adsorpt. Sci. Technol.* **2009**, *27*, 193–203. [CrossRef]
34. Duan, Y.; Bao, N.; Li, J. Fate characteristics of florasulam in soil environment. *Guizhou Agric. Sci.* **2018**, *46*, 66–70.
35. Zhang, X.H.; Wan, T.; Cheng, W. Effects of quinolones and sulfonamides on the growth of green algae. *J. Water Resour. Water Engin.* **2018**, *29*, 115–120.

36. Peng, Q.C.; Song, J.M.; Li, X.G. Biogeochemical characteristics and ecological risk assessment of pharmaceutically active compounds (PhACs) in the surface seawaters of Jiaozhou Bay, North China. *Environ. Pollut.* **2001**, *255*, 113247. [CrossRef]
37. Bergmann, T. Synergy of light and nutrients on the photosynthetic efficiency of phytoplankton populations from the Neuse River Estuary, North Carolina. *J. Plankton Res.* **2002**, *24*, 923–933. [CrossRef]
38. Hoagand, R.E. Herbicidal properties of the antibiotic monensin. *J. Sci.Food Agric.* **1996**, *70*, 373–379. [CrossRef]
39. National standards of People's Republic of China. In *GB/T 27861-2011 (Chemicals-Fish Acute Toxicity Test)*; Standardization Administration of the People's Republic of China: Beijing, China, 2011.
40. Liu, P.; Wang, M.Z.; Zhong, W.X. Stress-responsive genes (*hsp70* and *mt*) and genotoxicity elicited by roxarsone exposure in *Carassius auratus*. *Environ. Toxicol. Pharmacol.* **2018**, *62*, 132–139.
41. Cai, M.T.; Hou, G.Q.; Xi, H. Combined toxicity of co-exposure of typical antibiotic and heavy metal copper on freshwater green algae and zebrafish. *J. Zhejiang Shuren Univ.* **2018**, *18*, 11–15.
42. Wenshyg, C.; Kuolung, C.; Bi, Y. Effects of roxarsone on performance, toxicity, tissue accumulation and residue of eggs and excreta in laying hens. *J. Sci. Food Agric.* **1997**, *74*, 229–236.
43. Otvos, L., Jr. Antibacterial peptides and proteins with multiple cellular targets. *J. Pept Sci.* **2005**, *11*, 697–706. [CrossRef]
44. Wang, F.M.; Chen, Z.L.; Zhang, L. Arsenic uptake and accumulation in rice (*Oryza sativa* L.) at different growth stages following soil incorporation of roxarsone and arsanilic acid. *Plant. Soil.* **2006**, *285*, 359–367. [CrossRef]
45. Tasho, R.P.; Cho, J.Y. Veterinary antibiotics in animal waste, its distribution in soil and uptake by plants: A review. *Sci. Total Environ.* **2016**, *563–564*, 366–376. [CrossRef]
46. National standards of People's Republic of China. In *GB/T 21851-2008 (Chemicals-Adsorption-desorption Using a Batch Equilibrium Method), GB/T 21805-2008 (Chemicals-Alga Growth Inhibition Test), GB/T 21830-2008 (Chemicals-Daphnia sp., Acute Immobilisation Test)*; Standardization Administration of the People's Republic of China: Beijing, China, 2008.
47. *OECD Guidelines for the Testing of Chemicals, No.207 "Earthworm, Acute Toxicity Tests"*; Organisation for Economic Cooperation and Development: Paris, France, 1984.
48. Helling, C.S.; Turner, B.C. Pesticide mobility: Determination by soil thin-layer chromatography. *Science* **1968**, *162*, 562. [CrossRef]
49. Yan, J. Determination of the content of lincomycin hydrochloride injection by spectrophotometry. *Shanghai Med. Pharm. J.* **1994**, *12*, 33–34.
50. Zeng, Z.G.; Liu, B.; Chen, Y.H. Determination of monensin by spectrophotometry. *Chin. J. Vet. Drug* **2007**, *41*, 19–21.
51. Zhang, F.F.; Wang, W.; Yuan, S.J. Biodegradation and speciation of roxarsone in an anaerobic granular sludge system and its impacts. *J. Hazard. Mater.* **2014**, *279*, 562–568. [CrossRef] [PubMed]
52. Deng, X.R.; Qian, Q.M.; Sun, C. Comparative study on the toxic effects of lindane and chlorpyrifos on freshwater algae. *Ecol. Environ.* **2014**, *3*, 472–476.

Sample Availability: Samples of the compounds are available from the authors.

Review

Selected Pharmaceuticals in Different Aquatic Compartments: Part I—Source, Fate and Occurrence

André Pereira *, Liliana Silva, Célia Laranjeiro, Celeste Lino and Angelina Pena

LAQV, REQUIMTE, Laboratory of Bromatology and Pharmacognosy, Faculty of Pharmacy, University of Coimbra, Polo III, Azinhaga de St^a Comba, Coimbra, 3000-548, Portugal; ljgsilva@hotmail.com (L.S.); celialaranjeiro@gmail.com (C.L.); cmlino@ci.uc.pt (C.L.); apena@ci.uc.pt (A.P.)

* Correspondence: andrepereira@ff.uc.pt

Academic Editor: Jolanta Kumirska

Received: 29 January 2020; Accepted: 21 February 2020; Published: 25 February 2020

Abstract: Potential risks associated with releases of human pharmaceuticals into the environment have become an increasingly important issue in environmental health. This concern has been driven by the widespread detection of pharmaceuticals in all aquatic compartments. Therefore, 22 pharmaceuticals, 6 metabolites and transformation products, belonging to 7 therapeutic groups, were selected to perform a systematic review on their source, fate and occurrence in different aquatic compartments, important issues to tackle the Water Framework Directive (WFD). The results obtained evidence that concentrations of pharmaceuticals are present, in decreasing order, in wastewater influents (WWIs), wastewater effluents (WWEs) and surface waters, with values up to 14 mg L^{-1} for ibuprofen in WWIs. The therapeutic groups which presented higher detection frequencies and concentrations were anti-inflammatories, antiepileptics, antibiotics and lipid regulators. These results present a broad and specialized background, enabling a complete overview on the occurrence of pharmaceuticals in the aquatic compartments.

Keywords: environmental contaminants; pharmaceuticals occurrence; pharmaceuticals; aquatic compartments

1. Introduction

Human pharmaceuticals, presenting different characteristics and, consequently, producing different environmental exposure profiles, represent a group of widely used chemicals that contaminate the aquatic environment. Albeit in trace amounts, they are of concern, since they are designed to perform a biological effect. Moreover, given their continuous introduction into the environment, their impact, both as stressors and as agents of change, is of great importance [1].

The main source of pharmaceuticals residues in the aquatic environment is human excretion, and consequently, the widespread presence of pharmaceuticals in environmental samples is most likely to occur from wastewater treatment plants (WWTPs), which incompletely remove these compounds. Pharmaceuticals are then released into the environment as parent compounds, metabolites, as well as transformation products [2], leading to the contamination of surface waters, seawaters, groundwater and even some drinking waters already identified by new analytical methodologies which allowed the detection at low ng L^{-1} [3–10].

Although no legal limits have been established in water, seven pharmaceuticals and one metabolite became part of the WFD watch list established by the Directive 2013/39/EU amended by the Commission Implementing Decision from the EU 2015/495 and the EU 2018/840. This list is dynamic, changing with the awareness on the persistence in the water cycle, and its validity in time is limited. Therefore, identifying and prioritizing new pharmaceuticals are important goals to be accomplished for future updates in order to minimize the aquatic environmental contamination by pharmaceuticals [11].

Additionally, as a part of the strategy implemented by the Directive 2013/39/EU, all member states shall monitor the substances in the watch list at the selected surface waters' representative monitoring stations.

Globally, heavy contamination pressures from extensive urban activities characterize the main rivers that might lead to high aquatic contamination levels and consequent environmental and human exposure. Although the concentrations of pharmaceuticals in influents (WWIs) and effluents (WWEs) of WWTPs and surface waters are routinely monitored in many countries, only in recent years there has been an increase in the number of studies concerning the occurrence of pharmaceuticals in the aquatic environment [12–16]. Additionally, other aquatic compartments such as seawater, groundwater, mineral water and drinking water have a lower amount of data available regarding this contamination. However, most of these studies are primarily focused on a small number of targeted compounds in localized areas. Therefore, there is a knowledge gap which demands a comprehensive and systematic evaluation of pharmaceuticals, its metabolites and transformation products in the aquatic environment.

Thus, a systematic review, in order to provide a clear insight on pharmaceuticals' contamination of the water compartment, should embrace, not only several parent compounds, but also metabolites and transformations products belonging to different therapeutic groups (Table 1).

The pharmaceuticals in study, key representatives of major classes of pharmaceuticals, were selected based on the EU watch list, their high consumption, pharmacokinetics, physicochemical properties, persistence, previous studies on the occurrence on the aquatic environment and their potential toxicological impact, both on humans and on the aquatic environment [11,17–20]. In this way, the complete scenario of the contamination of pharmaceuticals in the aquatic environment could be acquired, contributing to future improvements in minimization measures, calculation of the environmental risk assessment and legislation.

In a larger vision of future water resource management sustainability, with the escalating population growth and intensified agricultural and industrial activity, water scarcity will be a reality [21–23]. Therefore, there will be the need for water/wastewater recycling, and the contamination of water resources by pharmaceuticals gains yet another perspective. Therefore, it is important to obtain a better understanding of the context, concerning the source, fate and occurrence posed by pharmaceuticals in the aquatic environment.

2. Sources and Fate of Pharmaceuticals in the Environment

2.1. Sources

Pharmaceuticals are widely consumed throughout the world and can reach the aquatic environment, primarily through human excretion or by direct disposal of unused or expired drugs in toilets, being WWTPs are considered the primary sources of these contaminants into the water bodies [18,24]. Although they are administered within healthcare facilities, namely, hospitals, nursing, assisted living and independent living healthcare facilities, its contribution to the input of pharmaceuticals into the municipal WWTPs is quite low, since these facilities typically make a small contribution to the overall load [3,25,26]. The hospital contribution to the total load of pharmaceuticals in municipal WWTPs is for most compounds under 10% and, usually, even below 3% [9]. However, wastewaters from drug production can be a potential source of pharmaceuticals in certain locations, namely, in major production areas for the global bulk drug market [6]. Finally, veterinary medicines can also enter the environment; however, their environmental exposure routes and fate differ from human pharmaceuticals [19,27].

Thus, these drugs, their metabolites and/or transformation products may enter the environment via WWTPs effluents or by land application of biosolids, originating from WWTPs sludges, which, through runoff or leaching, can enter the aquatic environment, surface or groundwaters [28]. It is important to highlight that the EU banned disposal of sewage sludge at sea in 1998, and since then, its application rate to land has risen significantly [29].

Table 1. Selected pharmaceuticals.

Therapeutic Group	Compound and Chemical Structure			
Anxiolytics and Hypnotics (Anx)	Alprazolam (ALP) 	Lorazepam (LOR) 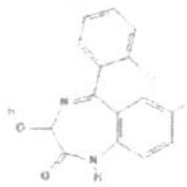	Zolpidem (ZOL) 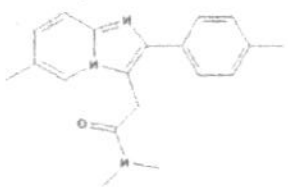	
Antibiotics (Antib)	Azithromycin (AZI) 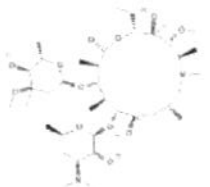	Ciprofloxacin (CIP) 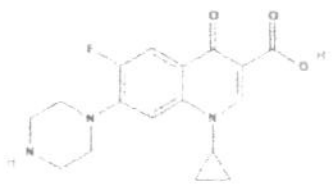	Clarithromycin (CLA) 	Erythromycin (ERY) 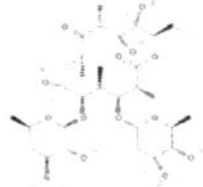
Lipid Regulators (Lip Reg)	Bezafibrate (BEZ) 	Gemfibrozil(GEM) 	Simvastatin (SIM) 	
Antiepileptic (Antiepi)	Carbamazepine (CAR) 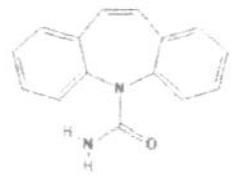			
Selective Serotonin Reuptake Inhibitors (SSRIs)	Citalopram (CIT) 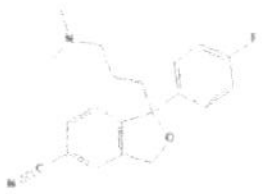	Desmethylcitalopram (N-Cit) (metabolite) 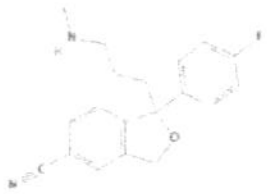	Escitalopram (ESC) 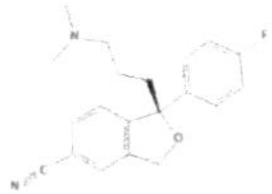	Fluoxetine (FLU) 
	Norfluoxetine (Nor-FLU) (metabolite) 	Paroxetine (PAR) 	Sertraline (SER) 	Desmethylsertraline (Nor-SER) (metabolite) 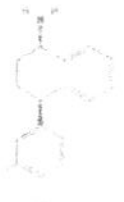

Table 1. *Cont.*

Therapeutic Group	Compound and Chemical Structure			
Anti-Inflammatories (Anti-inf)	Diclofenac (DIC) 	4-hydroxydiclofenac (4-OH-DIC) (metabolite) 	Ibuprofen(IBU) 	Naproxen (NAP) 
	Paracetamol (PARA) 	4-aminophenol (4-PARA) (transformation product) 		
Hormones (Horm)	Estrone (E1) (natural hormone/metabolite) 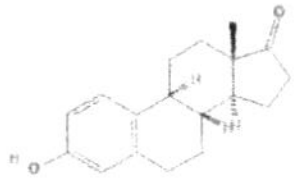	17β-estradiol (E2) 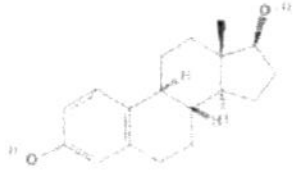	17α-ethinylestradiol (EE2) 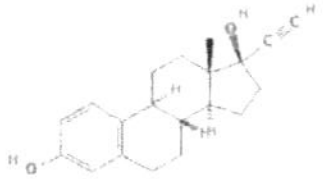	

2.2. Consumption Patterns

The presence of pharmaceuticals in the environment generally correlates well with the amount used in human medicine. Therefore, these data can be used to identify pharmaceuticals that may pose a risk to the environment [30]. An accurate estimate of the extent of drug exposure in a population is difficult in most countries, as precise consumption data are often lacking. In addition, the statistics frequently cover prescription drugs only and do not include over-the-counter medicines or hospital use of pharmaceuticals [31].

Nevertheless, for several reasons, consumption of pharmaceuticals is expected to increase and, thus, increase the burden of their presence in the environment. First, as the number of older people is rising, with frequent therapeutic regimes of five or more medicines, the extensive use of pharmaceuticals will also increase. In addition, with a rise in living standards and with a decrease in pharmaceuticals price, their usage will escalate throughout the world [9].

Bearing in mind the available data on antidepressants and lipid regulators provided by the Organization for Economic Cooperation and Development (OECD), in defined daily dose (DDD), which is calculated per 1000 inhabitants per day, the increased consumption from 2000 to 2015 is clear, with an increase of 30.7 to 60.6 DDD and of 28.1 to 100.7 DDD in antidepressants and lipid regulators, respectively [32].

However, the correlation between consumption data and environmental contamination is related to the amount consumed per year (kg y^{-1}), which may not correspond to a higher DDD, that varies widely between pharmaceuticals. For example, in 2000, approximately 100 million women worldwide were current users of combined hormonal contraceptives; however, since the DDD is very low for hormones, this will not correlate with the amount sold in kg [33].

When observing the pharmaceuticals consumption data on European countries (Table 2), namely, the amount consumed per year, we can realize that the amount used in Switzerland and Sweden is lower than the rest of the countries. This is explained by the fact that they have a significantly lower population when compared to the other countries referred to in Table 2 (Germany, France, Italy and Spain).

Table 2. International consumption of the selected pharmaceuticals.

Therapeutic Group	Pharmaceutical	DDD 1000 inh^{-1} d^{-1}	mg inh^{-1} y^{-1}	kg y^{-1}	Year	Country	Reference
Anx	ALP	17.64 [a]	6.4 [a]	302 [a]	2010	Spain	[27]
		NA	2.9	178	2004	France	
	LOR	19.67 [a]	17.9	844	2010	Spain	[27]
		NA	9.6	585	2004	France	
		13.3	NA	709	2010	Italy	[5]
Antib	AZI	0.9 [a]	98.6	4634 [a]	2010	Spain	[27]
		NA	67.1	4073	2004	France	
		NA	NA	13870	2010	Italy	[34]
		1.3	NA	13870	2010	Italy	[5]
	CIP	1.1 [a]	401.5	18870 [a]	2010	Spain	[27]
		NA	200.7	12186	2004	France	
		NA	NA	21672	2010	Italy	[34]
		1.0	NA	21672	2010	Italy	[5]
	CLA	0.6 [a]	231.0	10864 [a]	2010	Spain	[27]
		NA	150	12360	2010	Germany	
		NA	232.9	1700	2010	Switzerland	
		NA	276.1	16889	2010	France	
		NA	NA	64470	2010	Italy	[34]
		3.0	NA	64470	2010	Italy	[5]
	ERY	0.1 [a]	NA	1716a	2010	Spain	[27]
		NA	NA	0.12	2010	Italy	[34]

Table 2. *Cont.*

Therapeutic Group	Pharmaceutical	DDD 1000 inh^{-1} d^{-1}	mg inh^{-1} y^{-1}	kg y^{-1}	Year	Country	Reference
Lip reg	BEZ	0.6 [a]	133.0 [a]	6178 [a]	2010	Spain	
		NA	475.2	39158	2010	Germany	
		NA	215.6	1574	2010	Switzerland	[27]
		NA	343.4	20852	2004	France	
		NA	66.7	NA	2005	Sweden	
		NA	NA	7600	2001	Italy	[5]
	SIM	NA	282.7 [a]	13340 [a]	2010	Spain	[27]
		NA	114.3	6943	2004	France	
Antiepi	CAR	1.2 [a]	438.0	20595	2010	Spain	
		NA	1010.9	83299	2010	Germany	
		NA	857.5	6260	2010	Switzerland	[27]
		NA	554.3	33364	2010	France	
		NA	463.0	820	2005	Sweden	
		NA	NA	31190	2010	Italy	[34]
		NA	0.61–0.98	NA	2010	Europe	[35]
		NA	NA	31190	2010	Italy	[5]
		NA	NA	88000	2001	Germany	[1]
SSRIs	ESC	0.01 [a]	38.8	1824 [a]	2010	Spain	[27]
		NA	0.08	4.6	2004	France	
	FLU	0.02 [a]	62.0	2914 [a]	2010	Spain	[27]
		NA	61.6	3740	2004	France	
	PAR	0.02 [a]	69.4	3264 [a]	2010	Spain	[27]
		NA	90.8	5515	2004	France	
	SER	0.05 [a]	102.1	4800 [a]	2010	Spain	[27]
		NA	102.5	6224	2004	France	
Anti-inf	DIC	7.9 [a]	369.9	17395 [a]	2010	Spain	
		NA	953.6	78579	2010	Germany	
		NA	934.1	6819	2010	Switzerland	[27]
		NA	370.1	22640	2010	France	
		NA	375.9	NA	2005	Sweden	
		NA	60–880	NA	2009	Europe	[35]
		4.5	NA	9602	2010	Italy	[5]
		NA	NA	345000	2001	Germany	[1]
	IBU	NA	4647.5	218527	2010	Spain	
		NA	3043.6	250792	2010	Germany	
		NA	3078.2	22471	2010	Switzerland	[27]
		NA	953.8	58353	2010	France	
		NA	NA	7864	2005	Sweden	
		NA	NA	622000	2001	Germany	[1]
	NAP	5.15 [a]	1205.9	56700 [a]	2010	Spain	[27]
		NA	614.7	37332	2004	France	
	PARA	NA	22667.7	1065835	2010	Spain	[27]
		NA	54389.5	3303077	2004	France	
		NA	NA	836000	2001	Germany	[1]
Horm	E2	0.894 [a]		12.6 [a]	2010	Spain	[27]
	EE2	1.1969 [a]	0.03	1.2 [a]	2010	Spain	
		NA	0.58	48.2	2001	Germany	[27]
		NA	0.54	4.0	2000	Switzerland	
		NA	0.11	NA	2005	Sweden	

Anx—anxiolytics, Antib—antibiotics, Lip reg—lipid regulators, Antiepi—antiepileptics, Anti-inf—anti-inflammatories, Horm—hormones, NA—not available, DDD—defined daily dose and SSRIs—selective serotonin reuptake inhibitors. [a] Estimated consumption. Data on ZOL, GEM and CIT was not possible to obtain.

Besides the differences in population, different patterns are also observed between countries, even within each therapeutic group; however, some trends are clear regarding the global consumption of therapeutic groups. Anti-inflammatories are clearly the group with higher consumption (in kg), being PARA the pharmaceutical with the highest consumption. This group is followed by the antiepileptic CAR, with particularly high values in Germany. Antibiotics and lipid regulators have similar consumption patterns; nonetheless, these groups have great variations within them, showing distinct trends in different countries. Anxiolytics, SSRIs and hormones, in decreasing order, were the therapeutic groups with the lowest consumptions.

One should note that there are often discrepancies between pharmaceuticals sold and those actually consumed, due to delays between sales and actual use of medication. Moreover, patterns of local consumption might differ from those observed on a national scale [34,35].

2.3. Mechanism of Action, Metabolization and Excretion

Pharmaceuticals have different mechanisms of action resulting in several therapeutical indications, which differ between therapeutic groups. However, within each group, some variations can also occur, since there is more than one class of pharmaceuticals in each group.

The therapeutic group of anxiolytics include pharmaceuticals from the class of benzodiazepines like ALP and LOR, which are used for numerous indications, including anxiety, insomnia, muscle relaxation, relief from spasticity caused by central nervous system pathology and epilepsy. They act by binding to gamma-aminobutyric acid, increasing its activity, reducing the excitability of neurons and promoting a calming effect on the brain [36]. Although the hypnotic ZOL is not a benzodiazepine, it also acts on gamma-aminobutyric acid, promoting a shorter effect than benzodiazepines [37].

The selected antibiotics belong to two different classes, fluoroquinolones (CIP) and macrolides (AZI, CLA and ERY), which inhibit bacterial growth. Fluoroquinolones act by inhibiting bacterial DNA synthesis, and macrolides link to the bacterial ribosomes, inhibiting protein biosynthesis [38,39].

Lipid regulators drugs are used to treat dyslipidaemias; primarily, raised cholesterol. Statins like SIM have the capacity to reduce the endogenous cholesterol synthesis by inhibiting the principal enzyme involved. The fibrates (BEZ and GEM) increase the expression of some proteins in the liver, which results in a substantial decrease in plasma triglycerides and is usually associated with a moderate decrease in cholesterol concentrations [40,41].

The antiepileptic CAR has been extensively used in the treatment of epilepsy, as well as in the treatment of neuropathic pain and affective disorders, mainly due to the inhibition of sodium channel activity [42].

The SSRIs (CIT, ESC, FLU, PAR and SER) are antidepressants that, via inhibition of the serotonin reuptake mechanism, induce an increase in serotonin concentration within the central nervous system [43]. It should be noticed that CIT is a racemic mixture of *R*-citalopram and *S*-citalopram enantiomers with different potencies, but since *S*-citalopram is more potent, it is also marketed as the single *S*-enantiomer formulation ESC [44].

The anti-inflammatories DIC, IBU and NAP are non-steroids, and their mechanism of action is through inhibition of cyclooxygenase (1 and 2) in the periphery and central nervous system, reducing pain and inflammation but also other physiologic processes [45]. As for PARA, it acts on cyclooxygenase (2 and 3) in the central nervous system and only reduces pain and fever [46].

Finally, the hormones E1 and E2 are estrogen sex hormones, mainly female, and although they regulate the reproductive system, they also act in very different endocrine systems. As pharmaceuticals, E2 is mostly used in hormone replacement therapy, and EE2, a synthetic hormone more potent than E2, is primarily used in oral contraception [47,48].

According to other authors, pharmacokinetic data could provide a better knowledge of the environmental fate of pharmaceuticals, especially in the water compartment [30,49].

After consumption, pharmaceuticals are metabolized and primarily excreted in urine and feces as a mixture of the parent compound and its metabolites. The elimination in urine and/or feces is driven by two mechanisms, Phase I and Phase II metabolites. The first one uses the hepatic metabolism and, through biochemical oxidations, reductions and hydrolysis, increases the polarity and water solubility of the metabolites. Phase II metabolites are produced by a biochemical reaction through a conjugation step (i.e., glucuronidation and sulphation), where polar groups are transferred to parent compounds or metabolites, allowing these conjugated metabolites to become enough hydrophilic and water soluble to be eliminated through urine and/or feces [1,50,51]. These processes usually promote the loss of pharmaceutical activity of the compound. However, there are pharmaceuticals that are only active after metabolic activation by enzymatic system(s) of the parent compound (pro-drugs) to metabolite(s) [1].

To determine this pharmacokinetic feature, the proportion of the unchanged active molecule excreted in urine and/or in feces and the proportion of the parent molecule excreted as conjugates (glucuronide and sulphate) was included when available [52,53] (Table 3). The excretion rate, in addition to the consumption data, contributes to either a greater or lesser environmental impact and is related to the reported occurrence of the parent compound and its metabolites in the aquatic compartment [30,54]. Therefore, the excretion features were revised and are presented in Table 3.

Table 3. Excretion rates of the selected pharmaceuticals.

Therapeutic Group	Pharmaceutical	Excretion Results	References
Anx	ALP	20	[55]
	LOR	72.5	[56]
	ZOL	0.75	[57]
Antib	AZI	12	[56]
	CIP	60/83.7	[1]
		70	[5]
		70	[56]
	CLA	25	[58]
		25	[25]
	ERY	25	[49]
		10	[58]
		5	[59]
Lip reg	BEZ	72	[60]
		69	[5]
		47.5	[1]
		50	[61]
		45	[62]
	GEM	50	[63]
	SIM	12.5	[1]
		12.5	[62]
Antiepi	CAR	33	[25]
		5	[64]
		3	[29]
		3	[59]
SSRIs	CIT	23	[56]
		12/20	[65]
	ESC	9	[66]
	FLU	5/10/11	[65]
		10	[28]
	SER	0.2	[56]
		0.2	[28]
		0.2	[65]
	PAR	3	[56]
		3	[28]
		3	[65]
Anti-inf	DIC	39	[5]
		15	[1]
		15	[63]
		15	[60]
		12.5	[62]
	IBU	15	[67]
		10	[68]
		10	[61]
		5	[1]
	NAP	10	[25]
		<1	[59]
	PARA	80	[69]
		75	[56]
Horm	E2	5.6	[70]
	EE2	22/26/27/35/42/53/66/68	[71]

Anx—anxiolytics, Antib—antibiotics, Lip reg—lipid regulators, Antiepi—antiepileptics, SSRIs—selective serotonin reuptake inhibitors, Anti-inf—anti-inflammatories and Horm—hormones.

While several publications are available on the metabolism of pharmaceuticals, the results of these studies can vary. The observed differences are probably explained by genomically distinct metabolizing capacities, as well as differences in race, sex, age and health status of the studied subjects, which are all known to affect the route and rate of metabolism [54,72]. SSRIs are clearly the therapeutic group with lower excretion rates, ranging from 0.2% to 23%, whereas the other groups present higher variability. The compounds with higher excretion rates are CIP (84%), PARA (80%), LOR (73%), BEZ (72%), E2 (68%) and GEM (50%).

3. Physicochemical Properties and Fate

3.1. Physicochemical Properties

The fate and persistence of the excreted pharmaceuticals and/or metabolites in the aquatic environment depend upon their physicochemical properties and the chemical and biological characteristics of the receiving water compartment. Several important chemical measurements of the pharmaceuticals in study, such as pKa (acid dissociation constant), log K_{ow} (octanol-water partitioning coefficient), log D_{ow} (the pH-dependent n-octanol-water distribution ratio), log Koc (soil organic carbon-water partitioning coefficient) and solubility, are presented in Table S1 (supporting information). These features can provide strong evidence of the ionization state of the compounds, their hydrophobicity and can help determining whether they will partition into water, biosolids, sediment and/or biological media [28,73].

Some authors defend that the log K_{ow} and log K_{oc} approaches are excessive restrictive models of pharmaceuticals distribution in the environment. In complex natural water and wastewater samples, partitioning due to hydrophobicity/lipophilicity is not the only physicochemical force of attraction operating between molecules. Electrostatic interactions, chemical bounding and nonspecific forces between ionized molecules and dissolved organic matter are neglected through exclusive log K_{ow} and K_{oc} approaches. Some studies have illustrated that water pH could play an important role in the interactions between organic matter and pH-depending pharmaceuticals, since there is a great variability between these compounds as regard to their pKa (4.0–18.3) [1]. Therefore, the log D_{ow} and log K_{oc} values presented in Table S1 (supporting information) are specific for pH 7.4, a value close to the ones usually observed in the water compartments (wastewater and surface water) [29,73,74].

With a log D_{ow} superior to 1, the likelihood of predominance of the chemical in the aqueous phase decreases logarithmically, whereas below a log D_{ow} of -1, the likelihood of predominance of the chemical in the aqueous phase increases logarithmically. Therefore, compounds having log D_{ow} values between -1 to +1 could be anticipated to be distributed in both the water and organic phases [73].

As seen in Table S1 (supporting information), the physicochemical properties of pharmaceuticals show a high variability. For example, the log D_{ow} ranges from -2.23 to 4.6, the log K_{oc} varies between 0 and 3.88 and even solubility goes from 0.1 to 101,200 (mg L^{-1}). These variations are not only observed between different therapeutic groups but also within each group, since, as previously referred, this pharmaceuticals grouping does not correspond to similar chemical structures and there are more than one class per group. This can be seen especially for antibiotics, lipid regulators and anti-inflammatories, where greater fluctuations in these parameters are reported.

In summary, although pharmaceuticals present different physicochemical properties, some are expected to be more lipophilic and others to sorb to soils and sediments, they all have relatively high water solubility, having the potential to contaminate the aquatic environment [75].

3.2. Fate in Wastewater Treatment Plants

After excretion, pharmaceuticals are transported to WWTPs through the sewer system, and no significant removal occurs during transport in sewer pipes to WWTPs [76]. As hotspots of aquatic contamination, WWTPs play an important role in the life cycle of pharmaceuticals, since many are incompletely removed by conventional treatment processes and behave as persistent organic micropollutants [77].

The removal of pharmaceuticals in WWTPs is a complex phenomenon with many plausible mechanisms; additionally, these facilities are generally not equipped to deal with complex pharmaceuticals, as they were built and upgraded with the principal aim of removing easily or moderately biodegradable carbon, nitrogen and phosphorus compounds and microbiological organisms [18,78]. The main mechanisms involved in the removal of pharmaceuticals by WWTPs are filtration; biodegradation (e.g., oxidation, hydrolysis, demethylation and cleavage of glucuronide conjugates); sorption to sludge or particulate matter (by hydrophobic or electrostatic interactions) and chemical oxidation. Loss by volatilization can be considered as negligible [79–81].

WWTPs employ a primary, a secondary and an optional tertiary treatment process, being the last one is always associated with a high treatment cost. During primary treatment, physical removal of solids is achieved through a sieve, regularly followed by coagulation-flocculation processes for the removal of particulate matter, as well as colloids and some dissolved substances; however, this process is ineffective for the elimination of pharmaceuticals [82]. In the secondary treatment, usually with activated sludges, pharmaceuticals are subjected to a range of processes, including dispersion, dilution, partition, biodegradation and abiotic transformation, being biodegradation and sorption to solids are the main removal pathways of pharmaceuticals during this biological treatment. Afterwards, some WWTPs possess tertiary treatments like advanced oxidation processes, ultraviolet radiation (UV) or ozonation [82,83]. Most of the WWTPs in northern Europe comprise tertiary wastewater treatment; however, in other countries, they are less frequent [18].

Besides the type of wastewater treatment, WWTPs' efficiency in removing pharmaceuticals is influenced by operational and environmental conditions, namely, the hydraulic retention time (HRT) (high HRT allows reactions like biodegradation and sorption mechanisms to occur); solid retention time (SRT) (which controls the size and diversity of the microbial community, and higher SRT will facilitate the build-up of slowly growing bacteria enhancing removal); environmental temperature (since higher temperatures reflect superior removal efficiencies) and pH conditions (affecting on the degradation kinetics of the compounds) [50,78,82,84,85].

As previously mentioned, the physicochemical characteristics of the pharmaceuticals also affect their removal in WWTPs. Since a significant part of the removal process is through sorption or biodegradation in sludge, the ability to interact with solid particles plays a major role. Thus, compounds with low sorption coefficients tend to remain in the aqueous phase, favoring their mobility through the WWTPs and into the receiving waters [86,87]. Independently of their physicochemical characteristics, some authors state that the portion of some pharmaceuticals in the treated sludge is negligible (<20%) when compared to the aqueous fraction for NAP, DIC, BEZ, GEM, LOR and CAR, although higher sorption removals were noted for selected compounds (AZI, CIP, IBU, PAR and PARA) [29,85].

Generally, during secondary treatment, compounds with log D_{ow} higher than 3, which indicates high sorption potential, tend to be removed through sorption onto sewage sludge, while compounds with log D_{ow} between 1.5 and 3 are removed mainly by biodegradation. The remaining pharmaceuticals with log D_{ow} inferior to 1.5 tend to remain dissolved [50,80,82,88]. Therefore, it is expected that the removal efficiency of substances with higher log D_{ow} are more influenced by SRT, while compounds with low log D_{ow} are more influenced by HRT [78]. During the secondary treatment, besides sorption to sludges, another removal mechanism is through microbial degradation, where nitrifiers are the most important group. This mechanism has been described as the main removal pathway for polar acidic pharmaceuticals; however, they are also sensitive to inhibitors, and some pharmaceuticals can have this effect on these microorganism [89,90].

Currently, besides the conventional treatments, new methodologies have been applied as tertiary treatments with higher removal efficiencies, but some of these new methods have high construction, maintenance and energy costs associated [77]. Advanced oxidation processes that include UV, ozone and hydrogen peroxide, among others, can also be used. UV treatment has been shown to partially remove some pharmaceuticals; however, it does not completely eliminate them [49,64,91,92]. Ozonation alone promotes the partial oxidation of pharmaceuticals, and to overcome this drawback, this process has been combined with heterogeneous catalysts or membrane technologies, such as nanoparticles of titanium dioxide, a known photocatalyst [11,77,82]. Adsorption by activated carbon is another methodology that proves to be effective in removing pharmaceuticals, with powdered activated carbon and granular activated carbon widely used in these adsorption processes. Generally, efficient removals are obtained when the compounds have nonpolar characteristics, as well as matching pore size/shape requirements. The main advantage of using activated carbon to remove pharmaceuticals is that it does not generate toxic or pharmacologically active products [82,93].

More recently, the growing trend of improving sustainability and reducing energy demands in WWTPs have encouraged alternative methods, such as algae ponds for secondary effluent polishing, with promising results [29].

As previously referred, metabolization in the human body can lead to elimination of pharmaceuticals conjugates. However, these phase II metabolites can be converted back into the parent compound, especially in WWTPs, being infrequently found in surface waters. One of the mechanisms used is the action of a β-glucuronidase enzyme produced by *Escherichia coli* capable of deconjugating the β-glucuronated pharmaceuticals excreted by the human body, resulting in the release of the active pharmaceutical into the wastewater [29,50,89,94,95]. On the other hand, the WWTPs processes responsible for pharmaceuticals elimination do not commonly lead to their complete mineralization; instead, breakdown products can emerge, which can also be toxic to the environment. In general, there is still a knowledge gap concerning the generation of metabolites and transformation products of known contaminants, which can potentially be as hazardous, or even more, than the parent compounds and can be present in different aquatic bodies at a higher concentration than parent compounds [90,96–98].

Naturally, the type of treatment can affect not only the removal efficiencies but also the metabolites and transformation products generated.

This supports the need for the evaluation of metabolites and transformation products and the further development of new treatment techniques to achieve complete mineralization of emerging contaminants [90,97]. Besides the fact that some of the new treatments, like advanced oxidation processes, can originate toxic transformation products, they have higher efficiencies when compared to traditional treatments [77,82,99,100].

Data from 52 publications were collected, and removal efficiencies of the selected pharmaceuticals are summarized in Figure 1. One should note that, although we are comparing the fate of pharmaceuticals in WWTPs, there are some countries with inadequate wastewater and collection infrastructures or even functional WWTPs. For example, in Ghana and India, only 7.9% and 30.7% of the wastewaters are treated, which anticipates that the presence of pharmaceuticals in the aquatic environment in these countries should represent an even bigger problem [101].

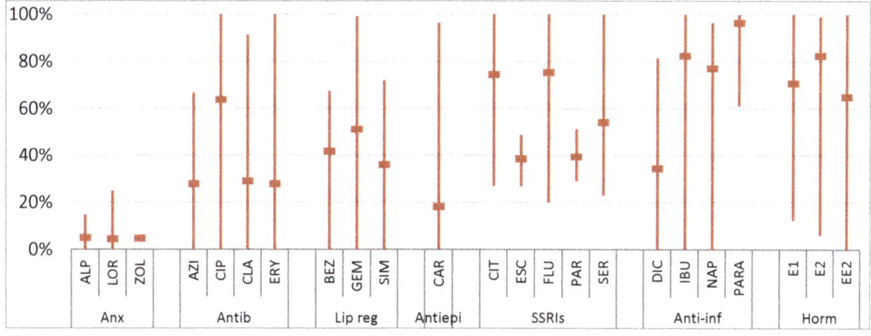

Figure 1. Minimum, maximum and average removal efficiencies in WWTPs (%). Anx—anxiolytics, Antib—antibiotics, Lip reg—lipid regulators, Antiepi—antiepileptics, SSRIs—selective serotonin reuptake inhibitors, Anti-inf—anti-inflammatories and Horm—hormones [3,5,13,16,18,51,59,63,67,68,71,78–82,85,87,88,92,99,102–132].

Although, as mentioned, some studies indicate that physicochemical properties set the efficiency of removal of pharmaceuticals in WWTPs, the literature review performed showed that the target compounds present very different removal rates, ranging between negative and high removal rates, and no obvious pattern in behavior was observed, even within the same therapeutic group, implying that factors other than compound-specific properties affect removal efficiency [68,85]. Negative values for some compounds have been reported and may reflect deconjugation of metabolites during the treatment process or changes in the adsorption to particles during treatment [133]. Generally, what becomes evident is that the elimination of most pharmaceuticals is incomplete, and it is not exclusively related neither to the physicochemical properties nor to the type of treatment processes. Additionally, most pharmaceuticals have always one report that shows no removal [16,18,85,88].

Concerning the removal efficiencies of each therapeutic group, anxiolytics present the lowest average, having a small variation due to their similar physicochemical properties, with values ranging from 0% to 25%. Although their log D_{ow} (from 2.49 to 3.06), higher than most of the selected pharmaceuticals, predicted large sorption to sludge and higher removal rates, this was not observed in real removal data.

As for antibiotics, the range observed in the removal efficiencies was from 0% to 100%, similar to anti-inflammatories and hormones. The average removal rates for AZI, CLA and ERY (macrolides) are near 30%, whereas CIP presented higher removal rates (64%). Despite the lower log D_{ow} for CIP (-2.23) sorption to sludges, it has been suggested as the primary removal mechanism for fluoroquinolones, whereas, for macrolides, limited sorption to sludge is observed [108,132,134].

Although the therapeutic group of lipid regulators encloses a statin (SIM) and fibrates (BEZ and GEM) and their removals vary between 0% and 99%, their averages are similar, ranging from 36% to 51%, being also found in sludges [33].

For CAR, although presenting a lower log D_{ow} (2.28) than anxiolytics and a wide range of removal efficiencies, it is one of the most persistent compounds and is averagely reduced by only 18.1% [135,136]. This pharmaceutical is very resistant to wastewater treatments, since it has low biological degradation and sorption and has only higher removal rates with the use of advanced treatments such as ozonation together with the usage of the photocatalyst titanium dioxide [134,135].

Regarding SSRIs, even though they all belong to the same group, the average removal efficiencies range from 39% to 75%, with ESC, PAR and SER presenting lower values, below 55%, when compared to CIT and FLU that present higher removal rates, 75%.

The most investigated therapeutic group in WWTPs are anti-inflammatories, and despite their high variability, average removal rates are above 77% and up to 96% (PARA), with the exception for DIC (34%) [82,135]. Excluding DIC, anti-inflammatories undergo sorption to sludges and biological and photolytic degradation [33,82,89,96,137]. As for DIC, sorption to sludge and biodegradability have been reported but to a lower extent, translating into low elimination rates during wastewater treatment; moreover, a low removal efficiency of 4-OH-DIC has been reported in WWTPs [89]. Advanced oxidation processes are described as highly efficient for DIC removal, since it rapidly decomposes by direct photo-oxidation, indicating that this pathway is one of its main degradation mechanisms. However, ozonation alone is not completely effective, but the O_3/H_2O_2 system shows high efficacy [11,135]. On the other hand, PARA, which has the higher removal rate during wastewater treatment, can generate different transformation products, being 4-PARA was identified as the main one, and its presence in wastewater samples was already reported. However, there are other possible sources, since it is also widely used in industrial applications and is a known transformation product from pesticides. Furthermore, 4-PARA was also described as the primary degradation product of PARA during storage [138].

Hormones are the therapeutic group with higher log Dow and high average removal efficiency, which ranges from 65% to 82%. This low variation was expected, since the molecules have similar physicochemical properties [82]. Although most hormone conjugates are degraded in the WWTPs, some are still observed in WWEs representing less than 33% of the parent compound (E1 and E2), which can be reconverted back into the parent compound in the environment [50,139]. It is also possible that E2 can be converted in E1 in the WWTPs, possibly explaining the higher removal rate for this pharmaceutical [71]. Once again, advanced oxidation processes are described as highly efficient processes in hormone removal [11].

As observed, the WWTPs are unable to completely remove the pharmaceuticals, and through direct discharge of WWEs in surface water or by land application of WWTPs' sludge or through leaching, these facilities are the major sources of pharmaceuticals in the environment [29,59,79,140,141].

Optimization of wastewater treatment still remains a task of high priority. Biological treatment is commonly unable to remove pharmaceuticals; however, its efficacy can be improved under favorable conditions. Although advanced treatment technologies, such as membrane and advanced oxidation processes, have been promising for pharmaceuticals removal, high operation costs and formation of degradation products still remain an issue [82].

3.3. Fate in Surface Waters

Since WWTPs are not able to completely remove pharmaceuticals, they are disseminated through their WWEs and sludges, mostly, into surface waters. In the aquatic environment, the fate and concentration of pharmaceuticals can be reliant on the receiving water body flow rate, partitioning to sediments, biological entities and consequent degradation, uptake by biota, volatilization, photodegradation or transformation through other abiotic mechanisms, such as hydrolysis [29,74,134,142].

When WWEs reach the surface waters, the dilution effect varies significantly due to different flows in different rivers; however, this effect can be relatively low, especially in arid or semi-arid regions due to water scarcity, like some Iberian rivers, where other processes gain relative importance [143,144]. Although multiple biotic and abiotic routes could transform pharmaceuticals once they reach the surface water, the predominant pathways to remove pharmaceuticals are photodegradation and sorption [77,143].

The fate of different pharmaceuticals has already been studied in surface waters by several authors using estimates of mass loading, dilution and in-stream attenuation, here understood as the reduction of the concentration of pharmaceuticals along the river segment by processes different from dilution [28,74,98,141,143].

Overall, it is expected that the log D_{ow} of a given compound influences its in-stream attenuation; in the case of hydrophobic compounds (with higher log D_{ow}), sorption to suspended particles and sediments is a dominant process leading to in-stream attenuation by reducing the concentration in the aqueous phase along the river segment [74]. In this way, these compounds become less exposed to other biotic (biotransformation) and abiotic (photolysis and volatilization) transformation processes and, therefore, become less affected by the variation of environmental conditions between river segments. Therefore, it is expected that compounds with low log D_{ow} show not only more differences in attenuation rates between sites but also more temporal differences (i.e., seasonal and day–night) within each site [143]. This sorption mechanism in the aquatic environment represents an important sink for pharmaceuticals, as it has been suggested that strong pharmaceutical interactions may act as a long-term storage of pharmaceuticals that will increase their persistence, while their bioavailability in the environment is reduced, being recalcitrant to microbial degradation [28,33]. In fact, the sediments could be a source of contaminants in downstream river segments if resuspension of fine-grained bedded sediments occurs, for instance, during seasonal increases in flow rate or during flood events [143]. Moreover, the activity of benthic invertebrate in sediments can result in an increased desorption, leading to improved bioavailability in the water compartment [29]. Additionally, sorption to colloids can also provide an important sink for the pharmaceuticals in the aquatic environment, increasing their persistence while reducing their bioavailability. In general, sorption may result in a biased risk estimation [9].

As already referred, in complex natural waters, electrostatic interactions, chemical bounding and nonspecific forces between ionized molecules and dissolved organic matter can also occur, meaning that we cannot generalize the attenuation of a compound based on its physicochemical properties alone [98,143]. However, the different log D_{ow} of pharmaceuticals influence the variability of rates among rivers, likely due to its effect on sorption to sediments and suspended particles, and therefore, influence the balance between the different attenuation mechanisms (biotransformation, photolysis and sorption) [143].

The attenuation of pharmaceuticals was evaluated in surface water in Spain where the total concentration of pharmaceuticals (CLA, DIC, IBU, BEZ, GEM, CAR and CIT) decreased about 40% in less than 5 km, although the number of compounds detected only decreased 13% [74]. Studies also reported that GEM is a quite persistent compound in surface water, with half-lives ranging from 70 to 288 days [137]. As for CIP, photodegradation is reported to be the main mechanism of attenuation [90]. However, for CAR, there are reports evaluated in a Swedish lake where no attenuation was observed and with an estimated half-life of 780-5700 days [98]. This was also supported by other studies that revealed that CAR and IBU were stable against sunlight, while PARA suffers moderate photodegradation and DIC was rapidly photodegraded in surface water [90,145]. Accordingly, another study noticed that no biodegradation of IBU was observed in a sterile river, but in river water and using microbial biofilms, biodegradation occurred in a few hours, evidencing that although its transformation is a complex process, microorganisms play an important role in IBU degradation [137]. Concerning SSRIs, which have high sorption coefficients, they have proven to be persistent compounds, and FLU demonstrated that it was far more resistant to photolysis than the other SSRIs, with a half-life of 122 days [28].

Besides the presence of the parent compounds in surface waters, sulphate conjugates of E1 and E2 have already been observed. Although these conjugates no longer possess a significant biological activity, they can act as precursor steroid reservoirs that might be converted into free estrogens [128,139]. Even though the synthetic hormone EE2 has lower solubility than E2, it is also considerably more persistent in the aquatic environment, with an estimated half-life in surface water between 1.5 and 17 days [146].

In addition to the parent compounds, some studies also addressed the contribution of WWTPs for pharmaceuticals transformation products in surface waters and confirmed that these facilities were a major source of contamination to the recipients [74,98].

In summary, on one hand, the emissions from WWEs vary widely because of differences in regional usage of the compounds and efficiency of WWTPs. On the other hand, the processes that drive in-stream attenuation (i.e., biotransformation, photolysis, sorption and volatilization) depend on the different pharmaceutical characteristics, as well as on a series of physicochemical and biological parameters of the river, such as river flow rate, temperature, the vertical hydrological exchange between surface and subsurface compartments, turbidity, dissolved oxygen concentration, biofilm biomass and pH [143]. The magnitude of the measured attenuation rates urges scientists to consider them as important as dilution when aiming to predict concentrations in freshwater ecosystems. Since pharmaceuticals are continuously introduced in surface waters and are not completely removed, they eventually will reach groundwater, seawater, mineral water and drinking water, contaminating all aquatic compartments [98].

4. Occurrence

Along with advances in analytical instruments and techniques, trace levels of various pharmaceuticals and their metabolites have been detected in the aquatic compartment since the latter half of the 1970s [145].

A literature review on worldwide monitoring programs in recent years, presented in Figures 2–5 and Tables S2–S5 (supporting information), clearly reveals the ubiquitous distribution of pharmaceuticals in different aquatic environment compartments, including WWIs, WWEs and surface waters, with concentrations up to mg L^{-1} [145,147]. Usually, this occurrence is related to the gross domestic product per capita of each country and is presented as the shape of an inverted-U; i.e., pollution worsens as the economy of countries starts to grow (increased consumption of pharmaceuticals), and then it improves when countries reach a higher stage of economic growth (improved WWTPs) [101].

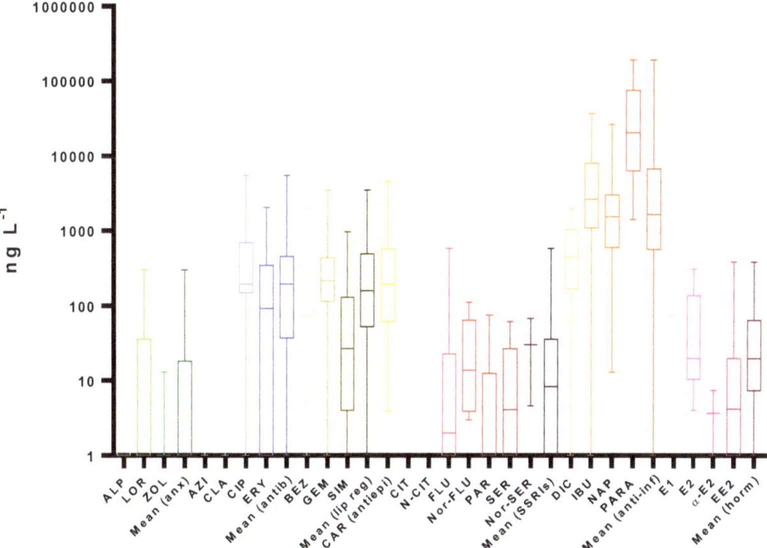

Figure 2. Boxplots with median, maximum and minimum average concentrations of pharmaceuticals in wastewater influents (WWIs). Anx—anxiolytics, Antib—antibiotics, Lip reg—lipid regulators, Antiepi—antiepileptics, SRRIs—selective serotonin uptake inhibitors, Anti-inf—anti-inflammatories and Horm—hormones [3,13,15,16,18,29,34,59,63,67,68,71,78,79,82,83,86,87,94,96,100,102,107–109,111–115,117–120,122,123,126,128,130–133,140,147–170].

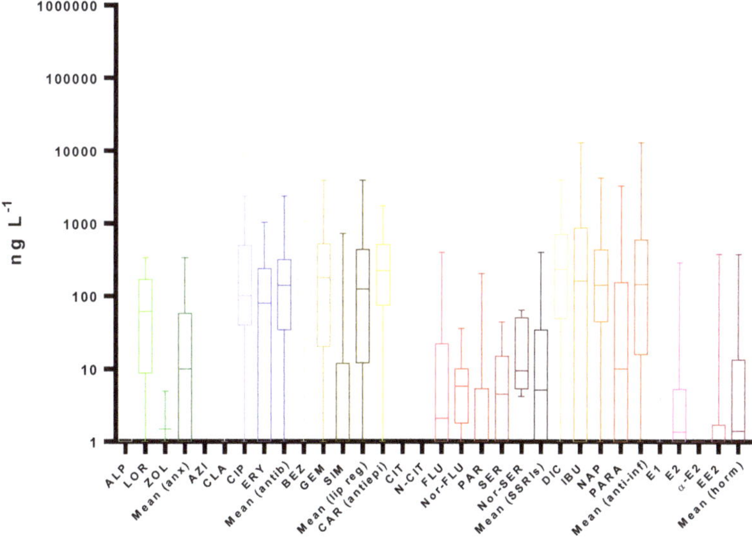

Figure 3. Boxplots with median, maximum and minimum average concentrations of pharmaceuticals in wastewater effluents (WWEs). Anx—anxiolytics, Antib—antibiotics, Lip reg—lipid regulators, Antiepi—antiepileptics, SSRIs—selective serotonin reuptake inhibitors, Anti-inf—anti-inflammatories, Horm—hormones and different letters represent significant statistical differences) [3,5,6,8,13,15,16,18,29,34,59,63,67,68,71,78,79,82,83,86,87,91,94,96,100,102,103,107–109,111–115,117–120,122,126,128–133,136,139–141,146,147,149,150,152–183].

Beside the aspects previously referred, several others can influence the concentration of pharmaceuticals in the different aquatic compartments, promoting a great variability in the detected concentrations. In WWTPs, other aspects that can influence the detected concentrations are the flow rate, the time of the year, the temperature, the type of WWTPs, day and the type of sampling, etc. [105]. As for surface waters, the flow rate, temperature, sunlight, time of the year, day and the type of sampling are also parameters that can influence pharmaceuticals concentrations [23]. Moreover, some of these parameters can also influence the detected concentrations in other water bodies.

4.1. Wastewater

4.1.1. Wastewater Influents

Figure 2 and Table S2 (supporting information) summarizes the median, averages and maximum concentrations of the targeted pharmaceuticals in the WWIs across the world, collected from 66 references. These concentrations are likely to be influenced by both consumption data and excretion rates.

All investigated pharmaceuticals were frequently detected in WWIs, with PARA, CIT, IBU, CAR, BEZ, CLA and α-E2 (E2 isomer) presenting detection frequencies higher than 88%. As for the different therapeutic groups, antiepileptics and anti-inflammatories were the ones with higher detection frequencies, above 86%, followed by lipid regulators (75%) and hormones (74%). Anxiolytics were the group with lower values (31%), much different from the other groups. The highest median concentration (1.7 $\mu g\ L^{-1}$) was observed in the anti-inflammatories group, with statistical differences for all of the other therapeutic groups, being the maximum individual concentration observed for IBU (700 $\mu g\ L^{-1}$) [78]. Antibiotics, lipid regulators and the antiepileptics had median concentrations between 160 and 196 $ng\ L^{-1}$, followed by the other groups, with medians under 20 $ng\ L^{-1}$.

Although anxiolytics were the group with the lower detection frequency and median, ALP had concentrations up to 4.7 $\mu g\ L^{-1}$. Additionally, the highest detection frequency belonged to LOR, with 38% [150]. These results are in line with data already mentioned, such as the low consumption and low excretion rates observed for this therapeutic group. The anxiolytic with the highest excretion rates and consumption is LOR, which is reflected on the occurrence reported.

Antibiotics were the most homogenous group, with median concentrations ranging from 93 to 324 $ng\ L^{-1}$ and with all detection frequencies above 65%. Although some discrepancies in excretion rates, with higher values for CIP, both CIP and CLA have higher consumptions, being this pattern was observed in the occurrence data.

Lipid regulators occurrence data was comparable to that of antibiotics, mostly because of similar consumption and excretion rates. Within this group, we can observe that the one with the highest consumption in most countries, SIM, had the lowest detection frequency and median concentration in WWIs. This can be due to a significant difference in excretion data, where BEZ have clearly higher rates than SIM, with excretion values up to 72% and 12.5%, respectively [1,60]. Therefore, it is shown that a pharmaceutical with low consumption can reach relatively high detection frequencies and median concentration in WWIs (89% and 271 $ng\ L^{-1}$, respectively).

The antiepileptic CAR with excretion rates up to 33%, and whose consumption is only surpassed by anti-inflammatories, had a detection frequency of 89% and concentrations up to 22 $\mu g\ L^{-1}$ [25,111].

Like anxiolytics, SSRIs also had low consumption and excretion rates, which reflected also in low concentrations in the WWIs, with a median concentration of 8 $ng\ L^{-1}$. However, this group presented some peculiarities, SER being one of them. This SSRI has the highest consumption in European countries. Nonetheless, due to its very low excretion rate (0.2%), this compound and its metabolite (Nor-SER) present lower median concentrations than CIT and Nor-FLU [56]. On the other hand, despite the low consumption data for CIT, its higher excretion rate explains the fact that this SSRI and its metabolite (N-CIT) are the ones with the highest concentrations within this therapeutic group, followed by FLU and its metabolite (Nor-FLU), that also present higher excretion rates (up to 11%) [65].

As referred, anti-inflammatories were the group with higher concentrations in WWIs, not only due to their high consumption but also to significant excretion rates (up to 80%), with median concentrations of 450, 1550, 2680 and 20 601 ng L^{-1} for DIC, NAP, IBU and PARA, respectively [69].

In the hormones group, although there were lower excretion rates observed for E2, its higher consumption (2.5 kg y^{-1}) when compared to EE2 (0.7 kg y^{-1}) resulted in higher concentrations even for its metabolite E1, being even present in the enantiomer of E2 (α-E2) up to 10 μg L^{-1} [155]. As previously mentioned, one should also take into account that both E1 and E2 are produced in the human body and can be excreted naturally [71,128].

These data highlight that pharmaceutical compounds with low excretion rates are not necessarily present at low levels in WWIs, because this could be offset by the massive use of these compounds [82]. Additionally, it was also observed that, in general, the mean pharmaceutical concentrations could vary between 1 to 3 orders of magnitude from one sampling day or week to the next. Diurnal trends were also observed, and peak concentrations were highly unpredictable [150].

4.1.2. Wastewater Effluents

The first report of human pharmaceuticals in WWEs is from 1976, and subsequent studies have confirmed the presence of pharmaceuticals in this aquatic compartment [170]. After passing through WWTPs and being submitted to the different treatments already discussed, it would be expected that WWEs presented lower concentrations than the influent, with a decrease proportional to the removal efficiency of the WWTPs [18].

Data regarding 87 references were collected and summarized in Figure 3 and Table S3 (supporting information). In the effluents, the median concentrations of the therapeutic groups varied from 1.4 ng L^{-1}, for hormones, to 226 ng L^{-1}, for antiepileptics, and, in general, significantly lower concentrations were found when comparing to influent samples, as shown in Figure 2. However, since concentrations in WWIs, as well as removal efficiencies, have a wide variability, the range of concentrations in WWEs is still high [78].

In general, regarding the median concentrations, antiepileptics were followed by anti-inflammatories (146 ng L^{-1}), antibiotics (142 ng L^{-1}) and lipid regulators (126 ng L^{-1}), a similar pattern to that in WWIs but with no statistical significance between them. The remaining three groups had lower medians, with 10, 5.2 and 1.4 ng L^{-1} for anxiolytics, SSRIs and hormones, respectively. The highest individual mean concentration observed was for DIC 233 ng L^{-1}; however, the maximum concentration regarded CIP, 14 mg L^{-1}. This high value, along with others that are completely offset, were observed in the effluents of pharmaceutical industries and hospitals [25,26,111,183].

Anxiolytics were the only therapeutic group with a clear higher median and individual concentrations in WWEs than in WWIs and surpassed the mean concentration of hormones and SSRIs. This is justified by the fact that anxiolytics have the lowest removal efficiencies, and, in some cases, even negative values are found. This increased concentration in WWEs is related to the transformation of metabolites and/or transformation products back into the parent compounds during wastewater treatment [80,82]. Since all the three compounds have similar removal efficiencies, LOR, with the highest concentration in WWIs, presented again the highest values in WWEs, both median (61 ng L^{-1}) and individual (438 ng L^{-1}) levels [94].

As indicated in Table S3 (supporting information), CLA was once again the antibiotic more frequently detected in WWEs (87%), and this group remained the most homogenic, with median concentrations ranging from 80 to 200 ng L^{-1}. The extremely high value found for CIP was observed in the effluent of a pharmaceutical industry [111].

As regard to the antiepileptic CAR, the fact that it does not adsorb to soils and has low removal efficiencies in WWTPs results in a small increased median from WWIs to WWEs, from 193 to 226 ng L^{-1}, respectively [184].

Lipid regulators having removal efficiencies analogous to those observed for antibiotics present an occurrence pattern in WWEs comparable to that of WWIs, again with SIM presenting the lowest median concentration (1 ng L^{-1}).

The therapeutic group SSRIs had also the same pattern observed in WWIs, with CIT and N-CIT presenting the higher median concentrations of 73 and 107 ng L^{-1}, respectively, and, once again, the metabolites (N-CIT, Nor-FLU and Nor-SER) concentrations were in the same range or higher as the parent compounds [118]. The highest value regarded CIT with 430 μg L^{-1}, which was also detected in a pharmaceutical industry effluent [111].

Anti-inflammatories had one of the highest removal efficiencies, only comparable to hormones, and although they remain with a high median concentration, the difference to the other therapeutic groups (antiepileptics, lipid regulators and antibiotics) was significantly reduced. Within this therapeutic group, DIC presented the highest median concentration, followed by IBU, NAP and PARA, with 163, 142 and 10 ng L^{-1}, respectively, meaning that PARA shifted from the highest median concentration in WWIs to the fourth in WWEs, mainly due to the high removal average (96%) presented.

As for hormones, with average removal efficiencies above 60%, concentrations were also significantly reduced, with the highest median concentration belonging to E1 (14 ng L^{-1}) and the lowest to α-E2 (0.4 ng L^{-1}); the highest individual value was also for α-E2 (4.7 μg L^{-1}), observed in only one study [155].

Despite these concentrations, it is possible that some conjugates, which were not evaluated, enter surface waters, where they can be reconverted back to the parent compound, increasing the pharmaceuticals contamination burden [29].

As expected, some positive correlation could be observed between the concentrations found in WWIs and in WWEs with removal efficiencies. Nonetheless, even at relatively low population densities and low industrial and hospital activity, human pharmaceuticals are present at quantifiable levels in WWEs [170].

4.2. Surface Water

The release of WWEs into surface water, in comparison to other sources, has been considered the main cause of the presence of pharmaceuticals in this water body [59,184].

As previously discussed, following the treatment processes in WWTPs, pharmaceuticals are subjected to different degrees of natural attenuation. These conditions can promote a variation higher than one order of magnitude in the same sampling location and even higher between different rivers [19]. Due to these factors, pharmaceutical compounds are expected to occur in surface waters at lower levels than in WWEs [82,98,185].

Since 1970, the issue regarding the presence of chemicals in surface waters has been addressed by the EU. Nowadays, the chemical quality of surface waters is controlled under the WFD (Directive 2000/60/EC of the European Parliament and of the council of 23 October 2000, establishing a framework for community action in the field of water policy), transposed into the Portuguese legal system by the Law N 58/2005 of 29 December 2005 (the Water law). Within this framework, the key strategy adopted was the establishment of priority substances or groups of substances due to their persistence, toxicity, bioaccumulation, widespread use and detection in rivers, lakes, transitional and coastal waters. Additionally, a list of environmental quality standards have been issued for these substances, to ensure adequate protection of the aquatic environment and human health [8]. Although no pharmaceutical belongs to this list, their environmental presence in surface waters is a growing problem that must be tackled and was addressed by the WFD in order to minimize their aquatic environmental contamination and support future prioritization measures. Despite this awareness, legal limits have not yet been set for pharmaceuticals in surface water, although a watch list that includes seven pharmaceuticals (E2, EE2, AZI, CLA, ERY, amoxicilin and CIP) and one metabolite (E1) has been recently established [17,34,186,187]. IBU has also been proposed to enter this list; however, its inclusion was rejected in January 2012 owing to a lack of sufficient evidence of significant risks to aquatic environments [9].

According to the Directive 2013/39/EU strategy, all member states shall monitor each substance in the watch list at selected surface waters representative monitoring stations at least once per year. The number of monitoring stations varies within each member state, taking into account the population and area of each country. About 40% of European water bodies still have an unknown chemical status, as not even the monitoring of the EU priority substances have been performed [21].

After reviewing 88 scientific references, as expected, lower median concentrations (ten times lower) were found in surface waters than in WWEs (Figure 4 and Table S4 (supporting information).

We can observe similar patterns in WWEs, with the same four therapeutic groups presenting higher median concentrations, anti-inflammatories (34 ng L^{-1}), antiepileptics (28 ng L^{-1}), antibiotics (20 ng L^{-1}) and lipid regulators (16 ng L^{-1}). These four therapeutic groups had statistically significant higher median concentrations than SSRIs and hormones. SSRIs, hormones and anxiolytics, with notably lower values, had the lowest median concentrations of 0.8, 0.4 and 0 ng L^{-1}, respectively. The highest values observed were reported for CIP in India, with a maximum concentration of 650 μg L^{-1} for CIP [6].

Regarding anxiolytics, only LOR and ALP were found in surface waters, with a detection frequency of 30%. ZOL was evaluated in only one study, which did not detect it [188].

As above mentioned, antibiotics were one of the therapeutic groups with high median concentrations (20 ng L^{-1}). It also presented two extremely high average concentrations detected for CIP in surface waters near pharmaceutical industries in Pakistan (1.3 μg L^{-1}) and in India (164 μg L^{-1}); however, all the other average concentrations were below 108 ng L^{-1} [6,189]. Comparing the antibiotics concentrations with WWEs, a very similar pattern was observed, with a tendency for a relative higher detection frequency and concentration for ERY, probably revealing a higher persistency in the environment.

Lipid regulators presented similar patterns than in WWE, with SIM being the one with the lower median concentration. BEZ, apparently, presented higher persistence, since its detection frequency and median concentration, 67% and 22 ng L^{-1}, respectively, surpassed those of GEM, 51% and 19 ng L^{-1}, respectively.

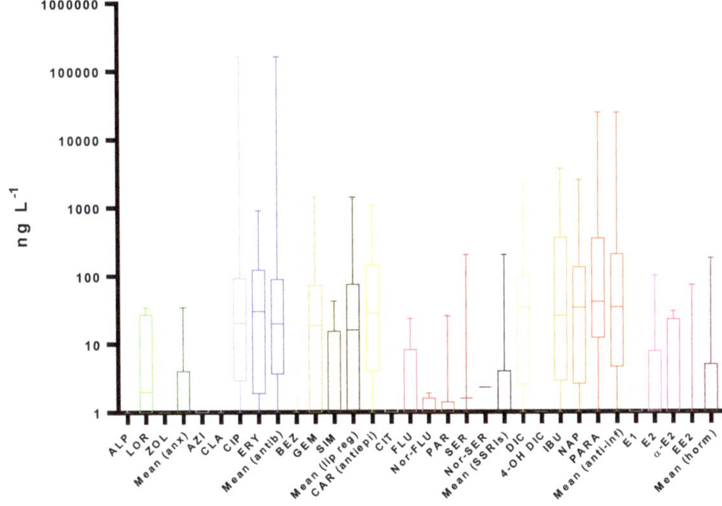

Figure 4. Boxplots with median, maximum and minimum average concentrations of pharmaceuticals in surface waters. Anx—anxiolytics, Antib—antibiotics, Lip reg—lipid regulators, Antiepi—antiepileptics, SSRIs—selective serotonin reuptake inhibitors, Anti-inf—anti-inflammatories, Horm—hormones and different letters represent significant statistical differences [1,5,6,29,34,58,59,73,74,82,96,98,101,113,119, 122,128,129,132,134,136,138–141,145–147,154,155,157–159,161,163,164,166,167,169–171,174–176,178–182,184–186,189–224].

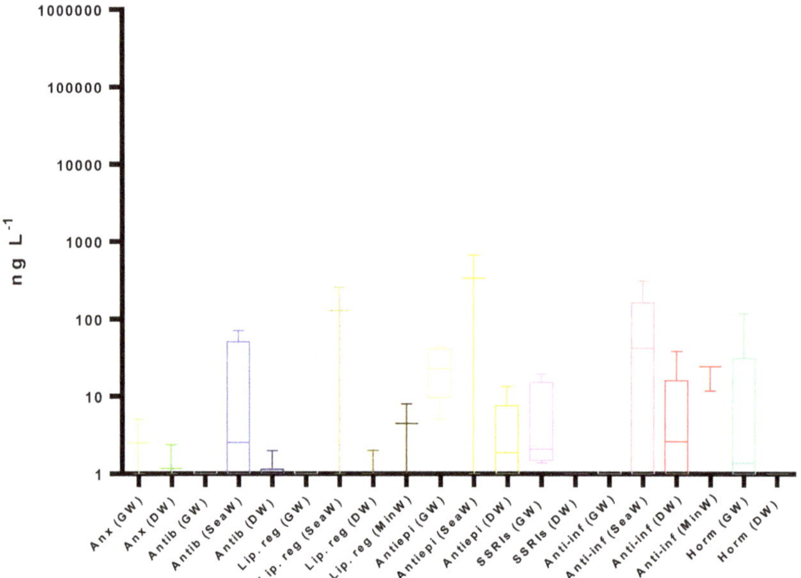

Figure 5. Boxplots with median, maximum and minimum average concentrations of pharmaceuticals in other water bodies. GW—groundwaters, SeaW—seawaters, DW—drinking waters, MinW—mineral waters, Anx—anxiolytics, Antib—antibiotics, Lip reg—lipid regulators, Antiepi—antiepileptics, SSRIs selective serotonin reuptake inhibitors, Anti-inf—anti-inflammatories, Horm—hormones and different letters represent significant statistical differences [1,6,82,91,113,128,136,141,145,166,167,174,179,189,193,200,206,209,213,215,217,225–231].

As previously noted, CAR continued among the most frequently detected pharmaceutical compounds in surface waters (78%) and presented concentrations up to 12 μg L^{-1}, reflecting, as expected, the recalcitrant nature of this molecule, given its high half-life [190]. In fact, it is also one of the most frequently detected pharmaceuticals in European surface waters [137].

The SSRIs values decreased from WWEs to surface waters in median concentrations (from 5.2 to 0.8 ng L^{-1}) and in detection frequencies from 55% to 26%. The highest concentration regarded CIT (76 μg L^{-1}); however, it was found, once again, near a pharmaceutical industry in India [6]. The metabolites suffer even a higher reduction than the parent compounds.

Anti-inflammatories presented, once again, higher concentrations when comparing with other therapeutic groups [170]. PARA presented the higher median concentration (41 ng L^{-1}), followed by DIC and NAP (34 ng L^{-1}) and IBU (26 ng L^{-1}). The results of PARA, higher than the ones in WWE, move PARA to values in the same range as the other anti-inflammatories. Looking at the detection frequencies, they all fall in the same range, from 52% to 59%. In this group, another extremely high concentration was observed for PARA in Kenya 107 μg L^{-1} [166]. Although, in wastewaters, no study on 4-OH-DIC was reviewed, in surface waters, two studies were found and 40 ng L^{-1} was the highest concentration found for this metabolite [191]. The average concentration observed for DIC (221 ng L^{-1}) was twice the purposed value of 100 ng L^{-1} for the environmental quality standard in 2012-2013. The high values in surface waters possibly raised some issues regarding the establishment of this standard.

Within the hormones group, E1 presented the higher median concentration (2.1 ng L^{-1}), and the highest average value was detected in China, 180 ng L^{-1}, whereas its detection frequency was slightly decreased (from 57% to 54%) [192]. Contrary to what was previously mentioned, namely that EE2 was more persistent than E2, EE2 registered a higher decrease in detection frequency (from 25% to

2%) than E2 (from 43% to 22%). In surface waters, conjugates of both E1 and E2 were also found in a concentration range from a quarter to half of the parent compound [139,193].

As above mentioned, lower concentrations of pharmaceuticals (ten times lower) were found in surface waters than in WWEs. Surface waters showed an overall trend of higher concentrations in sites influenced by the location of WWTPs [104,194].

4.3. Other Water Bodies

As discussed earlier, the concentrations of pharmaceuticals decrease from the WWIs to WWEs and to surface waters through different mechanisms. However, data collected from 28 references showed that pharmaceuticals can reach groundwaters, seawaters and even mineral waters and drinking waters (Figure 5 and Table S5 (supporting information)). Regarding groundwaters, it is important to underline that this is an important resource of water supply in the world, and it is especially vulnerable to contamination, although soil provides a big inertia to propagation of the contamination, and for that same reason, once contaminated, the effects can hardly ever be reverted [225].

The concentrations in remaining waters bodies should be lower than the previous ones, since they suffer attenuation mechanisms similar to surface water. Additionally, drinking water has dedicated treatment plants. However, these facilities do not completely remove pharmaceuticals and can also produce transformation products that can be toxic [145,173,199].

Although susceptible to degradation or transformation, pharmaceuticals' continuous introduction into the aquatic environment confers some degree of pseudo-persistence, reaching, at extremely low concentrations, all aquatic compartments all over the world, even drinking waters [64,91]. However, it is unlikely that pharmaceuticals pose significant threats to human health at the concentrations that may occur in drinking waters [145,231].

In Figure 5, we observe that, once again, antibiotics, lipid regulators, antiepileptics and anti-inflammatories had higher detection frequencies and median concentrations; however, CAR stands out from the others with a higher detection frequency and average concentration of 45% and 60 ng L^{-1}, respectively. Groundwater and seawater were the water bodies with higher detection frequencies and concentrations, and the highest concentration found was of 14 μg L^{-1} for CIP in groundwater [6]. No statistical significance was observed between the different therapeutic group averages.

5. Final Remarks

A careful literature review was conducted in order to understand the sources, fate and occurrence of pharmaceuticals in the aquatic environment. In this context, a broad and specialized background was obtained, enabling a complete overview of the state-of-the-art in these subjects.

The data provided in this review evidenced that WWTPs are the major source of pharmaceuticals contamination. It is also noteworthy that pharmaceuticals belonging to the same therapeutic group can have distinct physicochemical properties, resulting in different behaviours both in WWTPs and in the aquatic environment.

The concentrations of pharmaceuticals found in the aquatic bodies were, in decreasing order, WWIs, WWEs, surface water and other water bodies.

Overall, these results present a global picture of the pharmaceuticals' contamination, an important input for setting prioritizing measures and sustainable strategies to minimize their impact in the aquatic environment.

Supplementary Materials: The following are available online: Table S1: Physicochemical properties of the selected pharmaceuticals (adapted from Chemspider, Drugbank, Pubchem and ECOSARv1.11), Table S2: Occurrence of pharmaceuticals in wastewaters influents (WWIs), Table S3: Occurrence of pharmaceuticals in wastewater efluents (WWEs), Table S4: Occurrence of pharmaceuticals in surface waters (SWs) and Table S5: Occurrence of pharmaceuticals in seawaters (SeaW), groundwaters (GWs), drinking waters (DWs) and mineral waters (DWs).

Author Contributions: Conceptualization, A.P., L.S. and A.P.; methodology, A.P., L.S. and A.P.; formal analysis, A.P.; writing—original draft preparation, A.P.; writing—review and editing, A.P., L.S. and C.L. (Célia Laranjeiro); supervision, C.L. (Celeste Lino) and A.P. and project administration, A.P. All authors have read and agreed to the published version of the manuscript.

Funding: The work was supported by UIDB/50006/2020, with funding from FCT/MCTES through national funds.

Conflicts of Interest: The authors declare no conflicts of interest.

References

1. Mompelat, S.; Le Bot, B.; Thomas, O. Occurrence and fate of pharmaceutical products and by-products, from resource to drinking water. *Environ. Int.* **2009**, *35*, 803–814. [CrossRef] [PubMed]
2. Petrovic, M.; Barceló, D. LC-MS for identifying photodegradation products of pharmaceuticals in the environment. *TrAC Trends Anal. Chem.* **2007**, *26*, 486–493. [CrossRef]
3. Seifrtová, M.; Pena, A.; Lino, C.M.; Solich, P. Determination of fluoroquinolone antibiotics in hospital and municipal wastewaters in Coimbra by liquid chromatography with a monolithic column and fluorescence detection. *Anal. Bioanal. Chem.* **2008**, *391*, 799–805. [CrossRef] [PubMed]
4. Focazio, M.J.; Kolpin, D.W.; Furlong, E.T. Occurrence of human pharmaceuticals in water resources of the United States: A review. In *Pharmaceuticals in the Environment: Sources, Fate, Effects and Risks*; Springer: Berlin/Heidelberg, Germany, 2004; pp. 91–105.
5. Al Aukidy, M.; Verlicchi, P.; Jelic, A.; Petrovic, M.; Barcelò, D. Monitoring release of pharmaceutical compounds: Occurrence and environmental risk assessment of two WWTP effluents and their receiving bodies in the Po Valley, Italy. *Sci. Total Environ.* **2012**, *438*, 15–25. [CrossRef] [PubMed]
6. Fick, J.; Söderström, H. Contamination of surface, ground, and drinking water from pharmaceutical production. *Environ. Toxicol. Chem.* **2009**, *28*, 2522–2527. [CrossRef] [PubMed]
7. Jelic, A. *Occurrence and Fate of Pharmaceuticals in Wastewater Treatment Processes*; Universitat de Barcelona: Barcelona, Spain, 2012.
8. Bueno, M.J.M.; Gomez, M.J.; Herrera, S.; Hernando, M.D.; Agüera, A.; Fernández-Alba, A.R. Occurrence and persistence of organic emerging contaminants and priority pollutants in five sewage treatment plants of Spain: two years pilot survey monitoring. *Environ. Pollut.* **2012**, *164*, 267–273. [CrossRef] [PubMed]
9. Kümmerer, K. Pharmaceuticals in the Environment. *Annu. Rev. Environ. Resour.* **2010**, *35*, 57–75. [CrossRef]
10. Nikolaou, A.; Meric, S.; Fatta, D. Occurrence patterns of pharmaceuticals in water and wastewater environments. *Anal. Bioanal. Chem.* **2007**, *387*, 1225–1234. [CrossRef]
11. Ribeiro, A.R.; Nunes, O.C.; Pereira, M.F.R.; Silva, A.M.T. An overview on the advanced oxidation processes applied for the treatment of water pollutants defined in the recently launched Directive 2013/39/EU. *Environ. Int.* **2015**, *75*, 33–51. [CrossRef]
12. Seifrtová, M.; Aufartová, J.; Vytlacilová, J.; Pena, A.; Solich, P.; Nováková, L. Determination of fluoroquinolone antibiotics in wastewater using ultra high-performance liquid chromatography with mass spectrometry and fluorescence detection. *J. Sep. Sci.* **2010**, *33*, 2094–2108. [CrossRef]
13. Sousa, M.A.; Gonçalves, C.; Cunha, E.; Hajšlová, J.; Alpendurada, M.F. Cleanup strategies and advantages in the determination of several therapeutic classes of pharmaceuticals in wastewater samples by SPE-LC-MS/MS. *Anal. Bioanal. Chem.* **2011**, *399*, 807–822. [CrossRef] [PubMed]
14. Loos, R.; Carvalho, R.; Comero, S.; António, D.; Ghiani, M.; Lettieri, T.; Locoro, G.; Paracchini, B.; Tavazzi, S.; Gawlik, B.; et al. *EU Wide Monitoring Survey on Waste Water Treatment Plant Effluents*; European Comission Joint Research Centre Institute for Environment and Sustainability: Ispra, Italy, 2012; ISBN 978-92-79-26784-0.
15. Salgado, R.; Noronha, J.P.; Oehmen, A.; Carvalho, G.; Reis, M.A.M. Analysis of 65 pharmaceuticals and personal care products in 5 wastewater treatment plants in Portugal using a simplified analytical methodology. *Water Sci. Technol.* **2010**, *62*, 2862. [CrossRef] [PubMed]
16. Santos, L.H.M.L.M.; Gros, M.; Rodriguez-Mozaz, S.; Delerue-Matos, C.; Pena, A.; Barceló, D.; Montenegro, M.C.B.S.M. Contribution of hospital effluents to the load of pharmaceuticals in urban wastewaters: identification of ecologically relevant pharmaceuticals. *Sci. Total Environ.* **2013**, *461–462*, 302–316. [CrossRef] [PubMed]
17. European Commission Comission Implementing Decision (EU) 2018/840 of 5 June 2018 establishing a watch list of substances for Union-wide monitoring in the field of water policy pursuant to Directive 2008/105/EC of the European Parliament and of the Council and repealing 495. *Off. J. Eur. Union* **2018**, *L141*, 9–12.
18. Papageorgiou, M.; Kosma, C.; Lambropoulou, D. Seasonal occurrence, removal, mass loading and environmental risk assessment of 55 pharmaceuticals and personal care products in a municipal wastewater treatment plant in Central Greece. *Sci. Total Environ.* **2016**, *543*, 547–569. [CrossRef]

19. ter Laak, T.L.; Van der Aa, M.; Houtman, C.J.; Stoks, P.G.; Van Wezel, A.P. Relating environmental concentrations of pharmaceuticals to consumption: A mass balance approach for the river Rhine. *Environ. Int.* **2010**, *36*, 403–409. [CrossRef]
20. Bade, R.; Rousis, N.I.; Bijlsma, L.; Gracia-Lor, E.; Castiglioni, S.; Sancho, J.V.; Hernandez, F. Screening of pharmaceuticals and illicit drugs in wastewater and surface waters of Spain and Italy by high resolution mass spectrometry using UHPLC-QTOF MS and LC-LTQ-Orbitrap MS. *Anal. Bioanal. Chem.* **2015**, *407*, 8979–8988. [CrossRef]
21. Altenburger, R.; Ait-Aissa, S.; Antczak, P.; Backhaus, T.; Barceló, D.; Seiler, T.-B.; Brion, F.; Busch, W.; Chipman, K.; de Alda, M.L.; et al. Future water quality monitoring — Adapting tools to deal with mixtures of pollutants in water resource management. *Sci. Total Environ.* **2015**, *512–513*, 540–551. [CrossRef]
22. Radjenović, J.; Petrović, M.; Barceló, D. Fate and distribution of pharmaceuticals in wastewater and sewage sludge of the conventional activated sludge (CAS) and advanced membrane bioreactor (MBR) treatment. *Water Res.* **2009**, *43*, 831–841. [CrossRef]
23. Pereira, A.M.P.T.; Silva, L.J.G.; Laranjeiro, C.S.M.; Meisel, L.M.; Lino, C.M.; Pena, A. Human pharmaceuticals in Portuguese rivers: The impact of water scarcity in the environmental risk. *Sci. Total Environ.* **2017**, *609*, 1182–1191. [CrossRef]
24. Daughton, C.G.; Ruhoy, I.S. Environmental footprint of pharmaceuticals: the significance of factors beyond direct excretion to sewers. *Environ. Toxicol. Chem.* **2009**, *28*, 2495–2521. [CrossRef] [PubMed]
25. Mendoza, A.; Aceña, J.; Pérez, S.; López de Alda, M.; Barceló, D.; Gil, A.; Valcárcel, Y. Pharmaceuticals and iodinated contrast media in a hospital wastewater: A case study to analyse their presence and characterise their environmental risk and hazard. *Environ. Res.* **2015**, *140*, 225–241. [CrossRef] [PubMed]
26. Herrmann, M.; Olsson, O.; Fiehn, R.; Herrel, M.; Kümmerer, K. The significance of different health institutions and their respective contributions of active pharmaceutical ingredients to wastewater. *Environ. Int.* **2015**, *85*, 61–76. [CrossRef] [PubMed]
27. Ortiz de García, S.; Pinto Pinto, G.; García Encina, P.; Irusta Mata, R. Consumption and occurrence of pharmaceutical and personal care products in the aquatic environment in Spain. *Sci. Total Environ.* **2013**, *444*, 451–465. [CrossRef] [PubMed]
28. Silva, L.J.G.; Lino, C.M.; Meisel, L.M.; Pena, A. Selective serotonin re-uptake inhibitors (SSRIs) in the aquatic environment: an ecopharmacovigilance approach. *Sci. Total Environ.* **2012**, *437*, 185–195. [CrossRef] [PubMed]
29. Petrie, B.; Barden, R.; Kasprzyk-Hordern, B. A review on emerging contaminants in wastewaters and the environment: Current knowledge, understudied areas and recommendations for future monitoring. *Water Res.* **2014**, *72*, 3–27. [CrossRef] [PubMed]
30. Almeida, A.; Duarte, S.; Nunes, R.; Rocha, H.; Pena, A.; Meisel, L. Human and Veterinary Antibiotics Used in Portugal—A Ranking for Ecosurveillance. *Toxics* **2014**, *2*, 188–225. [CrossRef]
31. Grung, M.; Källqvist, T.; Sakshaug, S.; Skurtveit, S.; Thomas, K.V. Environmental assessment of Norwegian priority pharmaceuticals based on the EMEA guideline. *Ecotoxicol. Environ. Saf.* **2008**, *71*, 328–340. [CrossRef]
32. OECD Pharmaceutical consumption. *Heal. glanceOECD Indic.* **2012**, 88–89.
33. Zenker, A.; Cicero, M.R.; Prestinaci, F.; Bottoni, P.; Carere, M. Bioaccumulation and biomagnification potential of pharmaceuticals with a focus to the aquatic environment. *J. Environ. Manage.* **2014**, *133*, 378–387. [CrossRef]
34. Verlicchi, P.; Al Aukidy, M.; Jelic, A.; Petrović, M.; Barceló, D. Comparison of measured and predicted concentrations of selected pharmaceuticals in wastewater and surface water: a case study of a catchment area in the Po Valley (Italy). *Sci. Total Environ.* **2014**, *470–471*, 844–854. [CrossRef] [PubMed]
35. Oosterhuis, M.; Sacher, F.; Ter Laak, T.L. Prediction of concentration levels of metformin and other high consumption pharmaceuticals in wastewater and regional surface water based on sales data. *Sci. Total Environ.* **2013**, *442*, 380–388. [CrossRef] [PubMed]
36. Iii, C.E.G.; Kaye, A.M.; Pharm, D.; Bueno, F.R.; Kaye, A.D. Benzodiazepine pharmacology and central nervous system – Mediated effects. *Ochsner J.* **2013**, 214–223.
37. Crestani, F.; Martin, J.R.; Möhler, H.; Rudolph, U. Mechanism of action of the hypnotic zolpidem in vivo. *Br. J. Pharmacol.* **2000**, *131*, 1251–1254. [CrossRef] [PubMed]
38. Tenson, T.; Lovmar, M.; Ehrenberg, M. The mechanism of action of macrolides, lincosamides and streptogramin B reveals the nascent peptide exit path in the ribosome. *J. Mol. Biol.* **2003**, *330*, 1005–1014. [CrossRef]
39. Blondeau, J.M. Fluoroquinolones: mechanism of action, classification, and development of resistance. *Surv. Ophthalmol.* **2004**, *49*, S73–S78. [CrossRef] [PubMed]

40. Staels, B.; Dallongeville, J.; Auwerx, J.; Schoonjans, K.; Leitersdorf, E.; Fruchart, J.C. Mechanism of action of fibrates on lipid and lipoprotein metabolism. *Circulation* **1998**, *98*, 2088–2093. [CrossRef]
41. Stancu, C.; Sima, A. Statins: mechanism of action and effects. *J. Cell. Mol. Med.* **2001**, *5*, 378–387. [CrossRef]
42. Ambrósio, A.F.; Soares-da-Silva, P.; Carvalho, C.M.; Carvalho, A.P. Mechanisms of Action of Carbamazepine and Its Derivatives, Oxcarbazepine, BIA 2-093, and BIA 2-024. *Neurochem. Res.* **2002**, *27*, 121–130. [CrossRef]
43. Kreke, N.; Dietrich, D.R. Physiological endpoints for potential SSRI interactions in fish. *Crit. Rev. Toxicol.* **2008**, *38*, 215–247. [CrossRef] [PubMed]
44. Kosjek, T.; Heath, E. Tools for evaluating selective serotonin re-uptake inhibitor residues as environmental contaminants. *TrAC Trends Anal. Chem.* **2010**, *29*, 832–847. [CrossRef]
45. Vane, J.R.; Botting, R.M. Anti-inflammatory drugs and their mechanism of action. *Inflamm. Res.* **1998**, *47*, 78–87. [CrossRef] [PubMed]
46. Aronoff, D.M.; Oates, J.A.; Boutaud, O. New insights into the mechanism of action of acetaminophen: Its clinical pharmacologic characteristics reflect its inhibition of the two prostaglandin H2 synthases. *Clin. Pharmacol. Ther.* **2006**, *79*, 9–19. [CrossRef] [PubMed]
47. Brann, D.W.; Hendry, L.B.; Mahesh, V.B. Emerging diversities in the mechanism of action of steroid hormones. *J. Steroid Biochem. Mol. Biol.* **1995**, *52*, 113–133. [CrossRef]
48. Rivera, R.; Yacobson, I.; Grimes, D. The mechanism of action of hormonal contraceptives and intrauterine contraceptive devices. *Am. J. Obstet. Gynecol.* **1999**, *181*, 1263–1269. [CrossRef]
49. WHO. *Pharmaceuticals in drinking water*; World Health Organization: Geneva, Switzerland, 2011.
50. Evgenidou, E.N.; Konstantinou, I.K.; Lambropoulou, D.A. Occurrence and removal of transformation products of PPCPs and illicit drugs in wastewaters: A review. *Sci. Total Environ.* **2015**, *505*, 905–926. [CrossRef]
51. Leclercq, M.; Mathieu, O.; Gomez, E.; Casellas, C.; Fenet, H.; Hillaire-Buys, D. Presence and fate of carbamazepine, oxcarbazepine, and seven of their metabolites at wastewater treatment plants. *Arch. Environ. Contam. Toxicol.* **2009**, *56*, 408–415. [CrossRef]
52. Straub, J. An Environmental Risk Assessment for Human-Use Trimethoprim in European Surface Waters. *Antibiotics* **2013**, *2*, 115–162. [CrossRef]
53. Coetsier, C.M.; Spinelli, S.; Lin, L.; Roig, B.; Touraud, E. Discharge of pharmaceutical products (PPs) through a conventional biological sewage treatment plant: MECs vs PECs? *Environ. Int.* **2009**, *35*, 787–792. [CrossRef]
54. Pereira, A.M.P.T.; Silva, L.J.G.; Lino, C.M.; Meisel, L.M.; Pena, A. A critical evaluation of different parameters for estimating pharmaceutical exposure seeking an improved environmental risk assessment. *Sci. Total Environ.* **2017**, *603C–604C*, 226–236. [CrossRef]
55. Fraser, A.D.; Bryan, W.; Isner, A.F. Urinary Screening for Alprazolam and its Major Metabolites by the Abbott ADx and TDx Analyzers with Confirmation by GC/MS. *J. Anal. Toxicol.* **1991**, *15*, 25–29. [CrossRef] [PubMed]
56. Medical Products Agency–Sweden Medical Products Agency–Sweden. Available online: https://lakemedelsverket.se/english/ (accessed on 1 January 2015).
57. Hempel, G.; Blaschke, G. Direct determination of zolpidem and its main metabolites in urine using capillary electrophoresis with laser-induced fluorescence detection. *J. Chromatogr. B. Biomed. Appl.* **1996**, *675*, 131–137. [CrossRef]
58. Calamari, D.; Zuccato, E.; Castiglioni, S.; Bagnati, R.; Fanelli, R. Strategic Survey of Therapeutic Drugs in the Rivers Po and Lambro in Northern Italy. *Environ. Sci. Technol.* **2003**, *37*, 1241–1248. [CrossRef]
59. Kasprzyk-Hordern, B.; Dinsdale, R.M.; Guwy, A.J. The removal of pharmaceuticals, personal care products, endocrine disruptors and illicit drugs during wastewater treatment and its impact on the quality of receiving waters. *Water Res.* **2009**, *43*, 363–380. [CrossRef] [PubMed]
60. Ternes, T. Occurrence of drugs in German sewage treatment plants and rivers. *Water Res.* **1998**, *32*. [CrossRef]
61. Zuccato, E.; Castiglioni, S.; Fanelli, R. Identification of the pharmaceuticals for human use contaminating the Italian aquatic environment. *J. Hazard. Mater.* **2005**, *122*, 205–209. [CrossRef]
62. Jjemba, P.K. Excretion and ecotoxicity of pharmaceutical and personal care products in the environment. *Ecotoxicol. Environ. Saf.* **2006**, *63*, 113–130. [CrossRef]
63. Bendz, D.; Paxéus, N.A.; Ginn, T.R.; Loge, F.J. Occurrence and fate of pharmaceutically active compounds in the environment, a case study: Höje River in Sweden. *J. Hazard. Mater.* **2005**, *122*, 195–204. [CrossRef]
64. Czech, B.; Buda, W. Photocatalytic treatment of pharmaceutical wastewater using new multiwall-carbon nanotubes/TiO2/SiO2 nanocomposites. *Environ. Res.* **2015**, *137*, 176–184. [CrossRef]

65. Calisto, V.; Esteves, V.I. Psychiatric pharmaceuticals in the environment. *Chemosphere* **2009**, *77*, 1257–1274. [CrossRef]
66. Rao, N. The clinical pharmacokinetics of escitalopram. *Clin. Pharmacokinet.* **2007**, *46*, 281–290. [CrossRef] [PubMed]
67. Gómez, M.J.; Martínez Bueno, M.J.; Lacorte, S.; Fernández-Alba, A.R.; Agüera, A. Pilot survey monitoring pharmaceuticals and related compounds in a sewage treatment plant located on the Mediterranean coast. *Chemosphere* **2007**, *66*, 993–1002. [CrossRef] [PubMed]
68. Nakada, N.; Tanishima, T.; Shinohara, H.; Kiri, K.; Takada, H. Pharmaceutical chemicals and endocrine disrupters in municipal wastewater in Tokyo and their removal during activated sludge treatment. *Water Res.* **2006**, *40*, 3297–3303. [CrossRef] [PubMed]
69. Kasprzyk-Hordern, B.; Dinsdale, R.M.; Guwy, A.J. Multi-residue method for the determination of basic/neutral pharmaceuticals and illicit drugs in surface water by solid-phase extraction and ultra performance liquid chromatography-positive electrospray ionisation tandem mass spectrometry. *J. Chromatogr. A* **2007**, *1161*, 132–145. [CrossRef] [PubMed]
70. Beer, T.; Gallagher, T.F. Excretion of estrogen metabolistes by humans II. The fate of large doses of estradiol-17beta after intramuscular and oral administration. *J. Biol. Chem.* **1955**, *214*, 315–364.
71. Johnson, A.C.; Williams, R.J. A Model To Estimate Influent and Effluent Concentrations of Estradiol, Estrone, and Ethinylestradiol at Sewage Treatment Works. *Environ. Sci. Technol.* **2004**, *38*, 3649–3658. [CrossRef] [PubMed]
72. Boxall, A.B.A.; Keller, V.D.J.; Straub, J.O.; Monteiro, S.C.; Fussell, R.; Williams, R.J. Exploiting monitoring data in environmental exposure modelling and risk assessment of pharmaceuticals. *Environ. Int.* **2014**, *73*, 176–185. [CrossRef] [PubMed]
73. Jones-Lepp, T.L.; Sanchez, C.; Alvarez, D.A.; Wilson, D.C.; Taniguchi-Fu, R.L. Point sources of emerging contaminants along the Colorado River Basin: Source water for the arid Southwestern United States. *Sci. Total Environ.* **2012**, *430*, 237–245. [CrossRef] [PubMed]
74. Huerta, B.; Rodriguez-Mozaz, S.; Nannou, C.; Nakis, L.; Ruhí, A.; Acuña, V.; Sabater, S.; Barcelo, D. Determination of a broad spectrum of pharmaceuticals and endocrine disruptors in biofilm from a waste water treatment plant-impacted river. *Sci. Total Environ.* **2016**, *540*, 241–249. [CrossRef]
75. Cunningham, V.L.; Binks, S.P.; Olson, M.J. Human health risk assessment from the presence of human pharmaceuticals in the aquatic environment. *Regul. Toxicol. Pharmacol.* **2009**, *53*, 39–45. [CrossRef]
76. Jelic, A.; Rodriguez-Mozaz, S.; Barceló, D.; Gutierrez, O. Impact of in-sewer transformation on 43 pharmaceuticals in a pressurized sewer under anaerobic conditions. *Water Res.* **2014**, *8*. [CrossRef] [PubMed]
77. He, Y.; Sutton, N.B.; Rijnaarts, H.H.H.; Langenhoff, A.A.M. Degradation of pharmaceuticals in wastewater using immobilized TiO2 photocatalysis under simulated solar irradiation. *Appl. Catal. B Environ.* **2016**, *182*, 132–141. [CrossRef]
78. Verlicchi, P.; Al Aukidy, M.; Zambello, E. Occurrence of pharmaceutical compounds in urban wastewater: removal, mass load and environmental risk after a secondary treatment–a review. *Sci. Total Environ.* **2012**, *429*, 123–155. [CrossRef] [PubMed]
79. Miège, C.; Choubert, J.M.; Ribeiro, L.; Eusèbe, M.; Coquery, M. Fate of pharmaceuticals and personal care products in wastewater treatment plants-conception of a database and first results. *Environ. Pollut.* **2009**, *157*, 1721–1726. [CrossRef] [PubMed]
80. Blair, B.; Nikolaus, A.; Hedman, C.; Klaper, R.; Grundl, T. Evaluating the degradation, sorption, and negative mass balances of pharmaceuticals and personal care products during wastewater treatment. *Chemosphere* **2015**, *134*, 395–401. [CrossRef] [PubMed]
81. Salgado, R.; Marques, R.; Noronha, J.P.; Carvalho, G.; Oehmen, A.; Reis, M.A.M. Assessing the removal of pharmaceuticals and personal care products in a full-scale activated sludge plant. *Environ. Sci. Pollut. Res. Int.* **2012**, *19*, 1818–1827. [CrossRef] [PubMed]
82. Luo, Y.; Guo, W.; Ngo, H.H.; Nghiem, L.D.; Hai, F.I.; Zhang, J.; Liang, S.; Wang, X.C. A review on the occurrence of micropollutants in the aquatic environment and their fate and removal during wastewater treatment. *Sci. Total Environ.* **2014**, *473–474*, 619–641. [CrossRef]
83. Gao, P.; Ding, Y.; Li, H.; Xagoraraki, I. Occurrence of pharmaceuticals in a municipal wastewater treatment plant: mass balance and removal processes. *Chemosphere* **2012**, *88*, 17–24. [CrossRef]

84. Santos, J.L.; Aparicio, I.; Callejón, M.; Alonso, E. Occurrence of pharmaceutically active compounds during 1-year period in wastewaters from four wastewater treatment plants in Seville (Spain). *J. Hazard. Mater.* **2009**, *164*, 1509–1516. [CrossRef]
85. Jelic, A.; Gros, M.; Ginebreda, A.; Cespedes-Sánchez, R.; Ventura, F.; Petrovic, M.; Barcelo, D. Occurrence, partition and removal of pharmaceuticals in sewage water and sludge during wastewater treatment. *Water Res.* **2011**, *45*, 1165–1176. [CrossRef]
86. Rosal, R.; Rodríguez, A.; Perdigón-Melón, J.A.; Petre, A.; García-Calvo, E.; Gómez, M.J.; Agüera, A.; Fernández-Alba, A.R. Occurrence of emerging pollutants in urban wastewater and their removal through biological treatment followed by ozonation. *Water Res.* **2010**, *44*, 578–588. [CrossRef] [PubMed]
87. Behera, S.K.; Kim, H.W.; Oh, J.-E.; Park, H.-S. Occurrence and removal of antibiotics, hormones and several other pharmaceuticals in wastewater treatment plants of the largest industrial city of Korea. *Sci. Total Environ.* **2011**, *409*, 4351–4360. [CrossRef] [PubMed]
88. Gardner, M.; Jones, V.; Comber, S.; Scrimshaw, M.D.; Coello-Garcia, T.; Cartmell, E.; Lester, J.; Ellor, B. Performance of UK wastewater treatment works with respect to trace contaminants. *Sci. Total Environ.* **2013**, *456–457*, 359–369. [CrossRef] [PubMed]
89. Vieno, N.; Sillanpää, M. Fate of diclofenac in municipal wastewater treatment plant—A review. *Environ. Int.* **2014**, *69C*, 28–39. [CrossRef] [PubMed]
90. Kosjek, T.; Heath, E.; Pérez, S.; Petrović, M.; Barceló, D. Metabolism studies of diclofenac and clofibric acid in activated sludge bioreactors using liquid chromatography with quadrupole – time-of-flight mass spectrometry. *J. Hydrol.* **2009**, *372*, 109–117. [CrossRef]
91. McEneff, G.; Barron, L.; Kelleher, B.; Paull, B.; Quinn, B. A year-long study of the spatial occurrence and relative distribution of pharmaceutical residues in sewage effluent, receiving marine waters and marine bivalves. *Sci. Total Environ.* **2014**, *476–477*, 317–326. [CrossRef] [PubMed]
92. Roberts, P.H.; Thomas, K.V. The occurrence of selected pharmaceuticals in wastewater effluent and surface waters of the lower Tyne catchment. *Sci. Total Environ.* **2006**, *356*, 143–153. [CrossRef]
93. Rivera-Utrilla, J.; Sánchez-Polo, M.; Ferro-García, M.Á.; Prados-Joya, G.; Ocampo-Pérez, R. Pharmaceuticals as emerging contaminants and their removal from water. A review. *Chemosphere* **2013**, *93*, 1268–1287. [CrossRef]
94. Subedi, B.; Kannan, K. Occurrence and fate of select psychoactive pharmaceuticals and antihypertensives in two wastewater treatment plants in New York State, USA. *Sci. Total Environ.* **2015**, *514*, 273–280. [CrossRef]
95. Badia-Fabregat, M.; Lucas, D.; Gros, M.; Rodríguez-Mozaz, S.; Barceló, D.; Caminal, G.; Vicent, T. Identification of some factors affecting pharmaceutical active compounds (PhACs) removal in real wastewater. Case study of fungal treatment of reverse osmosis concentrate. *J. Hazard. Mater.* **2015**, *283*, 663–671. [CrossRef]
96. Ferrando-Climent, L.; Collado, N.; Buttiglieri, G.; Gros, M.; Rodriguez-Roda, I.; Rodriguez-Mozaz, S.; Barceló, D. Comprehensive study of ibuprofen and its metabolites in activated sludge batch experiments and aquatic environment. *Sci. Total Environ.* **2012**, *438*, 404–413. [CrossRef] [PubMed]
97. Boix, C.; Ibáñez, M.; Sancho, J.V.; Parsons, J.R.; de Voogt, P.; Hernández, F. Biotransformation of pharmaceuticals in surface water and during waste water treatment: Identification and occurrence of transformation products. *J. Hazard. Mater.* **2016**, *302*, 175–187. [CrossRef] [PubMed]
98. Li, Z.; Sobek, A.; Radke, M. Fate of pharmaceuticals and their transformation products in four small european rivers receiving treated wastewater. *Environ. Sci. Technol.* **2016**, *50*, 5614–5621. [CrossRef] [PubMed]
99. Schaar, H.; Clara, M.; Gans, O.; Kreuzinger, N. Micropollutant removal during biological wastewater treatment and a subsequent ozonation step. *Environ. Pollut.* **2010**, *158*, 1399–1404. [CrossRef] [PubMed]
100. Sui, Q.; Huang, J.; Deng, S.; Chen, W.; Yu, G. Seasonal variation in the occurrence and removal of pharmaceuticals and personal care products in different biological wastewater treatment processes. *Environ. Sci. Technol.* **2011**, *45*, 3341–3348. [CrossRef] [PubMed]
101. Segura, P.A.; Takada, H.; Correa, J.A.; El Saadi, K.; Koike, T.; Onwona-Agyeman, S.; Ofosu-Anim, J.; Sabi, E.B.; Wasonga, O.V.; Mghalu, J.M.; et al. Global occurrence of anti-infectives in contaminated surface waters: Impact of income inequality between countries. *Environ. Int.* **2015**, *80*, 89–97. [CrossRef]
102. Gracia-Lor, E.; Sancho, J.V.; Serrano, R.; Hernández, F. Occurrence and removal of pharmaceuticals in wastewater treatment plants at the Spanish Mediterranean area of Valencia. *Chemosphere* **2012**, *87*, 453–462. [CrossRef]

103. Verlicchi, P.; Galletti, A.; Petrovic, M.; Barceló, D.; Al Aukidy, M.; Zambello, E. Removal of selected pharmaceuticals from domestic wastewater in an activated sludge system followed by a horizontal subsurface flow bed—Analysis of their respective contributions. *Sci. Total Environ.* **2013**, *454–455*, 411–425. [CrossRef]
104. Pereira, A.M.P.T.; Silva, L.J.G.; Meisel, L.M.; Lino, C.M.; Pena, A. Environmental impact of pharmaceuticals from Portuguese wastewaters: geographical and seasonal occurrence, removal and risk assessment. *Environ. Res.* **2015**, *136*, 108–119. [CrossRef]
105. Pereira, A.M.P.T.; Silva, L.J.G.; Lino, C.M.; Meisel, L.M.; Pena, A. Assessing environmental risk of pharmaceuticals in Portugal: an approach for the selection of the Portuguese monitoring stations in line with Directive 2013/39/EU. *Chemosphere* **2016**, *144*, 2507–2515. [CrossRef]
106. Petrovic, M.; Gros, M.; Barcelo, D. Multi-residue analysis of pharmaceuticals in wastewater by ultra-performance liquid chromatography-quadrupole-time-of-flight mass spectrometry. *J. Chromatogr. A* **2006**, *1124*, 68–81. [CrossRef] [PubMed]
107. Dutta, K.; Lee, M.Y.; Lai, W.W.P.; Lee, C.H.; Lin, A.Y.C.; Lin, C.F.; Lin, J.G. Removal of pharmaceuticals and organic matter from municipal wastewater using two-stage anaerobic fluidized membrane bioreactor. *Bioresour. Technol.* **2014**, *165*, 42–49. [CrossRef] [PubMed]
108. Karthikeyan, K.G.; Meyer, M.T. Occurrence of antibiotics in wastewater treatment facilities in Wisconsin, USA. *Sci. Total Environ.* **2006**, *361*, 196–207. [CrossRef] [PubMed]
109. Lindberg, R.H.; Wennberg, P.; Johansson, M.I.; Tysklind, M.; Andersson, B.A.V. Screening of Human Antibiotic Substances and Determination of Weekly Mass Flows in Five Sewage Treatment Plants in Sweden. *Environ. Sci. Technol.* **2005**, *39*, 3421–3429. [CrossRef] [PubMed]
110. Castiglioni, S.; Bagnati, R.; Fanelli, R.; Pomati, F.; Calamari, D.; Zuccato, E. Removal of Pharmaceuticals in Sewage Treatment Plants in Italy. *Environ. Sci. Technol.* **2006**, *40*, 357–363. [CrossRef] [PubMed]
111. Sim, W.-J.; Lee, J.-W.; Lee, E.-S.; Shin, S.-K.; Hwang, S.-R.; Oh, J.-E. Occurrence and distribution of pharmaceuticals in wastewater from households, livestock farms, hospitals and pharmaceutical manufactures. *Chemosphere* **2011**, *82*, 179–186. [CrossRef] [PubMed]
112. Kosma, C.I.; Lambropoulou, D.A.; Albanis, T.A. Investigation of PPCPs in wastewater treatment plants in Greece: Occurrence, removal and environmental risk assessment. *Sci. Total Environ.* **2014**, *466–467*, 421–438. [CrossRef]
113. Carmona, E.; Andreu, V.; Picó, Y. Occurrence of acidic pharmaceuticals and personal care products in Turia River Basin: from waste to drinking water. *Sci. Total Environ.* **2014**, *484*, 53–63. [CrossRef]
114. Lee, H.-B.; Peart, T.E.; Svoboda, M.L. Determination of endocrine-disrupting phenols, acidic pharmaceuticals, and personal-care products in sewage by solid-phase extraction and gas chromatography-mass spectrometry. *J. Chromatogr. A* **2005**, *1094*, 122–129. [CrossRef]
115. Sun, Q.; Lv, M.; Hu, A.; Yang, X.; Yu, C.-P. Seasonal variation in the occurrence and removal of pharmaceuticals and personal care products in a wastewater treatment plant in Xiamen, China. *J. Hazard. Mater.* **2013**. [CrossRef]
116. Styrishave, B.; Halling-Sørensen, B.; Ingerslev, F. Environmental risk assessment of three selective serotonin reuptake inhibitors in the aquatic environment: a case study including a cocktail scenario. *Environ. Toxicol. Chem.* **2011**, *30*, 254–261. [CrossRef] [PubMed]
117. Vasskog, T.; Berger, U.; Samuelsen, P.-J.; Kallenborn, R.; Jensen, E. Selective serotonin reuptake inhibitors in sewage influents and effluents from Tromsø, Norway. *J. Chromatogr. A* **2006**, *1115*, 187–195. [CrossRef] [PubMed]
118. Vasskog, T.; Anderssen, T.; Pedersen-Bjergaard, S.; Kallenborn, R.; Jensen, E. Occurrence of selective serotonin reuptake inhibitors in sewage and receiving waters at Spitsbergen and in Norway. *J. Chromatogr. A* **2008**, *1185*, 194–205. [CrossRef] [PubMed]
119. Lajeunesse, A.; Gagnon, C.; Sauvé, S. Determination of Basic Antidepressants and Their N-Desmethyl Metabolites in Raw Sewage and Wastewater Using Solid-Phase Extraction and Liquid Chromatography-Tandem Mass Spectrometry. *Anal. Chem.* **2008**, *80*, 5325–5333. [CrossRef] [PubMed]
120. Yuan, S.; Jiang, X.; Xia, X.; Zhang, H.; Zheng, S. Detection, occurrence and fate of 22 psychiatric pharmaceuticals in psychiatric hospital and municipal wastewater treatment plants in Beijing, China. *Chemosphere* **2013**, *90*, 2520–2525. [CrossRef] [PubMed]
121. Silva, L.J.G.; Pereira, A.M.P.T.; Meisel, L.M.; Lino, C.M.; Pena, A. A one-year follow-up analysis of antidepressants in Portuguese wastewaters: occurrence and fate, seasonal influence, and risk assessment. *Sci. Total Environ.* **2014**, *490*, 279–287. [CrossRef] [PubMed]

122. Baker, D.R.; Kasprzyk-Hordern, B. Spatial and temporal occurrence of pharmaceuticals and illicit drugs in the aqueous environment and during wastewater treatment: New developments. *Sci. Total Environ.* **2013**, *454–455*, 442–456. [CrossRef]
123. Santos, L.H.M.L.M.; Araújo, A.N.; Fachini, A.; Pena, A.; Delerue-Matos, C.; Montenegro, M.C.B.S.M. Ecotoxicological aspects related to the presence of pharmaceuticals in the aquatic environment. *J. Hazard. Mater.* **2010**, *175*, 45–95. [CrossRef]
124. Hernando, M.D.; Mezcua, M.; Fernández-Alba, A.R.; Barceló, D. Environmental risk assessment of pharmaceutical residues in wastewater effluents, surface waters and sediments. *Talanta* **2006**, *69*, 334–342. [CrossRef]
125. Lin, A.Y.-C.; Yu, T.-H.; Lateef, S.K. Removal of pharmaceuticals in secondary wastewater treatment processes in Taiwan. *J. Hazard. Mater.* **2009**, *167*, 1163–1169. [CrossRef]
126. Tauxe-Wuersch, A.; De Alencastro, L.F.; Grandjean, D.; Tarradellas, J. Occurrence of several acidic drugs in sewage treatment plants in Switzerland and risk assessment. *Water Res.* **2005**, *39*, 1761–1772. [CrossRef] [PubMed]
127. Choi, K.; Kim, Y.; Park, J.; Park, C.K.; Kim, M.; Kim, H.S.; Kim, P. Seasonal variations of several pharmaceutical residues in surface water and sewage treatment plants of Han River, Korea. *Sci. Total Environ.* **2008**, *405*, 120–128. [CrossRef] [PubMed]
128. Zuehlke, S.; Duennbier, U.; Heberer, T. Determination of estrogenic steroids in surface water and wastewater by liquid chromatography-electrospray tandem mass spectrometry. *J. Sep. Sci.* **2005**, *28*, 52–58. [CrossRef] [PubMed]
129. Laganà, A.; Bacaloni, A.; De Leva, I.; Faberi, A.; Fago, G.; Marino, A. Analytical methodologies for determining the occurrence of endocrine disrupting chemicals in sewage treatment plants and natural waters. *Anal. Chim. Acta* **2004**, *501*, 79–88. [CrossRef]
130. Boisvert, M.; Fayad, P.B.; Sauvé, S. Development of a new multi-residue laser diode thermal desorption atmospheric pressure chemical ionization tandem mass spectrometry method for the detection and quantification of pesticides and pharmaceuticals in wastewater samples. *Anal. Chim. Acta* **2012**, *754*, 75–82. [CrossRef]
131. Dasenaki, M.E.; Thomaidis, N.S. Multianalyte method for the determination of pharmaceuticals in wastewater samples using solid-phase extraction and liquid chromatography-tandem mass spectrometry. *Anal. Bioanal. Chem.* **2015**, 4229–4245. [CrossRef]
132. Zuccato, E.; Castiglioni, S.; Bagnati, R.; Melis, M.; Fanelli, R. Source, occurrence and fate of antibiotics in the Italian aquatic environment. *J. Hazard. Mater.* **2010**, *179*, 1042–1048. [CrossRef]
133. Golovko, O.; Kumar, V.; Fedorova, G.; Randak, T.; Grabic, R. Seasonal changes in antibiotics, antidepressants/psychiatric drugs, antihistamines and lipid regulators in a wastewater treatment plant. *Chemosphere* **2014**, *111*, 418–426. [CrossRef]
134. Blair, B.D.; Crago, J.P.; Hedman, C.J.; Klaper, R.D. Pharmaceuticals and personal care products found in the Great Lakes above concentrations of environmental concern. *Chemosphere* **2013**, *93*, 2116–2123. [CrossRef]
135. Qian, F.; He, M.; Song, Y.; Tysklind, M.; Wu, J. A bibliometric analysis of global research progress on pharmaceutical wastewater treatment during 1994-2013. *Environ. Earth Sci.* **2015**, 4995–5005. [CrossRef]
136. Grujić, S.; Vasiljević, T.; Laušević, M. Determination of multiple pharmaceutical classes in surface and ground waters by liquid chromatography-ion trap-tandem mass spectrometry. *J. Chromatogr. A* **2009**, *1216*, 4989–5000. [CrossRef] [PubMed]
137. Caracciolo, A.B.; Topp, E.; Grenni, P. Pharmaceuticals in the environment: Biodegradation and effects on natural microbial communities. A review. *J. Pharm. Biomed. Anal.* **2015**, *106*, 25–36. [CrossRef] [PubMed]
138. Santos, L.H.M.L.M.; Paíga, P.; Araújo, A.N.; Pena, A.; Delerue-Matos, C.; Montenegro, M.C.B.S.M. Development of a simple analytical method for the simultaneous determination of paracetamol, paracetamol-glucuronide and p-aminophenol in river water. *J. Chromatogr. B. Anal. Technol. Biomed. Life Sci.* **2013**, *930*, 75–81. [CrossRef] [PubMed]
139. Isobe, T.; Shiraishi, H.; Yasuda, M.; Shinoda, A.; Suzuki, H.; Morita, M. Determination of estrogens and their conjugates in water using solid-phase extraction followed by liquid chromatography-tandem mass spectrometry. *J. Chromatogr. A* **2003**, *984*, 195–202. [CrossRef]
140. Al-Qaim, F.F.; Abdullah, M.P.; Othman, M.R.; Latip, J.; Zakaria, Z. Multi-residue analytical methodology-based liquid chromatography-time-of-flight-mass spectrometry for the analysis of pharmaceutical residues in surface water and effluents from sewage treatment plants and hospitals. *J. Chromatogr. A* **2014**, *1345*, 139–153. [CrossRef] [PubMed]

141. McEachran, A.D.; Shea, D.; Bodnar, W.; Nichols, E.G. Pharmaceutical occurrence in groundwater and surface waters in forests land-applied with municipal wastewater. *Environ. Toxicol. Chem.* **2016**, *35*, 898–905. [CrossRef] [PubMed]
142. Liu, J.-L.; Wong, M.-H. Pharmaceuticals and personal care products (PPCPs): A review on environmental contamination in China. *Environ. Int.* **2013**, *59*, 208–224. [CrossRef]
143. Acuña, V.; Schiller, D.; García-Galán, M.J.; Rodríguez-Mozaz, S.; Corominas, L.; Petrovic, M.; Poch, M.; Barceló, D.; Sabater, S. Occurrence and in-stream attenuation of wastewater-derived pharmaceuticals in Iberian rivers. *Sci. Total Environ.* **2015**, *503–504*, 133–141. [CrossRef]
144. Navarro-Ortega, A.; Acuña, V.; Bellin, A.; Burek, P.; Cassiani, G.; Choukr-Allah, R.; Dolédec, S.; Elosegi, A.; Ferrari, F.; Ginebreda, A.; et al. Managing the effects of multiple stressors on aquatic ecosystems under water scarcity. The GLOBAQUA project. *Sci. Total Environ.* **2015**, *503–504*, 3–9. [CrossRef]
145. Simazaki, D.; Kubota, R.; Suzuki, T.; Akiba, M.; Nishimura, T.; Kunikane, S. Occurrence of selected pharmaceuticals at drinking water purification plants in Japan and implications for human health. *Water Res.* **2015**, *76*, 187–200. [CrossRef]
146. Corcoran, J.; Winter, M.J.; Tyler, C.R. Pharmaceuticals in the aquatic environment: a critical review of the evidence for health effects in fish. *Crit. Rev. Toxicol.* **2010**, *40*, 287–304. [CrossRef] [PubMed]
147. Murata, A.; Takada, H.; Mutoh, K.; Hosoda, H.; Harada, A.; Nakada, N. Nationwide monitoring of selected antibiotics: Distribution and sources of sulfonamides, trimethoprim, and macrolides in Japanese rivers. *Sci. Total Environ.* **2011**, *409*, 5305–5312. [CrossRef] [PubMed]
148. Vergeynst, L.; Haeck, A.; De Wispelaere, P.; Van Langenhove, H.; Demeestere, K. Multi-residue analysis of pharmaceuticals in wastewater by liquid chromatography-magnetic sector mass spectrometry: Method quality assessment and application in a Belgian case study. *Chemosphere* **2015**, *119*, S2–S8. [CrossRef] [PubMed]
149. YUAN, S.-L.; LI, X.-F.; JIANG, X.-M.; ZHANG, H.-X.; ZHENG, S.-K. Simultaneous determination of 13 psychiatric pharmaceuticals in sewage by automated solid phase extraction and liquid chromatography-mass spectrometry. *Chinese J. Anal. Chem.* **2013**, *41*, 49–56. [CrossRef]
150. Salgado, R.; Marques, R.; Noronha, J.P.; Mexia, J.T.; Carvalho, G.; Oehmen, A.; Reis, M.A.M. Assessing the diurnal variability of pharmaceutical and personal care products in a full-scale activated sludge plant. *Environ. Pollut.* **2011**, *159*, 2359–2367. [CrossRef] [PubMed]
151. Ginebreda, A.; Jelić, A.; Petrović, M.; López de Alda, M.; Barceló, D. New indexes for compound prioritization and complexity quantification on environmental monitoring inventories. *Environ. Sci. Pollut. Res. Int.* **2012**, *19*, 958–970. [CrossRef] [PubMed]
152. Lindberg, R.H.; Östman, M.; Olofsson, U.; Grabic, R.; Fick, J. Occurrence and behaviour of 105 active pharmaceutical ingredients in sewage waters of a municipal sewer collection system. *Water Res.* **2014**, *58*, 221–229. [CrossRef]
153. Gros, M.; Petrović, M.; Barceló, D. Development of a multi-residue analytical methodology based on liquid chromatography-tandem mass spectrometry (LC-MS/MS) for screening and trace level determination of pharmaceuticals in surface and wastewaters. *Talanta* **2006**, *70*, 678–690. [CrossRef]
154. Gros, M.; Rodríguez-Mozaz, S.; Barceló, D. Rapid analysis of multiclass antibiotic residues and some of their metabolites in hospital, urban wastewater and river water by ultra-high-performance liquid chromatography coupled to quadrupole-linear ion trap tandem mass spectrometry. *J. Chromatogr. A* **2013**, *1292*, 173–188. [CrossRef]
155. Blair, B.D.; Crago, J.P.; Hedman, C.J.; Treguer, R.J.F.; Magruder, C.; Royer, L.S.; Klaper, R.D. Evaluation of a model for the removal of pharmaceuticals, personal care products, and hormones from wastewater. *Sci. Total Environ.* **2013**, *444*, 515–521. [CrossRef]
156. Sim, W.; Lee, J.; Oh, J. Occurrence and fate of pharmaceuticals in wastewater treatment plants and rivers in Korea. *Environ. Pollut.* **2010**, *158*, 1938–1947. [CrossRef] [PubMed]
157. Matongo, S.; Birungi, G.; Moodley, B.; Ndungu, P. Occurrence of selected pharmaceuticals in water and sediment of Umgeni River, KwaZulu-Natal, South Africa. *Environ. Sci. Pollut. Res.* **2015**, *22*, 10298–10308. [CrossRef] [PubMed]
158. Stumpf, M.; Ternes, T.A.; Wilken, R.D.; Rodrigues, S.V.; Baumann, W. Polar drug residues in sewage and natural waters in the state of Rio de Janeiro, Brazil. *Sci. Total Environ.* **1999**, *225*, 135–141. [CrossRef]

159. Sun, Q.; Li, M.; Ma, C.; Chen, X.; Xie, X.; Yu, C.P. Seasonal and spatial variations of PPCP occurrence, removal and mass loading in three wastewater treatment plants located in different urbanization areas in Xiamen, China. *Environ. Pollut.* **2016**, *208*, 371–381. [CrossRef]
160. Miao, X.-S.; Metcalfe, C.D. Determination of cholesterol-lowering statin drugs in aqueous samples using liquid chromatography–electrospray ionization tandem mass spectrometry. *J. Chromatogr. A* **2003**, *998*, 133–141. [CrossRef]
161. Koutsouba, V.; Heberer, T.; Fuhrmann, B.; Schmidt-Baumler, K.; Tsipi, D.; Hiskia, A. Determination of polar pharmaceuticals in sewage water of Greece by gas chromatography-mass spectrometry. *Chemosphere* **2003**, *51*, 69–75. [CrossRef]
162. Hernando, M.D.; Heath, E.; Petrovic, M.; Barceló, D. Trace-level determination of pharmaceutical residues by LC-MS/MS in natural and treated waters. A pilot-survey study. *Anal. Bioanal. Chem.* **2006**, *385*, 985–991. [CrossRef]
163. Pailler, J.-Y.; Krein, A.; Pfister, L.; Hoffmann, L.; Guignard, C. Solid phase extraction coupled to liquid chromatography-tandem mass spectrometry analysis of sulfonamides, tetracyclines, analgesics and hormones in surface water and wastewater in Luxembourg. *Sci. Total Environ.* **2009**, *407*, 4736–4743. [CrossRef]
164. Samaras, V.G.; Stasinakis, A.S.; Mamais, D.; Thomaidis, N.S.; Lekkas, T.D. Fate of selected pharmaceuticals and synthetic endocrine disrupting compounds during wastewater treatment and sludge anaerobic digestion. *J. Hazard. Mater.* **2013**, *244–245*, 259–267. [CrossRef]
165. Wu, M.; Xiang, J.; Que, C.; Chen, F.; Xu, G. Occurrence and fate of psychiatric pharmaceuticals in the urban water system of Shanghai, China. *Chemosphere* **2015**, *138*, 486–493. [CrossRef]
166. K'oreje, K.O.; Vergeynst, L.; Ombaka, D.; De Wispelaere, P.; Okoth, M.; Van Langenhove, H.; Demeestere, K. Occurrence patterns of pharmaceutical residues in wastewater, surface water and groundwater of Nairobi and Kisumu city, Kenya. *Chemosphere* **2016**, *149*, 238–244. [CrossRef] [PubMed]
167. Subedi, B.; Balakrishna, K.; Joshua, D.I.; Kannan, K. Mass loading and removal of pharmaceuticals and personal care products including psychoactives, antihypertensives, and antibiotics in two sewage treatment plants in southern India. *Chemosphere* **2017**, *167*, 429–437. [CrossRef] [PubMed]
168. Paíga, P.; Santos, L.H.M.L.M.; Ramos, S.; Jorge, S.; Silva, J.G.; Delerue-Matos, C. Presence of pharmaceuticals in the Lis river (Portugal): Sources, fate and seasonal variation. *Sci. Total Environ.* **2016**, *573*, 164–177. [CrossRef] [PubMed]
169. Rivera-Jaimes, J.A.; Postigo, C.; Melgoza-Alemán, R.M.; Aceña, J.; Barceló, D.; López de Alda, M. Study of pharmaceuticals in surface and wastewater from Cuernavaca, Morelos, Mexico: Occurrence and environmental risk assessment. *Sci. Total Environ.* **2018**, *613–614*, 1263–1274. [CrossRef] [PubMed]
170. Nebot, C.; Falcon, R.; Boyd, K.G.; Gibb, S.W. Introduction of human pharmaceuticals from wastewater treatment plants into the aquatic environment: a rural perspective. *Environ. Sci. Pollut. Res.* **2015**, *14*, 10559–10568. [CrossRef] [PubMed]
171. Kostich, M.S.; Batt, A.L.; Lazorchak, J.M. Concentrations of prioritized pharmaceuticals in effluents from 50 large wastewater treatment plants in the US and implications for risk estimation. *Environ. Pollut.* **2014**, *184*, 354–359. [CrossRef] [PubMed]
172. Loos, R.; Carvalho, R.; António, D.C.; Comero, S.; Locoro, G.; Tavazzi, S.; Paracchini, B.; Ghiani, M.; Lettieri, T.; Blaha, L.; et al. EU-wide monitoring survey on emerging polar organic contaminants in wastewater treatment plant effluents. *Water Res.* **2013**, *47*, 6475–6487. [CrossRef]
173. Li, W.C. Occurrence, sources, and fate of pharmaceuticals in aquatic environment and soil. *Environ. Pollut.* **2014**, *187C*, 193–201. [CrossRef]
174. Kim, S.D.; Cho, J.; Kim, I.S.; Vanderford, B.J.; Snyder, S.A. Occurrence and removal of pharmaceuticals and endocrine disruptors in South Korean surface, drinking, and waste waters. *Water Res.* **2007**, *41*, 1013–1021. [CrossRef]
175. Hilton, M.J.; Thomas, K.V. Determination of selected human pharmaceutical compounds in effluent and surface water samples by high-performance liquid chromatography–electrospray tandem mass spectrometry. *J. Chromatogr. A* **2003**, *1015*, 129–141. [CrossRef]
176. Bueno, M.J.M.; Agüera, A.; Hernando, M.D.; Gómez, M.J.; Fernández-Alba, A.R. Evaluation of various liquid chromatography-quadrupole-linear ion trap-mass spectrometry operation modes applied to the analysis of organic pollutants in wastewaters. *J. Chromatogr. A* **2009**, *1216*, 5995–6002. [CrossRef] [PubMed]

177. Verenitch, S.S.; Lowe, C.J.; Mazumder, A. Determination of acidic drugs and caffeine in municipal wastewaters and receiving waters by gas chromatography-ion trap tandem mass spectrometry. *J. Chromatogr. A* **2006**, *1116*, 193–203. [CrossRef] [PubMed]
178. Subedi, B.; Codru, N.; Dziewulski, D.M.; Wilson, L.R.; Xue, J.; Yun, S.; Braun-Howland, E.; Minihane, C.; Kannan, K. A pilot study on the assessment of trace organic contaminants including pharmaceuticals and personal care products from on-site wastewater treatment systems along Skaneateles Lake in New York State, USA. *Water Res.* **2014**, *72*, 28–39. [CrossRef] [PubMed]
179. Daneshvar, A.; Svanfelt, J.; Kronberg, L.; Prévost, M.; Weyhenmeyer, G.A. Seasonal variations in the occurrence and fate of basic and neutral pharmaceuticals in a Swedish river-lake system. *Chemosphere* **2010**, *80*, 301–309. [CrossRef] [PubMed]
180. Crouse, B.A.; Ghoshdastidar, A.J.; Tong, A.Z. The presence of acidic and neutral drugs in treated sewage effluents and receiving waters in the Cornwallis and Annapolis River watersheds and the Mill CoveSewage Treatment Plant in Nova Scotia, Canada. *Environ. Res.* **2012**, *112*, 92–99. [CrossRef]
181. Mijangos, L.; Ziarrusta, H.; Ros, O.; Kortazar, L.; Fernández, L.A.; Olivares, M.; Zuloaga, O.; Prieto, A.; Etxebarria, N. Occurrence of emerging pollutants in estuaries of the Basque Country: Analysis of sources and distribution, and assessment of the environmental risk. *Water Res.* **2018**, *147*, 152–163. [CrossRef]
182. Chiffre, A.; Degiorgi, F.; Buleté, A.; Spinner, L.; Badot, P.M. Occurrence of pharmaceuticals in WWTP effluents and their impact in a karstic rural catchment of Eastern France. *Environ. Sci. Pollut. Res.* **2016**, *23*, 25427–25441. [CrossRef]
183. Lin, A.Y.C.; Tsai, Y.T. Occurrence of pharmaceuticals in Taiwan's surface waters: Impact of waste streams from hospitals and pharmaceutical production facilities. *Sci. Total Environ.* **2009**, *407*, 3793–3802. [CrossRef]
184. Kim, J.-W.; Jang, H.-S.; Kim, J.-G.; Ishibashi, H.; Hirano, M.; Nasu, K.; Ichikawa, N.; Takao, Y.; Shinohara, R.; Arizono, K. Occurrence of Pharmaceutical and Personal Care Products (PPCPs) in Surface Water from Mankyung River, South Korea. *J. Heal. Sci.* **2009**, *55*, 249–258. [CrossRef]
185. Lv, M.; Sun, Q.; Hu, A.; Hou, L.; Li, J.; Cai, X.; Yu, C.-P. Pharmaceuticals and personal care products in a mesoscale subtropical watershed and their application as sewage markers. *J. Hazard. Mater.* **2014**, *280*, 696–705. [CrossRef]
186. European Commission Commission implementing decision (EU) 2015/495 of 20 March 2015 establishing a watch list of substances for Union-wide monitoring in the field of water policy pursuant to Directive 2008/105/EC of the European Parliament and of the Council. *Off. J. Eur. Union* **2015**.
187. European Parliament Directive 2013/39/EU of the European Parliament and of the Council of 12 August 2013 amending Directives 2000/60/EC and 2008/105/EC as regards priority substances in the field of water policy. *Off. J. Eur. Union* **2013**, *L 226*, 1–17.
188. Gonçalves, C.M.O.; Sousa, M.A.D.; Alpendurada, M.D.F.P. Analysis of acidic, basic and neutral pharmaceuticals in river waters: clean-up by 1°, 2° amino anion exchange and enrichment using an hydrophilic adsorbent. *Int. J. Environ. Anal. Chem.* **2013**, *93*, 1–22.
189. Khan, G.A.; Berglund, B.; Khan, K.M.; Lindgren, P.E.; Fick, J. Occurrence and Abundance of Antibiotics and Resistance Genes in Rivers, Canal and near Drug Formulation Facilities—A Study in Pakistan. *PLoS ONE* **2013**, *8*, 4–11. [CrossRef] [PubMed]
190. Fram, M.S.; Belitz, K. Occurrence and concentrations of pharmaceutical compounds in groundwater used for public drinking-water supply in California. *Sci. Total Environ.* **2011**, *409*, 3409–3417. [CrossRef] [PubMed]
191. López-Serna, R.; Petrović, M.; Barceló, D. Direct analysis of pharmaceuticals, their metabolites and transformation products in environmental waters using on-line TurboFlow^{TM} chromatography-liquid chromatography-tandem mass spectrometry. *J. Chromatogr. A* **2012**, *1252*, 115–129. [CrossRef]
192. Yang, L.; Luan, T.; Lan, C. Solid-phase microextraction with on-fiber silylation for simultaneous determinations of endocrine disrupting chemicals and steroid hormones by gas chromatography-mass spectrometry. *J. Chromatogr. A* **2006**, *1104*, 23–32. [CrossRef]
193. López-Roldán, R.; de Alda, M.L.; Gros, M.; Petrovic, M.; Martín-Alonso, J.; Barceló, D. Advanced monitoring of pharmaceuticals and estrogens in the Llobregat River basin (Spain) by liquid chromatography-triple quadrupole-tandem mass spectrometry in combination with ultra performance liquid chromatography-time of flight-mass spectrometry. *Chemosphere* **2010**, *80*, 1337–1344. [CrossRef]
194. Madureira, T.V.; Barreiro, J.C.; Rocha, M.J.; Rocha, E.; Cass, Q.B.; Tiritan, M.E. Spatiotemporal distribution of pharmaceuticals in the Douro River estuary (Portugal). *Sci. Total Environ.* **2010**, *408*, 5513–5520. [CrossRef]

195. Vulliet, E.; Cren-Olivé, C.; Grenier-Loustalot, M.-F. Occurrence of pharmaceuticals and hormones in drinking water treated from surface waters. *Environ. Chem. Lett.* **2009**, *9*, 103–114. [CrossRef]
196. Christian, T.; Schneider, R.J.; Färber, H.A.; Skutlarek, D.; Meyer, M.T.; Goldbach, H.E. Determination of antibiotic residues in manure, soil, and surface waters. *Acta Hydrochim. Hydrobiol.* **2003**, *31*, 36–44. [CrossRef]
197. Managaki, S.; Murata, A.; Takada, H.; Bui, C.T.; Chiem, N.H. Distribution of macrolides, sulfonamides, and trimethoprim in tropical waters: Ubiquitous occurrence of veterinary antibiotics in the Mekong Delta. *Environ. Sci. Technol.* **2007**, *41*, 8004–8010. [CrossRef] [PubMed]
198. Hoa, P.T.P.; Managaki, S.; Nakada, N.; Takada, H.; Shimizu, A.; Anh, D.H.; Viet, P.H.; Suzuki, S. Antibiotic contamination and occurrence of antibiotic-resistant bacteria in aquatic environments of northern Vietnam. *Sci. Total Environ.* **2011**, *409*, 2894–2901. [CrossRef]
199. Zhang, R.; Zhang, G.; Zheng, Q.; Tang, J.; Chen, Y.; Xu, W.; Zou, Y.; Chen, X. Occurrence and risks of antibiotics in the Laizhou Bay, China: Impacts of river discharge. *Ecotoxicol. Environ. Saf.* **2012**, *80*, 208–215. [CrossRef] [PubMed]
200. Focazio, M.J.; Kolpin, D.W.; Barnes, K.K.; Furlong, E.T.; Meyer, M.T.; Zaugg, S.D.; Barber, L.B.; Thurman, M.E. A national reconnaissance for pharmaceuticals and other organic wastewater contaminants in the United States—II) Untreated drinking water sources. *Sci. Total Environ.* **2008**, *402*, 201–216. [CrossRef] [PubMed]
201. Bu, Q.; Wang, B.; Huang, J.; Deng, S.; Yu, G. Pharmaceuticals and personal care products in the aquatic environment in China: A review. *J. Hazard. Mater.* **2013**, *262*, 189–211. [CrossRef] [PubMed]
202. Hughes, S.R.; Brown, L.E.; Kay, P. Global Synthesis and Critical Evaluation of Pharmaceutical Data Sets Collected from River Systems. *Environ. Sci. Technol.* **2013**, *47*, 661–677. [CrossRef]
203. Ginebreda, A.; Muñoz, I.; Alda, M.L.; Brix, R.; López-Doval, J.; Barceló, D. Environmental risk assessment of pharmaceuticals in rivers: Relationships between hazard indexes and aquatic macroinvertebrate diversity indexes in the Llobregat River (NE Spain). *Environ. Int.* **2010**, *36*, 153–162. [CrossRef]
204. Wu, C.; Huang, X.; Witter, J.D.; Spongberg, A.L.; Wang, K.; Wang, D.; Liu, J. Occurrence of pharmaceuticals and personal care products and associated environmental risks in the central and lower Yangtze river, China. *Ecotoxicol. Environ. Saf.* **2014**, *106*, 19–26. [CrossRef]
205. de Gaffney, V.J.; Almeida, C.M.M.; Rodrigues, A.; Ferreira, E.; Benoliel, M.J.; Cardoso, V.V. Occurrence of pharmaceuticals in a water supply system and related human health risk assessment. *Water Res.* **2015**, *72*, 199–208. [CrossRef]
206. Kolpin, D.W.; Furlong, E.T.; Meyer, M.T.; Thurman, E.M.; Zaugg, S.D.; Barber, L.B.; Buxton, H.T. Pharmaceuticals, hormones, and other organic wastewater contaminants in U.S. streams, 1999-2000: A national reconnaissance. *Environ. Sci. Technol.* **2002**, *36*, 1202–1211. [CrossRef] [PubMed]
207. Pena, A.; Chmielova, D.; Lino, C.M.; Solich, P. Determination of fluoroquinolone antibiotics in surface waters from Mondego River by high performance liquid chromatography using a monolithic column. *J. Sep. Sci.* **2007**, *30*, 2924–2928. [CrossRef] [PubMed]
208. Kleywegt, S.; Pileggi, V.; Yang, P.; Hao, C.; Zhao, X.; Rocks, C.; Thach, S.; Cheung, P.; Whitehead, B. Pharmaceuticals, hormones and bisphenol A in untreated source and finished drinking water in Ontario, Canada—occurrence and treatment efficiency. *Sci. Total Environ.* **2011**, *409*, 1481–1488. [CrossRef] [PubMed]
209. Ginebreda, A.; Kuzmanovic, M.; Guasch, H.; de Alda, M.L.; López-Doval, J.C.; Muñoz, I.; Ricart, M.; Romaní, A.M.; Sabater, S.; Barceló, D. Assessment of multi-chemical pollution in aquatic ecosystems using toxic units: compound prioritization, mixture characterization and relationships with biological descriptors. *Sci. Total Environ.* **2014**, *468–469*, 715–723. [CrossRef] [PubMed]
210. de Jongh, C.M.; Kooij, P.J.F.; de Voogt, P.; ter Laak, T.L. Screening and human health risk assessment of pharmaceuticals and their transformation products in Dutch surface waters and drinking water. *Sci. Total Environ.* **2012**, *427–428*, 70–77. [CrossRef]
211. Kunkel, U.; Radke, M. Fate of pharmaceuticals in rivers: Deriving a benchmark dataset at favorable attenuation conditions. *Water Res.* **2012**, *46*, 5551–5565. [CrossRef]
212. Benotti, M.J.; Trenholm, R.A.; Vanderford, B.J.; Holady, J.C.; Stanford, B.D.; Snyder, S.A. Pharmaceuticals and endocrine disrupting compounds in U.S. drinking water. *Environ. Sci. Technol.* **2009**, *43*, 597–603. [CrossRef]
213. Zhao, J.L.; Ying, G.G.; Wang, L.; Yang, J.F.; Yang, X.B.; Yang, L.H.; Li, X. Determination of phenolic endocrine disrupting chemicals and acidic pharmaceuticals in surface water of the Pearl Rivers in South China by gas chromatography-negative chemical ionization-mass spectrometry. *Sci. Total Environ.* **2009**, *407*, 962–974. [CrossRef]

214. Schriks, M.; Heringa, M.B.; Van der Kooi, M.M.E.; de Voogt, P.; Van Wezel, A.P. Toxicological relevance of emerging contaminants for drinking water quality. *Water Res.* **2010**, *44*, 461–476. [CrossRef]
215. Weigel, S.; Kallenborn, R.; Hühnerfuss, H. Simultaneous solid-phase extraction of acidic, neutral and basic pharmaceuticals from aqueous samples at ambient (neutral) pH and their determination by gas chromatography-mass spectrometry. *J. Chromatogr. A* **2004**, *1023*, 183–195. [CrossRef]
216. Vulliet, E.; Wiest, L.; Baudot, R.; Grenier-Loustalot, M.F. Multi-residue analysis of steroids at sub-ng/L levels in surface and ground-waters using liquid chromatography coupled to tandem mass spectrometry. *J. Chromatogr. A* **2008**, *1210*, 84–91. [CrossRef] [PubMed]
217. Na, T.W.; Kang, T.W.; Lee, K.H.; Hwang, S.H.; Jung, H.J.; Kim, K. Distribution and ecological risk of pharmaceuticals in surface water of the Yeongsan river, Republic of Korea. *Ecotoxicol. Environ. Saf.* **2019**, *181*, 180–186. [CrossRef] [PubMed]
218. Hossain, A.; Nakamichi, S.; Habibullah-Al-Mamun, M.; Tani, K.; Masunaga, S.; Matsuda, H. Occurrence and ecological risk of pharmaceuticals in river surface water of Bangladesh. *Environ. Res.* **2018**, *165*, 258–266. [CrossRef] [PubMed]
219. Cantwell, M.G.; Katz, D.R.; Sullivan, J.C.; Shapley, D.; Lipscomb, J.; Epstein, J.; Juhl, A.R.; Knudson, C.; O'Mullan, G.D. Spatial patterns of pharmaceuticals and wastewater tracers in the Hudson River Estuary. *Water Res.* **2018**, *137*, 335–343. [CrossRef]
220. Reis-Santos, P.; Pais, M.; Duarte, B.; Caçador, I.; Freitas, A.; Vila Pouca, A.S.; Barbosa, J.; Leston, S.; Rosa, J.; Ramos, F.; et al. Screening of human and veterinary pharmaceuticals in estuarine waters: A baseline assessment for the Tejo estuary. *Mar. Pollut. Bull.* **2018**, *135*, 1079–1084. [CrossRef]
221. Gumbi, B.P.; Moodley, B.; Birungi, G.; Ndungu, P.G. Detection and quantification of acidic drug residues in South African surface water using gas chromatography-mass spectrometry. *Chemosphere* **2017**, *168*, 1042–1050. [CrossRef]
222. Ivešić, M.; Krivohlavek, A.; Žuntar, I.; Tolić, S.; Šikić, S.; Musić, V.; Pavlić, I.; Bursik, A.; Galić, N. Monitoring of selected pharmaceuticals in surface waters of Croatia. *Environ. Sci. Pollut. Res.* **2017**, *24*, 23389–23400. [CrossRef]
223. Olatunji, O.S.; Fatoki, O.S.; Opeolu, B.O.; Ximba, B.J.; Chitongo, R. Determination of selected steroid hormones in some surface water around animal farms in Cape Town using HPLC-DAD. *Environ. Monit. Assess.* **2017**, *189*. [CrossRef]
224. Mokh, S.; El Khatib, M.; Koubar, M.; Daher, Z.; Al Iskandarani, M. Innovative SPE-LC-MS/MS technique for the assessment of 63 pharmaceuticals and the detection of antibiotic-resistant-bacteria: A case study natural water sources in Lebanon. *Sci. Total Environ.* **2017**, *609*, 830–841. [CrossRef]
225. López-Serna, R.; Jurado, A.; Vázquez-Suñé, E.; Carrera, J.; Petrović, M.; Barceló, D. Occurrence of 95 pharmaceuticals and transformation products in urban groundwaters underlying the metropolis of Barcelona, Spain. *Environ. Pollut.* **2013**, *174*, 305–315. [CrossRef]
226. Afonso-Olivares, C.; Torres-Padrón, M.; Sosa-Ferrera, Z.; Santana-Rodríguez, J. Assessment of the Presence of Pharmaceutical Compounds in Seawater Samples from Coastal Area of Gran Canaria Island (Spain). *Antibiotics* **2013**, *2*, 274–287. [CrossRef] [PubMed]
227. Barnes, K.K.; Kolpin, D.W.; Furlong, E.T.; Zaugg, S.D.; Meyer, M.T.; Barber, L.B. A national reconnaissance of pharmaceuticals and other organic wastewater contaminants in the United States—I) Groundwater. *Sci. Total Environ.* **2008**, *402*, 192–200. [CrossRef] [PubMed]
228. Magnér, J.; Filipovic, M.; Alsberg, T. Application of a novel solid-phase-extraction sampler and ultra-performance liquid chromatography quadrupole-time-of-flight mass spectrometry for determination of pharmaceutical residues in surface sea water. *Chemosphere* **2010**, *80*, 1255–1260. [CrossRef] [PubMed]
229. Sacher, F.; Lange, F.T.; Brauch, H.-J.; Blankenhorn, I. Pharmaceuticals in groundwaters. *J. Chromatogr. A* **2001**, *938*, 199–210. [CrossRef]
230. Riva, F.; Castiglioni, S.; Fattore, E.; Manenti, A.; Davoli, E.; Zuccato, E. Monitoring emerging contaminants in the drinking water of Milan and assessment of the human risk. *Int. J. Hyg. Environ. Health* **2018**, *221*, 451–457. [CrossRef] [PubMed]
231. Snyder, S.A. Occurrence, Treatment, and Toxicological Relevance of EDCs and Pharmaceuticals in Water. *Ozone Sci. Eng.* **2008**, *30*, 65–69. [CrossRef]

Review

Selected Pharmaceuticals in Different Aquatic Compartments: Part II—Toxicity and Environmental Risk Assessment

André Pereira *, Liliana Silva, Célia Laranjeiro, Celeste Lino and Angelina Pena

LAQV, REQUIMTE, Laboratory of Bromatology and Pharmacognosy, Faculty of Pharmacy, University of Coimbra, Polo III, Azinhaga de St Comba, 3000-548 Coimbra, Portugal; ljgsilva@hotmail.com (L.S.); celialaranjeiro@gmail.com (C.L.); cmlino@ci.uc.pt (C.L.); apena@ci.uc.pt (A.P.)

* Correspondence: andrepereira@ff.uc.pt

Received: 12 March 2020; Accepted: 8 April 2020; Published: 14 April 2020

Abstract: Potential risks associated with releases of human pharmaceuticals into the environment have become an increasingly important issue in environmental health. This concern has been driven by the widespread detection of pharmaceuticals in all aquatic compartments. Therefore, 22 pharmaceuticals, 6 metabolites and transformation products, belonging to 7 therapeutic groups, were selected to perform a review on their toxicity and environmental risk assessment (ERA) in different aquatic compartments, important issues to tackle the water framework directive (WFD). The toxicity data collected reported, with the exception of anxiolytics, at least one toxicity value for concentrations below 1 $\mu g\ L^{-1}$. The results obtained for the ERA revealed risk quotients (RQs) higher than 1 in all the aquatic bodies and for the three trophic levels, algae, invertebrates and fish, posing ecotoxicological pressure in all of these compartments. The therapeutic groups with higher RQs were hormones, antiepileptics, anti-inflammatories and antibiotics. Unsurprisingly, RQs values were highest in wastewaters, however, less contaminated water bodies such as groundwaters still presented maximum values up to 91,150 regarding 17α-ethinylestradiol in fish. Overall, these results present an important input for setting prioritizing measures and sustainable strategies, minimizing their impact in the aquatic environment.

Keywords: environmental contaminants; pharmaceuticals; pharmaceuticals toxicity; environmental risk assessment; aquatic compartments

1. Introduction

The environmental impact of medicinal products has been recognized worldwide, and as its use cannot be avoided, a sound risk assessment of their presence in the environment is a key issue that must be tackled to meet the European Union (EU) Water Framework Directive (WFD) [1]. The potential of human pharmaceuticals for negative ecotoxicological effects, even at sublethal concentrations, in the aquatic environment has been of concern since the issue was first brought to attention in 1985 [2]. Nonetheless, the ecotoxicological risks associated to the ubiquitous occurrence of pharmaceuticals in aquatic ecosystems are far from being fully known [3].

According to the European Medicines Agency (EMA) legislation, and since 2006, before a pharmaceutical obtains a marketing authorization approval, it must be demonstrated that it poses no risk to the environment through an environmental risk assessment (ERA). ERA compares the predicted environmental concentrations (PECs), with the predicted no effect concentrations (PNECs) of three trophic levels of aquatic organisms [4,5]. When the pharmaceutical is already on the market, instead of using the PEC, which predict the environmental concentration, we can use the measured environmental concentrations (MEC) reflecting the real concentration in the aquatic environment [6]. Therefore, for marketed pharmaceuticals, high-quality monitoring data, along with data on ecotoxicological and toxicological effects are crucial to perform the ERA, which associates the presence of pharmaceuticals with their impact on the aquatic ecosystem and human health, supporting the selection of possible new priority substances to be monitored [7–9].

Thus, a systematic review, in order to provide a clear insight on pharmaceuticals toxicology and on ERA, should embrace, not only several parent compounds, but also, metabolites and transformations products belonging to different therapeutic groups such as: the anxiolytics and hypnotics, further referred only as anxiolytics, alprazolam (ALP), lorazepam (LOR) and zolpidem (ZOL); the antibiotics azithromycin (AZI), ciprofloxacin (CIP), clarithromycin (CLA) and erythromycin (ERY); the lipid regulators bezafibrate (BEZ), gemfibrozil (GEM) and simvastatin (SIM); the antiepileptic carbamazepine (CAR); the selective serotonin reuptake inhibitors (SSRIs) citalopram (CIT) and its main metabolite desmethylcitalopram (N-CIT), escitalopram (ESC), fluoxetine (FLU) and its main metabolite norfluoxetine (Nor-FLU), paroxetine (PAR), sertraline (SER) and its main metabolite desmethylsertraline (Nor-SER); the anti-inflammatories and/or analgesics and antipyretics, further referred only as anti-inflammatories, diclofenac (DIC) and its main metabolite 4-hydroxydiclofenac (4-OH-DIC), ibuprofen (IBU), naproxen (NAP), paracetamol (PARA) and its transformation product 4-aminophenol (4-PARA); and the hormones 17β-estradiol (E2) and its main metabolite estrone (E1) and 17α-ethinylestradiol (EE2; Table 1).

Table 1. Selected pharmaceuticals.

Therapeutic Group	Compound and Chemical Structure			
Anxiolytics (Anx)	Alprazolam (ALP) 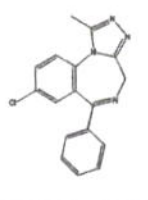	Lorazepam (LOR) 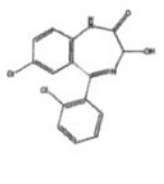	Zolpidem (ZOL) 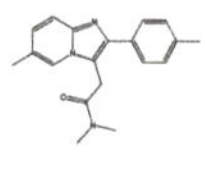	
Antibiotics (Antib)	Azithromycin (AZI) 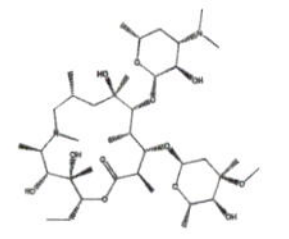	Ciprofloxacin (CIP) 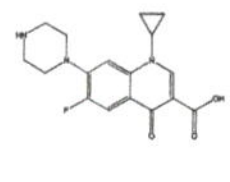	Clarithromycin (CLA) 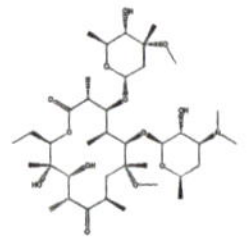	Erythromycin (ERY) 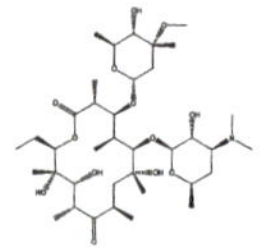
Lipid regulators (Lip Reg)	Bezafibrate (BEZ) 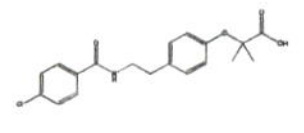	Gemfibrozil (GEM) 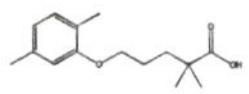	Simvastatin (SIM) 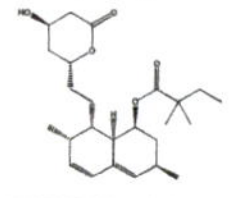	
Antiepileptic (Antiepi)	Carbamazepine (CAR)	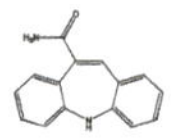		

Table 1. *Cont.*

Therapeutic Group	Compound and Chemical Structure			
Selective serotonin reuptake inhibitors (SSRIs)	Citalopram (CIT) 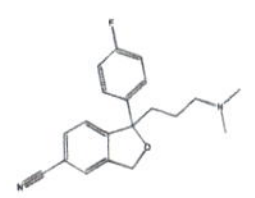	Desmethylcitalopram (N-Cit) (metabolite) 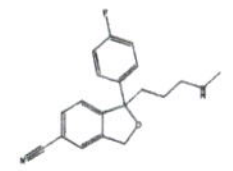	Escitalopram (ESC) 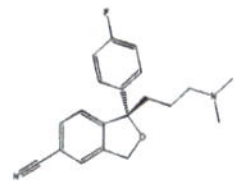	Fluoxetine (FLU) 
	Norfluoxetine (Nor-FLU) (metabolite) 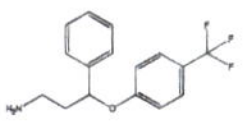	Paroxetine (PAR) 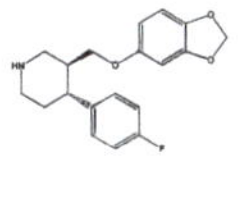	Sertraline (SER) 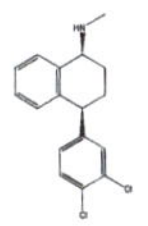	Desmethylsertraline (Nor-SER) (metabolite) 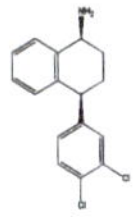
Anti-inflammatories (Anti-inf)	Diclofenac (DIC) 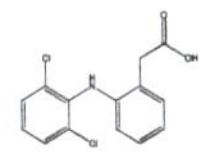	4-hydroxydiclofenac (4-OH-DIC) (metabolite) 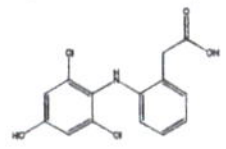	Ibuprofen (IBU) 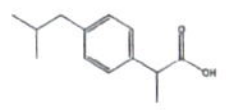	Naproxen (NAP) 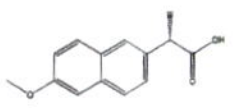
	Paracetamol (PARA) 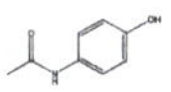	4-aminophenol (4-PARA) (transformation product) 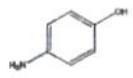		
Hormones (Horm)	Estrone (E1) (natural hormone/metabolite) 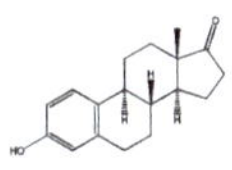	17β-estradiol (E2) 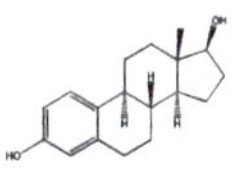	17α-ethinylestradiol (EE2) 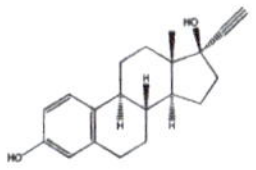	

The pharmaceuticals in study, key representatives of major classes of pharmaceuticals, were selected based on the EU watch list, their high consumption, pharmacokinetics, physicochemical properties, persistence, previous studies on the occurrence on the aquatic environment, and their potential toxicological impact, both on humans and on the aquatic environment [10–14]. This review will provide a more realistic water quality assessment contributing for a more integrative approach to rank and prioritize pharmaceuticals, based on an integrated assessment of ERA and exposure in the aquatic environment.

"Water is not a commercial product like any other but, rather, a heritage which must be protected, defended and treated as such", the claim by the EU WFD contrasts with the poor ecological status of European freshwater bodies, where only 43% achieve a good ecological status. In addition, and despite the enormous efforts, the picture that emerges regarding ecological status is still incomplete, fragmented and with contradictory assessments of the situation. Therefore, it is important to obtain a better understanding of the regional and global context, concerning the environmental risk posed by pharmaceuticals in the aquatic environment.

2. Toxicity

Since pharmaceuticals are continuously introduced into the aquatic environment, they can promote toxic effects on living organisms, even when present at concentrations on the ng L^{-1} level [15]. This potential for negative effects of pharmaceuticals even at sublethal concentrations, namely for aquatic organisms, has been of concern since the issue was first brought to attention in 1985 [2]. Therefore, their presence poses a threat to the quality of water resources [5,16].

Pharmaceuticals have a relatively clear mode of action in target organisms, and given that fish and invertebrates share more drug targets with humans, it would be expected that they would also respond to pharmaceuticals in a similar way. However, when non-target-species are exposed, unknown effects and potential risks need to be assessed. One example is the impact of EE2 in the feminization of fish [17–19]. Nonetheless, all the ecotoxicological risks associated to the ubiquitous occurrence of pharmaceuticals in aquatic ecosystems are far from known [3].

Sorption to sediments is one factor that influences toxicity of pharmaceuticals, although higher sorption to sediments results in an apparent reduction of bioavailability and toxicity, the activity of benthic invertebrate in sediments results in a higher exposure for these organisms [20].

Moreover, bioaccumulation and biomagnification should also be accounted for since they can increase toxicity [17]. These parameters are also related to log $D_{ow,}$ since compounds with values higher than 3 have a tendency for bioaccumulation [17,21], which means that the ionization state can influence the toxicity of pharmaceuticals, and that the pH variability in surface water should also be taken into account [17].

A bibliographic search of the scientific literature was conducted on Google Scholar using the search terms "ecotoxicology" and each of the selected compounds. All the publications that presented ecotoxicological studies on the selected compounds, referring to the concentrations, were included. Below, the ecotoxicological data in the aquatic biota was reviewed, presenting the toxicity data obtained from 120 exposure studies of three trophic levels of non-target organisms, algae (Figure 1), invertebrates (Figure 2) and fish (Figure 3), Table S1 (Supporting information). The data was divided by the different endpoints found in the literature: no observed effect concentrations (NOEC), lowest observed effect concentrations (LOEC), effective concentration (EC50) and lethal concentration (LC50). These endpoints are expected to have increasing concentrations, since they were organized from the more susceptible endpoint (NOEC) to the less one (LC50). However, each endpoint encloses various species of the same trophic level and different toxicological tests like immobilization, growth, luminescence, reproduction, morphology, behavior, etc. When no experimental data was available, L(E)C50 values were estimated with ECOSAR 2.0. This program estimates data on acute toxicity through the molecule structure, sometimes underestimating toxic effects. The data was also divided in acute and chronic toxicity, depending on the time of exposure and trophic level. For algae, acute toxicity was considered when the toxicity tests lasted until 4 days (96 h), longer exposures were considered chronic toxicity. Regarding invertebrates, with the exception of *Brachionus calyciflorus* (were 2 days was considered chronic data, since it has a shorter life cycle), acute toxicity was accounted when the exposure took place until 2 days (48 h) and chronic toxicity when it was equal or longer than 7 days. For fish, tests until 4 days (96 h) were included in acute toxicity data and exposures equal or above 7 days entered the chronic toxicity data. These criteria were based on OECD tests for each trophic level [22].

Figure 1. Median, maximum and minimum concentration values reported for acute (**A**) and chronic (**B**) toxicity concerning algae. (Anx—anxiolytics; Antib—antibiotics; Lip reg—lipid regulators; Antiepi—antiepileptics; SSRIs—Selective serotonin reuptake inhibitors; Anti-inf—anti-inflammatories; Horm—hormones) [23–52].

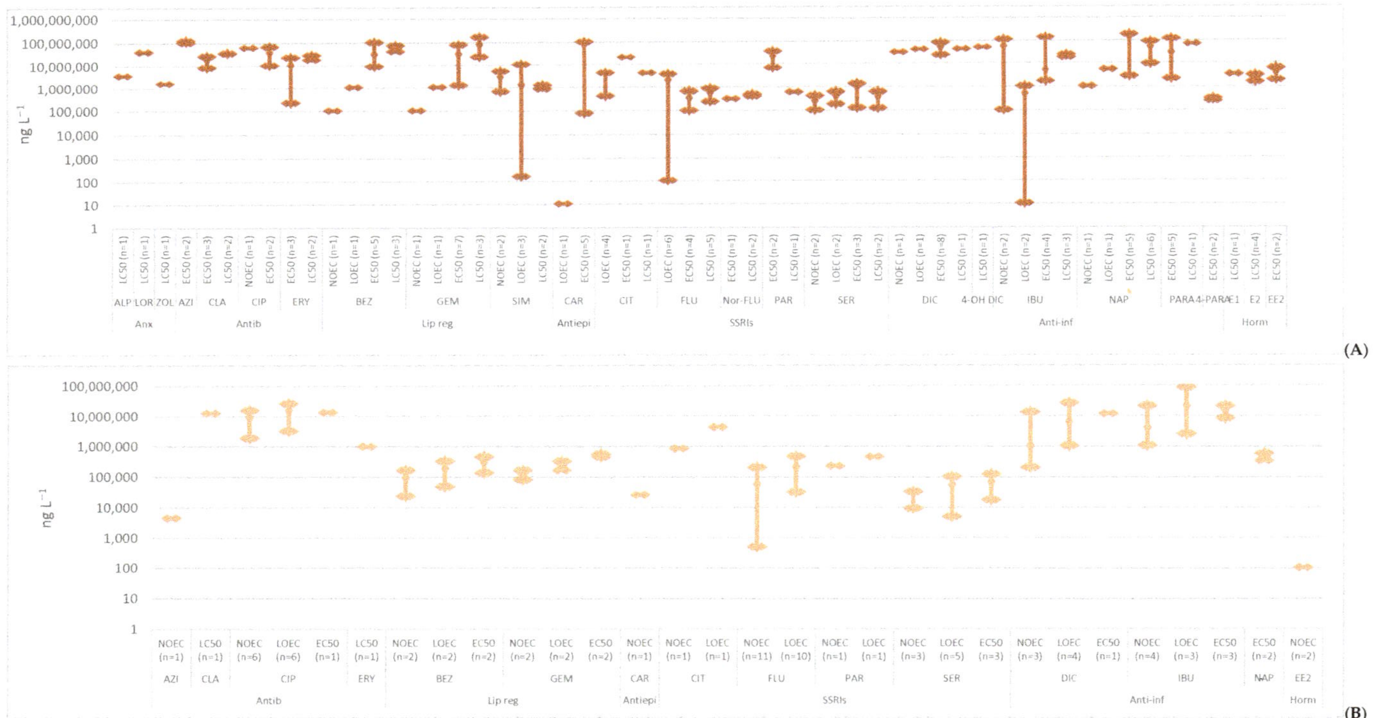

Figure 2. Median, maximum and minimum concentration values reported for acute (**A**) and chronic (**B**) toxicity data concerning invertebrates. (Anx—anxiolytics; Antib—antibiotics; Lip reg—lipid regulators; Antiepi—antiepileptics; SSRIs—Selective serotonin reuptake inhibitors; Anti-inf—anti-inflammatories; Horm—hormones) [23,27–32,35,36,40–42,44,46,51–84].

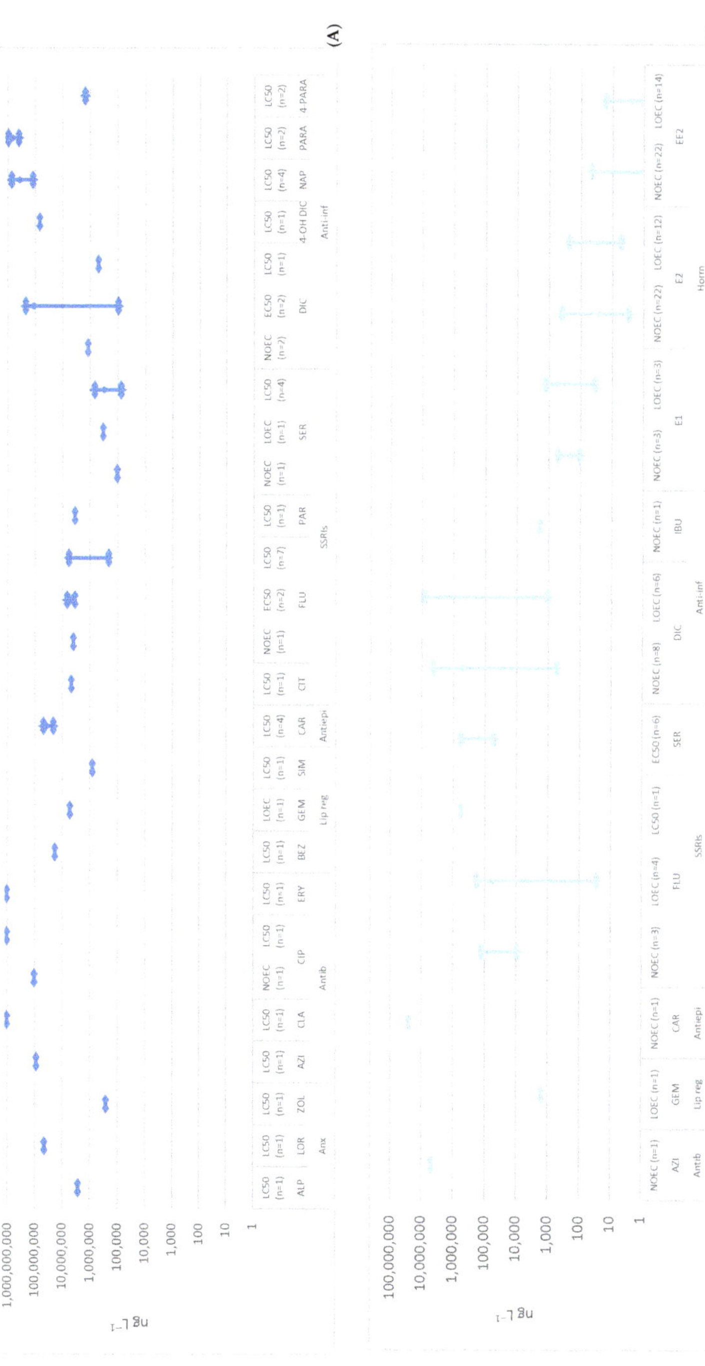

Figure 3. Median, maximum and minimum concentration values reported for acute (**A**) and chronic (**B**) toxicity data concerning fish. (Anx—anxiolytics; Antib—antibiotics; Lip reg—lipid regulators; Antiepi—antiepileptics; SSRIs—Selective serotonin reuptake inhibitors; Anti-inf—anti-inflammatories; Horm—hormones) [23,29,31,32,35,36,42,44,58,59,70,74,82,83,85–137].

Although, as expected, some therapeutic groups presented higher toxicity, such as hormones, which can promote endocrine modifications, all therapeutic groups presented toxicity at low concentrations, highlighting the ecotoxicity of the selected pharmaceuticals [138]. Overall, considering all trophic levels, all therapeutic groups with the exception of anxiolytics, had at least one toxicity report for concentrations below 1 $\mu g\ L^{-1}$, near the concentrations found in the aquatic environment.

Considering the toxicity of the selected pharmaceuticals in all trophic levels, we could observe that the most sensitive one, with the lowest concentrations promoting toxic effects was fish, followed by invertebrates and algae. The limitation of this analysis is that, regarding fish, there were also toxicity data obtained through cell line or tissue testing, which can be difficult to extrapolate to the entire organism. The therapeutic group with higher toxicity, mainly chronic toxicity in fish and invertebrates, are hormones. Additionally, the pharmaceutical that presented higher toxicity, with the lowest concentration promoting toxic effects, was EE2 at 0.1 ng L^{-1} in fish (NOEC, chronic toxicity) [123]. The highest concentrations promoting toxicity were detected in fish (LC50, acute toxicity), for CLA, CIP and ERY (1 g L^{-1}), [23,123,126].

Ecotoxicological chronic studies on pharmaceuticals are lacking, meaning that many questions about the threat to the environment of pharmaceuticals remain unanswered. Additionally, the actual exposure scenario regards multiple pharmaceuticals, posing uncertainty regarding toxicology in long-term exposure. If many pharmaceuticals are present and share the same mode of action, then the toxicity of this mixture could be higher than if only one pharmaceutical is present, being usually considered the concept of concentration addition, although antagonistic and synergistic effects may also occur. This could result in risk underestimation, as the typical exposure is toward multicomponent chemicals [139–142].

One example of mixture effects was observed when using a mixture of anti-inflammatories (DIC, IBU and NAP). In this case, the acute toxicity was detected at concentrations where little or no effect was observed for the chemicals individually [20]. Even in mixtures with pharmaceuticals belonging to different therapeutic groups, additive and synergistic effects were reported. A mixture with E2 and FLU promoted a decrease in the reproductive success of *D. magna* more significantly than either chemical compounds alone [143]. Another example was provided by exposing *D. magna* to a mixture of CAR and a lipid lowering agent, which exhibited stronger effects during immobilization tests than the single compounds at the same concentration [20].

Taking into account mixture effects, some research has already been developed focusing on toxic effects, and not on specific pharmaceuticals. This was already used to evaluate wastewater treatment plants (WWTPs) removal efficiencies, by evaluating and comparing the toxicity (androgenecity, cytotoxicity, anti-estrogenicity and *L. variegatus* decrease in reproduction and biomass) both in wastewater influents (WWIs) and efluents (WWEs) [144,145].

Additionally to the active compounds of pharmaceuticals, excipients and additives are also present in medicines, that may contain endocrine disrupting chemical excipients and additives [138].

The measured concentrations of some of the selected pharmaceuticals reported for surface water all over the world surpassed the concentrations here described for toxicity, which suggests that the aquatic biota could be vulnerable to the presence of pharmaceuticals in their environment, and that toxic effects are expected to occur with unexpected outcomes [146].

It is unlikely that pharmaceuticals present in drinking water may pose a risk to the human health through chronic exposure, however, the toxicological implications are not clear [147]. Furthermore, studies have shown that infants may have difficulty in metabolizing drugs therefore, being more vulnerable to the toxic effects of these compounds [17].

As referred, many pharmaceuticals have the potential for bioaccumulation and biomagnification, and chronic effects on ecosystems cannot be ignored for animals at the higher end of the food web [148]. Thus, the health hazard of human exposure by ingestion of contaminated foods should also be taken into account [17].

2.1. Anxiolytics

No ecotoxicological data was found in literature for ALP, LOR and ZOL, and for that reason, all the results for this therapeutic group were obtained from ECOSAR 2.0 [18]. In decreasing order, the more toxic was ZOL, followed by ALP and LOR. The trophic level with the lowest reported concentrations producing toxicity was algae (from 0.211 to 6.07 mg $L^{-1)}$, followed by fish (from 0.248 to 43.1 mg L^{-1}) and invertebrates (from 1.55 to 39.4 mg L^{-1}).

2.2. Antibiotics

Observing the acute toxicity for antibiotics, since there is little data on chronic endpoints, the pattern for the three trophic levels was similar for all antibiotics, with algae being more susceptible at lower concentrations (from 0.0018 to 20.6 mg L^{-1}), followed by invertebrates (from 0.22 to 120 mg L^{-1}) and fish (from 84 to 1000 mg L^{-1}). If we compare each antibiotic, concerning invertebrates, it can be observed that CLA and CIP presented similar results, but when compared with ERY, lower concentrations (220 μg L^{-1}) of this antibiotic can produce the same toxic effects, in this case growth inhibition [23].

In this therapeutic class, in addition to direct toxicological risks, concern has been raised about the potential for the antibiotic residues in water, since they are typically found in the aquatic environment at subtherapeutic concentrations, promoting the emergence of resistant bacteria and subsequent development of more resistant and virulent pathogens [149]. These bacterial resistances, through horizontal gene transfer, may end up in human pathogens, raising questions on human health and the stability of the ecosystem [150–154].

This emergence of bacterial resistance presents one of the major emerging threats to human health and is by far the highest risk for humans of having medicinal products residues in the environment [155]. Furthermore, historical evidence appears to indicate that in the aquatic environment resistance might be acquired faster than in the terrestrial environment [156].

Corroborating the effects on bacteria, changes in biomass and growth rate were reported at concentrations above 5.7 μg L^{-1} [47]. This therapeutic class can also induce immunotoxicity in the freshwater mussel at low concentrations, between 2 ng L^{-1} and 1100 ng L^{-1} [157].

2.3. Lipid Regulators

In this group, the pattern observed with both previous therapeutic groups was not so clear, with median concentrations similar in all trophic levels for acute toxicity. Observing these data, SIM was clearly the pharmaceutical, which promoted toxicity at lower concentrations for invertebrates (160 ng L^{-1}) and fish (765 μg L^{-1}) [26,56]. However, data on chronic toxicity, only available for GEM in two trophic levels, showed that the highest toxicity regarded fish (1.5 μg L^{-1}), followed by invertebrates (78.0 μg L^{-1}) [51,85].

2.4. Antiepileptics

For CAR, once again, the pattern of acute data, was similar to that registered for anxiolytics and antibiotics, with the lowest concentrations promoting toxicity at 10.0 μg L^{-1}, 20 000 μg L^{-1} and 0.01 μg L^{-1} for algae, fish and invertebrates, respectively [27,44,57]. Considering the chronic data, similar concentrations were found to produce toxicity in invertebrates and fish trophic levels, ranging from 25 to 25,000 μg L^{-1} [27,29].

2.5. SSRIs

This therapeutic group has the peculiarity that the phylogenetically ancient and highly conserved neurotransmitter and neurohormone serotonin has been found in invertebrates and vertebrates, although its specific physiological role and mode of action is unknown for many species [158]. Many biological functions within invertebrates, such as reproduction, metabolism, molting and behavior, are under the control of serotonin [159]. Therefore, the pharmaceuticals in this therapeutic group could have tremendous effects on these and other organisms [160]. These facts are in agreement with those found in acute toxicity data found, since for all trophic levels this group had globally the lowest concentrations, which promoted toxic effects, being some of these on reproduction, survival and behavior [161].

When observing these data, the most sensitive trophic level was the invertebrates (0.1 μg L^{-1}), followed by algae (12.1 μg L^{-1}) and fish (72.0 μg L^{-1}) [33,57,92]. In invertebrates, the pharmaceuticals with higher toxicity were FLU (100 ng L^{-1}) and its metabolite Nor-FLU (300 μg L^{-1}) and SER (100 μg L^{-1}). On the other side, PAR was the one with lower toxicity [33,36,57,69]. In algae, the pharmaceutical with highest toxicity was SER, however, in invertebrates, FLU surpassed SER toxicity.

The only metabolite referred in the literature concerning toxicity studies was Nor-FLU, with data for algae and invertebrates. When comparing with FLU (algae and invertebrates), it is clear that the median concentrations inducing toxicity were always lower [162].

Studies performed on SER and FLU demonstrated the influence of pH on toxicity, since the uncharged drug can pass easier through the membrane and act inside the cells, showing a tenfold increased toxicity when shifting the pH closest to their pKa, increasing the nonionized form, from 6.5 to 8.5 and from 7.8 to 9, respectively [17,89,92].

2.6. Anti-Inflammatories

Most anti-inflammatories induce the nonspecific inhibition of prostaglandins. This, in turn, means that there is the potential for effects on any of the normal physiological functions mediated by prostaglandins. In fish, for instance, prostaglandins influence mechanisms of behavior and reproduction and, therefore, they can act as endocrine disruptors or modulators, because they can exert their effects by mimicking or antagonizing the effects of hormones, alter their pattern of synthesis and metabolism and modify hormone receptor levels, leading to possible adverse effects [7,163–165]. However, different and unexpected toxicity effects were also observed. One of the first was reported in Pakistan, where a catastrophic decline in the Oriental White-backed Vulture population (95%) originated from the exposure to DIC contaminated live-stock carcasses, which promoted fatal renal disease [98,138].

Overall, excepting anxiolytics, anti-inflammatories were less toxic than the other therapeutic groups. Regarding the lowest concentrations that produced acute toxicity in the three trophic levels, invertebrates had the lowest value (10 ng L^{-1}), followed by algae (10 μg L^{-1}) and fish (90 μg L^{-1}), however, when using median values, the differences become less clear [37,44,57,96]. As for chronic data, higher toxicity was observed in fish (500 ng L^{-1}) and invertebrates (200 μg L^{-1}), when compared with algae (4.01 mg L^{-1}), which is in line with the already referred anti-inflammatories mode of action [70,95].

Data for each anti-inflammatory showed no clear pattern, nonetheless, except for invertebrates, NAP and PARA seemed to have lower toxicity than DIC and IBU. When performing a comparison between DIC and its metabolite (4-OH-DIC) in invertebrates and fish, one could observe that they have similar toxicities. Conversely, PARA transformation product (4-PARA) presented higher toxicity than the parent molecule in all three trophic levels.

2.7. Hormones

Although hormones like E1, E2 and EE2 are mainly used for contraception purposes, the physiological effects are not restricted to effects on reproductive and sexual development, and can target mitochondrial function, energy metabolism and cell cycle control [165].

For acute toxicity, there is only data on algae and invertebrates, and algae presented higher toxicity since the lowest concentration promoting toxic effects was at 162 μg L^{-1}, lower than the 1500 μg L^{-1} observed in invertebrates [43,78]. Nonetheless, the toxicity promoted by this therapeutic group is mainly expected to be detected through chronic toxicity, however, these data could only be obtained for invertebrates and fish. Considering chronic data, in these two trophic levels, hormones presented higher toxicity than the other therapeutic groups, since the lowest concentrations reported were of 100 ng L^{-1} and 0.1 ng L^{-1}, for invertebrates and fish, respectively [81,82,123]. It should also be noted that, the highest concentration found that promoted toxicity for fish was also very low (1188 ng L^{-1}) [102].

Individually, there were no differences observed between E1 and E2 toxicity, while EE2 seems the most toxic compound regarding chronic toxicity in invertebrates and especially in fish, where the 36 results available presented concentrations below 44 ng L^{-1} [135]. Namely, when two different fish species were exposed to EE2 at 3 ng L^{-1} and 4 ng L^{-1} they suffered sex gender reversal, from male to female, which can strongly unbalance the aquatic ecosystem [130,136,138].

3. Environmental Risk Assessment

The data regarding occurrence and toxicity already presented is crucial in order to perform the ERA, and can be used to select the pharmaceuticals that are more prone to induce toxic effects in aquatic biota [166]. The risk assessment, mentioned in the EMA guideline on the ERA of medicinal products for human use [4], is performed through the risk quotient (RQ) calculation, dividing the PEC by the PNEC for each pharmaceutical, observing three different trophic levels (algae, invertebrates and fish). If RQ is equal or above 1, there is a potential environmental risk situation, whereas when values are lower than 1, no risk is expected. However, a certain risk could be expected for the substances with a RQ between 0.1 and 1 [167,168]. However, this guideline is only applied for marketing authorizations and for pharmaceuticals marketed after 2006. Additionally, it does not constitute a valid criterion upon which to base the refusal of a market authorization of medicinal products for human use in the EU [6]. Our evaluation of the potential ecotoxicological risk posed for the aquatic compartment was based on a dual approach: one using the worst case scenario, as stated by the EMA guideline on the ERA [4], where the maximum individual concentrations of pharmaceuticals found in the respective aquatic compartment were used as MEC [15,169,170], and another using the median concentrations for each pharmaceutical as MEC [171]. This evaluation can also be an important tool to suggest the inclusion or removal of pharmaceuticals in the watch list of the Directive 2013/39/EU.

As discussed, some concentrations compiled in surface water are higher than the levels that induce toxicity, not applying any uncertainty factor (UF) for the PNEC calculation. Additionally, some studies have indicated that concentrations of several pharmaceuticals belonging to different therapeutic groups can promote toxic effects on negatively impacted aquatic biota, presenting RQ higher than 1 [11,18,139,157,172,173].

As referred, aquatic biota inhabiting the receiving environment are unintentionally exposed throughout their lifetime to a complex mixture of residual pharmaceuticals and these mixtures can exhibit a greater effect than individual compounds [20,174,175]. Therefore, it is a challenge to address the concerns related to the chronic effect, low-level exposure to these compounds, including exposure of sensitive subpopulations to pharmaceutical mixtures [17,174].

3.1. Predicted No-Effect Concentration (PNECs)

Based on the toxicity data (Figures 4–7), Table 2 presents the PNECs for the selected pharmaceuticals. These values were calculated by applying an UF of 100 and 10 to the long-term EC50 and NOEC values, and an UF of 50 and 1000 to the short-term LOEC and L(E)C50 values, respectively, available in the literature. The UF is an expression of the degree of uncertainty in the extrapolation from the test data on a limited number of species to the actual environment [4]. As referred, when no experimental data are available, L(E)C50 values were estimated through ECOSAR 2.0.

Table 2. Predicted no-effect concentrations of the selected pharmaceuticals for algae, invertebrates and fish for the studied pharmaceuticals.

Therapeutic Group	Pharmaceutical	PNEC (ng L^{-1}) Algae	PNEC (ng L^{-1}) Invertebrates	PNEC (ng L^{-1}) Fish
Anx	ALP	892 ^{a,b}	3590 ^{b,c}	2540 ^{b,c}
	LOR	6070 ^{a,b}	39,400 ^{b,c}	43,100 ^{b,c}
	ZOL	211 ^{a,b}	1550 ^{b,c}	248 ^{b,c}
Antib	AZI	1.8 ^{b} [176]	440 ^{e,g} [83]	84,000 ^{b}
	CLA	2 ^{b} [23]	8160 ^{b} [23]	1,000,000 ^{b} [23]
	CIP	5 ^{b} [35]	10,000 ^{b} [44]	1,000,000 ^{b} [44]
	ERY	20 ^{b} [23]	220 ^{b} [23]	1,000,000 ^{b} [23]
Lip reg	BEZ	4870 ^{a,b}	1300 ^{e,f} [51]	17,600 ^{b,c}
	GEM	15,190 ^{b} [51]	1180 ^{b} [53]	150 ^{e,g} [85]
	SIM	22,800 ^{b} [26]	3.2 ^{d} [56]	765 ^{b,c}
Antiepi	CAR	31.6 ^{b} [27]	0.2 ^{d} [57]	20,000 ^{b} [44]
SSRIs	CIT	1600 ^{b} [30]	3900 ^{b} [58]	4470 ^{b,c}
	FLU	44.99 ^{b} [33]	2 ^{d} [57]	2.8 ^{e,g} [177]
	Nor-FLU (M)	189 ^{b} [34]	300 ^{b} [61]	n.a
	PAR	140 ^{b} [30]	580 ^{b} [58]	3290 ^{b,c}
	SER	12.10 ^{b} [33]	120 ^{b} [58]	72 ^{b} [92]
Anti-inf	DIC	200 ^{d} [37]	20,000 ^{e,g} [70]	50 ^{e,g} [95]
	4-OH-DIC (M)	660,300 ^{e,f} [38]	48,200 ^{b,c}	65,200 ^{b,c}
	IBU	40,100 ^{e,f} [39]	0.2 ^{d} [57]	180 ^{e,g}
	NAP	31,820 ^{b} [40]	2620 ^{b} [53]	115,200 ^{b} [99]
	PARA	134,000 ^{b} [59]	2040 ^{b} [73]	378,000 ^{b} [42]
	4-PARA (TP)	11,300 ^{a,b}	240 ^{b} [77]	1430 ^{b} [100]
Horm	E1 (NH/M)	355 ^{a,b}	3160 ^{b,c}	3.4 ^{e,g} [103]
	E2	162 ^{a,b}	1500 ^{b} [78]	0.29 ^{e,g} [113]
	EE2	730 ^{b} [43]	10 ^{e,g} [81,82]	0.01 ^{e,g} [123]

M—metabolite; TP—transformation product; NH—natural hormone. ^{a}: EC50 was estimated with ECOSAR. ^{b}: UF = 1000. ^{c}: LC50 was estimated with ECOSAR. ^{d}: UF = 50 (uncertainty factor used for lowest observed effect concentrations (LOEC) and no observed effect concentrations (NOEC) in acute toxicity). ^{e}: long-term data. ^{f}: UF = 100. ^{g}: UF = 10 (uncertainty factor used for LOEC and NOEC in chronic toxicity).

It should be taken into account that the choice of toxicity data can obviously affect the outcome [15]. However, the results obtained for the PNECs were directly related to the toxicity data, and a similar pattern to the toxicity data was observed, with the therapeutic groups and pharmaceuticals with higher toxicity presenting the lowest PNEC values.

3.2. Risk Assessment

Using the occurrence data obtained from Part I of this review (Table S2, Supporting information) and the PNECs previously calculated (Table 2), RQs were deemed for all the selected pharmaceuticals in the different aquatic compartments and are presented in Figures 4–7 [146].

In general, the results revealed that RQs higher than 1 could be observed for all the aquatic bodies, posing ecotoxicological pressure in all of these compartments.

3.2.1. Wastewater Influents

The RQs observed in WWIs were the highest from all the aquatic compartments, as well as the concentrations of the selected pharmaceuticals (Figure 4). The highest value (274,816) and median (13,400) were observed for IBU in invertebrates. Anti-inflammatories were the therapeutic group with the highest RQs median, both for the maximum and median values, followed by antiepileptic and hormones.

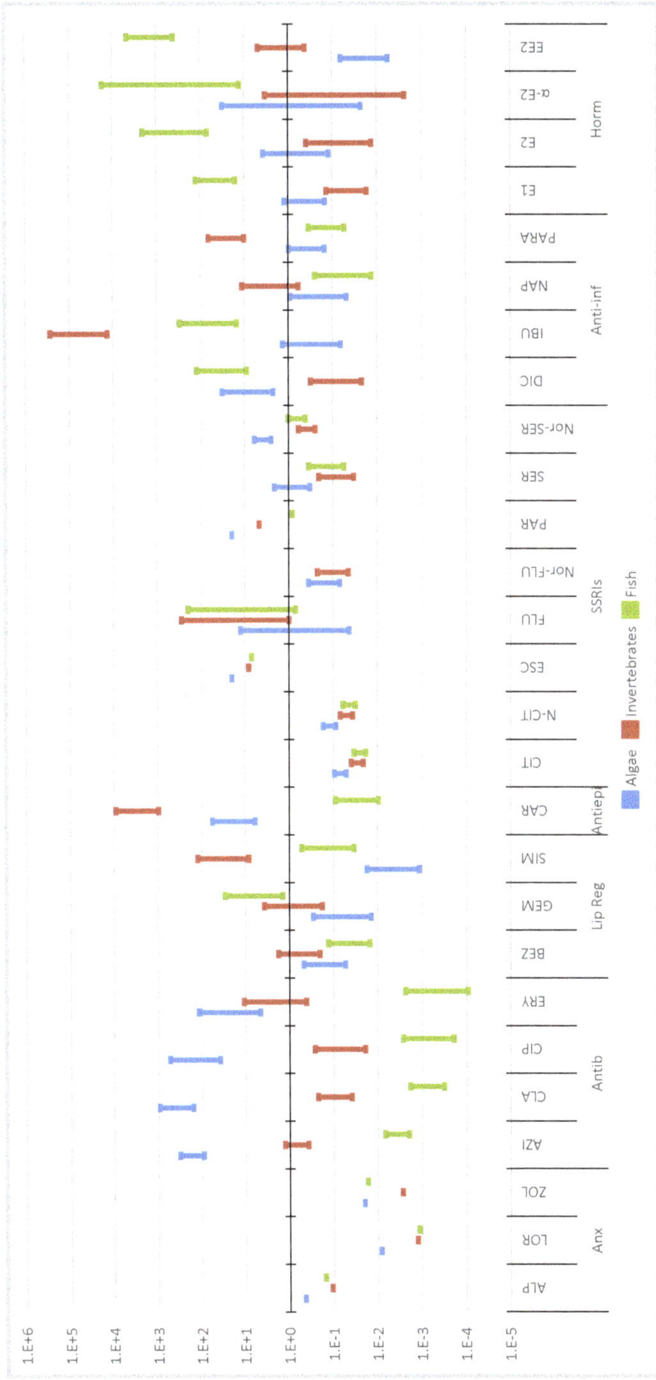

Figure 4. Median and maximum risk quotients of pharmaceuticals in WWIs for the three trophic levels. (Anx—anxiolytics; Antib—antibiotics; Lip reg—lipid regulators; Antiepi—antiepileptics; SSRIs—Selective serotonin reuptake inhibitors; Anti-inf—anti-inflammatories; Horm—hormones).

With the exception of anxiolytics, all the other therapeutic groups presented RQs > 1 for at least two trophic levels, being SSRIs and anxiolytics the only groups that did not present risk RQs > 1 for all pharmaceuticals.

For the anxiolytics and antibiotics, the algae were clearly the most susceptible trophic level, presenting higher RQs. Another clear pattern was observed for the hormones, where for fish all the RQs (median and maximum) were higher than 1 and with the maximum values between 129 and 17 271

These results demonstrated that the concentrations reaching the WWTPs could clearly endanger all the trophic levels that might be exposed to this aquatic matrix.

3.2.2. Wastewater Effluents

This aquatic compartment presents lower RQs than the WWIs (Figure 5). Antibiotics along with antiepileptic, anti-inflammatories and hormones were the therapeutic groups with highest RQs values. When considering median values alone, antibiotics have lower RQs than the other three therapeutic groups. The values observed for antibiotics can promote an even bigger problem than the direct toxicity to aquatic organism: the emergence of bacterial resistance. Nonetheless, the ERA approach does not address this issue [155].

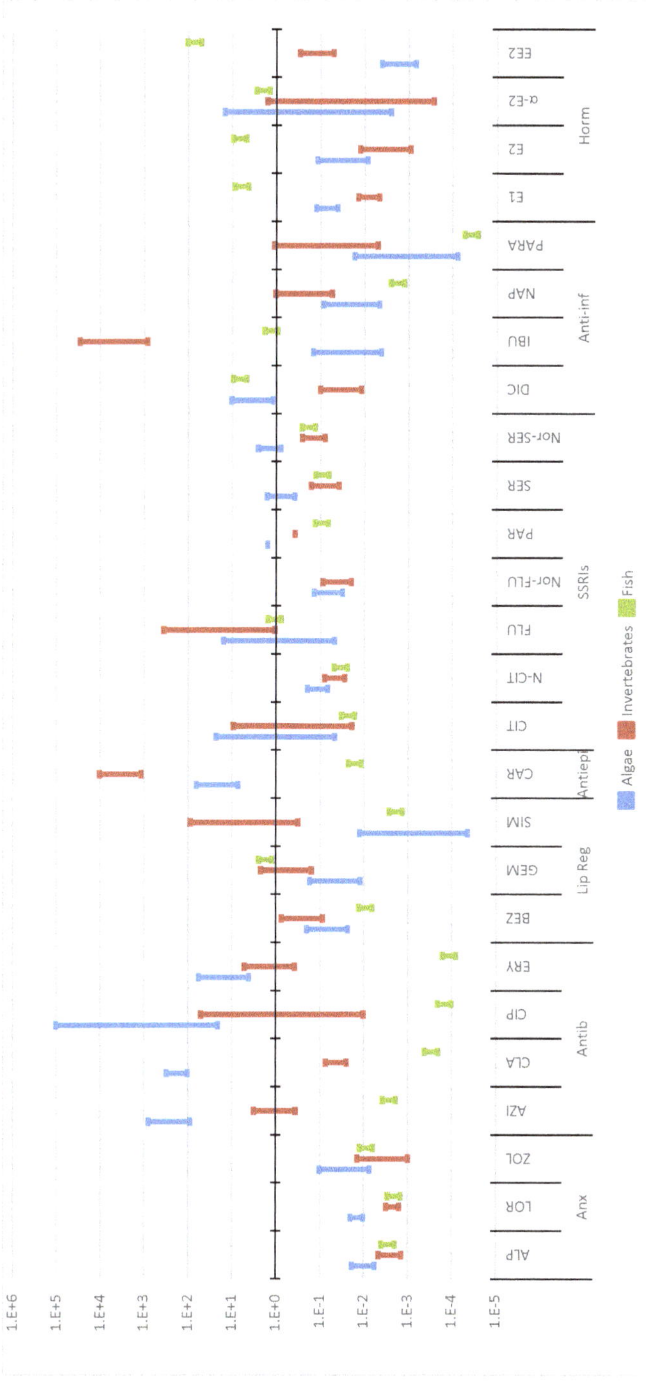

Figure 5. Median and maximum risk quotients of pharmaceuticals in WWEs for the three trophic levels. (Anx—anxiolytics; Antib—antibiotics; Lip reg—lipid regulators; Antiepi—antiepileptics; SSRIs—Selective serotonin reuptake inhibitors; Anti-inf—anti-inflammatories; Horm—hormones).

Anxiolytics continued to present RQs values lower than 1. Antibiotics, on the other hand, had the highest value observed for CIP in algae (100,258). As in WWI, antiepileptic continued to present the highest median RQs.

In WWI, some metabolites of the SSRIs presented similar or slightly higher RQs than the parent compounds (N-CIT and Nor-SER), in WWE a similar pattern was observed. These results highlight the fact that parent compounds and metabolites reached WWEs, and that the concentrations found in this matrix were able to promote toxic effects in the aquatic biota. This fact suggests that metabolites and transformation products should also be monitored in the environment.

Regarding the anti-inflammatories therapeutic group, DIC and IBU stand out from the other pharmaceuticals presenting clearly higher median and maximum RQs. As for hormones and observing the fish trophic level, as in the WWI, all RQs were higher than one.

3.2.3. Surface Waters

This aquatic body was clearly more problematic to the environment when compared to WWIs and WWEs since here is where most of aquatic life inhabits. However, like in the previous water compartments, RQs higher than 1 were observed for all trophic levels and therapeutic groups, with the exception of the anxiolytics. Antiepileptic, anti-inflammatories, antibiotics and hormones remained the therapeutic groups with the highest RQs (Figure 6).

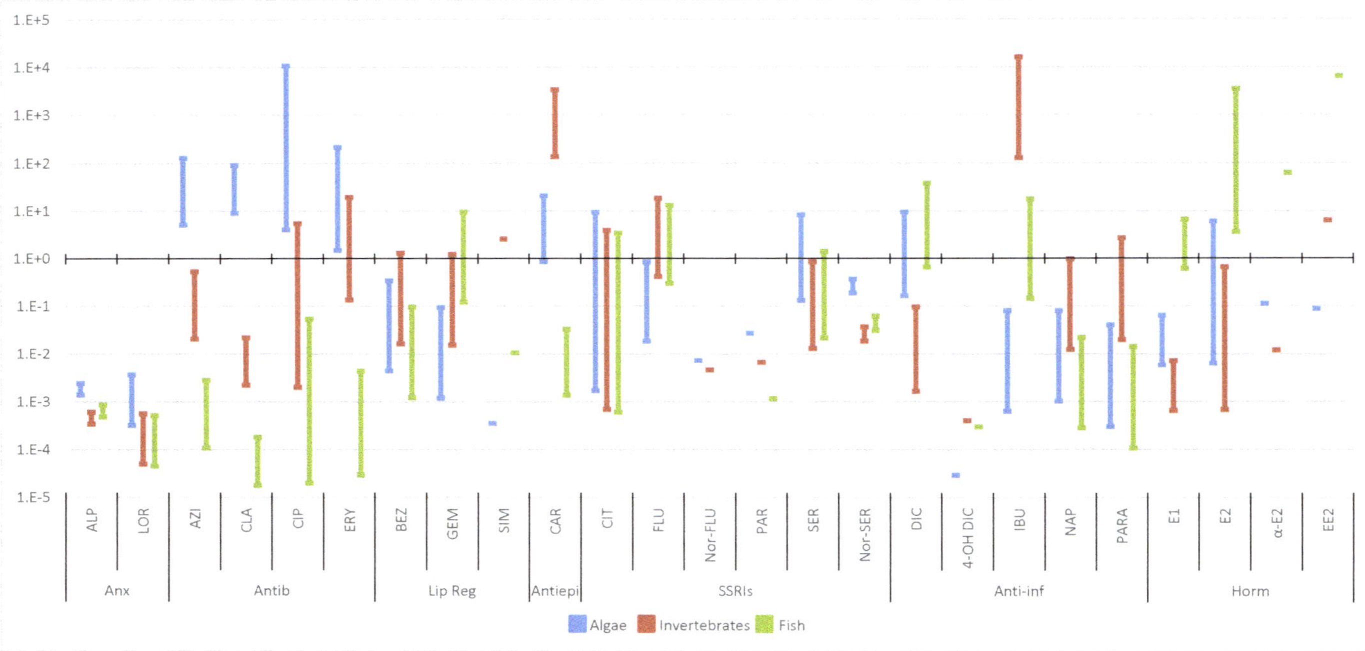

Figure 6. Median and maximum risk quotients of pharmaceuticals in surface waters for the three trophic levels. (Anx—anxiolytics; Antib—antibiotics; Lip reg—lipid regulators; Antiepi—antiepileptics; SSRIs—Selective serotonin reuptake inhibitors; Anti-inf—anti-inflammatories; Horm—hormones).

The highest maximum and median values regarded IBU (16,327) and CAR (138) in invertebrates. There were still eighteen pharmaceuticals (AZI, CLA, CIP, ERY, BEZ, GEM, SIM, CAR, CIT, FLU, SER, DIC, IBU, PARA, E1, E2, αE2 and EE2) with RQs above 1.

Antibiotics still presented all of their maximum RQs higher than 1 for algae, whereas lipid regulators presented the same pattern for invertebrates. As for the antiepileptic, their maximum values were above one for algae and invertebrates. In SSRIs therapeutic group, CIT, FLU and SER were the ones that presented RQs higher than 1, contributing to the possible risk posed by this group. The hormones, regarding fish, still presented all median RQs higher than 1, with the exception of E1, with the EE2 obtaining the highest RQs for this trophic level.

As already mentioned, this water body, encompassing rivers and lakes, should be free from risk. Nonetheless, from the 28 detected pharmaceuticals, 18 presented maximum RQs above 1 and even 8 had median RQs superior to 1, posing a threat to all the aquatic organisms.

3.2.4. Other Water Bodies

The results obtained for the RQs of other water bodies are presented in Figure 7. The RQs, in decreasing order, were groundwater, seawater, mineral water and drinking water. Since drinking water is usually obtained through surface and groundwaters, these results suggest that the sources used to produce drinking water were the ones with lower pharmaceutical contaminations or that water treatment plants were removing the selected pharmaceuticals. The results obtained in seawater can be biased, since fresh water organisms were used to evaluate the risk and there are reports that marine organism can be more vulnerable, increasing the risk in this water matrix [178,179].

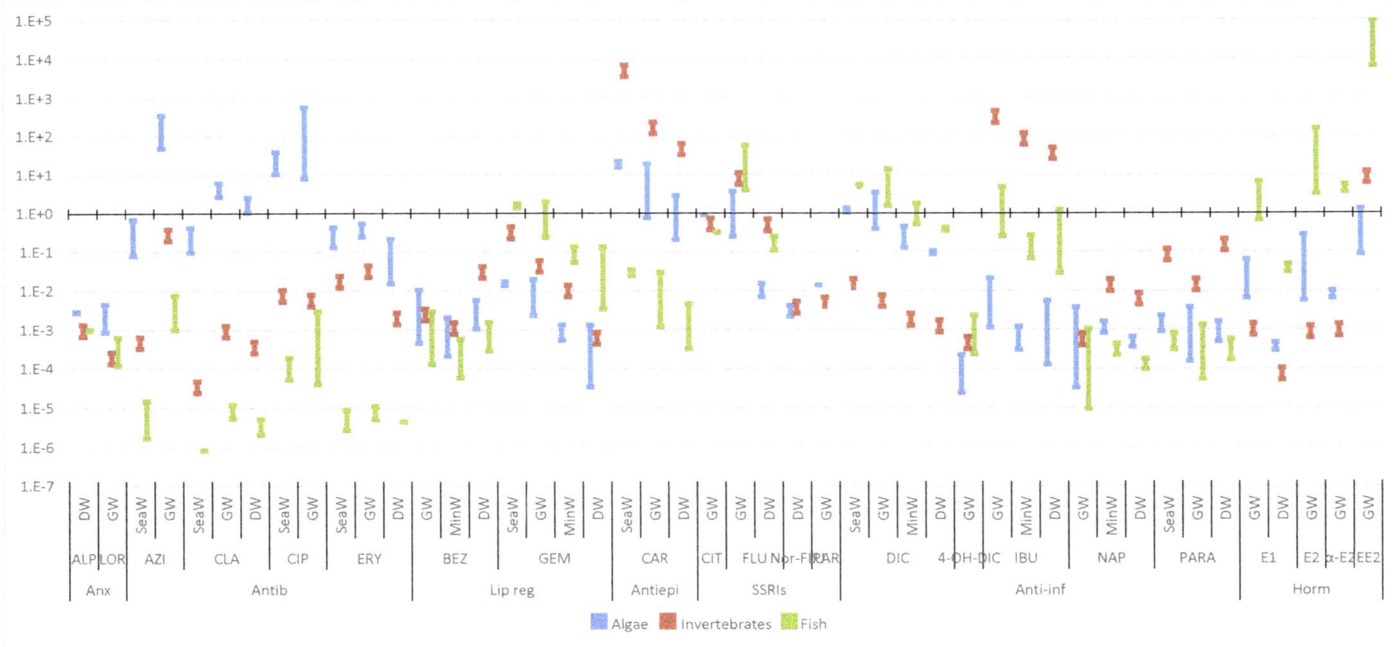

Figure 7. Median and maximum risk quotients of pharmaceuticals in other water bodies for the three trophic levels. (DW—drinking water; GW groundwater; SeaW—seawater; MinW—mineral water; Anx—anxiolytics; Antib—antibiotics; Lip reg—lipid regulators; Antiepi—antiepileptics; SSRIs—Selective serotonin reuptake inhibitors; Anti-inf—anti-inflammatories; Horm—hormones).

All of these compartments presented values above 1 for all trophic levels, with the exception of mineral water in algae, highlighting, once again, the pressure sustained by the aquatic organisms in all the aquatic compartments. This indicates that there is also a possible risk for humans. The therapeutic groups with higher maximum and median RQs are hormones, antiepileptic, antibiotics and anti-inflammatories, being the same groups as in WWIs.

With the exception of anxiolytics, all therapeutic groups still present RQ medians above 1, with 12 pharmaceuticals with maximum values above 1, and 11 with medians also higher than 1. If we use the threshold of 0.1, where some risk might be expected, we can find 16 of these compounds. Some RQs are still extremely high, with EE2 presenting values up to 91,150 and a median of 6091 for fish in groundwater. The higher RQs observed in these compartments were for CAR and IBU in invertebrates, and for hormones in fish.

Viewing these results, it is possible to observe not only high risk for aquatic organism in wastewaters but, despite the RQ reduction, several pharmaceuticals still promote risk in other supposedly cleaner aquatic matrices. Additionally, additive or even synergistic effects can occur, especially in the pharmaceuticals with the same mechanism of action [139–142].

3.3. Mitigation Measures

The RQs obtained for all the water compartments, particularly the RQs higher than 1 in surface waters and in the other water bodies, raise not only the issue of toxicity for the aquatic environment but also for humans using these aquatic bodies as a source of drinking water and also for whom eating animals living in these contaminated environments. Therefore, mitigation measures should be implemented to prevent high RQs in these important water resources. These measures should begin through the awareness of the problem. For example, in Sweden, an environmental classification system for drugs has been established through collaboration between producers, authorities and the public health care. This system assesses the environmental risk of pharmaceuticals being publicly available, therefore, the market can demand for medicines with less environmental impact, stimulating producers to design more environmentally friendly medicines [17]. This includes the concept of green pharmacy, where the design of pharmaceutical products focus also on their high metabolization and environmental degradation, reducing the environmental burden and improving environmental safety and health impacts [138].

Additionally, the ERA guideline on human pharmaceuticals should be revised in order to: enter the risk benefit analysis; impose its revision every five years with the new available data; incorporate the pharmaceuticals marketed before 2006 and include metabolites and transformation products [6].

Another issue already referred is the possible improvement of WWTPs removal efficiencies and the improvement of toxicity studies. The main challenges regarding the assessment of ecotoxicity are the scarce information available for some of the selected pharmaceuticals, namely chronic data and effects on multiple life stages or even multiple generations, which mimics the environmental exposure. Behavior studies are also lacking, before death and other major toxicity effects there can be diminished sexual interest, fear or activity, increased curiosity, etc. Although this seems like small behavior changes, it can be enough to unbalance an ecosystem eliminating one or more species by being unable to escape from predators or by lack of reproduction. This leads to another point that ecotoxicity studies should also be performed in ecosystems, because some effects previously referred are more evident in this type of studies. It should also be considered the increase of species tested and mixture effects, not only between different pharmaceuticals of the same therapeutic group but also from different groups and with other substances, like heavy metals. Given the pharmaceutical environmental presence in mixtures and with other substances, additive or even synergistic effects occur; therefore, the real hazard may be greater than that calculated [24,33,180,181]. Another issue that is neglected with the traditional ecotoxicity studies is the emergence of bacterial resistance. We are testing antibiotics in algae, invertebrates and fish but probably the biggest impact will be on bacteria, namely on the emergence of resistant bacteria that can reach not only aquatic animals but also humans. Additionally, the emergence of bacterial resistance is a major concern involving the presence of pharmaceuticals in the aquatic compartment, which is more prone in this environment [6,182].

This could generate benefits in water resource management, by providing the means for cross-compliance measures in environmental regulation and providing an adequate risk assessment for pharmaceuticals mixtures [183]. In this way, the complete scenario of the contamination of pharmaceuticals in the aquatic environment and their risk could be performed, contributing to future improvements in minimization measures and legislation.

4. Final Remarks

A literature review was conducted in order to understand the toxicity and ERA of pharmaceuticals in the aquatic environment. In this context, a broad and specialized background was obtained, enabling an overview of the state of the art in these subjects.

Regarding the toxicity data, although the differences observed between different therapeutic groups and within each therapeutic group, all therapeutic groups with the exception of anxiolytics, had at least one toxicity report for concentrations below 1 $\mu g\ L^{-1}$. The trophic level with the lowest concentrations promoting toxic effects was fish, followed by invertebrates and algae, emphasizing that fish, the trophic level closer to humans, are more prone to toxicity effects from the selected pharmaceuticals.

The results also show that pharmaceuticals with higher RQs are not the ones with higher occurrence and that proper toxicity data is important to a correct evaluation of the ERA.

The ERA performed for the pharmaceuticals in the different aquatic compartments revealed, with the exception of anxiolytics, RQs higher than 1, not only for WWIs but also for all aquatic compartments, for all trophic levels and therapeutic groups.

The therapeutic groups with the highest RQs in all aquatic compartments are hormones, antiepileptics, anti-inflammatories and antibiotics and the pharmaceuticals with the highest values are the EE2, CAR, IBU, CIP and AZI in all aquatic compartments. Highlighting threat to all the aquatic organisms exposed, namely on the feminization of fish by EE2 and its impact on aquatic ecosystems. Additionally, two antibiotics are among the pharmaceuticals with higher RQs and the ERA does not evaluate the emergence of bacterial resistance. If this issue was also evaluated it would probably confirm why they are considered the therapeutic group with the highest risk for humans, regarding the residues of medicinal products in the environment.

Unfortunately, the pressure of pharmaceuticals on aquatic bodies will continue to rise, and, therefore, mitigation measures and changes in legislation must be implemented.

Supplementary Materials: The following are available online: Table S1: Ecotoxicological data on the selected pharmaceuticals, Table S2: Occurrence data from the different aquatic compartments for environmental risk assessment evaluation.

Author Contributions: Conceptualization, A.P. (André Pereira), L.S. and A.P. (Angelina Pena); Methodology, A.P. (André Pereira), L.S. and A.P. (Angelina Pena); Formal analysis, A.P. (André Pereira); Writing—original draft preparation, A.P. (André Pereira); Writing—review and editing, A.P. (André Pereira), L.S. and C.L. (Celia Laranjeiro); Supervision, C.L. (Celeste Lino) and A.P. (Angelina Pena); Project administration, A.P. (Angelina Pena). All authors have read and agreed to the published version of the manuscript.

Funding: The work was supported by UIDB/50006/2020 with funding from FCT/MCTES through national funds.

Conflicts of Interest: The authors declare no conflict of interest.

References

1. Robles-Molina, J.; Lara-Ortega, F.J.; Gilbert-López, B.; García-Reyes, J.F.; Molina-Díaz, A. Multi-residue method for the determination of over 400 priority and emerging pollutants in water and wastewater by solid-phase extraction and liquid chromatography-time-of-flight mass spectrometry. *J. Chromatogr. A* **2014**, *1350*, 30–43. [CrossRef] [PubMed]
2. Silva, L.J.G.; Lino, C.M.; Meisel, L.M.; Pena, A. Selective serotonin re-uptake inhibitors (SSRIs) in the aquatic environment: An ecopharmacovigilance approach. *Sci. Total Environ.* **2012**, *437*, 185–195. [CrossRef] [PubMed]
3. Gonzalez-Rey, M.; Bebianno, M.J. Does selective serotonin reuptake inhibitor (SSRI) fluoxetine affects mussel Mytilus galloprovincialis? *Environ. Pollut.* **2013**, *173*, 200–209. [CrossRef]
4. *European Medicines Agency Guideline on the Environmental Risk Assessment of Medicinal Products for Human Use*; European Medicines Agency: London, UK, 2006.
5. Taylor, D.; Senac, T. Human pharmaceutical products in the environment—The "problem" in perspective. *Chemosphere* **2014**, *115*, 95–99. [CrossRef]
6. Pereira, A.M.P.T.; Silva, L.J.G.; Lino, C.M.; Meisel, L.M.; Pena, A. A critical evaluation of different parameters for estimating pharmaceutical exposure seeking an improved environmental risk assessment. *Sci. Total Environ.* **2017**, *603–604*, 226–236. [CrossRef]
7. Mompelat, S.; Le Bot, B.; Thomas, O. Occurrence and fate of pharmaceutical products and by-products, from resource to drinking water. *Environ. Int.* **2009**, *35*, 803–814. [CrossRef]
8. Van Doorslaer, X.; Dewulf, J.; Van Langenhove, H.; Demeestere, K. Fluoroquinolone antibiotics: An emerging class of environmental micropollutants. *Sci. Total Environ.* **2014**, *500–501*, 250–269. [CrossRef]
9. Boxall, A.B.A.; Keller, V.D.J.; Straub, J.O.; Monteiro, S.C.; Fussell, R.; Williams, R.J. Exploiting monitoring data in environmental exposure modelling and risk assessment of pharmaceuticals. *Environ. Int.* **2014**, *73*, 176–185. [CrossRef]
10. European Commission Implementing Decision (EU) 2018/840 of 5 June 2018 establishing a watch list of substances for Union-wide monitoring in the field of water policy pursuant to Directive 2008/105/EC of the European Parliament and of the Council and repealing 495. *Off. J. Eur. Union* **2018**, *L141*, 9–12.
11. Papageorgiou, M.; Kosma, C.; Lambropoulou, D. Seasonal occurrence, removal, mass loading and environmental risk assessment of 55 pharmaceuticals and personal care products in a municipal wastewater treatment plant in Central Greece. *Sci. Total Environ.* **2016**, *543*, 547–569. [CrossRef]
12. Ter Laak, T.L.; Van der Aa, M.; Houtman, C.J.; Stoks, P.G.; Van Wezel, A.P. Relating environmental concentrations of pharmaceuticals to consumption: A mass balance approach for the river Rhine. *Environ. Int.* **2010**, *36*, 403–409. [CrossRef] [PubMed]
13. Bade, R.; Rousis, N.I.; Bijlsma, L.; Gracia-Lor, E.; Castiglioni, S.; Sancho, J.V.; Hernandez, F. Screening of pharmaceuticals and illicit drugs in wastewater and surface waters of Spain and Italy by high resolution mass spectrometry using UHPLC-QTOF MS and LC-LTQ-Orbitrap MS. *Anal. Bioanal. Chem.* **2015**, *407*, 8979–8988. [CrossRef] [PubMed]
14. Ribeiro, A.R.; Nunes, O.C.; Pereira, M.F.R.; Silva, A.M.T. An overview on the advanced oxidation processes applied for the treatment of water pollutants defined in the recently launched Directive 2013/39/EU. *Environ. Int.* **2015**, *75*, 33–51. [CrossRef] [PubMed]
15. Pereira, A.M.P.T.; Silva, L.J.G.; Meisel, L.M.; Lino, C.M.; Pena, A. Environmental impact of pharmaceuticals from Portuguese wastewaters: Geographical and seasonal occurrence, removal and risk assessment. *Environ. Res.* **2015**, *136*, 108–119. [CrossRef]
16. He, Y.; Sutton, N.B.; Rijnaarts, H.H.H.; Langenhoff, A.A.M. Degradation of pharmaceuticals in wastewater using immobilized TiO2 photocatalysis under simulated solar irradiation. *Appl. Catal. B Environ.* **2016**, *182*, 132–141. [CrossRef]
17. Zenker, A.; Cicero, M.R.; Prestinaci, F.; Bottoni, P.; Carere, M. Bioaccumulation and biomagnification potential of pharmaceuticals with a focus to the aquatic environment. *J. Environ. Manag.* **2014**, *133*, 378–387. [CrossRef]
18. Vergeynst, L.; Haeck, A.; De Wispelaere, P.; Van Langenhove, H.; Demeestere, K. Multi-residue analysis of pharmaceuticals in wastewater by liquid chromatography-magnetic sector mass spectrometry: Method quality assessment and application in a Belgian case study. *Chemosphere* **2015**, *119*, S2–S8. [CrossRef]

19. Johnson, A.C.; Sumpter, J.P. Putting pharmaceuticals into the wider context of challenges to fish populations in rivers. *Philos. Trans. R. Soc. B Biol. Sci.* **2014**, *369*, 20130581. [CrossRef]
20. Petrie, B.; Barden, R.; Kasprzyk-Hordern, B. A review on emerging contaminants in wastewaters and the environment: Current knowledge, understudied areas and recommendations for future monitoring. *Water Res.* **2014**, *72*, 3–27. [CrossRef]
21. Lavén, M.; Alsberg, T.; Yu, Y.; Adolfsson-Erici, M.; Sun, H. Serial mixed-mode cation- and anion-exchange solid-phase extraction for separation of basic, neutral and acidic pharmaceuticals in wastewater and analysis by high-performance liquid chromatography-quadrupole time-of-flight mass spectrometry. *J. Chromatogr. A* **2009**, *1216*, 49–62. [CrossRef]
22. *OECD Guidance Document on the Use of the Harmonised System for the Classification of Chemicals Which Are Hazardous for the Aquatic Environment*; OECD: Paris, France, 2001.
23. Isidori, M.; Lavorgna, M.; Nardelli, A.; Pascarella, L.; Parrella, A. Toxic and genotoxic evaluation of six antibiotics on non-target organisms. *Sci. Total Environ.* **2005**, *346*, 87–98. [CrossRef] [PubMed]
24. Yang, L.-H.; Ying, G.-G.; Su, H.-C.; Stauber, J.L.; Adams, M.S.; Binet, M.T. Growth-inhibiting effects of 12 antibacterial agenst and their mixtures on the freshwater microalga Pseudokirchneriella subcapitata. *Environ. Toxicol. Chem.* **2008**, *27*, 1201. [CrossRef] [PubMed]
25. Farré, M.; Ferrer, I.; Ginebreda, A.; Figueras, M.; Olivella, L.; Tirapu, L.; Vilanova, M.; Barceló, D. Determination of drugs in surface water and wastewater samples by liquid chromatography–mass spectrometry: Methods and preliminary results including toxicity studies with Vibrio fischeri. *J. Chromatogr. A* **2001**, *938*, 187–197. [CrossRef]
26. DeLorenzo, M.E.; Fleming, J. Individual and mixture effects of selected pharmaceuticals and personal care products on the marine phytoplankton species Dunaliella tertiolecta. *Arch. Environ. Contam. Toxicol.* **2008**, *54*, 203–210. [CrossRef] [PubMed]
27. Ferrari, B.; Mons, R.; Vollat, B.; Fraysse, B.; Paxéus, N.; Lo Giudice, R.; Pollio, A.; Garric, J. Environmental risk assessment of six human pharmaceuticals: Are the current environmental risk assessment procedures sufficient for the protection of the aquatic environment? *Environ. Toxicol. Chem.* **2004**, *23*, 1344–1354. [CrossRef]
28. Cleuvers, M. Aquatic ecotoxicity of pharmaceuticals including the assessment of combination effects. *Toxicol. Lett.* **2003**, *142*, 185–194. [CrossRef]
29. Ferrari, B.; Paxéus, N.; Giudice, R.L.; Pollio, A.; Garric, J. Ecotoxicological impact of pharmaceuticals found in treated wastewaters: Study of carbamazepine, clofibric acid, and diclofenac. *Ecotoxicol. Environ. Saf.* **2003**, *55*, 359–370. [CrossRef]
30. Christensen, A.M.; Faaborg-Andersen, S.; Ingerslev, F.; Baun, A. Mixture and single-substance toxicity of selective serotonin reuptake inhibitors toward algae and crustaceans. *Environ. Toxicol. Chem.* **2007**, *26*, 85. [CrossRef]
31. Brooks, B.W.; Turner, P.K.; Stanley, J.K.; Weston, J.J.; Glidewell, E.A.; Foran, C.M.; Slattery, M.; La Point, T.W.; Huggett, D.B. Waterborne and sediment toxicity of fluoxetine to select organisms. *Chemosphere* **2003**, *52*, 135–142. [CrossRef]
32. Brooks, B.W.; Foran, C.M.; Richards, S.M.; Weston, J.; Turner, P.K.; Stanley, J.K.; Solomon, K.R.; Slattery, M.; La Point, T.W. Aquatic ecotoxicology of fluoxetine. *Toxicol. Lett.* **2003**, *142*, 169–183. [CrossRef]
33. Johnson, D.J.; Sanderson, H.; Brain, R.A.; Wilson, C.J.; Solomon, K.R. Toxicity and hazard of selective serotonin reuptake inhibitor antidepressants fluoxetine, fluvoxamine, and sertraline to algae. *Ecotoxicol. Environ. Saf.* **2007**, *67*, 128–139. [CrossRef]
34. Neuwoehner, J.; Fenner, K.; Escher, B.I. Physiological modes of action of fluoxetine and its human metabolites in algae. *Environ. Sci. Technol.* **2009**, *43*, 6830–6837. [CrossRef]
35. Halling-Sorensen, B. Environmental risk assessment of antibiotics: Comparison of mecillinam, trimethoprim and ciprofloxacin. *J. Antimicrob. Chemother.* **2000**, *46*, 53–58. [CrossRef]
36. Minagh, E.; Hernan, R.; O'Rourke, K.; Lyng, F.M.; Davoren, M. Aquatic ecotoxicity of the selective serotonin reuptake inhibitor sertraline hydrochloride in a battery of freshwater test species. *Ecotoxicol. Environ. Saf.* **2009**, *72*, 434–440. [CrossRef] [PubMed]
37. Lawrence, J.R.; Swerhone, G.D.W.; Topp, E.; Korber, D.R.; Neu, T.R.; Wassenaar, L.I. Structural and functional responses of river biofilm communities to the nonsteroidal anti-inflammatory diflofenac. *Environ. Toxicol. Chem.* **2007**, *26*, 573. [CrossRef]

38. Ortiz de García, S.; Pinto, G.P.; García-Encina, P.A.; Mata, R.I. Ranking of concern, based on environmental indexes, for pharmaceutical and personal care products: An application to the Spanish case. *J. Environ. Manage.* **2013**, *129*, 384–397. [CrossRef] [PubMed]
39. Pomati, F.; Netting, A.G.; Calamari, D.; Neilan, B.A. Effects of erythromycin, tetracycline and ibuprofen on the growth of Synechocystis sp. and Lemna minor. *Aquat. Toxicol.* **2004**, *67*, 387–396. [CrossRef] [PubMed]
40. Isidori, M.; Lavorgna, M.; Nardelli, A.; Parrella, A.; Previtera, L.; Rubino, M. Ecotoxicity of naproxen and its phototransformation products. *Sci. Total Environ.* **2005**, *348*, 93–101. [CrossRef]
41. Cleuvers, M. Mixture toxicity of the anti-inflammatory drugs diclofenac, ibuprofen, naproxen, and acetylsalicylic acid. *Ecotoxicol. Environ. Saf.* **2004**, *59*, 309–315. [CrossRef]
42. Henschel, K.-P.; Wenzel, A.; Diedrich, M.; Fliedner, A. Environmental Hazard Assessment of Pharmaceuticals. *Regul. Toxicol. Pharmacol.* **1997**, *25*, 220–225. [CrossRef] [PubMed]
43. Onogbosele, C.O. Bioavailability of organic contaminants in rivers. Ph.D. Thesis, Brunel University, London, UK, 2015.
44. Li, Y.; Zhang, L.; Liu, X.; Ding, J. Ranking and prioritizing pharmaceuticals in the aquatic environment of China. *Sci. Total Environ.* **2019**, *658*, 333–342. [CrossRef] [PubMed]
45. Machado, M.D.; Soares, E.V. Sensitivity of freshwater and marine green algae to three compounds of emerging concern. *J. Appl. Phycol.* **2019**, *31*, 399–408. [CrossRef]
46. Martins, N.; Pereira, R.; Abrantes, N.; Pereira, J.; Gonçalves, F.; Marques, C.R. Ecotoxicological effects of ciprofloxacin on freshwater species: Data integration and derivation of toxicity thresholds for risk assessment. *Ecotoxicology* **2012**, *21*, 1167–1176. [CrossRef] [PubMed]
47. Ebert, I.; Bachmann, J.; Kühnen, U.; Küster, A.; Kussatz, C.; Maletzki, D.; Schlüter, C. Toxicity of the fluoroquinolone antibiotics enrofloxacin and ciprofloxacin to photoautotrophic aquatic organisms. *Environ. Toxicol. Chem.* **2011**, *30*, 2786–2792. [CrossRef]
48. Brain, R.A.; Johnson, D.J.; Richards, S.M.; Sanderson, H.; Sibley, P.K.; Solomon, K.R. Effects of 25 pharmaceutical compounds to Lemna gibba using a seven-day static-renewel test. *Environ. Toxicol. Chem.* **2004**, *23*, 371–382. [CrossRef]
49. Robinson, A.A.; Belden, J.B.; Lydy, M.J. Toxicity of fluoroquinolone antibiotics to aquatic organisms. *Environ. Toxicol. Chem.* **2005**, *24*, 423. [CrossRef]
50. Nie, X.; Wang, X.; Chen, J.; Zitko, V.; An, T. Response of the freshwater Alga chlorella vulgaris to trichloroisocyanuric acid and ciprofloxacin. *Environ. Toxicol. Chem.* **2008**, *27*, 168–173. [CrossRef]
51. Isidori, M.; Nardelli, A.; Pascarella, L.; Rubino, M.; Parrella, A. Toxic and genotoxic impact of fibrates and their photoproducts on non-target organisms. *Environ. Int.* **2007**, *33*, 635–641. [CrossRef]
52. Zurita, J.L.; Repetto, G.; Jos, A.; Salguero, M.; López-Artíguez, M.; Cameán, A.M. Toxicological effects of the lipid regulator gemfibrozil in four aquatic systems. *Aquat. Toxicol.* **2007**, *81*, 106–115. [CrossRef]
53. Quinn, B.; Gagné, F.; Blaise, C. An investigation into the acute and chronic toxicity of eleven pharmaceuticals (and their solvents) found in wastewater effluent on the cnidarian, Hydra attenuata. *Sci. Total Environ.* **2008**, *389*, 306–314. [CrossRef]
54. Han, G.H.; Hur, H.G.; Kim, S.D. Ecotoxicological risk of pharmaceuticals from wastewater treatment plants in Korea: Occurrence and toxicity to Daphnia magna. *Environ. Toxicol. Chem.* **2006**, *25*, 265–271. [CrossRef]
55. Key, P.B.; Hoguet, J.; Reed, L.A.; Chung, K.W.; Fulton, M.H. Effects of the statin antihyperlipidemic agent simvastatin on grass shrimp, Palaemonetes pugio. *Environ. Toxicol.* **2008**, *23*, 153–160. [CrossRef]
56. Dahl, U.; Gorokhova, E.; Breitholtz, M. Application of growth-related sublethal endpoints in ecotoxicological assessments using a harpacticoid copepod. *Aquat. Toxicol.* **2006**, *77*, 433–438. [CrossRef] [PubMed]
57. De Lange, H.J.; Noordoven, W.; Murk, A.J.; Lürling, M.; Peeters, E.T.H.M. Behavioural responses of Gammarus pulex (Crustacea, Amphipoda) to low concentrations of pharmaceuticals. *Aquat. Toxicol.* **2006**, *78*, 209–216. [CrossRef] [PubMed]
58. Kim, J.-W.; Ishibashi, H.; Yamauchi, R.; Ichikawa, N.; Takao, Y.; Hirano, M.; Koga, M.; Arizono, K. Acute toxicity of pharmaceutical and personal care products on freshwater crustacean (Thamnocephalus platyurus) and fish (Oryzias latipes). *J. Toxicol. Sci.* **2009**, *34*, 227–232. [CrossRef] [PubMed]
59. Kim, Y.; Choi, K.; Jung, J.; Park, S.; Kim, P.-G.; Park, J. Aquatic toxicity of acetaminophen, carbamazepine, cimetidine, diltiazem and six major sulfonamides, and their potential ecological risks in Korea. *Environ. Int.* **2007**, *33*, 370–375. [CrossRef]

60. Fong, P.P.; Molnar, N. Antidepressants cause foot detachment from substrate in five species of marine snail. *Mar. Environ. Res.* **2013**, *84*, 24–30. [CrossRef]
61. Nałecz-Jawecki, G. Evaluation of the in vitro biotransformation of fluoxetine with HPLC, mass spectrometry and ecotoxicological tests. *Chemosphere* **2007**, *70*, 29–35. [CrossRef]
62. Nentwig, G. Effects of pharmaceuticals on aquatic invertebrates. Part II: The antidepressant drug fluoxetine. *Arch. Environ. Contam. Toxicol.* **2007**, *52*, 163–170. [CrossRef]
63. Péry, A.R.R.; Gust, M.; Vollat, B.; Mons, R.; Ramil, M.; Fink, G.; Ternes, T.; Garric, J. Fluoxetine effects assessment on the life cycle of aquatic invertebrates. *Chemosphere* **2008**, *73*, 300–304. [CrossRef]
64. Gust, M.; Buronfosse, T.; Giamberini, L.; Ramil, M.; Mons, R.; Garric, J. Effects of fluoxetine on the reproduction of two prosobranch mollusks: Potamopyrgus antipodarum and Valvata piscinalis. *Environ. Pollut.* **2009**, *157*, 423–429. [CrossRef]
65. Hazelton, P.D.; Cope, W.G.; Mosher, S.; Pandolfo, T.J.; Belden, J.B.; Barnhart, M.C.; Bringolf, R.B. Fluoxetine alters adult freshwater mussel behavior and larval metamorphosis. *Sci. Total Environ.* **2013**, *445–446*, 94–100. [CrossRef]
66. Cunningham, V.L.; Constable, D.J.C.; Hannah, R.E. Environmental Risk Assessment of Paroxetine. *Environ. Sci. Technol.* **2004**, *38*, 3351–3359. [CrossRef]
67. Lamichhane, K.; Garcia, S.N.; Huggett, D.B.; DeAngelis, D.L.; La Point, T.W. Exposures to a selective serotonin reuptake inhibitor (SSRI), sertraline hydrochloride, over multiple generations: Changes in life history traits in Ceriodaphnia dubia. *Ecotoxicol. Environ. Saf.* **2014**, *101*, 124–130. [CrossRef]
68. Cleuvers, M. Chronic Mixture Toxicity of Pharmaceuticals to Daphnia—The Example of Nonsteroidal Anti-Inflammatory Drugs. In *Pharmaceuticals in the Environment*; Kümmerer, K., Ed.; Springer: Berlin, Heidelberg, 2006; pp. 277–284. ISBN 978-3-540-74664-5.
69. Nalecz-Jawecki, G.; Persoone, G. Toxicity of selected pharmaceuticals to the Anostracan crustacean Thamnocephalus platyurus—Comparison of sublethal and lethal effect levels with the 1 h Rapidtoxkit and the 24 h Thamnotoxkit microbiotests. *Environ. Sci. Pollut. Res. Int.* **2006**, *13*, 22–27. [CrossRef]
70. Haap, T.; Triebskorn, R.; Köhler, H.-R. Acute effects of diclofenac and DMSO to Daphnia magna: Immobilisation and hsp70-induction. *Chemosphere* **2008**, *73*, 353–359. [CrossRef]
71. Heckmann, L.-H.; Callaghan, A.; Hooper, H.L.; Connon, R.; Hutchinson, T.H.; Maund, S.J.; Sibly, R.M. Chronic toxicity of ibuprofen to Daphnia magna: Effects on life history traits and population dynamics. *Toxicol. Lett.* **2007**, *172*, 137–145. [CrossRef]
72. Pounds, N.; Maclean, S.; Webley, M.; Pascoe, D.; Hutchinson, T. Acute and chronic effects of ibuprofen in the mollusc Planorbis carinatus (Gastropoda: Planorbidae). *Ecotoxicol. Environ. Saf.* **2008**, *70*, 47–52. [CrossRef]
73. Dave, G.; Herger, G. Determination of detoxification to Daphnia magna of four pharmaceuticals and seven surfactants by activated sludge. *Chemosphere* **2012**, *88*, 459–466. [CrossRef]
74. Sherer, J.T. Pharmaceuticals in the environment. *Am. J. Heal. Pharm.* **2006**, *63*, 174–178. [CrossRef]
75. El-Bassat, R.; Touliabah, H.; Harisa, G. Toxicity of four pharmaceuticals from different classes to isolated plankton species. *African J. Aquat. Sci.* **2012**, *37*, 71–80. [CrossRef]
76. Li, M.-H. Acute toxicity of 30 pharmaceutically active compounds to freshwater planarians, Dugesia japonica. *Toxicol. Environ. Chem.* **2013**, *95*, 1157–1170. [CrossRef]
77. Kühn, R.; Pattard, M.; Pernak, K.; Winter, A. Results of the harmful effects of selected water pollutants (anilines, phenols, aliphatic compounds) to Daphnia magna. *Water Res.* **1989**, *23*, 495–499. [CrossRef]
78. Li, M. Acute toxicity of industrial endocrine-disrupting chemicals, natural and synthetic sex hormones to the freshwater planarian, Dugesia japonica. *Toxicol. Environ. Chem.* **2013**, *95*, 984–991. [CrossRef]
79. Jaser, W. Effects of 17α-ethinylestradiol on the reproduction of the cladoceran species Ceriodaphnia reticulata and Sida crystallina. *Environ. Int.* **2003**, *28*, 633–638. [CrossRef]
80. Caldwell, D.J.; Mastrocco, F.; Hutchinson, T.H.; Länge, R.; Heijerick, D.; Janssen, C.; Anderson, P.D.; Sumpter, J.P. Derivation of an Aquatic Predicted No-Effect Concentration for the Synthetic Hormone, 17α-Ethinyl Estradiol. *Environ. Sci. Technol.* **2008**, *42*, 7046–7054. [CrossRef]
81. Vandenbergh, G.F.; Adriaens, D.; Verslycke, T.; Janssen, C.R. Effects of 17α-ethinylestradiol on sexual development of the amphipod Hyalella azteca. *Ecotoxicol. Environ. Saf.* **2003**, *54*, 216–222. [CrossRef]
82. Jobling, S.; Casey, D.; Rodgers-Gray, T.; Oehlmann, J.; Schulte-Oehlmann, U.; Pawlowski, S.; Baunbeck, T.; Turner, A.; Tyler, C. Comparative responses of molluscs and fish to environmental estrogens and an estrogenic effluent. *Aquat. Toxicol.* **2004**, *66*, 207–222. [CrossRef]

83. Sidhu, H.; O'Connor, G.; McAvoy, D. Risk assessment of biosolids-borne ciprofloxacin and azithromycin. *Sci. Total Environ.* **2019**, *651*, 3151–3160. [CrossRef]
84. Henry, T.B.; Kwon, J.W.; Armbrust, K.L.; Black, M.C. Acute and chronic toxicity of five selective serotonin reuptake inhibitors in Ceriodaphnia dubia. *Environ. Toxicol. Chem.* **2004**, *23*, 2229–2233. [CrossRef]
85. Mimeault, C.; Trudeau, V.L.; Moon, T.W. Waterborne gemfibrozil challenges the hepatic antioxidant defense system and down-regulates peroxisome proliferator-activated receptor beta (PPARbeta) mRNA levels in male goldfish (Carassius auratus). *Toxicology* **2006**, *228*, 140–150. [CrossRef]
86. Raldúa, D.; André, M.; Babin, P.J. Clofibrate and gemfibrozil induce an embryonic malabsorption syndrome in zebrafish. *Toxicol. Appl. Pharmacol.* **2008**, *228*, 301–314.
87. Caminada, D.; Escher, C.; Fent, K. Cytotoxicity of pharmaceuticals found in aquatic systems: Comparison of PLHC-1 and RTG-2 fish cell lines. *Aquat. Toxicol.* **2006**, *79*, 114–123. [CrossRef]
88. Richards, S.M.; Cole, S.E. A toxicity and hazard assessment of fourteen pharmaceuticals to Xenopus laevis larvae. *Ecotoxicology* **2006**, *15*, 647–656. [CrossRef]
89. Stanley, J.K.; Ramirez, A.J.; Chambliss, C.K.; Brooks, B.W. Enantiospecific sublethal effects of the antidepressant fluoxetine to a model aquatic vertebrate and invertebrate. *Chemosphere* **2007**, *69*, 9–16. [CrossRef]
90. Henry, T.B.; Black, M.C. Acute and chronic toxicity of fluoxetine (selective serotonin reuptake inhibitor) in western mosquitofish. *Arch. Environ. Contam. Toxicol.* **2008**, *54*, 325–330. [CrossRef]
91. Nakamura, Y.; Yamamoto, H.; Sekizawa, J.; Kondo, T.; Hirai, N.; Tatarazako, N. The effects of pH on fluoxetine in Japanese medaka (Oryzias latipes): Acute toxicity in fish larvae and bioaccumulation in juvenile fish. *Chemosphere* **2008**, *70*, 865–873. [CrossRef]
92. Valenti, T.W.; Hurtado, P.P.; Chambliss, C.K.; Brooks, B.W. Aquatic toxicity of sertraline to Pimephales promelas at environmentally relevant surface water pH. *Environ. Toxicol. Chem.* **2009**, *28*, 2685–2694. [CrossRef]
93. Schwaiger, J.; Ferling, H.; Mallow, U.; Wintermayr, H.; Negele, R.D. Toxic effects of the non-steroidal anti-inflammatory drug diclofenac. Part I: Histopathological alterations and bioaccumulation in rainbow trout. *Aquat. Toxicol.* **2004**, *68*, 141–150. [CrossRef]
94. Triebskorn, R.; Casper, H.; Heyd, A.; Eikemper, R.; Köhler, H.-R.; Schwaiger, J. Toxic effects of the non-steroidal anti-inflammatory drug diclofenac. Part II: Cytological effects in liver, kidney, gills and intestine of rainbow trout (Oncorhynchus mykiss). *Aquat. Toxicol.* **2004**, *68*, 151–166. [CrossRef]
95. Hoeger, B.; Köllner, B.; Dietrich, D.R.; Hitzfeld, B. Water-borne diclofenac affects kidney and gill integrity and selected immune parameters in brown trout (Salmo trutta f. fario). *Aquat. Toxicol.* **2005**, *75*, 53–64. [CrossRef] [PubMed]
96. Memmert, U.; Peither, A.; Burri, R.; Weber, K.; Schmidt, T.; Sumpter, J.P.; Hartmann, A. Diclofenac: New data on chronic toxicity and bioconcentration in fish. *Environ. Toxicol. Chem.* **2013**, *32*, 442–452. [CrossRef] [PubMed]
97. Mehinto, A.C.; Hill, E.M.; Tyler, C.R. Uptake and biological effects of environmentally relevant concentrations of the nonsteroidal anti-inflammatory pharmaceutical diclofenac in rainbow trout (Oncorhynchus mykiss). *Environ. Sci. Technol.* **2010**, *44*, 2176–2182. [CrossRef]
98. Oaks, J.L.; Gilbert, M.; Virani, M.Z.; Watson, R.T.; Meteyer, C.U.; Rideout, B.A.; Shivaprasad, H.L.; Ahmed, S.; Iqbal Chaudhry, M.J.; Arshad, M.; et al. Diclofenac residues as the cause of vulture population decline in Pakistan. *Nature* **2004**, *427*, 630–633. [CrossRef] [PubMed]
99. Li, Q.; Wang, P.; Chen, L.; Gao, H.; Wu, L. Acute toxicity and histopathological effects of naproxen in zebrafish (Danio rerio) early life stages. *Environ. Sci. Pollut. Res.* **2016**, *23*, 18832–18841. [CrossRef]
100. Sun, L.W.; Qu, M.M.; Li, Y.Q.; Wu, Y.L.; Chen, Y.G.; Kong, Z.M.; Liu, Z.T. Toxic effects of aminophenols on aquatic life using the Zebrafish embryo test and the comet assay. *Bull. Environ. Contam. Toxicol.* **2004**, *73*, 628–634. [CrossRef] [PubMed]
101. Metcalfe, C.D.; Metcalfe, T.L.; Kiparissis, Y.; Koenig, B.G.; Khan, C.; Hughes, R.J.; Croley, T.R.; March, R.E.; Potter, T. Estrogenic potency of chemicals detected in sewage treatment plant effluents as determined by in vivo assays with Japanese medaka (Oryzias latipes). *Environ. Toxicol. Chem.* **2001**, *20*, 297–308. [CrossRef]
102. Imai, S.; Koyama, J.; Fujii, K. Effects of estrone on full life cycle of Java medaka (Oryzias javanicus), a new marine test fish. *Environ. Toxicol. Chem.* **2007**, *26*, 726–731. [CrossRef]

103. Thorpe, K.L.; Benstead, R.; Hutchinson, T.H.; Tyler, C.R. Associations between altered vitellogenin concentrations and adverse health effects in fathead minnow (Pimephales promelas). *Aquat. Toxicol.* **2007**, *85*, 176–183. [CrossRef]
104. Kang, I.J.; Yokota, H.; Oshima, Y.; Tsuruda, Y.; Yamaguchi, T.; Maeda, M.; Imada, N.; Tadokoro, H.; Honjo, T. Effect of 17β-estradiol on the reproduction of Japanese medaka (Oryzias latipes). *Chemosphere* **2002**, *47*, 71–80. [CrossRef]
105. Caldwell, D.J.; Mastrocco, F.; Anderson, P.D.; Länge, R.; Sumpter, J.P. Predicted-no-effect concentrations for the steroid estrogens estrone, 17β-estradiol, estriol, and 17α-ethinylestradiol. *Environ. Toxicol. Chem.* **2012**, *31*, 1396–1406. [CrossRef] [PubMed]
106. Kramer, V.J.; Miles-Richardson, S.; Pierens, S.L.; Giesy, J.P. Reproductive impairment and induction of alkaline-labile phosphate, a biomarker of estrogen exposure, in fathead minnows (Pimephales promelas) exposed to waterborne 17β-estradiol. *Aquat. Toxicol.* **1998**, *40*, 335–360. [CrossRef]
107. Shappell, N.W.; Elder, K.H.; West, M. Estrogenicity and nutrient concentration of surface waters surrounding a large confinement dairy operation using best management practices for land application of animal wastes. *Environ. Sci. Technol.* **2010**, *44*, 2365–2371. [CrossRef]
108. Seki, M.; Fujishima, S.; Nozaka, T.; Maeda, M.; Kobayashi, K. Comparison of response to 17β-estradiol and 17β-trenbolone among three small fish species. *Environ. Toxicol. Chem.* **2006**, *25*, 2742. [CrossRef] [PubMed]
109. Brion, F.; Tyler, C.; Palazzi, X.; Laillet, B.; Porcher, J.; Garric, J.; Flammarion, P. Impacts of 17β-estradiol, including environmentally relevant concentrations, on reproduction after exposure during embryo-larval-, juvenile- and adult-life stages in zebrafish (Danio rerio). *Aquat. Toxicol.* **2004**, *68*, 193–217. [PubMed]
110. Van der Ven, L.T.M.; Van den Brandhof, E.-J.; Vos, J.H.; Wester, P.W. Effects of the estrogen agonist 17β-estradiol and antagonist tamoxifen in a partial life-cycle assay with with zebrafish (danio rerio). *Environ. Toxicol. Chem.* **2007**, *26*, 92. [CrossRef] [PubMed]
111. Nash, J.P.; Kime, D.E.; Van der Ven, L.T.M.; Wester, P.W.; Brion, F.; Maack, G.; Stahlschmidt-Allner, P.; Tyler, C.R. Long-term exposure to environmental concentrations of the pharmaceutical ethynylestradiol causes reproductive failure in fish. *Environ. Health Perspect.* **2004**, *112*, 1725–1733. [CrossRef]
112. Hirai, N.; Nanba, A.; Koshio, M.; Kondo, T.; Morita, M.; Tatarazako, N. Feminization of Japanese medaka (Oryzias latipes) exposed to 17β-estradiol: Formation of testis–ova and sex-transformation during early-ontogeny. *Aquat. Toxicol.* **2006**, *77*, 78–86. [CrossRef]
113. Seki, M.; Yokota, H.; Maeda, M.; Kobayashi, K. Fish full life-cycle testing for 17beta-estradiol on medaka (Oryzias latipes). *Environ. Toxicol. Chem.* **2005**, *24*, 1259–1266. [CrossRef]
114. Jukosky, J.A.; Watzin, M.C.; Leiter, J.C. The effects of environmentally relevant mixtures of estrogens on Japanese medaka (Oryzias latipes) reproduction. *Aquat. Toxicol.* **2008**, *86*, 323–331. [CrossRef]
115. Imai, S.; Koyama, J.; Fujii, K. Effects of 17β-estradiol on the reproduction of Java-medaka (Oryzias javanicus), a new test fish species. *Mar. Pollut. Bull.* **2005**, *51*, 708–714. [CrossRef] [PubMed]
116. Cripe, G.M.; Hemmer, B.L.; Goodman, L.R.; Fournie, J.W.; Raimondo, S.; Vennari, J.C.; Danner, R.L.; Smith, K.; Manfredonia, B.R.; Kulaw, D.H.; et al. Multigerational exposure of the estuarine Sheepshead minnow (cyprinodon variegatus) to 17β-estradiol. I. Organism-level effects over three generations. *Environ. Toxicol. Chem.* **2009**, *28*, 2397. [CrossRef] [PubMed]
117. Toft, G.; Baatrup, E. Altered sexual characteristics in guppies (Poecilia reticulata) exposed to 17β-estradiol and 4-tert-octylphenol during sexual development. *Ecotoxicol. Environ. Saf.* **2003**, *56*, 228–237. [CrossRef]
118. Liao, T.; Guo, Q.L.; Jin, S.W.; Cheng, W.; Xu, Y. Comparative responses in rare minnow exposed to 17β-estradiol during different life stages. *Fish Physiol. Biochem.* **2009**, *35*, 341–349. [CrossRef] [PubMed]
119. Robinson, C.D.; Brown, E.; Craft, J.A.; Davies, I.M.; Megginson, C.; Miller, C.; Moffat, C.F. Bioindicators and reproductive effects of prolonged 17β-oestradiol exposure in a marine fish, the sand goby (Pomatoschistus minutus). *Aquat. Toxicol.* **2007**, *81*, 397–408. [CrossRef]
120. Pollino, C.A.; Georgiades, E.; Holdway, D.A. Use of the Australian crimson-spotted rainbowfish (Melanotaenia fluviatilis) as a model test species for investigating the effects of endocrine disruptors. *Environ. Toxicol. Chem.* **2007**, *26*, 2171. [CrossRef] [PubMed]

121. Pawlowski, S.; van Aerle, R.; Tyler, C.; Braunbeck, T. Effects of 17α-ethinylestradiol in a fathead minnow (Pimephales promelas) gonadal recrudescence assay. *Ecotoxicol. Environ. Saf.* **2004**, *57*, 330–345. [CrossRef]
122. Örn, S.; Holbech, H.; Madsen, T.H.; Norrgren, L.; Petersen, G.I. Gonad development and vitellogenin production in zebrafish (Danio rerio) exposed to ethinylestradiol and methyltestosterone. *Aquat. Toxicol.* **2003**, *65*, 397–411. [CrossRef]
123. Zha, J.; Sun, L.; Zhou, Y.; Spear, P.A.; Ma, M.; Wang, Z. Assessment of 17α-ethinylestradiol effects and underlying mechanisms in a continuous, multigeneration exposure of the Chinese rare minnow (Gobiocypris rarus). *Toxicol. Appl. Pharmacol.* **2008**, *226*, 298–308. [CrossRef]
124. Schäfers, C.; Teigeler, M.; Wenzel, A.; Maack, G.; Fenske, M.; Segner, H. Concentration- and time-dependent effects of the synthetic estrogen, 17α-ethinylestradiol, on reproductive capabilities of the Zebrafish, Danio rerio. *J. Toxicol. Environ. Heal. Part A* **2007**, *70*, 768–779. [CrossRef]
125. Van den Belt, K.; Berckmans, P.; Vangenechten, C.; Verheyen, R.; Witters, H. Comparative study on the in vitro/in vivo estrogenic potencies of 17β-estradiol, estrone, 17α-ethynylestradiol and nonylphenol. *Aquat. Toxicol.* **2004**, *66*, 183–195. [CrossRef] [PubMed]
126. Xu, H.; Yang, J.; Wang, Y.; Jiang, Q.; Chen, H.; Song, H. Exposure to 17α-ethynylestradiol impairs reproductive functions of both male and female zebrafish (Danio rerio). *Aquat. Toxicol.* **2008**, *88*, 1–8. [CrossRef] [PubMed]
127. Soares, J.; Coimbra, A.M.; Reis-Henriques, M.A.; Monteiro, N.M.; Vieira, M.N.; Oliveira, J.M.A.; Guedes-Dias, P.; Fontaínhas-Fernandes, A.; Parra, S.S.; Carvalho, A.P. Disruption of zebrafish (Danio rerio) embryonic development after full life-cycle parental exposure to low levels of ethinylestradiol. *Aquat. Toxicol.* **2009**, *95*, 330–338. [CrossRef]
128. Parrott, J.L.; Blunt, B.R. Life-cycle exposure of fathead minnows (Pimephales promelas) to an ethinylestradiol concentration below 1 ng/L reduces egg fertilization success and demasculinizes males. *Environ. Toxicol.* **2005**, *20*, 131–141. [CrossRef]
129. Länge, R.; Hutchinson, T.H.; Croudace, C.P.; Siegmund, F.; Schweinfurth, H.; Hampe, P.; Panter, G.H.; Sumpter, J.P. Effects of the synthetic estrogen 17 alpha-ethinylestradiol on the life-cycle of the fathead minnow (Pimephales promelas). *Environ. Toxicol. Chem.* **2001**, *20*, 1216–1227. [CrossRef]
130. Fenske, M.; Maack, G.; Schäfers, C.; Segner, H. An environmentally relevant concentration of estrogen induces arrest of male gonad development in zebrafish, Danio rerio. *Environ. Toxicol. Chem.* **2005**, *24*, 1088–1098. [CrossRef]
131. Scholz, S. 17-α-ethinylestradiol affects reproduction, sexual differentiation and aromatase gene expression of the medaka (Oryzias latipes). *Aquat. Toxicol.* **2000**, *50*, 363–373. [CrossRef]
132. Balch, G.C.; Mackenzie, C.A.; Metcalfe, C.D. Alterations of gonadal development and reproductive success in Japanese medaka (Oryzias latipes) exposed to 17α-ethinylestradiol. *Environ. Toxicol. Chem.* **2004**, *23*, 782. [CrossRef]
133. Schultz, I.R.; Skillman, A.; Nicolas, J.-M.; Cyr, D.G.; Nagler, J.J. Short-term exposure to 17α-ethynylestradiol decreases the fertility of sexually maturing male rainbow trout (Oncorhynchus mykiss). *Environ. Toxicol. Chem.* **2003**, *22*, 1272–1280. [CrossRef] [PubMed]
134. Zillioux, E.J.; Johnson, I.C.; Kiparissis, Y.; Metcalfe, C.D.; Wheat, J.V.; Ward, S.G.; Liu, H. The sheepshead minnow as an in vivo model for endocrine disruption in marine teleosts: A partial life-cycle test with 17α-ethynylestradiol. *Environ. Toxicol. Chem.* **2001**, *20*, 1968–1978. [CrossRef] [PubMed]
135. Kristensen, T.; Baatrup, E.; Bayley, M. 17α-Ethinylestradiol Reduces the Competitive Reproductive Fitness of the Male Guppy (Poecilia reticulata)1. *Biol. Reprod.* **2005**, *72*, 150–156. [CrossRef] [PubMed]
136. Lange, A.; Katsu, Y.; Ichikawa, R.; Paull, G.C.; Chidgey, L.L.; Coe, T.S.; Iguchi, T.; Tyler, C.R. Altered sexual development in roach (Rutilus rutilus) exposed to environmental concentrations of the pharmaceutical 17α-Ethinylestradiol and associated expression dynamics of aromatases and estrogen receptors. *Toxicol. Sci.* **2008**, *106*, 113–123. [CrossRef] [PubMed]
137. Yang, X.; Xu, X.; Wei, X.; Wan, J.; Zhang, Y. Biomarker effects in carassius auratus exposure to ofloxacin, sulfamethoxazole and ibuprofen. *Int. J. Environ. Res. Public Health* **2019**, *16*, 1628. [CrossRef] [PubMed]
138. Kümmerer, K. Pharmaceuticals in the Environment. *Annu. Rev. Environ. Resour.* **2010**, *35*, 57–75. [CrossRef]
139. Oliveira, T.S.; Murphy, M.; Mendola, N.; Wong, V.; Carlson, D.; Waring, L. Characterization of Pharmaceuticals and Personal Care products in hospital effluent and waste water influent/effluent by direct-injection LC-MS-MS. *Sci. Total Environ.* **2015**, *518–519*, 459–478. [CrossRef] [PubMed]

140. Backhaus, T.; Faust, M. Predictive environmental risk assessment of chemical mixtures: A conceptual framework. *Environ. Sci. Technol.* **2012**, *46*, 2564–2573. [CrossRef]
141. Escher, B.I.; Baumgartner, R.; Koller, M.; Treyer, K.; Lienert, J.; McArdell, C.S. Environmental toxicology and risk assessment of pharmaceuticals from hospital wastewater. *Water Res.* **2011**, *45*, 75–92. [CrossRef]
142. Jones-Lepp, T.L.; Sanchez, C.; Alvarez, D.A.; Wilson, D.C.; Taniguchi-Fu, R.L. Point sources of emerging contaminants along the Colorado River Basin: Source water for the arid Southwestern United States. *Sci. Total Environ.* **2012**, *430*, 237–245. [CrossRef]
143. McEachran, A.D.; Shea, D.; Bodnar, W.; Nichols, E.G. Pharmaceutical occurrence in groundwater and surface waters in forests land-applied with municipal wastewater. *Environ. Toxicol. Chem.* **2016**, *35*, 898–905. [CrossRef]
144. Stalter, D.; Magdeburg, A.; Wagner, M.; Oehlmann, J. Ozonation and activated carbon treatment of sewage effluents: Removal of endocrine activity and cytotoxicity. *Water Res.* **2011**, *45*, 1015–1024. [CrossRef]
145. Magdeburg, A.; Stalter, D.; Oehlmann, J. Whole effluent toxicity assessment at a wastewater treatment plant upgraded with a full-scale post-ozonation using aquatic key species. *Chemosphere* **2012**, *88*, 1008–1014. [CrossRef] [PubMed]
146. Pereira, A.; Silva, L.; Laranjeiro, C.; Lino, C.; Pena, A. Selected Pharmaceuticals in Different Aquatic Compartments: Part I—Source, Fate and Occurrence. *Molecules* **2020**, *25*, 1026. [CrossRef] [PubMed]
147. Benotti, M.J.; Trenholm, R.A.; Vanderford, B.J.; Holady, J.C.; Stanford, B.D.; Snyder, S.A. Pharmaceuticals and endocrine disrupting compounds in U.S. drinking water. *Environ. Sci. Technol.* **2009**, *43*, 597–603. [CrossRef] [PubMed]
148. Crouse, B.A.; Ghoshdastidar, A.J.; Tong, A.Z. The presence of acidic and neutral drugs in treated sewage effluents and receiving waters in the Cornwallis and Annapolis River watersheds and the Mill CoveSewage Treatment Plant in Nova Scotia, Canada. *Environ. Res.* **2012**, *112*, 92–99. [CrossRef] [PubMed]
149. Tuševljak, N.; Dutil, L.; Rajić, A.; Uhland, F.C.; McClure, C.; St-Hilaire, S.; Reid-Smith, R.J.; McEwen, S.A. Antimicrobial Use and Resistance in Aquaculture: Findings of a Globally Administered Survey of Aquaculture-Allied Professionals. *Zoonoses Public Health* **2013**, *60*, 426–436. [CrossRef]
150. Kostich, M.S.; Batt, A.L.; Lazorchak, J.M. Concentrations of prioritized pharmaceuticals in effluents from 50 large wastewater treatment plants in the US and implications for risk estimation. *Environ. Pollut.* **2014**, *184*, 354–359. [CrossRef] [PubMed]
151. Khan, G.A.; Berglund, B.; Khan, K.M.; Lindgren, P.E.; Fick, J. Occurrence and Abundance of Antibiotics and Resistance Genes in Rivers, Canal and near Drug Formulation Facilities—A Study in Pakistan. *PLoS ONE* **2013**, *8*, e62712. [CrossRef]
152. Zuccato, E.; Castiglioni, S.; Bagnati, R.; Melis, M.; Fanelli, R. Source, occurrence and fate of antibiotics in the Italian aquatic environment. *J. Hazard. Mater.* **2010**, *179*, 1042–1048. [CrossRef]
153. Rizzo, L.; Manaia, C.; Merlin, C.; Schwartz, T.; Dagot, C.; Ploy, M.C.; Michael, I.; Fatta-Kassinos, D. Urban wastewater treatment plants as hotspots for antibiotic resistant bacteria and genes spread into the environment: A review. *Sci. Total Environ.* **2013**, *447*, 345–360. [CrossRef]
154. Fick, J.; Söderström, H. Contamination of surface, ground, and drinking water from pharmaceutical production. *Environ. Toxicol. Chem.* **2009**, *28*, 2522–2527. [CrossRef]
155. *EAHC Study on the Environmental Risks of Medicinal Products Executive Agency for Health and Consumers Document Information*; Executive Agency for Health and Consumers: Luxembourg, 2013.
156. Cabello, F.C. Heavy use of prophylactic antibiotics in aquaculture: A growing problem for human and animal health and for the environment. *Environ. Microbiol.* **2006**, *8*, 1137–1144. [CrossRef] [PubMed]
157. Segura, P.A.; Takada, H.; Correa, J.A.; El Saadi, K.; Koike, T.; Onwona-Agyeman, S.; Ofosu-Anim, J.; Sabi, E.B.; Wasonga, O.V.; Mghalu, J.M.; et al. Global occurrence of anti-infectives in contaminated surface waters: Impact of income inequality between countries. *Environ. Int.* **2015**, *80*, 89–97. [CrossRef]
158. Kreke, N.; Dietrich, D.R. Physiological endpoints for potential SSRI interactions in fish. *Crit. Rev. Toxicol.* **2008**, *38*, 215–247. [CrossRef] [PubMed]
159. Bossus, M.C.; Guler, Y.Z.; Short, S.J.; Morrison, E.R.; Ford, A.T. Behavioural and transcriptional changes in the amphipod Echinogammarus marinus exposed to two antidepressants, fluoxetine and sertraline. *Aquat. Toxicol.* **2014**, *151*, 46–56. [CrossRef] [PubMed]

160. Lajeunesse, A.; Gagnon, C.; Sauvé, S. Determination of Basic Antidepressants and Their N-Desmethyl Metabolites in Raw Sewage and Wastewater Using Solid-Phase Extraction and Liquid Chromatography-Tandem Mass Spectrometry. *Anal. Chem.* **2008**, *80*, 5325–5333. [CrossRef] [PubMed]
161. Weinberger, J.; Klaper, R. Environmental concentrations of the selective serotonin reuptake inhibitor fluoxetine impact specific behaviors involved in reproduction, feeding and predator avoidance in the fish Pimephales promelas (fathead minnow). *Aquat. Toxicol.* **2014**, *151*, 77–83. [CrossRef] [PubMed]
162. Celiz, M.D.; Tso, J.; Aga, D.S. Pharmaceutical metabolites in the environment: Analytical challenges and ecological risks. *Environ. Toxicol. Chem.* **2009**, *28*, 2473. [CrossRef]
163. Laganà, A.; Bacaloni, A.; De Leva, I.; Faberi, A.; Fago, G.; Marino, A. Analytical methodologies for determining the occurrence of endocrine disrupting chemicals in sewage treatment plants and natural waters. *Anal. Chim. Acta* **2004**, *501*, 79–88. [CrossRef]
164. Fent, K.; Weston, A.A.; Caminada, D. Ecotoxicology of human pharmaceuticals. *Aquat. Toxicol.* **2006**, *76*, 122–159. [CrossRef]
165. Corcoran, J.; Winter, M.J.; Tyler, C.R. Pharmaceuticals in the aquatic environment: A critical review of the evidence for health effects in fish. *Crit. Rev. Toxicol.* **2010**, *40*, 287–304. [CrossRef]
166. Pereira, A.M.P.T.; Silva, L.J.G.; Laranjeiro, C.S.M.; Meisel, L.M.; Lino, C.M.; Pena, A. Human pharmaceuticals in Portuguese rivers: The impact of water scarcity in the environmental risk. *Sci. Total Environ.* **2017**, *609*, 1182–1191. [CrossRef] [PubMed]
167. Bound, J.P.; Voulvoulis, N. Predicted and measured concentrations for selected pharmaceuticals in UK rivers: Implications for risk assessment. *Water Res.* **2006**, *40*, 2885–2892. [CrossRef] [PubMed]
168. Holm, G.; Snape, J.R.; Murray-Smith, R.; Talbot, J.; Taylor, D.; Sörme, P. Implementing ecopharmacovigilance in practice: Challenges and potential opportunities. *Drug Saf.* **2013**, *36*, 533–546. [CrossRef] [PubMed]
169. Santos, L.H.M.L.M.; Gros, M.; Rodriguez-Mozaz, S.; Delerue-Matos, C.; Pena, A.; Barceló, D.; Montenegro, M.C.B.S.M. Contribution of hospital effluents to the load of pharmaceuticals in urban wastewaters: Identification of ecologically relevant pharmaceuticals. *Sci. Total Environ.* **2013**, *461–462*, 302–316. [CrossRef] [PubMed]
170. Vazquez-Roig, P.; Andreu, V.; Blasco, C.; Picó, Y. Risk assessment on the presence of pharmaceuticals in sediments, soils and waters of the Pego-Oliva Marshlands (Valencia, eastern Spain). *Sci. Total Environ.* **2012**, *440*, 24–32. [CrossRef]
171. Pereira, A.M.P.T.; Silva, L.J.G.; Lino, C.M.; Meisel, L.M.; Pena, A. Assessing environmental risk of pharmaceuticals in Portugal: An approach for the selection of the Portuguese monitoring stations in line with Directive 2013/39/EU. *Chemosphere* **2016**, *144*, 2507–2515. [CrossRef]
172. Zhang, R.; Zhang, G.; Zheng, Q.; Tang, J.; Chen, Y.; Xu, W.; Zou, Y.; Chen, X. Occurrence and risks of antibiotics in the Laizhou Bay, China: Impacts of river discharge. *Ecotoxicol. Environ. Saf.* **2012**, *80*, 208–215. [CrossRef]
173. Acuña, V.; Schiller, D.; García-Galán, M.J.; Rodríguez-Mozaz, S.; Corominas, L.; Petrovic, M.; Poch, M.; Barceló, D.; Sabater, S. Occurrence and in-stream attenuation of wastewater-derived pharmaceuticals in Iberian rivers. *Sci. Total Environ.* **2015**, *503–504*, 133–141. [CrossRef]
174. Simazaki, D.; Kubota, R.; Suzuki, T.; Akiba, M.; Nishimura, T.; Kunikane, S. Occurrence of selected pharmaceuticals at drinking water purification plants in Japan and implications for human health. *Water Res.* **2015**, *76*, 187–200. [CrossRef]
175. Henry, T.B.; Black, M.C. Mixture and single-substance acute toxicity of selective serotonin reuptake inhibitors in Ceriodaphnia dubia. *Environ. Toxicol. Chem.* **2007**, *26*, 1751–1755. [CrossRef]
176. Vestel, J.; Caldwell, D.J.; Constantine, L.; D'Aco, V.J.; Davidson, T.; Dolan, D.G.; Millard, S.P.; Murray-Smith, R.; Parke, N.J.; Ryan, J.J.; et al. Use of acute and chronic ecotoxicity data in environmental risk assessment of pharmaceuticals. *Environ. Toxicol. Chem.* **2016**, *35*, 1201–1212. [CrossRef] [PubMed]
177. Schultz, M.M.; Painter, M.M.; Bartell, S.E.; Logue, A.; Furlong, E.T.; Werner, S.L.; Schoenfuss, H.L. Selective uptake and biological consequences of environmentally relevant antidepressant pharmaceutical exposures on male fathead minnows. *Aquat. Toxicol.* **2011**, *104*, 38–47. [CrossRef] [PubMed]
178. Minguez, L.; Pedelucq, J.; Farcy, E.; Ballandonne, C.; Budzinski, H.; Halm-Lemeille, M.P. Toxicities of 48 pharmaceuticals and their freshwater and marine environmental assessment in northwestern France. *Environ. Sci. Pollut. Res.* **2016**, *23*, 4992–5001. [CrossRef] [PubMed]

179. Gao, P.; Li, Z.; Gibson, M.; Gao, H. Ecological risk assessment of nonylphenol in coastal waters of China based on species sensitivity distribution model. *Chemosphere* **2014**, *104*, 113–119. [CrossRef]
180. Richards, S.M.; Wilson, C.J.; Johnson, D.J.; Castle, D.M.; Lam, M.; Mabury, S.A.; Sibley, P.K.; Solomon, K.R. Effects of pharmaceutical mixtures in aquatic microcosms. *Environ. Toxicol. Chem.* **2004**, *23*, 1035–1042. [CrossRef]
181. Goolsby, E.W.; Mason, C.M.; Wojcik, J.T.; Jordan, A.M.; Black, M.C. Acute and chronic effects of diphenhydramine and sertraline mixtures in Ceriodaphnia dubia. *Environ. Toxicol. Chem.* **2013**, *32*, 2866–2869. [CrossRef]
182. Pereira, A.M.P.T.; Silva, L.J.G.; Meisel, L.M.; Pena, A. Fluoroquinolones and Tetracycline Antibiotics in a Portuguese Aquaculture System and Aquatic Surroundings: Occurrence and Environmental Impact. *J. Toxicol. Environ. Heal. Part A* **2015**, *78*, 959–975. [CrossRef]
183. Altenburger, R.; Ait-Aissa, S.; Antczak, P.; Backhaus, T.; Barceló, D.; Seiler, T.-B.; Brion, F.; Busch, W.; Chipman, K.; de Alda, M.L.; et al. Future water quality monitoring — Adapting tools to deal with mixtures of pollutants in water resource management. *Sci. Total Environ.* **2015**, *512–513*, 540–551. [CrossRef]

Review

The Influence of Ionic Liquids on the Effectiveness of Analytical Methods Used in the Monitoring of Human and Veterinary Pharmaceuticals in Biological and Environmental Samples—Trends and Perspectives

Natalia Treder [1], Tomasz Bączek [1], Katarzyna Wychodnik [2], Justyna Rogowska [2], Lidia Wolska [2] and Alina Plenis [1,*]

[1] Department of Pharmaceutical Chemistry, Medical University of Gdańsk, Hallera 107, 80-416 Gdańsk, Poland; natalia.treder@gumed.edu.pl (N.T.); tbaczek@amg.gda.pl (T.B.)

[2] Department of Environmental Toxicology, Faculty of Health Sciences with Institute of Maritime and Tropical Medicine, Medical University of Gdańsk, Dębowa 23 A, 80-204 Gdańsk, Poland; k.wychodnik@gumed.edu.pl (K.W.); justyna.rogowska@gumed.edu.pl (J.R.); lidia.wolska@gumed.edu.pl (L.W.)

* Correspondence: aplenis@gumed.edu.pl; Tel.: +48-58-349-12-36

Received: 28 November 2019; Accepted: 8 January 2020; Published: 10 January 2020

Abstract: Recent years have seen the increased utilization of ionic liquids (ILs) in the development and optimization of analytical methods. Their unique and eco-friendly properties and the ability to modify their structure allows them to be useful both at the sample preparation stage and at the separation stage of the analytes. The use of ILs for the analysis of pharmaceuticals seems particularly interesting because of their systematic delivery to the environment. Nowadays, they are commonly detected in many countries at very low concentration levels. However, due to their specific physiological activity, pharmaceuticals are responsible for bioaccumulation and toxic effects in aquatic and terrestrial ecosystems as well as possibly upsetting the body's equilibrium, leading to the dangerous phenomenon of drug resistance. This review will provide a comprehensive summary of the use of ILs in various sample preparation procedures and separation methods for the determination of pharmaceuticals in environmental and biological matrices based on liquid-based chromatography (LC, SFC, TLC), gas chromatography (GC) and electromigration techniques (e.g., capillary electrophoresis (CE)). Moreover, the advantages and disadvantages of ILs, which can appear during extraction and separation, will be presented and attention will be given to the criteria to be followed during the selection of ILs for specific applications.

Keywords: ionic liquids; green chemistry; environmental and biological samples; sample preparation; determination of pharmaceuticals; chromatographic methods; electromigration techniques

1. Introduction

Analytical chemistry focused on the development of methods for the qualitative and quantitative determination of compounds with different chemical structures is a huge, dynamically developing field of science. The number of available methods and techniques is impressive. However, in addition to successes, there are many limitations regarding the use of such approaches. Problems may appear already at the sample preparation stage. Inadequate selectivity, and the use of large volumes of harmful organic solvents with a high vapor pressure in liquid-liquid extraction (LLE) or solid-phase extraction (SPE) are some of the many reasons for the search for alternatives [1]. The introduction of microextraction combined with the reduction of organic solvents used, and the inclusion of additional physical and chemical factors (sonication, temperature) have brought enormous progress, but also have several difficulties. Microextraction into both solid and liquid phases is a time-consuming process, and the final results require the indication of many other conditions [2]. For example, in solid-phase microextraction (SPME), commercially available fibers are not always suitable for the target compounds, while for single-drop microextraction (SDME), the stability of the drop in the sample may be a problem [3,4]. These limitations, as well as the need for even greater process control by affecting the retention time, and improving the extraction efficiency and resolution of analytes, are responsible for the attempt to include new structures in the extraction process, which can help to achieve these goals [5]. Modifications, such as the introduction of additional processes in liquid-based sample preparation procedures or changes on the surface of sorbents in SPE-based extraction and microextraction procedures are a good direction in analytics, but often insufficient to achieve the expected effects.

Equally as crucial as sample preparation is the process of the separation and detection of the compounds of interest. Among the many available techniques, chromatography or electrophoresis are most often used for the determination of pharmaceuticals in different matrices. Chromatographic techniques exist in a variety of types: the oldest thin-layer chromatography (TLC), the commonly used high performance liquid chromatography (HPLC) and gas chromatography (GC) as well as the less popular supercritical fluid chromatography (SFC) techniques. These methods can be coupled to various types of detectors, including ultraviolet (UV), fluorescence (FL) or mass spectrometry (MS). There are many important parameters during the development and optimization of methods but the most important include the choice of the stationary phase (the place of separation of the analytes) and the mobile phase composition. If the analytes show excessive column adsorption, tailing of the chromatographic peaks occurs and their width is incorrect [6]. In turn, when choosing a mobile phase, problems can occur with obtaining separate peaks for specific compounds, a too long analysis time and low efficiency [7]. However, other chromatographic conditions, such as the column temperature and the flow rate of the mobile phase as well as the parameters of detection should be carefully selected. This is a particular challenge for pharmaceutical determinations because their diverse structures and rich (despite extraction) matrices, and the necessity to detect many analytes at the same time, are just some of the reasons for difficulties in their separation. In addition, it should be highlighted that the mobile phases in LC often contain large volumes of organic solvents which are highly toxic. An interesting alternative seems to be electromigration techniques such as capillary electrophoresis (CE), micellar electrokinetic chromatography (MEKC) or non-aqueous capillary electrophoresis (NACE). These analytical approaches have been considered to be powerful separation methods due to low sample and reagent consumption, high efficiency, and simplicity. On the other hand, CE-based methods have relatively low sensitivity which makes their application difficult in real clinical and environmental studies. Thus, the above examples show that each stage in the development of an analytical method (both sample preparation and further analysis) can cause problems in performing experiments or in achieving reliable results.

Ionic liquids (ILs) are a relatively new class of compounds that became an object of special attention in the 21st century. Their simple cationic-anionic structure provides unusual and unparalleled properties. Therefore, it should not be surprising that their potential is exploited in many unrelated areas of science, for example, as a catalyst in chemical reactions [8], in drug delivery systems [9], in electroplating processes [10], in treating harmful compounds in wastewater [11], as matrices for analysis by matrix-assisted laser desorption/ionization mass spectrometry (MALDI-MS) [12] and many others. Scientists have also become interested in "designer solvents" in response to the constant demand for developing new and better methods, and improving the results obtained. The literature data show that their application is focused on sample preparation by extraction or microextraction as well as chromatography (adding ILs to the mobile phase or to prepare the stationary phase) and electrophoretic techniques (Figure 1).

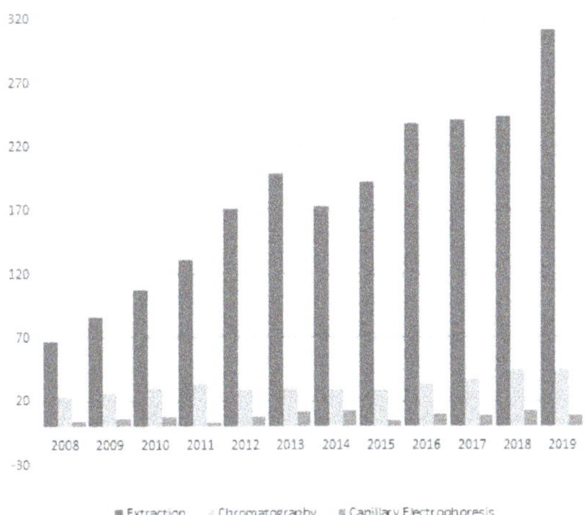

Figure 1. Number of publications on the use of ILs in sample preparation (extraction and microextraction) and chromatographic and electrophoretic techniques in 2008–2019 (the authors own elaboration according to ScienceDirect data).

In pharmaceutical sciences ILs can be used for a variety of purposes: as active pharmaceutical substances (API-IL) [13], to determine the solvent residues and impurities in drug quality testing [14] or as a source of information about the presence of pharmaceuticals in biological and environmental samples [15,16]. An important argument supporting their use was also the introduction by Anastas in 1999 of the 12 principles of green chemistry [17]. Attention was drawn to the excessive use of organic solvents and the need to eliminate or reduce environmentally harmful factors. The search for alternatives resulted in the inclusion of ILs in experiments. Negligible vapor pressure, non-flammability, thermal stability, and the possibility of reuse are just some of the properties that have allowed ILs to be described as more environmentally-friendly [18]. It should be highlighted that as newer compounds their literature data are incomplete. However, this does not preclude their use at various stages of analytical testing, from sample preparation to detection and the improvement of results, even for difficult to determine analytes, including the quantification of pharmaceuticals in biological and environmental samples. These substances, with different pharmacokinetic activity, can be delivered directly and indirectly (animal-derived foods) to the human body in very low concentrations. Moreover, pharmaceutical concentrations in urine or bile are different from those in blood or saliva [19]. For this reason, it is necessary to develop a method that will be adequate for the specific biological sample. In the treatment of patients, combination therapy is often used, which results in the presence in the matrix sample of many drugs with different physical and chemical properties making it difficult to choose the best extraction and separation conditions. It should also be remembered that these are not always stable compounds, and to obtain information on their concentrations, it may also be necessary to determine the degradation products and/or metabolites in the presence of many endogenous matrix compounds [20]. Similar considerations can be made in the field of drug determination in environmental samples. According to the data reported in the literature, the sources of pharmaceuticals in wastewater, river waters, lake waters and others are improper drug disposal, hospital wastewater or animal feces. If they occur in an unchanged form, they may cause the risk of typical side effects after they enter the body. One group of drugs often identified in environmental samples are antibiotics, which may be responsible for the development of antibiotic-resistant bacteria [21]. As in biological samples, pharmaceuticals are present in the environment in very low concentrations. Sample purification, the isolation of analytes or the possibility of enriching the sample are crucial and influence the final efficiency of a method.

As already mentioned, pharmaceuticals are compounds with high biological activity, so it is also important to develop simple, reproducible, quick methods, without the need to introduce additional steps to improve the safety of analysts [22]. The inclusion of ILs in their analyses not only improves safety due to the reduction of the use of organic solvent, but also, as confirmed by research, helps to overcome the mentioned difficulties in the analysis of drugs and to improve the validation parameters and efficiency. Therefore, the monitoring of these substances in both the environment and animal and human samples using IL-based environmentally-friendly analytical methods, which also offer reliability, and the qualitative and quantitative sensitivity and selectivity of the compounds of interest is one of the main tasks of modern analytics and chemistry.

The growing number of research papers on ILs has also increased interest in this topic in review articles. Their wide spectrum of possibilities is also clearly visible in the huge variety of subjects of such works. Some of them focused on IL in the context of "green chemistry", pointing to their great potential, but also disadvantages (the need to remove them from the environment, multi-stage synthesis) [23,24]. The reviews very often summarized their applications in sample preparation, especially solid phase microextraction. Most commonly, polymeric ionic liquids (PILs) were evaluated in such applications [25–28]. Some articles considered all the possibilities for using ILs, both at the extraction and detection stages [29–32]. However, the publication selection criteria in the review papers most often concerned analytical methods and techniques or the type of ILs and did not focus on the specific type of analytes or matrices. In addition, it should be noted that the dynamic development of analytical methods using IL requires continuous monitoring of current scientific reports and providing the latest information in current reviews papers. Thus, the purpose of this review was to summarize achievements in the use of ILs for the determination of drugs in biological and environmental samples. In order to properly understand the popularity of ILs in the modern laboratory, the section "Ionic Liquids" presents their history, with the inclusion of their most important features and properties. The basic criteria for choosing articles for the review was the use of ILs during the sample preparation procedure or in the chromatographic/electrophoretic separation of synthetic drugs quantified in biological and environmental samples. The review did not include endogenous compounds, substances responsible for addiction (e.g., nicotine and others) and herbal medicines, except for IL-applications in GC, TLC and SFC. This extension was made in order to fully present the capabilities of ILs and show current trends in the determination of different active biological substances.

2. Ionic Liquids

Regarding the history of ILs, and events that are responsible for their presence in many fields of science, it is first of all necessary to define the criteria used in the presentation of this subject. Considering the period of their greatest popularity, that is, the last two decades, we can accept the work of scientists who in their publications focused primarily on modifications of compounds in order to obtain the desired properties and identify their applications. However, to acquire information about the discovery of compounds that were ILs, although no one was aware of this and such a definition was not used, we should return to the mid 19th century. At that time, a by-product known as "red oil" was obtained in the Friedel-Crafts reaction. As shown later, this was the first recorded IL [33]. In the following years, Gabriel and Warner also made an important contribution to the development of ILs. In 1888, for the first time, they synthesized ethanol-ammonium nitrate [34]. Although all previous events were very important, the synthesis of ethylammonium nitrate by Walden in 1914 has most often appeared in publications in the context of the discovery of ILs [35]. Of course, it should be mentioned that in the case of ILs, as in all great discoveries, there are opinions that although Walden synthesized the compounds, he could not use them in practice and his success is over-emphasized [36]. Nevertheless, it was undoubtedly an important stage in the development of ILs. During the following years, there were further syntheses of the compounds and attempts to use them, among others by Yoke and his colleagues [37] and Koch and co-workers [38]. However, in more modern times, with the current compounds that are used in research, it is necessary to focus on analytical methods and extraction techniques. Considering the application of ILs, Pool's research should be mentioned, in which, using current knowledge, ILs were used in GC as stationary phases [39]. The results of this study prompted the beginning of their further development in this field and became the inspiration for subsequent publications. The 1980s were also important due to the synthesis of ILs based on the imidazolium cation, which are currently widely used in laboratories [40]. This event was important because the existing compounds of ILs had significant limitations in their application, while the imidazolium group provided new opportunities for researchers. Following the trend, subsequent years of research into the use of ILs increased knowledge about them, and consequently led to the introduction of some standards in this area. At the turn of the 20th century, ILs began to function under the name Task-specific Ionic Liquids (TSILs) [41] and companies marketing the first commercially available ILs appeared [42]. Increased access and the positive opinion of the scientific community prompted attempts to apply them to novel projects. In 1998, for the first time, ILs were used as extractants for LLE [43], and in 2005, they were used to coat SPME fibers [44]. Recent years have seen a period of their participation in advanced research, but this will be discussed in detail in subsequent sections. However, it should be highlighted that the most important factor responsible for the rich and long history of ILs is their specific structures, illustrated in Figure 2, which provide the enormous possibilities of these compounds. The cation-anion combinations, described in most definitions, create many possibilities for structure modification, and can thus change the properties of the designed compounds. The cations may have one or more nitrogen, sulfur, phosphorus or oxygen atom in the structure, described as ammonium, sulfonium, phosphonium or oxonium cations, respectively, but in most cases, they are large organic aromatic moieties: pyridinium, piperidinium and the most widely used imidazolium cations. In turn, anions are much smaller and can be both organic and inorganic. In research, tetrafluoroborate ([BF_4]), hexafluorophosphate ([PF_6]) and halogen anions, and many other compositions appear.

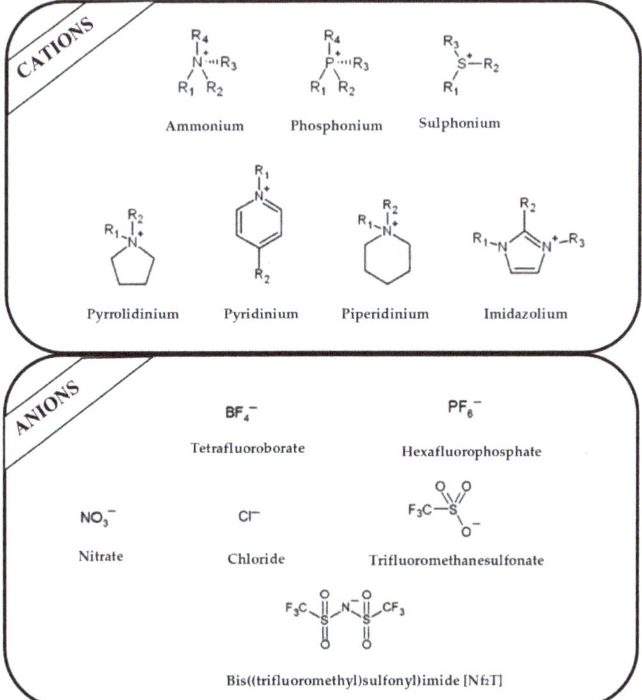

Figure 2. Examples of popular anions and cations of ILs used in analytical methods.

Besides the selection of the cation and anion, an important aspect that affects further results is the substituents on the cation, and especially the alkyl chain, the length, branching and position of which have a huge influence on applications of ILs [45]. To fully understand the unique properties of an IL, it is also necessary to pay attention to Coulombic interactions occurring in the molecules, dipole-dipole interactions, Van der Waals forces and hydrogen bonds [46]. It is estimated that the number of available combinations may allow up to 1018 different ILs to be obtained [47]. The differences in the size of the cation and anion, the asymmetry in the structure as well as the mentioned interactions mean that they have no regular, crystalline structure and the delocalization of the cation and anion composition is very possible. Thanks to this, their melting temperature does not exceed 100 °C, and in many cases it is close to room temperature (RTIL) [48]. This feature distinguishes ILs from typical inorganic salts, which, due to the much stronger Coulombic and hydrogen interactions, have a melting point of even above 400 °C. Equally as interesting as their melting point is the viscosity of ILs, which is at a higher level than that of organic solvent. Knowledge of these parameters is necessary when an IL is used in separation and detection techniques. The electrostatic interactions in alkyl chain cations have an enormous impact on viscosity. Coulomb forces, H-bonding and π-π dipole lead to increases in the flow resistance, and additionally, the presence of van der Waals interactions between the cation and anion, depending on the size of the molecule, also causes interactions in the same direction. This property can be modified by changing the temperature or adding an organic solvent [49–52]. Viscosity also influences another property, namely electrochemical conductivity. Thanks to their ionic structure, ILs can carry a charge, but this possibility is not the same for all compounds. When the flow resistance increases, conductivity becomes more difficult. However, increasing the temperature and mixing with organic solvents improves the results. Furthermore, the size of molecules can hinder access to the charge, so it is necessary to select the appropriate cations, which are large ions [53]. ILs are widely used in sample preparation techniques because they can be created as both hydrophilic and hydrophobic compounds that mix with water, and/or organic solvents [54]. It has been proven that the change in the position of the methyl group in the cation determines the change in the acid-base character, and therefore the C2 position is strongly acidic, which affects the interaction with other compounds [55]. The thermal stability of ILs is also important. As studies have shown, the majority of popular ILs are stable even above 300 °C, which is of great importance during GC analysis, where a high temperature is required. As with previous properties, the size and type of ions, pKa, chain length and electrostatic interactions determine the stability of individual ILs. Halogen anions, probably due to their nucleophilic character, have less stability than other inorganic anions, while the most stable is bis(trifluoromethanesulfonyl)imide ([Nf2T]). In turn, among the cations, the stability of pyrrolidinium and piperidinium is lower than that of imidazolium, regardless of the anion used [56–58]. An interesting property is also the insignificant vapor pressure which occurs at elevated temperatures. Zaitaus et al. confirmed the influence of the structure of ILs on vapor pressure. In their study, the absolute vapor pressures for a series of $[C_nMIM][BF_4]$ ionic liquids with (n = 2, 4, 6, 8, and 10) were measured. The results of experiments confirmed that an increase in the number of carbon atoms in the alkyl chain in the imidazolium cation caused a decrease in absolute vapor pressures. However, this effect was different for the homologies of $[C_nMIM][BF_4]$ and $[C_nMIM][Nf_2T]$. Moreover, it was observed that the volatility for $[C_nMIM][BF_4]$ was significantly lower in comparison to $[C_nMIM][Nf_2T]$. In addition, ILs added to organic solvents also reduced their evaporation [59–61].

It should be noted that there are a huge variety of IL combinations, so it is difficult to establish a clear classification. The most popular approach concerns the structure of these compounds (Table 1). A more detailed description concerns three generations of ILs in view of the anion or cation used. The first includes molecules with specific physical properties, which are described in the previous paragraph. The second generation includes ILs for which it is possible to tune their chemical and physical properties and then to use them for a specific purpose, while the last group are compounds with biological activity [62]. From the point of view of analytical applications, it seems reasonable to focus attention on a large and diverse IL group referred to as Task-specific Ionic Liquids. The results of subsequent tests confirmed that apart from typical ILs, it is necessary to design more specific molecules to achieve a specific goal. This led to the use of ILs in polymerization processes. ILs as monomers can form combinations with other molecules, thus improving the results.

Table 1. List of ILs used in modern laboratory for drug analysis.

Cation Structure	Full Name		Abbreviations
	Cations	Anions	
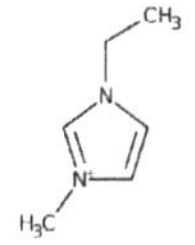 CH_3, N, N^+, H_3C	1-ethyl-3-methylimidazolium	hexafluorophosphate, tetrafluoroborate, chloride, bromide, bis(trifluoromethylsulfonyl)imide, methyl sulfate	$[C_2MIM][PF_6]$, $[C_2MIM][BF_4]$, $[C_2MIM][Cl]$, $[C_2MIM][Br]$, $[C_2MIM][Nf_2T]$, $[C_2MIM][CH_3(SO_4)]$
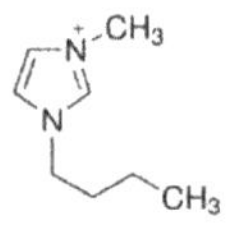 N^+, CH_3, N, CH_3	1-butyl-3-methylimidazolium	hexafluorophosphate, tetrafluoroborate, chloride, bromide, bis(trifluoromethylsulfonyl)imide, nitrate, methyl sulfate, octyl sulfate, trifluoromethanesulfonate, dimethyl phosphate, hydroxide	$[C_4MIM][PF_6]$, $[C_4MIM][BF_4]$, $[C_4MIM][Cl]$, $[C_4MIM][Br]$, $[C_4MIM][Nf_2T]$, $[C_4MIM][NO_3]$, $[C_4MIM][CH_3(SO_4)]$, $[C_4MIM][C_8H_{17}(SO_4)]$, $[C_4MIM][CF_3SO_4]$, $[C_4MIM][DMP]$, $[C_4MIM][OH]$
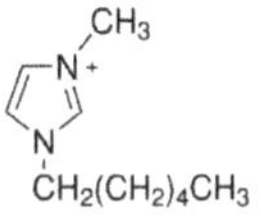 CH_3, N^+, N, $CH_2(CH_2)_4CH_3$	1-hexyl-3-methylimidazolium	hexafluorophosphate, tetrafluoroborate, chloride, bromide, bis(trifluoromethylsulfonyl)imide, tris(pentafluoroethyl)trifluoro-phosphate	$[C_6MIM][PF_6]$, $[C_6MIM][BF_4]$, $[C_6MIM][Cl]$, $[C_6MIM][Br]$, $[C_6MIM][Nf_2T]$, $[C_6MIM][TFP]$
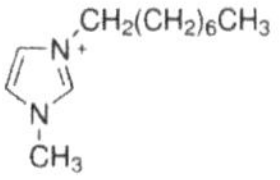 $CH_2(CH_2)_6CH_3$, N^+, N, CH_3	1-octyl-3-methylimidazolium	hexafluorophosphate, tetrafluoroborate, chloride, bromide, bis(trifluoromethylsulfonyl)imide	$[C_8MIM][PF_6]$, $[C_8MIM][BF_4]$, $[C_8MIM][Cl]$, $[C_8MIM][Br]$, $[C_8MIM][Nf_2T]$
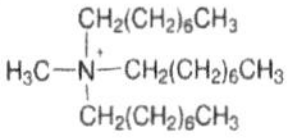 $CH_2(CH_2)_6CH_3$, $H_3C-N^+-CH_2(CH_2)_6CH_3$, $CH_2(CH_2)_6CH_3$	methyltrioctylammonium	tetrachloroferrate, tetrachloromanganate(II)	$[C_8MAmm][FeCl_4]$, $[C_8MAmm][MnCl_4{}^{2-}]$
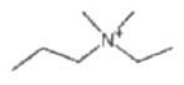 N^+	ethyl-dimethyl-propylammonium	bis(trifluoromethylsulfonyl)imide	$[NEMMP][Nf_2T]$
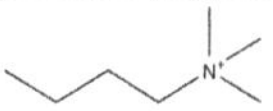 N^+	1-butyl-3-methylammonium	bis(trifluoromethylsulfonyl)imide	$[C_4M_3Amm][Nf_2T]$
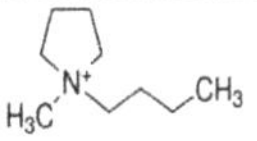 N^+, H_3C, CH_3	1-butyl-1-methylpyrrolidinium	bis(trifluoromethylsulfonyl)imide, tetracyanoborate, tris(pentafluoroethyl)trifluoro-phosphate	$[C_4MPyrr][Nf_2T]$, $[C_4MPyrr][B(CN)_4]$, $[C_4MPyrr][TFP]$
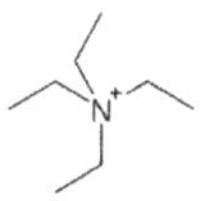 N^+	tetraethylammonium	tetrafluoroborate	$[(C_2H_5)_4N][BF_4]$
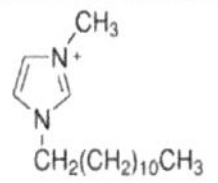 CH_3, N^+, N, $CH_2(CH_2)_{10}CH_3$	1-dodecyl-3-methylimidazolium	chloride	$[C_{12}MIM][Cl]$
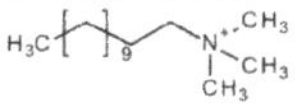 H_3C, 9, N^+, CH_3, CH_3, CH_3	1-dodecyl-3-methylammonium	chloride	$[C_{12}MAmm][Cl]$
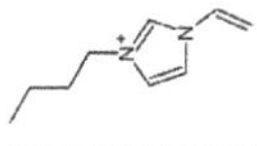	1-vinyl-3-butylimidazolium	chloride	$[ViC_4MIM][Cl]$

Table 1. *Cont.*

Cation Structure	Full Name		Abbreviations
	Cations	Anions	
[structure]	1-allyl-3-ethylimidazolium	bromide	$[AC_2MIM][Br]$
[structure]	1,3-dimethylimidazolium	methyl sulfate	$[MMIM]][CH_3(SO_4)]$
[structure]	1-(6-amino-hexyl)-1-methylpyrrolidinium	tris(pentafluoroethyl)trifluoro-phosphate	$[C_6NH_2MPyrr][TFP]$
[structure]	1-ethoxycarbonyl-methyl-1-methyl-pyrrolidinium	tris(pentafluoroethyl)trifluoro-phosphate	[ECMMPyrr][TFP]
[structure]	methoxyethyl-dimethylethyl-ammonium	tris(pentafluoroethyl)trifluoro-phosphate	[MOEDEAmm][TFP]
[structure]	1-methoxyethyl-3-methylimidazolium	tris(pentafluoroethyl)trifluoro-phosphate	[MOEMIM]][TFP]
[structure]	1-methoxyethyl-1-methylmorpholinium	tris(pentafluoroethyl)trifluorophosphate	[MOEMMO][TFP]
[structure]	1-methoxypropyl-1-methylpiperidinum	tris(pentafluoroethyl)trifluorophosphate	[MOPMPP][TFP]
$H_3C(H_2C)_5-P^+-(CH_2)_{13}CH_3$, $(CH_2)_5CH_3$, $(CH_2)_5CH_3$	trihexyltetradecylphosphonium	tetrachloromanganate(II), dicyanamide, bis(trifluoromethanesulfonyl) imide	$[P_{6,6,6,14}{}^+]_2[MnCl_4{}^{2-}]$, $[P_{6,6,6,14}{}^+][N(CN)_2]$, $[P_{6,6,6,14}{}^+][Nf_2T]$
$CH_3(CH_2)_{12}CH_2-P^+$, CH_3, CH_3, H_3C	tributyl(tetradecyl)phosphonium	*p*-dodecylbenzenesulfonate	$[P_{4,4,4,14}{}^+][DDBS])$
$CH_3(CH_2)_3-P^+-(CH_2)_3CH_3$, CH_2CH_3, $(CH_2)_3CH_3$	tributyl(ethyl)phosphonium	diethylphosphate	$[P_{2,4,4,4}{}^+][(2O)_2PO_2]$

The currently applied analytical methods use the structure and properties of ILs to create polymeric connections with cyclodextrins (CDs) [63] or magnetic imprinted nanoparticles (MILs) [64]. In addition, polymeric ionic liquids (PILs) can also be synthesized by a co-polymerization process [65]. Their participation in the molecular imprinting technique used to develop sorbents of monolithic columns has also been noted. Another, large subclass of ILs are chiral ionic liquids (CILs). Recent scientific reports show that amino acids can be used for the synthesis of CILs. Their carboxyl or amine functional groups determine the chiral nature and function in the structure (cation or anion). The use of amino acids results from the trend of reducing toxicity and the use of natural compounds [66]. In addition, these "designer molecules" are also used as chiral selectors in aqueous two-phase systems (ATPS) [67].

3. Sample Preparation

As it was earlier mentioned, the sample preparation procedure is still one of the most important stages in the development of analytical methods. The variety of biological and environmental samples makes them very complicated with regard to gathering all information about the sample preparation stage. Both types of samples are complex analytical matrices, and the stage of their preparation for analysis is multifactorial. It usually requires the performance of various operations and activities both in situ and in the laboratory. Due to the very low concentrations in real samples, the extraction method should have the highest possible recovery. In addition, the sample handling method largely depends on the chosen final determination technique. Knowing the chemical properties of the drug (or drugs) sought in the analyzed matrix, makes it possible to properly select the organic solvents in order to carry out a successful extraction from the sample, followed by purification, sometimes by back extraction. Considering all these aspects, it is necessary to search for new directions in sample pretreatment procedures. One of these is the use of ILs at the preparation stage of biological and environmental samples to isolate the drugs potentially present in them, both with the use of liquid-liquid based extraction and solid-phase based extraction procedures [68]. ILs are used as liquid phases, extractors, intermediate solvents, mediators and desorption solvents [68–111]. Exemplary applications of individual types of drug extraction from biological and environmental samples with IL-modifications are presented in Table 2. These summarized data clearly indicate that despite the determination of low pharmaceutical concentrations in both types of samples, IL-based extraction procedures go in a different direction. If the matrix is biological fluid, the most common problem is the distribution of peaks, selectivity, shape and of course performance. In turn, environmental samples most often focus on the need to improve extraction efficiency [69,70]. Matrix influence, peak shape and distribution are not the main reasons for using ILs in extraction. A difference also occurs in the volume of the analyzed sample, being much larger for environmental samples [69]. This factor is especially important during the formation of two phases with the participation of ILs, in which the proper volume ratios (aqueous phase, organic phase, IL and others) are needed for the proper phase separation and the subsequent separation [67,112–114]. In the publications presented below, it can also be seen that for environmental samples, there was a much greater variety in the choice of extraction method, especially in the area of dispersive liquid–liquid microextraction (DLLME). In the case of biological samples, they were also extracted by DLLME, but the modifications were much smaller in number.

3.1. Liquid-Phase Based Extraction and Microextraction Procedures

3.1.1. Liquid-Liquid Extraction

LLE is the oldest method of extracting analytes. Unfortunately, despite the simplicity of performance, the method has many disadvantages. It is a very time- and work-intensive process, and the results depend on many additional factors, e.g., the physicochemical character of analytes, the type of extraction solvents, the extraction time and the temperature, which in turn cause reproducibility and repeatability problems. Furthermore, according to the current trend of designing more environmentally-friendly analytical methods, the use of toxic organic solvents should be reduced. As is well known, LLE does not meet this condition. In all probability, this was the reason why in regard to drug quantification in biological and environmental samples, traditional LLE extraction supported by IL modification was rarely considered. To the best of our knowledge, no reports have been published for biological applications, while only one paper can be found in the field of environmental investigations (Table 2).

Table 2. Summary of the IL applications in liquid-phase microextraction drugs from biological and environmental samples.

Drug(s)	Matrices	Tested Ionic Liquids	Extraction Solvent	Analytical Technique	LOD [ng/mL]	Efficiency [%]	Ref.
		LLE					
		Environmental samples					
Ranitidine, nizatidine	River water, wastewater	**[C_4MIM][Nf_2T] [C_4MIM][PF_6]**	MeOH	HPLC-UV	90 430	100.4 101.2	[68]
		IL-DLLME					
		Biological samples					
NSAIDs	Human urine	**[C_4MIM][PF_6]** [C_6MIM][PF_6] [C_8MIM][PF_6]	MeOH	HPLC-UV	8.3–32	36.8–42.3	[70]
Anti-hypertensive drugs	Rat serum	**[C_4MIM][PF_6]** [C_6MIM][PF_6] [C_4MIM][BF_4] [C_6MIM][Cl] [C_2MIM][$CH_3(SO_4)$]	Acetone	HPLC-PDA	15–20	92.8–98.5	[80]
Balofloxacin	Rat serum	**[C_4MIM][PF_6]** [C_6MIM][PF_6] [C_4MIM][BF_4] [C_6MIM][Cl] [C_4MIM][Br] [C_2MIM][$CH_3(SO_4)$]	ACN	HPLC-DAD	10	99.5	[81]
Rifaximin	Rat serum	**[C_4MIM][PF_6]** [C_4MIM][BF_4] [C_6MIM][Cl] [C_4MIM][Br] [C_2MIM][$CH_3(SO_4)$]	MeOH	HPLC-DAD	10	99.5	[82]
Ofloxacin	Human urine and plasma, tablets	**[C_6MIM][PF_6]**	Ethanol	SFIS	29	89.5–93	[86]
Tetracycline	Eggs	**[C_4MIM][PF_6]** [C_6MIM][PF_6] [C_8MIM][PF_6] [IMIM][PF_6]	ACN	HPLC-DAD	2–12	58.6–95.3	[87]
Fluoroquinolone	Chicken, pork and fish meat	**[C_4MIM][PF_6]** [C_6MIM][PF_6] [C_8MIM][PF_6]	ACN	HPLC-DAD	0.5–1.1	60.4–96.3	[88]
Nifurtimox, benznidazole	Human plasma	**[C_8MIM][PF_6]** [C_4MIM][PF_6] [C_6MIM][PF_6]	MeOH	HPLC-UV	15.7 3.66	98 79.8	[72]
Nifurtimox, benznidazole	Human breast milk	**[C_8MIM][PF_6]**	MeOH	HPLC-UV	90 60	89.7 77.5	[73]
Sildenafil, Vardenafil, Aildenafil	Human plasma	**[C_8MIM][PF_6]** [C_4MIM][PF_6]	MeOH	HPLC-UV	0.92–2.69	100.4–103.9	[74]
Ephedrine, Ketamine	Human urine	**[C_4MIM][PF_6]** [C_6MIM][PF_6] [C_8MIM][PF_6]	ACN	HPCE	210 390	79–90	[89]
Daclatasvir	Rat serum	**[C_4MIM][PF_6]** [C_6MIM][PF_6] [C_4MIM][BF_4] [C_4MIM][Br] [C_6MIM][Cl] [C_2MIM][$CH_3(SO_4)$]	ACN	HPLC-DAD	15	99.4	[90]

Table 2. *Cont.*

Drug(s)	Matrices	Tested Ionic Liquids	Extraction Solvent	Analytical Technique	LOD [ng/mL]	Efficiency [%]	Ref.
		Environmental samples					
NSAIDs, acetazolamide, caffeine, sulfonamides, carbamazepine, gemfibrozil	Tap water, creek water	**[C_6NH_2MPyrr][TFP]** [C_4MIM][Cl] [C_4MIM][Nf_2T] [ECMMPyrr][TFP] [MOEDEAmm][TFP] [MOEMIM][TFP] [MOEMMO][TFP] [MOEMPyrr][TFP] [MOPMPP][TFP]	MeOH/Acetone	HPLC-UV	0.1–55.1	91–110	[92]
Triclosan, triclocarbon	Tap water, wastewater	**[C_6MIM][PF_6]** [C_4MIM][BF_4]	MeOH	LC-MS/MS	0.040–0.58	70.0–103.5	[93]
		DLLME modifications					
		Biological samples					
		US-IL-DLLME					
Salmeterol	Dried blood spot	**[C_4MIM][PF_6]** [C_6MIM][PF_6] [C_8MIM][PF_6]	MeOH	HPLC-FL	0.30	90	[71]
Citalopram, nortriptyline	Human plasma	**[C_8MIM][PF_6]** [C_4MIM][PF_6] [C_6MIM][PF_6]		HPLC-PDA	10 6	90–92	[79]
Venlafaxine, amitriptyline	Human plasma	**[C_8MIM][PF_6]** [C_4MIM][PF_6] [C_6MIM][PF_6]		HPLC-DAD	0.5 0.8	91.4–92.6	[83]
Ulipristal	Mice serum, tablets	**[C_8MIM][PF_6]** [C_4MIM][PF_6] [C_6MIM][PF_6]		HPLC-UV	6.8 9.3	95	[75]
Benzodiazepines and benzodiazepine-like	Human blood, post-mortem human blood	**[C_4MIM][PF_6]** [C_6MIM][PF_6] [C_8MIM][PF_6]		LC-MS/MS	0.03–4.74	24.7–126.2	[15,76,77]
Antidepressants	Human blood	**[C_4MIM][PF_6]** [C_6MIM][PF_6] [C_8MIM][PF_6] [C_4MPyrr][Nf_2T] [C_4M_3Amm][Nf_2T]		LC-MS/MS	1–2	53.11–132.98	[78]
		DLLME (rapid shooting)					
Danazol	Mice serum, capsules	**[C_8MIM][PF_6]** [C_4MIM][PF_6] [C_6MIM][PF_6]		UV	55 54	90.5–103.4	[84]
		TCIL-DLPME					
Piroxicam	Human urine, plasma, and tablets	**[C_6MIM][PF_6]**		SFIS	46	95.2–104	[85]
		Environmental samples					
		US-IL-DLLME					
Lovastatin, simvastatin	Tap water, lake water, river water	**[C_6MIM][PF_6]**	MeOH	HPLC-UV	0.17 0.29	90.0–102.2, 80.5–112.0	[95]
β-Blockers NSAIDs	Wastewaters	**[C_8MIM][PF_6]** [C_4MIM][PF_6] [C_6MIM][PF_6]	ACN	LC-MS	0.0002–0.060	88–111	[97]
Fluoroquinolones	Groundwater	**[C_8MIM][PF_6]** [C_4MIM][PF_6] [C_6MIM][PF_6]	MeOH	HPLC-FL	0.0008–0.013	105–107	[98]
NSAIDs	Tap water, drinking water	**[C_8MIM][PF_6]**, [C_4MIM][PF_6]	MeOH	UHPSFC-PDA	0.62–7.69	81.4–107.5	[99]

Table 2. *Cont.*

Drug(s)	Matrices	Tested Ionic Liquids	Extraction Solvent	Analytical Technique	LOD [ng/mL]	Efficiency [%]	Ref.
		MADLLME					
Derivatization of sulfonamides	River water	**$[C_6MIM][PF_6]$**, $[C_4MIM][PF_6]$, $[C_8MIM][PF_6]$	MeOH	HPLC-FD	0.011–0.018	95.0–110.8	[100]
		MIL-DLLME					
Acetaminophen sulfamethoxypyridazine, phenacetin, ketoprofen	Lake water, river water	**$[P_{6,6,6,14}{}^{+}]_2[MnCl_4{}^{2-}]$** $[Aliquat^{+}]_2[MnCl_4{}^{2-}]$ $[C_8MAmm][MnCl_4{}^{2-}]$	ACN/MeOH	HPLC-UV	0.25–1.0	42.9–114.7	[94]
		IL-DLLME-μ-SPE					
Antidepressant drugs	Canal water	**$[C_6MIM][TFP]$** $[C_6MIM][Nf_2T]$	Methanol	HPLC-UV	0.3–1.0	94.3–114.7	[101]
		IL-DLLME-SDS					
Tetracyclines	River water, fishpond water, leaching water	**$[C_4MIM][PF_6]$** $[C_6MIM][PF_6]$ $[C_8MIM][PF_6]$	Methanol	UHPLC-TUV	0.031–0.079	55.1–96.3	[96]
		IL/IL-DLLME					
Triclocarbon, triclosan	Tap, river, snow, Lake water	Hydrophobic: $[C_4MIM][PF_6]$ $[C_6MIM][PF_6]$ **$[C_8MIM][PF_6]$** Hydrophilic: $[C_2MIM][BF_4]$ **$[C_4MIM][BF_4]$** $[C_4MIM][NO_3]$	-	LC-MS/MS	0.23 0.35	88–111	[102]
NSAIDs	Tap water, River water	**$[C_4MIM][BF_4]$ $[C_4MIM][PF_6]$** $[NEMMP][Nf_2T]$ $[MOEDEA][TFP]$	Methanol	HPLC-DADHPLC-FL	17–95	89–103	[103]
		Other liquid phase extraction					
		Biological samples					
		IL-SE-UE-ME					
Doxepin, perphenazine	Human urine	**$[C_6MIM][PF_6]$** $[C_6MIM][Nf_2T]$ $[C_4MIM][PF_6]$		HPLC-MWD	100 1000	89–98	[104]
		IL/IL LPME					
Sulfonamides	Human, chicken, rabbit, cow, pig blood	**$[C_4MIM][BF_4]$ $[C_6MIM][PF_6]$**		HPLC-UV	3.77–5.21	90–113	[105]
		dLPME					
NSAIDS	Human urine	**$[C_4MIM][PF_6]$**	ACN	HPLC-UV	38–70	72.8–90.3	[91]
Phenothiazines	Human urine	**$[C_4MIM][PF_6]$**	ACN	HPLC-UV	21–60	72–98	[106]
		Environmental samples					
		SADBME					
Diclofenac, ibuprofen	Wastewater treatment plant, river and lake water	**$[C_8MAmm][FeCl_4]$**	ACN	HPLC-DAD			[64]
		ILSVA-SME					
Glucocorticoids	Mineral water, lake water, tap water	**$[C_4MIM][PF_6]$**	ACN	HPLC-DAD	4.11–7.50	≥97.24	[107]

Table 2. *Cont.*

Drug(s)	Matrices	Tested Ionic Liquids	Extraction Solvent	Analytical Technique	LOD [ng/mL]	Efficiency [%]	Ref.
		MA-LLME-SIL					
Sulfonamides	Tap water, lake water, river water, pool water	**[C_2MIM][PF_6]**	ACN	HPLC-UV	0.33–0.85	75.1–115.8	[108]
Sulfonamides	Wastewater paddy water, river water	**[C_8MIM][PF_6]**		HPLC-UV	0.1–0.4		[109]
		(MBA)-LPME					
Glucocorticoids	Wastewater	**[C_4MIM][$CH_3(SO_4)$]** [C_4MIM]BF_4] [C_4MIM][Cl] [C_4MIM][PF_6] [C_4MPyrr][TFP] [C_6MIM][TFP]	MeOH	UHPLC-MS/MS	0.0128–0.0470	49.40–83.1	[110]
		IL-AF-μ-EME					
Antidepressants	Tap water, river water	**[C_6MIM][PF_6]** [C_8MIM][PF_6]	MeOH/ACN	HPLC-UV	0.4	88.2–111.4, 90.9–107	[111]
		IL-IDME					
Amitriptyline	Hospital wastewater	**[C_6MIM][PF_6]**	MeOH/ACN	HPLC-UV	4.0	85.12	[16]
		ATPS					
		Biological samples					
Sulfonamides	Milk (from supermarket)	**[C_4MIM][BF_4]** [C_2MIM][BF_4] [C_6MIM][BF_4] [C_8MIM][BF_4]	ACN	HPLC-UV	2.04–2.84	72.3–108.9	[112]
Sulfonamides	Pig, rabbit, cow, chicken and human blood	**[C_6MIM][Cl]** [C_4MIM][Cl] [C_8MIM][Cl]		HPLC-UV	2.45–4.13	85.5–110.9	[67]
		Environmental samples					
		IL-ATPF					
Chlorampheni-col	Lake water, feed water	**[C_4MIM][Cl]** [C_8MIM][Cl] [C_4MIM][BF_4]		HPLC-UV	0.1	97.1–101.9	[113]
		MILATPs					
Chloramphenicol	River water	**[TMG][TEMPO-OSO_3]**		HPLC-UV	0.14	94.6–99.7	[114]

[C_2MIM]: 1-Ethyl-3-methylimidazolium; [C_4MIM]: 1-butyl-3-methylimidazolium; [C_6MIM]: 1-hexyl-3-methylimidazolium; [C_8MIM]: 1-octyl-3-methylimidazolium; [MMIM]: 1,3-dimethylimidazolium; [C_4MPyrr]: 1-butyl-1-methylpyrrolidinium; [C_4M_3Amm]: 1-butyl-3-methylammonium; [IMIM]: 1-isopentene-3-methylimidazolium; [Nf_2T]: bis(trifluoromethylsulfonyl)imid; [BF_4]: tetrafluoroborate; [PF_6]: hexafluorophosphate; [Cl]: chloride; [$CH_3(SO_4)$]: methylsulfate; HPCE: high performance capillary electrophoresis; HPLC-FL: high-performance liquid chromatography and fluorescence detection; HPLC-UV: high-performance liquid chromatography and ultraviolet detection; HPLC-DAD: high-performance liquid chromatography and diode array detection; LC-MS/MS: liquid chromatography—tandem mass spectrometry; UV: ultraviolet spectroscopy; HPLC-MWD: high-performance liquid chromatography and multiple wavelength detector; HPLC-PDA: high-performance liquid chromatography and photo-diode array detector; SFIS: stopped-flow injection spectrofluorimetry; TCIL-DLPME: temperature-controlled ionic liquid dispersive liquid phase microextraction; [C_6NH_2MPyrr]: 1-(6-aminohexyl)-1-methylpyrrolidinium; [TFP]: tris(pentafluoroethyl)trifluorophosphate; [ECMMPyrr]: 1-ethoxycarbonylmethyl-1-methylpyrrolidinium; [MOEDEAmm]: methoxyethyl-dimethyl-ethylammonium; [MOEMIM]: 1-methoxyethyl-3-methylimidazolium; [MOEMMO]: 1-methoxyethyl-1-methylmorpholinium; [MOPMPP]: 1-methoxypropyl-1-methylpiperidinum; [$Aliquat^+$]$_2$[$MnCl_4^{2-}$]: aliquat tetrachloromanganate(II); [C_8MAmm]: methyltrioctylammonium; [$P_{6,6,6,14}^+$]: trihexyltetradecylphosphonium; TUV: tunable ultraviolet detection; UHPLC: ultra-high pressure liquid chromatography; [MOEDEAmm]: ethyl-dimethyl-(2-methoxyethyl)ammonium; [NEMMP]: ethyl-dimethyl-propylammonium; MADLLME: microwave-assisted dispersive liquid–liquid microextraction; [TMG]: 1,1,3,3-tetramethylguanidine; [TEMPO]: 2,2,6,6-tetramethyl piperidine 1-oxyl free radical; IL-AF-μ-EME: Ionic liquid-impregnated agarose film two-phase micro-electrodriven membrane extraction; IL-ATPF: Ionic liquid–salt aqueous two-phase flotation; [DI-SPME]: direct-immersion solid-phase microextraction.

Environmental Samples

Kiszkiel et al. [68] were the only researchers who tested the ionic liquids [C_4MIM][PF_6] and [C_4MIM][Nf_2T]) for LLE, showing their ability to selectively isolate nizatidine and ranitidine from wastewater and river waters (Table 2). Based on preliminary studies, an IL with a different anion was selected for each analyte. In the case of nizatidine extraction, [C_4MIM][PF_6] was used, while for ranitidine—[C_4MIM][Nf_2T]. Their application allowed methanol consumption to be reduced (1.0 mL for nizatidine and 1.5 mL for ranitidine). During optimization, the appropriate volume of IL was selected and the impact of additional factors such as the effect and mixing time or pH was assessed. The ultimately optimized and validated method allowed over 100% recovery, a wide range of linearity and low LOD values to be obtained for both analytes. Thus, in this paper, the parameter values confirmed that LLE using ILs allows satisfactory results to be achieved.

3.1.2. Dispersive Liquid-Liquid Microextraction and Modifications

ILs have more often been applied in new solutions based on liquid-phase microextraction (LPME), and particularly in the increasingly used dispersive liquid-liquid microextraction method (DLLME), introduced for the first time by Rezaee [69]. The most important elements of the most popular microextraction methods are two solvents: extractant and disperser. After their quick injection into the sample, the disperser solvent causes the dispersion of the extraction solvent in the form of fine droplets. The large surface contact with the analyte helps in its adsorption. Then the two formed immiscible layers can be easily separated from each other. The method has many advantages, above all, lower consumption of organic solvents, a faster process and greater sample enrichment. Thus, subsequently, positive properties introduced further modifications leading to even better results. As it was mentioned above, one of them was the use of ILs.

Biological Samples

Cruz Vera and his colleagues [70] were among the first to use ILs in the DLLME of drugs from biological samples. In one-step in-syringe extraction of non-steroidal anti-inflammatory drugs (NSAIDs) from human urine, they used ILs as the extraction solvent and methanol as the disperser solvent. During optimization, they took into account not only the extraction efficiency, but also the enrichment factor and repeatability. Subsequent publications using IL-DLLME are modifications of the matrices and pharmaceuticals (Table 2). However, several repetitive elements of the study can be observed. First, the same group of molecules with the [PF_6] anion and the imidazolium cation were most often used to select the ionic liquid with the best results. Differences were related to the length of the cation alkyl chain (1-butyl-3-methylimidazolium ([C_4MIM]), 1-hexyl-3-methylimidazolium ([C_6MIM]) and 1-octyl-3-methylimidazolium ([C_8MIM])) [15,70–72,75–79,84,87–89]. Most often, ILs with a butyl or octyl substituent were qualified for further testing. Probably, the reason for choosing the C4 alkyl chain was the reduced viscosity and the resulting greater transfer of analytes to the IL (compared to C6 and C8) [15,70,71,76,78,80–82,87–90]. On the other hand, as the alkyl chain length increases, solubility in aqueous solutions decreases, and the analyte availability increases. This seems to be the reason for good results for [C_8MIM][PF_6] IL [72–75,79,83,84] (Figure 3).

However, it should be highlighted that despite the knowledge of the structure-properties, using only one criterion when choosing an IL is impossible. Thus, there is also the opinion that the structures of analytes should influence their choice. Moreover, the volume of ILs is an important factor. In many studies, it has been confirmed that as the volume increases, the efficiency and enrichment factors increase. However, the trend changes at some point and when the volume is too large, the results decrease. Probably, a large volume of ILs reduces the concentration of the analytes. On the other hand, if the volume is too small, the extraction and collection of the IL-analyte phase from the system is also problematic [71]. Therefore, this parameter should also be estimated in each study.

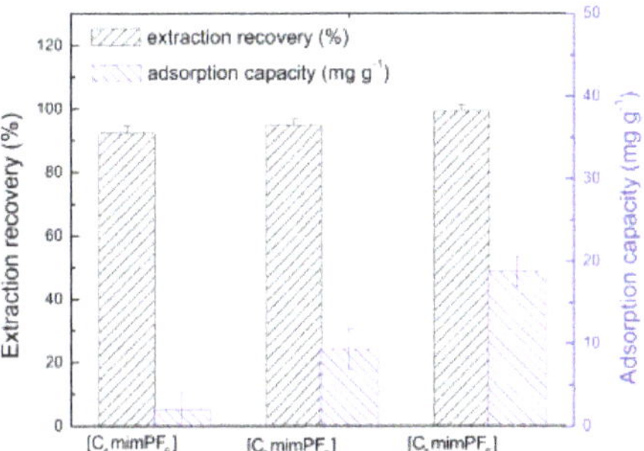

Figure 3. Effect of the kind of extraction solvents on ER of UPA and adsorption capacity. Extraction conditions: sample volume, 10.0 mL; sample amount, 10.0 μg; pH, 8.0; ultrasonic temperature, 313 K; ultrasonic time, 10 min; cooling temperature, 278 K; cooling time, 15 min; centrifugation time, 5 min. The error bars were standard deviation. Figure adopted from the reference [75] with copyright permission.

Besides those mentioned above, scientists also tried to include in the study ILs consisting of chloride ([Cl]), bromide ([Br]), [Nf_2T] and methyl sulfate ([$CH_3(SO_4)$]) anions [78,80–82]. Only in one publication was there an attempt to replace the imidazolium cation with 1-butyl-1-methyl-pyrrolidinium ([C_4MPyrr]) and 1-butyl-3-methylammonium ([C_4M_3Amm]) [78]. An important factor presented in the literature is the combination of ILs with organic solvents. Most often they are the dispenser solvent, but they can also be the solvent in back extraction [15,70,74–79,83–86,89]. It is known that ILs are highly viscous compounds. This property can hinder the chromatographic separation and detection of compounds, so the use of acetonitrile, methanol or ethanol is necessary. In some extractions, organic solvents are completely eliminated using instead sonication, controlled temperature, or intensive mixing. Their application helps to disperse ILs and gives as good results as organic solvents [15,74,76–79,83–86]. Gong et al. [75], during the determination of ulipristal acetate, completely eliminated the organic solvent as a dispersing agent. They used ultrasound energy without an organic solvent to disperse, and obtained an extraction recovery over of 95%. The addition of inorganic salts is also used in many works. The salting out process may affect the final results due to ionic strength and associated reactions with H_2O molecules [71,89,91]. The choice of pH is also important, the goal being to have analytes in neutral form, because in ionic form there is less availability for ILs, and the final extraction efficiency decreases [85]. As mentioned before, in addition to several constant elements, there are also several variables, such as analytes and matrices. Studies usually extract drugs commonly used to treat humans and animals, including antibiotics [81,82,86–88], antidepressants [78,79,83], benzodiazepines [15,76,77] and NSAIDs [70,85]. The matrices are most often human urine, plasma and serum (Table 2). An extraction method for a unique kind of matrix was developed by De Boeck and co-workers [15,76–78]. As the authors of several articles related to the use of ILs in DLLME, they started from choosing the best extraction and detection conditions for the determination of benzodiazepines, benzodiazepine-like hypnotics and antidepressants in whole human blood by LC-MS/MS. Then they transferred the optimized conditions for the analysis of postmortem blood samples. Both the matrix type and LC-MS/MS were first used in an IL-based analytical method for determining pharmaceuticals. Drugs used in veterinary medicine were determined in milk, eggs and the meat of pigs, cows, chickens and fish [87,88].

Environmental Samples

Similar to biological samples, two methods of the DLLME procedure can be observed: traditional, using only an IL (extractant) and organic solvent (dispersant) [92,93] or modified, using additional steps, such as ultrasound and others [94–103]. The traditional method was used by Yao et al. [92], who, by performing analyses with different ILs, drew attention to the impact of the character of the analyte on the final results. An IL with the [Nf_2T] anion and basic properties allows higher efficiency extraction for acidic compounds, whereas for compounds containing tertiary amines, the [TFP] anion was better. To further explain this phenomenon, the effect of surfactants on the results was also investigated. The use of the popular sodium dodecyl sulfate (SDS) without a primary amine did not improve the extraction efficiency, but after using a surfactant having such a moiety the result improved significantly. DLLME without modification also allowed the determination of triclosan and triclocarban by Zhoe et al. [93]. During optimization, $[C_6MIM][PF_6]$ was chosen for the analysis because of the higher solubility in water and worse efficiency of the $[C_4MIM]$ cation. The researchers also noted that the addition of an inorganic salt (most often NaCl), which changes the ionic strength, is responsible for two opposite effects. On the one hand, the addition of NaCl causes an increase in the solubility of an IL in water, thus increasing the volume sedimentation phase and consequently, the efficiency decreases, but on the other, there is an increase in analyte enrichment. Thus, the choice of this additive is not obvious.

DLLME modifications in the extraction of environmental samples are much more common. One of them is the use of ultrasound. Parrilla Vázquez and co-workers [97,98] focused on the optimization of this stage. They highlighted that the sonification time (too long may cause degradation) and sample cooling after the process have an impact on improving the results. Mao et al. [95] used high energy ultra-sound instead of normal ultrasound. In all US-IL-DLLME methods, the ILs for further analysis were selected from among the group with imidazolium cations and anions $[PF_6]$ in their structure. The best results were always obtained for ILs with the highest hydrophobicity, therefore the longest alkyl chain. Another modification was the inclusion of SDS in addition to the IL. The surfactant aimed to improve performance by reducing the adhesion of the IL to the walls of the tube. In addition, the novelty was heating the sample to 30 °C after the addition of the IL to completely dissolve the IL and then cooling to form two phases [96]. Yu et al. [94] used MIL to extract various compounds, including pharmaceuticals. They chose the best IL according to several criteria, such as magnetic susceptibility, HPLC compatibility, hydrophobicity needed for phase separation, minimal IL absorbance and minimal anion hydrolysis in the aqueous phase. These conditions were met by $[P_{6,6,6,14}{}^{+}]_2[MnCl_4]$. In order to achieve high efficiency, microwave energy was also used. However, its use could both improve and worsen the results, depending on the volume. Too high a temperature increases the contact of the IL with the aqueous phase and reduces the volume of the sedimentation phase, in consequence reducing the efficiency. The paper also discussed the influence of the dispersant on the final results. The choice of its volume is crucial as too large a volume causes an increase in the solubility of the IL in water, while too small hinders the formation of two phases [100]. Aimed at achieving environmentally-friendly procedures with the best results, methods using two ILs have also been proposed. Toledo-Neira et al. [103] used both $[C_4MIM][BF_4]$ and $[C_4MIM][PF_6]$ to change the polarity of the sample, and as an extractant, respectively. However, this work also uses an organic solvent as a dispersant. In another article with two ILs, one hydrophilic IL was used to disperse the other hydrophobic IL. Finally, only 50 μL of MeOH was used in the method to dissolve the sample prior to HPLC injection, thus the organic solvents were almost completely eliminated [102]. The last method of modification in the context of environmental samples was to combine DLLME with SPE. However the IL, as previously, was only applied as the extractant in DLLME. Among the tested ILs, the best result was obtained with $[C_6MIM][TFP]$ regarding the highest hydrophobicity, which was a constant trend in similar papers [101].

3.1.3. Other Liquid-Phase Extraction

Biological Samples

Good research results have encouraged researchers to further modify their extractions with ILs (Table 2). In 2015, doxepin and perphenazine were extracted according to a new procedure: ionic liquid-based surfactant emulsified microextraction accelerated by ultrasound radiation (IL-SE-UE-ME) [104]. Together with ultrasound applied to the surfactant, this led to the creation of an emulsion with the participation of ILs. The following year, Liu and co-workers [105] determined sulfonamides and used two ILs for extraction. They were the first to add [C_4MIM][PF_6] together with an inorganic salt, and after forming the precipitate, they added [C_6MIM][PF_6]. As a result, the analytes could be combined with the ionic liquid. Another equally effective extraction method is ionic liquid-based dynamic liquid-phase microextraction (dLPME). The method, for the extraction of phenothiazine and NSAID derivatives using the high density and viscosity of ILs, was developed by Cruz Vera [91,106]. The sample passes through the ionic liquid placed in a Pasteur pipette and the analytes are separated from the matrix. High viscosity is both an advantage and a limitation here as to perform the extraction it is necessary to reduce its value, therefore the addition of an organic solvent is also used.

Environmental Samples

In some publications, liquid-phase extraction procedures are very similar to DLLME procedures. One such procedure was proposed by Chatzimitakos et al. [64]. They used the potential of MIL [C_8MAmm][FeCl] to determine many analytes (including ibuprofen and diclofenac). The authors defined their novel method as stirring-assisted drop breakup microextraction (SADBME). Thus, the IL dispersion element was defined as drop breakup. Although the authors focused on the method itself, they showed that the application of MIL allows for the simplicity of extraction. Due to magnetic property, the separation of the IL-phase was possible by applying an external magnetic field. A similarity to DLLME can also be seen in synergistic centrifugal assisted ionic liquid assisted microextraction (ILSVA-SME). Faster formation of microemulsion (dispersion) with the IL and the surfactant used is achieved by vortex-assisted extraction. The method, as in other cases, allows for better efficiency of results [107]. Song et al. [108] also proposed a similar method of extraction to the above-described DLLME. They used a solid IL to extract sulfonamides and then they dissolved them by microwave energy namely microwave-assisted liquid-liquid microextraction (MA-LLME-SIL), and then cooled them again and dissolved them in acetonitrile. However, as the authors highlighted, this is a different method to DLLME, because a solid IL was used and an organic solvent was not necessary. The dispersion step is present here by shaking the molten IL sample. Thus, they do not define it as DLLME. In search of the best results, only one type of liquid was used, and its appropriate volume, duration of use and microwave power were chosen. Inorganic salt was also added but, as opposed to other works, it was not NaCl, but Na_2SO_4.

Another approach to minimize the amount of organic solvents was the modification and adaptation of LPME methods to determine analytes. The first work described the use of ILs in three-phase hollow fiber supported liquid-phase microextraction (HF-LPME). The procedure was based on the transfer of analytes from the donor phase to the acceptor phase through a membrane with an IL placed in the pores. Due to the good solubility of sulfonamides in water, their transfer based on passive diffusion can be difficult. For this purpose, the combination of an IL with tri-n-octylphosphine oxide (TOPO) was used to create a semi-liquid membrane and facilitate the transfer of the analyte to the acceptor phase. During optimization, the IL was compared with n-undecane and dihexyl ether (DHE). The IL, as the most polar compound, allowed the highest efficiency [109]. The second work describing the modification of LPME was related to the addition of an IL to the acceptor phase (IL/n-octanol) in membrane bag-assisted-liquid-phase microextraction ((MBA)-LPME). The extraction set was prepared by the author (a detailed description can be found in the original publication) [110]. The effect of using an IL was an increase in efficiency. Among the tested ILs, the results improved only after using [C_6MIM][TFP], which was explained by high hydrophobicity. Therefore, it should be noted that ILs, which are most relevant in DLLME, were not suitable for LPME-modification.

Extraction based on membranes was also proposed by Hanapi et al. [111] using an agarose membrane impregnated with an ionic liquid for electroconvulsive membrane extraction (IL-AF-μ-EME). An IL was used in both the membrane and the acceptor phase. During optimization, [C_6MIM][PF_6] and [C_8MIM][PF_6] were used. The cation with a hexyl substituent allowed for better performance. According to the authors, this is due to lower hydrophobicity, and therefore better solubility and conductivity. In addition to the type and the volume of the IL in the acceptor phase, other conditions (pH, ionic strength, mixing speed) were also optimized in experiments. The method was a fast process allowing for satisfactory validation parameters.

In one publication, ionic liquid-based immersed droplet microextraction (IL-IDME) was also used. Analytes, after transfer to IL droplets and in combination with a MeOH/ACN mixture, were analyzed by HPLC. Only one type of IL was used in the study, determining its optimal volume for analysis. During the optimization of other parameters, as in other papers, attention was paid to the effect of pH. Due to the determination of basic compounds, the samples were adjusted to an alkaline pH because analytes show greater affinity for ILs when in a non-ionized form [16].

3.1.4. Aqueous Two-Phase System

Biological Samples

In addition to DLLME extraction, ILs are also used in aqueous two-phase systems (ATPs) [67,112]. These consist of two immiscible water phases enriched with two different substances which affect their physical and chemical properties. These could be polymers, inorganic salts or surfactants. Unfortunately, there are also disadvantages in this process, such as interaction with analytes. Therefore, in searching for new solutions, it was decided to use the potential of ILs here. As in DLLME, ILs retain analytes in one of the phases and help in their separation. The group of tested ILs also remains constant (an imidazolium cation with a different alkyl chain length and a hexafluorophosphate anion); however, the selection criteria change. What is most important is the ability to separate the two phases. Shao et al. [112] chose [C_4MIM][PF_6] for further experiments. They also tested an IL with a 1-ethyl-3-methylimidazolium ([C_2MIM]) cation but then two phases were not formed. In longer alkyl chains, viscosity increased and analyte transfer decreased. In another publication with ATPs extraction, [C_6MIM][PF_6] was selected, the butyl substituent was too low in polarity and did not form separate phases with the SDS used, while an IL with C8 did not form a stable system. In addition, Yu et al. [67] checked how pH, IL volume, extraction time and the addition of inorganic salt affect ATPs. The results showed that all of the above factors are responsible for the total extraction effect. An increase in the volume of the IL caused an increase in the number of oil drops in the phase, the K_2HPO_4 used improved stability, while a change in pH and extraction time determined the final result of the efficiency. As we can also see, in this type of extraction the choice of IL is ambiguous and requires experimental testing. In addition, it is also important to choose a second substance that can determine the availability of analytes for the IL and the presence of two separate phases.

Environmental Samples

ATPs, which was described in the extraction of biological fluids, can also be used for environmental samples. Another form of ATPs, referred to as ionic liquid/salt aqueous two-phase flotation (IL-ATPF), was used to isolate chloramphenicol by Han et al. [113] (the solvent sublation apparatus was shown in the original work). The mechanism is based on the transfer of analytes into the IL droplets present in the upper surface phase of the system. As in other ATPs methods, the addition of inorganic salt was necessary and the best results were obtained through K_2HPO_4. The most appropriate IL was selected from three types: ([C_4MIM][Cl], [C_8MIM][Cl], [C_4MIM][BF_4]). [C_4MIM][Cl] was used in further analyses, because of the lowest viscosity and surface tension, which is crucial when analytes must be absorbed by the IL droplets. The particular novelty was also the use of MIL in this type of extraction (1,1,3,3-tetramethylguanidine and 2,2,6,6-tetramethylpiperidine 1-oxyl free radical [TMG][TEMPO-OSO3]). The common effect of MIL is to obtain a rapid extraction by easily collecting the IL-analyte complex with the help of an external magnesium field. The formation of MILATP requires the addition of inorganic salt (as already mentioned in paragraph 3.1.1). In this experiment, after the optimization and interpretation of results, the best addition was K_3PO_4. The choice of temperature was also important, as too high could cause an increase in the solubility of the IL in water, so finally room temperature was chosen [114].

3.2. Sorbent-Based Extraction Procedures

3.2.1. Solid-Phase Extraction

Solid-phase extraction (SPE) is a well-known sample pretreatment technique which ensures the simultaneous enrichment and purification of analytes [115]. In this technique, the compounds of interest and matrix interferences can be differentially desorbed from the SPE sorbent when water, an organic solvent or a mixture of organic solvent with water or salt solution are used as washing/eluting agents. It allows the analytes to be effectively extracted from the sample and the matrix interferences removed. In this extraction procedure, smaller amounts of organic solvent are required, and the risk of the formation of emulsions is decreased compared to LLE-based procedures. In effect, SPE is considered as a more environmentally-friendly method which is able to offer high analyte recoveries. Additionally, the SPE process is rapid and can be easily automated as an off-line SPE or on-line SPE system where direct coupling to chromatographic or electrophoretic separation systems is applied. In on-line SPE, a higher throughput and a more effective reduction of sample contamination or degradation can be obtained, while human exposure to potentially hazardous samples is decreased. On the other hand, the preconcentration and purification of the analytes in SPE may sometimes be ineffective because of the limited selectivity of conventional solid sorbents (e.g., modified silica-based sorbents). For this reason, new SPE materials are systematically developed and introduced to improve selectivity, including molecularly imprinted polymers (MIPs) as well as IL-based sorbents. In most investigations, ILs are immobilized by the covalent attachment of the imidazole group to the silica surface or polymeric support. These IL-based sorbents are considered to be interesting alternatives in SPE for different groups of pharmaceuticals from biological and environmental samples.

Biological Samples

Pang et al. [116] fabricated a polymer monolith column with 1-vinyl-3-hexylimidazolium bromide ([ViC$_6$MIM][Br]) IL which was used for the on-line SPE isolation of betamethasone, norgestrel, halcinonide, beclomethasone dipropionate and testosterone propionate from human plasma. The developed SPE-HPLC-UV method offered the effective extraction of the analytes (93–105%), which allowed the target compounds to be quantified with LODs of 1–2 ng/mL. In a study by Liu et al. [117] a poly(ionic liquid-glycidylmethacrylate-coethyleneglycol dimethacrylate) (IL-GMA-co-EDMA) monolithic column with 1-vinyl-3-butylimidazolium chloride ([ViC$_4$MIM][Cl]) was synthesized and applied as an SPE sorbent in the on-line SPE-HPLC-UV method for the determination of nifedipine, nitrendipine and felodipine in human plasma samples. The best extraction of the analytes and purification of the matrix sample was obtained when a methanol/water mixture was used as the eluting agent. It allowed the three antihypertensive drugs in human plasma samples to be determined with LODs of 2–3 ng/mL. Ferreira et al. [118] used 1-vinyl imidazole and 1,4-butane-sultane to create a silica-anchored IL-based material which was applied as a sorbent in an SPE system coupled online with HPLC-MS/MS for the quantification of the antibiotic ceftiofur in bovine milk samples. The extraction efficiency ranged from 70 to 130%, and the LOD was 0.1 μg/L. A sol–gel synthesis of three hybrid materials containing [C$_4$MIM][PF$_6$], [C$_6$MIM][PF$_6$] and [C$_8$MIM][PF$_6$], attached by covalent bonds, was published by da Silva and Mauro Lanças [119]. These IL-based hybrid materials were applied as the sorbents in off-line SPE for the isolation of five sulfonamides and trimethoprim from bovine milk samples. The results indicated that the extraction efficiency of the analytes systematically decreased when the alkyl chain of the IL increased from C4 to C8. This was probably caused by the reduction of the electron density and the steric hindrance from the methyl group on the three-substituent site of imidazole rings, which weakened the π–π interaction between the electron-rich benzene ring of the target compounds and the imidazole rings of the used ILs. The best efficiency was offered by an IL

(C4)-based sorbent which was applied for the isolation and preconcentration of sulfonamide in bovine milk by the on-line SPE-HPLC-MS method. The LODs for the method developed were in the range of 1.5–2.25 μg/mL, with extraction recoveries from 74 to 93%. Yan et al. [120] developed modified dummy molecularly imprinted microspheres (DMIMs) based on [AC_2MIM][Br] as the co-functional monomer and phenylephrine as the dummy template. These DMIMs were used as the SPE sorbent for the isolation of clenbuterol and clorprenaline from urine samples. The obtained results confirmed that they were able to more effectively extract the analytes and remove the matrix interferences than with other tested commercial sorbents such as HLB, PCX, C18 and SCX. For the DMIMs, the extraction efficiency ranged from 93.3 to 106%. The developed DMIMs-SPE-HPLC method allowed the analytes to be quantified with LODs of 0.19 and 0.070 μg/L for clorprenaline and clenbuterol, respectively. Ma and Row [1] synthesized a molecularly imprinted monolithic column using levofloxacin and ciprofloxacin as templates, 1-vinyl-3-ethylimidazolium bromide ([ViC_2MIM][Br] as the functional monomer, and graphene oxide (GO) as the core material. When the efficiency of the IL-based imprinted monolithic column was tested as the SPE sorbent for the extraction of levofloxacin and ciprofloxacin from human urine, the best results were achieved using water as the washing agent, and a mixture of ethanol/acetic acid (7:3 *v/v*) for the elution of the analytes. The main advantages of the developed SPE protocol were the effective purification of the matrix sample, and the good extraction recovery of the analytes (89.5% and 92.5% for levofloxacin and ciprofloxacin, respectively). However, relatively low sensitivity of the developed SPE-HPLC-UV method was also observed (LODs from 0.06 to 0.27 μg/mL). Wu et al. [121] used an SPE procedure based on hemimicelles and admicelles (mixed hemimicelles) supported by an IL for the simultaneous extraction of five cephalosporins from biological samples. In this technique, the sorbent possesses adsorbed ionic surfactants on the surface of mineral oxides (e.g., SDS or IL) which enables two mechanisms to occur for the retention of the analytes—hydrophobic and electrostatic interactions. In effect, the extraction efficiency can be improved. The authors tested seven different surfactants, such as SDS, cetyltrimethylammonium bromide (CTAB), [ViC_6MIM][Br], [C_4MIM][Br], [$C_{12}MIM$][PF_6], [$C_{16}MIM$][Br] and [$C_{12}MIM$][Br]. The best recoveries were obtained for the long-chain IL [$C_{16}MIM$][Br], which confirms data presented in (Figure 4).

The imidazolium-based IL with a longer alkyl side chain was probably able to strengthen the directionality of hydrogen bonds and van der Waals forces. In consequence, the interactions between the mixed hemimicelles and the hydrophobic regions of target compounds were more intensive and the efficiency increased. Taghvimi et al. [122] prepared mixed hemimicelle magnetic dispersive solid-phase extraction (MHMDSPE) based on carbon-coated magnetic nanoparticles and supported by the IL (IL-C/MNPs) for the extraction of tramadol from urine samples. In this study, MHMDSPE conditions were optimized, including both the selection of the adsorbent type and the solvent used as a desorbing agent. The results indicated that the IL-C/MNPs with [C_6MIM][PF_6] was more effective than that based on Fe_3O_4 NPs.

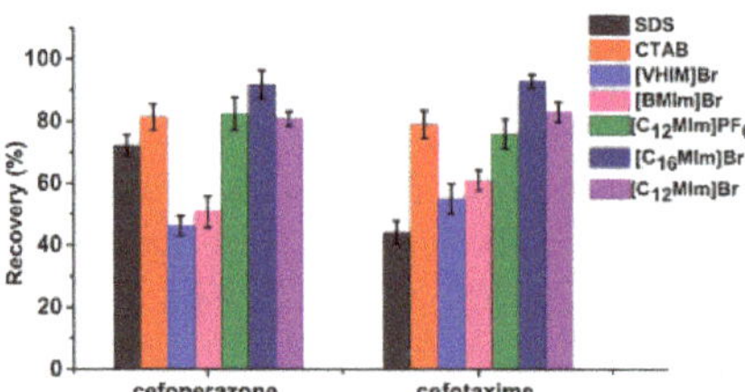

Figure 4. Comparison of the types of surfactants on the extraction efficiency of cefoperazone and cefotaxime. Figure adopted from the reference [121] with copyright permission.

This was probably related to the presence of carboxyl and hydroxyl groups on the surface of IL-C/MNPs, which improved the dispersion of the magnetic nano-adsorbent in the urine medium. In effect, stronger interactions between the analyte and the magnetic nano-adsorbent occurred, which improved the extraction efficiency. The best desorbing solvent was acetone, which allowed a recovery of 94% to be obtained. Yan et al. [123] prepared IL-modified magnetic polymer microspheres (ILMPM) based on Fe_3O_4 NPs and [C_4MIM][PF_6] used as a magnetic adsorbent of MDSPE for the determination of sulfamonomethoxine sodium and sulfachloropyrazine sodium in urine samples. The developed ILMPM-SPE sorbent provided a higher purification ability and extraction recovery of the tested analytes compared with magnetic polymers based on using 4-vinyl pyridine, methacrylic acid and acrylamide as monomers. A report was also published describing matrix solid-phase dispersion coupled with homogeneous ionic liquid microextraction (MSPE-HILME) applied for the extraction of sulfamerazine, sulfathiazole, sulfamethazine, sulfadoxine, sulfachloropyridazine, sulfaphenazole and sulfisoxazole from animal tissues [124]. In the study, three kinds of hydrophilic ILs, including [C_4MIM][BF_4], [C_6MIM][BF_4], and [C_8MIM][BF_4] were tested in MSPD and HILME simultaneously. The results confirmed that higher extraction recoveries of the analytes were obtained with the C_4 IL than those observed with C_6 and C_8 ILs. This was related to the significant loss of C6 and C8 ILs in MSPD, which resulted in a small volume of the IL phase and low extraction yields of the target analytes. Compared to C_6 and C_8 ILs, the C_4 IL possesses higher water miscibility and lower viscosity, which facilitates the transfer of target analytes from the sample matrix to the extraction solvent. In this study, this effect was predominant in respect to the extraction capacity of the IL, which often increases with the increase in the alkyl chain length of the IL [125]. Finally, water was selected as the elution solvent in MSPD because of the more effective extraction of sulfonamides, which are water-soluble polar compounds. In this procedure, the C4 IL was mixed with the dispersant and the sample before introduction to the MSPD column, and the IL phase was collected after HILM. When the MSPD-HILME method was coupled to HPLC-UV, the recoveries of the sulfonamides ranged from 85.4 to 118.0%. The LODs for the analytes were 4.3–13.4 g/kg. The application of magnetic core-shell nanoparticles (mag-NPs) of SiO_2@Fe_3O_4 type, covalently modified with the IL (dimethyl octadecyl [3-(trimethoxysilyl propyl)]ammonium chloride) as the MSPE material for the extraction of tolmetin, indometacin and naproxen from blood samples was also described in the literature [126]. The synthesized mag-NPs were applied as the adsorbent in MSPE according to the protocol presented in Figure 5.

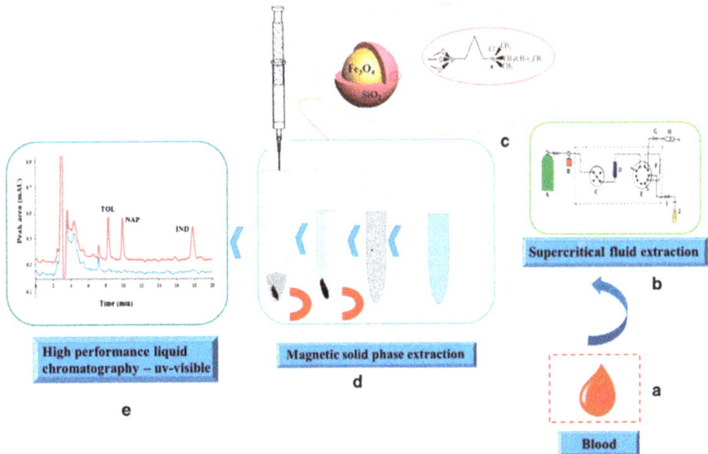

Figure 5. Schematic illustration of extraction procedure for tolmetin (TOL), indomethacin (IND) and naproxen (NAP) from blood samples. Figure adopted from the reference [126] with copyright permission.

The results of the study showed that the IL addition provided a more effective extraction of the NSAIDs probably due to an increase in both hydrophobic and π-π dipole or electrostatic interactions between the adsorbent surface and the analytes. On the other hand, the adsorption of the cationic molecules onto the sorbent was limited because of the repulsion interaction with the adsorbent surface. In consequence, a better purification of the sample was also achieved. The optimized MSPE was coupled to HPLC-UV and used alone or after supercritical fluid extraction (SFE) before HPLC separation. These protocols resulted in LODs between 0.1 and 0.3 μg/L for MSPE-HPLC and 0.2 to 0.3 mg/kg for SFE-MSPE-HPLC, respectively.

Environmental Samples

Fontanals et al. [127] synthesized and applied crosslinked polymer-supported imidazolium trifluoroacetate salt $[MI^+][CF3COO^-]$ as the SPE sorbent for the extraction of salicylic acid, carbamazepine, nalidixic acid, flumequine, gemfibrozil and four NSAIDs from aqueous samples. In the study, the developed IL-sorbent was tested under weak anion exchange (WAX), strong anion exchange (SAX) and strong cation exchange (SCX) as well as reversed-phase (RP) SPE conditions. The best purification and extraction results of acidic pharmaceuticals from different water samples (ultrapure, tap, river water and effluent wastewater) were obtained when the IL-based SAX material was applied. In the next study, two new imidazolium supported IL phases possessing different anions such as $[CF_3(SO_3)]$ and $[BF_4]$, were synthesized and applied as SPE-SAX sorbents for the isolation of acidic pharmaceuticals from water samples [128]. The obtained data indicated that $[MI^+][CF_3(SO_3)]$ and the previously developed $[MI^+][CF_3COO^-]$-SAX sorbent gave comparable results, whereas $[MI^+][BF_4]$ was not able to effectively extract and purify the acidic pharmaceuticals from environmental samples. On the other hand, the application of $[MI^+][CF_3(SO_3)]$ allowed only comparable efficiency to be obtained and calculated after using the commercially available Oasis MAX column, whereas $[MI^+][CF_3COO^-]$ was slightly more effective. Hydrophilic ciprofloxacin molecularly imprinted polymer material containing 1-allyl-3-vinylimidazole chloride ([AViMIM][Cl]) IL and 2-hydroxyethyl methacrylate as a bifunctional monomer was synthesized by Zhu and co-workers [129]. This MIP material was able to create strong hydrogen bonds, and electrostatic and π-π dipole interactions with ciprofloxacin in an aqueous solution. It offered excellent molecular recognition for common quinolone antibiotics (ciprofloxacin, levofloxacin and pefloxacin mesylate) in aqueous matrices as well as the selective isolation and separation of trace amounts of ciprofloxacin in real water, soil and pork samples, with recoveries of 87.3–102.5%.

3.2.2. Solid-Phase Microextraction

Solid-phase microextraction (SPME), developed by Pawliszyn and his co-workers in the 1990s [130], is a fast, solvent less-extraction technique for the sampling, cleaning-up and pre-concentration of analytes, which also offers the introduction of the sample to chromatography in a single solvent-free step. The SPME sorbents can be applied in both the headspace mode and the immersion mode. The simplicity of the SPME technique and other advantages, such as high selectivity and effective purification, the relatively low cost of equipment and the possibility of automation, mean that SPME is a powerful tool for the extraction of a wide range of compounds from different matrices. Moreover, new sorbents for SPME based on ILs are also synthesized. Several publications have described the results of their application for improving the extraction efficiency of different groups of pharmaceuticals from biological and environmental samples.

Biological Samples

A paper can be found in the literature describing the use of $SiO_2@Fe_3O_4$ functionalized with $[C_4MIM][PF_6]$ IL for the microextraction of four β-blockers (propranolol, metoprolol, atenolol and alprenolol) from human plasma [131]. In the study, two types of hydrophobic ILs, ($[C_4MIM][PF_6]$ and $[C_8MIM][PF_6]$), were tested. The results show that $[C_4MIM][PF_6]$ offered an extraction efficiency of 75 to 91%, while for $[C_8MIM][PF_6]$ these values were significantly lower (about 40%). This can be explained by the higher hydrophobicity of the long-chain IL, which leads to poor dispersion in the aqueous sample. Moreover, $[C_8MIM][PF_6]$ cannot be completely recovered by MNP, which can additionally decrease the extraction efficiency [132]. In the developed sample preparation procedure, an effervescent powder composed of sodium dihydrogen phosphate and sodium bicarbonate was also applied for the enhancement of the interaction between the magnetic sorbent and the analytes. When this protocol was coupled with LC-MS/MS, the developed method for the analysis of β-blockers in human plasma was able to monitor the compounds of interest with LODs from 0.03 to 0.62 ng/mL.

Environmental Samples

Serrano et al. [133] published the synthesis of GO functionalized with covalently attached 1-butyl-3-aminopropyl imidazolium chloride IL to GO sheets, and its application as an adsorbent for the dispersive micro SPE of six β-blockers and four anabolic steroids from aqueous samples prior to HPLC separation. It was observed that hydrophobic attraction between the compounds and the GO-IL was the predominant adsorption mechanism of steroids, while for β-blockers, their interactions with the adsorbent were more complicated. For them, both hydrophobic and electrostatic interactions can occur as well as the existence of interactions of electron-donor-acceptor type, which are dependent on the pH used in the extraction process. These mechanisms were more intense on the GO-IL sorbent, which was confirmed by the recovery results for the analytes (87–98%), which were found to be significantly higher than those observed with GO alone and graphene. Yu et al. [65] prepared six neat crosslinked polymeric ionic liquid (PIL) sorbent coatings for the SPME of selected phenolics, insecticides and pharmaceuticals, including phenacetin, ketoprofen, fenoprofen calcium, diclofenac sodium and ibuprofen, from environmental water samples (tap water and lake water). These PIL sorbents were prepared using various IL monomers such as 1-vinylbenzyl-3-hexadecyl-imidazolium chloride ([ViBC$_{16}$IM][Cl]), 1-vinylbenzyl-3-hexadecylimidazolium bis[(trifluoro-methyl)sulfonyl]imide ([ViBC$_{16}$IM][Nf$_2$T]), 1-vinyl-3-(2-hydroxyethyl)imidazolium bromide ([ViC$_2$OHIM][Br]), 1-vinyl-3-(10-hydroxydecyl)imidazolium chloride ([ViC$_{10}$OHIM][Cl]), 1-vinyl-3-(10-hydroxydecyl)imidazolium bis[(trifluoromethyl)sulfonyl]imide ([ViC$_{10}$OHIM][Nf$_2$T]), 1-vinyl-3-(9-carboxynonyl) imidazolium bromide ([ViC$_9$COOHIM][Br]), and crosslinkers like 1,12-di(3-vinyl-benzylimidazolium) dodecane dichloride [(ViBIM)$_2$C$_{12}$]2[Cl]), and 1,12-di(3-vinylbenzyl imidazolium)dodecane dibis[(trifluoromethyl) sulfonyl]imide ([(ViBIM)$_2$C$_{12}$]2[Nf$_2$T]). Next, they were tested in different experimental SPME conditions. The results indicated that all the developed PIL sorbent coatings were stable when the extraction was carried out under an acidic pH using various organic desorption solvents (e.g., methanol, acetonitrile, acetone). However, the best extraction results were obtained using the PIL-based sorbent coating polymerized from the IL monomer [VC$_{10}$OHIM][Cl] and the IL crosslinker [(VBIM)$_2$C$_{12}$]2[Cl]. The extraction efficiencies of pharmaceutical drugs and phenolics were higher when the film thickness of the PIL-based sorbent coating increased from 23 μm to 89 μm, whereas these values were largely unaffected for insecticides. This analysis allowed LODs to be obtained ranging from 0.2 to 2 g/L for the target compounds. A report presenting the synthesis of four different crosslinked PIL-based sorbent coatings by UV polymerization onto nitinol wires was also published in the literature [134]. These PIL coatings possessed either vinylbenzyl or vinyl alkyl imidazolium-based (ViBCnIM- or ViCnIM-) IL monomers with different types of anions, and various dicationic IL crosslinkers. They were used in a direct-immersion solid-phase microextraction (DI-SPME) method for the extraction of a group of polar analytes and non-polar analytes (10 different compounds), including gemfibrozil and carbamazepine. Two studied fibers, such as the polymers PIL–1a from the IL monomer [ViBC$_{16}$IM–Nf$_2$T] and IL crosslinker [(ViBIM)$_2$C$_{12}$–2Nf$_2$T], and PIL–2 based on the IL monomer [ViC$_{16}$IM–Nf$_2$T] and IL crosslinker [(ViIM)$_2$C$_{12}$–2Nf$_2$T] were used for the extraction of the analytes from real tap and river water samples. The results confirmed that these PIL-based fibers offered reproducible and effective extraction of most of the tested analytes from real samples. The extraction can be carried out many times (up to 100 extraction-desorption steps), and at low pH values.

3.3. Stir Bar Sorptive Extraction

In recent years, a sample preparation procedure based on stir bar sorptive extraction (SBSE) has been developed for the extraction of compounds occurring in matrices at trace levels. It should be noted that the extraction mechanism and the benefits of SBSE are identical to SPME. However, the enrichment factor obtained in SBSE can be significantly higher compared to SPME (~100 times). In SBSE, a glass tube with a magnetic core, coated with a layer of special polydimethylsiloxane (PDMS) tubing is applied to stir aqueous samples. After a certain time, the molecules captured on the bars can be desorbed either thermally for GC or into a solvent for LC. One drawback of SBSE is the low availability of different types of coatings. It should be noted that PDMS, mainly in SBSE, possesses a high affinity to extract non-polar compounds, while polar ones are poorly isolated. To overcome this limitation, new polymeric coatings are introduced, including poly (methyl methacrylate/ ethyleneglycol dimethacrylate) (PA-EG), and IL-based sorbents in order to improve the extraction efficiency of more polar compounds. Another problem of SBSE is the presence of the memory effect (carryover) during the desorption step using an organic solvent. According to the literature data, Talebpour et al. [135], in a comparative study, reported the application of a PA-EG polymeric phase and PDMS-coated stir bar supported by an IL for the extraction of carvedilol in human serum samples. In this investigation, $[C_8MIM][BF_4]$ IL was tested as a modifier in the desorption solvent (methanol) for checking whether better extraction efficiency and the elimination of carryover can be obtained. The results confirmed that carvedilol has a better affinity for the PA-EG phase than for PDMS. Moreover, the addition of $[C_8MIM][BF_4]$ at a concentration of 0.1 M to methanol significantly increased the recovery of carvedilol. Additionally, no carryover effect was observed, whereas it was detected when methanol was used without the IL (about 11% of the initial desorption step). Unfortunately, to the best of our knowledge, no report describing the use of IL-based sorbents for the SBSE extraction of pharmaceuticals from environmental samples has been published.

3.4. PASsive Sampling with Ionic Liquids

Extractions described so far can be classified as extractions with active sampling, because additional mechanisms, such as pressure and so on, are used for the flow of samples through the sorbent. However, the isolation of analytes is also possible in another way. Extraction using passive samplers can be used for the long-term monitoring of pharmaceuticals [136]. A significant difference in this method compared to procedures traditionally used in laboratories is the ability to estimate the time-weighted average concentration (TWAC) of analytes in ecosystems. Currently, the most popular Polar Organic Chemical Integrative Sampler (POCIS) techniques have been enriched with the new PASsive Sampling with Ionic Liquids (PASSIL) technique, developed by a team of scientists from the University of Gdansk. To carry this out, a dosimeter consisting of two disks and a membrane covered with the acceptor phase is necessary. Various ILs or their combinations with other sorbents are used as the acceptor, here.

Caban et al. [137] compared the results of the isolation of analytes (diclofenac, carbamazepine and two sulfonamide antibiotics) using dosimeters in which the membrane was covered only with an IL or a combination of IL and colloidal silica obtained from C18 SPE extraction columns. In the experiment, they tested four ILs using not only the popular imidazolium cation but also the phosphonium cation ($[C_6MIM][Tf_2N]$, $[P_{6,6,6,14}{}^+][N(CN)_2]$, $[P_{4,4,4,14}{}^+][DDBS]$, and $[P_{2,4,4,4}{}^+][(2O)_2PO_2]$). The most important and desirable property of such sorbents was water insolubility. The content of the IL transferred into the donor phase was determined by testing the pH, conductivity and recovery of the phase. In order to select the best extraction conditions, the extraction efficiencies were calculated for all experiments. The results confirmed that when the IL alone was applied (independent of the type of IL), it did not improve the efficiency, and sometimes lower extraction parameters were calculated than those using traditional C18 sorbents (carbamazepine). In contrast, by using the combination of the IL and C18, the efficiency increased, the acceptor phase stability was improved and less IL consumption was possible. The developed method was used to extract analytes from saline water. The use of the matrix, which caused changes in the properties of the IL and analytes due to pH modifications, proved that the final result is a consequence of many components, not only choosing the right sorbent at the stage of method optimization. The same effects were observed in the study using a similar procedure to assess the effect of pH and salinity on the extraction efficiency of β-blockers, NSAIDs and sulfonamides using the PASSIL technique. It was interesting that the results for samples taken from the donor phase by a dosimeter with the same IL were different depending on the analyte. The extraction of β-blockers was impossible when an IL was used as the sorbent, even after changing the pH. In contrast, for NSAIDs and sulfonamides, the extraction efficiency improved after the appropriate pH modification (Figure 6).

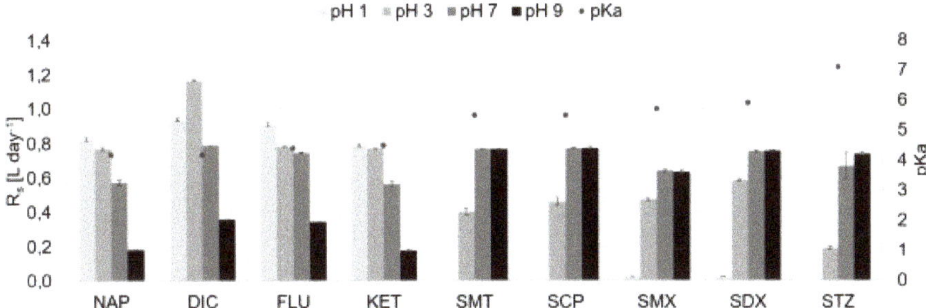

Figure 6. Dependency between sampling rate (Rs) values and the salinity of the donor solution for selected sulfonamides and NSAIDs (the pKa values of target compounds are specified by the black dots). Figure adopted from the reference [138] with copyright permission.

The authors suggest that this situation results from the presence of β-blockers in a neutral or cationic form in the solution which cannot be adsorbed on the membrane surface to large $[P_{6,6,6,14}{}^{+}]$IL cations. In turn, the increase in salinity caused a decrease in the efficiency of analyte extraction due to their competition with the ions of salts present in saline water [138]. Męczykowska et al. [139] also assessed the effect of humic acids, temperature and mixing on the final extraction results of various pharmaceuticals using the PASSIL technique. The results indicated that each of these parameters can decide on the final results. Moreover, the importance was emphasized of polarity or hydrophobic properties as factors affecting these parameters.

4. Chromatographic Techniques

4.1. High Performance Liquid Chromatography

4.1.1. IL Additives to the Mobile Phase

Liquid chromatography is the most commonly used technique for determining pharmaceuticals. Most of them are basic and their separation takes place in a reversed-phase using a silica-based column [140–142]. Unfortunately, this involves several serious problems during the analysis. The literature data indicate the main reason to be the presence of free silanol groups, which are negatively charged and can interact with positively charged basic analytes in an ion exchange reaction. Based on experimental research, it can be observed that this is often associated with problems with the resolution and shape of chromatographic peaks or a high retention factor. To prevent or minimize these deleterious effects, a mobile phase is used with additives for blocking free silanols [140]. The most popular additives are various types of amines, such as triethylamine (TEA), dimethyl-octylamine (DMOA) or buffers. The first researchers who noticed that ILs may also have suppressing properties against silanol groups were Kaliszan et al. [143]. In 2005 they published a paper in which they used an additive IL to the mobile phase in drug detection by thin layer chromatography (TLC) and reversed-phase liquid chromatography (RPLC) techniques. Since then, new publications have appeared systematically on similar topics (Table 3). However, considering the topic of ILs in drug determination, it should be highlighted that these works mainly focus on explaining the function of ILs in the suppression process and the drugs are less important as analytes. In addition, only a few works use biological [140,141,144–146] or environmental [142] samples as the matrices; in one, tablets were analyzed [147], but most often they were aqueous solutions [6,143,147–154].

Table 3. Summary of the HPLC methods for the determination of pharmaceuticals in biological and environmental samples supported by the addition of ILs to the mobile phase.

Drug(s)	Matrice(s)	Tested Ionic Liquids	Apparatures	Stationary Phase	Ref.
Psychotropic drugs	Human serum	**[C_4MIM][BF_4]**	HPLC-DAD λ = 240 nm	Synergi Polar RP 80A (150 × 4.6 mm, 5 μm)	[140]
Fluoroquinolone antibiotics	Bovine, ovine and caprine milk	**[C_2MIM][BF_4]**, [C_4MIM][BF_4]	HPLC-FL λ_{ex} = 280 nm λ_{em} = 450 nm	RP Nova-Pak C18 (150 × 3.9 mm, 4 μm)	[141]
Fluoroquinolone antibiotics	Mineral and tap water	[C_2MIM][BF_4], **[C_4MIM][BF_4]**, [C_6MIM][BF_4], [C_8MIM][BF_4], [$(C_2H_5)_4$N][BF_4]	HPLC-FL λ_{ex} = 280 nm λ_{em} = 450 nm	RP Nova-Pak C18 (150 × 3.9 mm, 4 μm)	[142]
Antiretroviral drugs	Rats plasma	**[C_4MIM][BF_4]**, [C_4MIM][Br], [C_4MIM][$C_8H_{17}(SO_4)$], **[C_2MIM][$CH_3(SO_4)$]**, [C_6MIM][Cl]	HPLC-DAD	monolithic RP-18e column (250 × 4.6 mm, porous material)	[144]
Antidepressants	Urine samples	[C_4MIM][BF_4], [C_4MIM][PF_6], **[C_4MIM][CF_3SO_4]**,	HPLC-UV λ = 254 nm	RP Eclipse X-DB-C8 (150 × 4.6 mm)	[145]
Ofloxacin, sparfloxacin, moxifloxacin, levofloxacin, *p*-amino-salicylic acid, ketoprofen, ibuprofen	Human plasma	[C_4MIM][Cl], **[C_6MIM][Cl]**, [C_8MIM][Cl], [C_{12}MIM][Cl], [C_6MIM][BF_4]	HPLC-DAD λ = 235–375 nm	Luna C18(150 × 4.6 mm, 5 μm)	[146]
Tricyclic antidepressants	Tablets	**[C_6MIM][Cl]**, [C_6MIM][BF_4]	HPLC-UV λ = 254 nm	Zorbax Eclipse XDB C18 and C8 (150 × 4.6 mm, 5 μm)	[147]
β-Blockers		[C_2MIM][Cl], [C_4MIM][Cl], [C_6MIM][Cl]	HPLC-UV λ = 254 nm	Zorbax Eclipse XDB (150 × 4.6 mm, 5 μm)	[6]
β-Blockers		**[C_6MIM][Cl]**	HPLC-UV λ = 254 nm, λ = 300 nm (timolol)	Zorbax Eclipse XDB C18 (150 × 4.6 mm, 5 μm)	[7]
Quinine, fluphenazine, thioridazine, chlorpromazine, trifluopromazine, phenazoline, tiamenidine, naphazoline propiomazine		[C_2MIM][BF_4], [MMIM][$CH_3(SO_4)$]	HPLC-DAD λ = 254 nm	LiChrospher RP-18 (250 × 4.6 mm, 5 μm)	[143]
β-Blockers		[C_4MIM][BF_4], **[C_6MIM][BF_4]**	HPLC-UV λ = 254 nm	Zorbax SB C18 X-Terra MS C18 Kromasil Lichrospher, Nucleosil, Spherisorb	[148]
β-Blockers		[C_2MIM][Cl], [C_4MIM][Cl], **[C_6MIM][Cl]**, [C_2MIM][BF_4], [C_4MIM][BF_4], **[C_6MIM][BF_4]**, [C_2MIM][PF_6]	HPLC-DADλ = 254 nm, λ = 300 nm (timolol)	Kromasil C18 (150 × 4.6 mm, 5 μm)	[149]
β-Blockers		[C_2MIM][PF_6], [C_4MIM][PF_6], [C_4MIM][BF_4], **[C_6MIM][BF_4]**	HPLC-UV λ = 254 nm, λ = 300 nm (timolol)	Kromasil C18 (150 × 4.6 mm, 5 μm)	[150]
Tricyclic antidepressants		**[C_4MIM][PF_6], [C_4MIM][Cl],**	HPLC-DAD λ = 254 nm	Gemini-NX C18 (150 x 4.6 mm, 5 μm)	[151]
β-Lactam antibiotics		**[C_4MIM][BF_4]**, [C_6MIM][BF_4], [C_8MIM][BF_4]	HPLC-UV λ = 254 nm	RS Tech C18 (250 × 4.6 mm, 5 μm)	[152]
β-Blockers		**[C_4MIM][BF_4]**, [C_8MIM][BF_4], [C_4MIM][PF_6]	HPLC-UV λ = 254 nm	RP Kromasil C18 (150 × 4.6 mm, 5μm)	[153]
Neuroleptic Drugs		[C_2MIM][PF_6], **[C_4MIM][PF_6]**, [C_4MIM][Cl]	HPLC-DAD	Zorbax Extend-C18 (150 × 4.6 mm, 5 μm)	[154]
Naphazoline, phenazoline, chlorpromazine, fluphenazine; propiomazine, thioridazine		[C_6MIM][BF_4], [C_8MIM][BF_4], [MMIM][$CH_3(SO_4)$],	HPLC-DAD λ = 254 nm	LiChrospher RP-18 (250 × 4.6 mm, 5 μm)	[155]

[$(C_2H_5)_4$N]: tetraethylammonium; [$C_8H_{17}(SO_4)$]: octylsulfate; [C_{12}MIM]: 1-dodecyl-3-methylimidazolium. (The other abbreviations explained under Table 2).

Focusing on the addition of an IL to the mobile phase in LC, it should be noted that the interpretation of the results requires consideration of the influence of both the IL anion and cation. Although the use of the term IL suggests that one large molecule is responsible for the effect, it should be remembered that in the mobile phase the IL dissociates into both the cation and anion, so their combined effect determines the final results. It should also be highlighted that despite the involvement of other physical and chemical factors in the separation process, the largest changes in the chromatogram can be seen when using different kinds of ILs [142]. Their basic mechanism during pharmaceutical analysis is the reaction of IL cations with free silanol groups, the repulsion of IL cations with cations of basic analytes, as well as the reaction of IL anions with cations of analytes [149]. Depending on their type, the mechanism may be a little different than described. The choice of cation, as in the case of extraction (see the Section 3.2) focuses on the selection of an appropriate imidazolium cation with a different alkyl chain length. In one work, the analysis of the imidazolium cation with two methyl substituents was also carried out [143]. The effect of the cation was studied by Herrera et al. [142]. They performed analyses for ILs with the same anion [BF_4] and different cations (Table 3). The results showed that an IL with a longer alkyl chain causes a decrease in the retention factor and an increase in efficiency. The effect of changing the retention time is similar in all analyses of basic analytes. The explanation for this effect may be an increase in hydrophobicity along with an increase in the length of the alkyl chain [144]. In turn, in a publication concerning the analysis of β-lactam antibiotics, an increase in the length of the alkyl chain caused an increase in retention. Han et al. [152] highlighted that a different effect may be the result of weak acidic properties and large analyte structures (the decrease in retention in other publications concerned basic analytes). The ester moiety of the antibiotic competed more strongly with the IL used for adsorption, and therefore despite the use of the long alkyl chain of the cation, retention increased. Ubeda-Torres et al. [149] also suggested that the size of the cation is more important than its nature. To study the effects of the IL anion, in other experiments, the same cation but a different anion was used during optimizing the IL selection. The number of anions tested is much greater than cations. The most commonly used are [PF_6], [Cl] and [BF_4] anions, but the less popular [$CH_3(SO_4)$], octylsulfate ([$C_8H_{17}(SO_4)$]) and [Nf_2T] have also been tested (Table 3). The analysis provided several important facts. First, the [PF_6] anion showed very strong adsorption on the column and had a stronger effect on the parameters than the present cation. This is probably the result of its strongly chaotropic character [148]. The [BF_4] anion is also a chaotropic ion, but with less adsorption than [PF_6]. For this reason [BF_4] more often qualified for further parts of the experiment. The next popular anion [Cl] belongs to strongly hydrated ions, and does not react with the analyte and the stationary phase; in its presence, the cation is mainly responsible for the mechanism [151]. Although the literature data provide information on the effects of the use of individual anions and cations, their choice is not obvious, not only because of the often antagonistic effect of ions. The use of an ionic liquid, which significantly reduces the retention time, is often associated with a poorer peak shape or resolution. In addition, too short a retention time for biological samples is not recommended because of the interference of analytes and background signals. In turn, improving the shape of the peak is possible at the expense of a higher retention factor. Figure 7 shows the change in retention after the use of two different ILs. Despite the shortening of the retention time by [C_6MIM][BF_4], an IL with [Cl] was chosen for the study due to better resolution. Therefore, the choice of IL is a kind of compromise, and the choice depends on many factors, including the type and number of analytes, and the type of matrices [147]. The results also show the influence of other factors on the final results. One of the most important modifications is the change in IL concentration. It was observed that a higher concentration leads to an improvement in the shape of the peak and reduces the retention time. However, this effect is more complex. First of all, the crucial factor here is whether the anion or cation has a stronger impact. For example, if the [PF_6] anion is used, which has strong adsorption on the stationary phase, the retention time increases, but if the low affinity [Cl] anion and the long alkyl chain cation are used, the retention time decreases. However, it was noted that both with increasing and decreasing retention the effect occurs already at a very low IL concentration, and occurs until the column is completely filled with the IL. When column saturation occurs, a further increase in the IL concentration in the mobile phase has less of an effect on the results. The retention time is constant or the effect is the opposite to the current one. The mechanism of action of the aforementioned [PF_6] in such a situation is explained by the reaction in the stationary phase until the column is saturated, and the reaction of this ion in the mobile phase after its saturation, and consequently, to a decrease in retention time [148,150]. The effect of pH on ILs was also analyzed, and it was found that at a lower pH the retention time is high because a larger number of [H^+] ions react with the IL anions and the elution power decreases. This can be both a disadvantage and an advantage, because on the one hand, the separation improves, but on the other hand, the analysis time is too long [152]. In another study, the purpose of which was to assess the effect of buffers on separation parameters in the presence and absence of ILs, it was observed that the IL is mainly responsible for the retention time, while the buffers more strongly affect the final effect without the addition of the IL. However, it should be mentioned that the IL [C_6MIM] with a strongly adsorbing cation was used in the experiment [7]. Another publication also suggests that retention is affected by the ratio of unprotonated to protonated silanols [6]. As mentioned above, ILs are an alternative to other mobile phase additives. For this reason, the results of studies with the addition of these compounds and with the addition of ILs are compared. ILs are better than other additives in all tests, but it must be highlighted that TEA the most popular compound, also gives good separation results [151]. In addition, the competitive advantage of ILs over other additives is the lack of effect on pH and, as shown in the literature, the involvement of both cations and anions in suppressing the silanol interaction and improving the results. In addition, the use of ILs is also possible in hydrophilic interaction liquid chromatography (HILIC). The increase in the stationary phase surface polarity obtained after the addition of an IL is responsible for improving the retention and efficiency parameters [146]. The verification of the positive effect of ILs was also presented in studies focusing more on the kind of columns used.

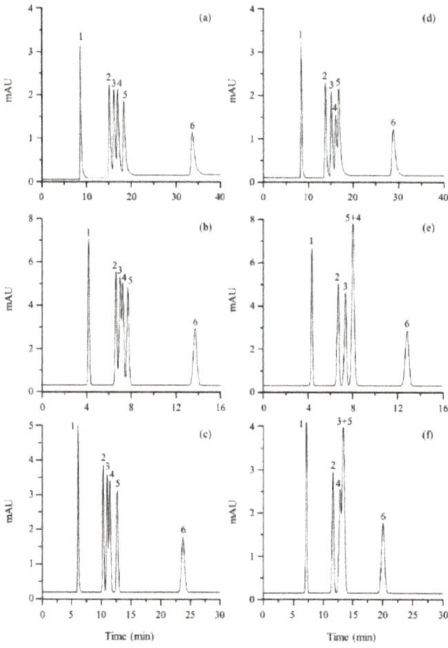

Figure 7. Simulated chromatograms for mixtures of the six TCAs using the C18 (**a–c**) and C8 (**d–f**) columns. Mobile phase composition: (**a**,**d**) 30% acetonitrile, (**b**,**e**) 30% acetonitrile/10 mM HMIM·Cl, and (**c** and **f**) 30% acetonitrile/10 mM HMIM·BF_4. Peak identity: (1) doxepin, (2) imipramine, (3) nortryptiline, (4) maprotiline, (5) amitryptiline, and (6) clomipramine. Figure adopted from reference [147] with permission of the copyright holder.

The analyses were performed on monolithic columns [144], popular C_8 and C_{18} columns (Figure 7) [147] and six commercially available stationary phases [148]. In each analysis, the addition of ILs improved the results. It was also noted that the results depend on the production process of columns, which decide about the number of free silanol groups. Thus, the best results during the application of ILs in LC are obtained for columns for which the result was worse when using a traditional mobile phase without the addition of an IL [148]. As already mentioned, the application of ILs in LC focuses on the reaction mechanism, and drug determination is not essential here. Apart from a small number of analyses for real samples, the quality of the developed methods is not confirmed by determining the validation parameters. Therefore, the aspects of linearity, repeatability or reproducibility are ignored. To our knowledge, only one work has performed validation [147]. Moreover, only a few anions and cations have been tested in the analyses. There is no information on the effects of less common ILs. In addition, not all works compare the results obtained for ILs with other popular additives. The application of ILs also has several limitations. Although they extend the life of the column by protecting the surface of the stationary phase, conditioning is necessary for several hours to remove adsorbed IL ions and return the column to the starting position [143]. Due to the involvement of both anions and cations in the separation mechanism, other unknown interactions with their participation may occur. In addition, the choice of detector is an important issue during the application of ILs to the mobile phase. The following detectors can be used: FL, UV or *diode array detector* (DAD), but it should be remembered that ILs have a natural ability for ultraviolet absorption, which may affect the final results or prevent the selection of the optimal wavelength for analytes [148,155]. Moreover, the use of mass spectrometry is very problematic, here. However, despite the inconveniences described above, the popularity of ILs is constantly increasing and they are being used in subsequent experiments

4.1.2. Ionic Liquid Stationary Phases

The application of ILs as an addition to mobile phases is not the only way to use them in liquid chromatography. In 2004, the first stationary phase appeared with ILs immobilized on the silica surface [156]. However, despite progress in this area, the use of IL stationary phases is much less popular than IL additives to mobile phases. Several-stage binding reactions are the first step of column preparation, producing as a final product modified IL-silica adsorbents, which finally coat the stationary phase (detailed reaction descriptions can be found in the original papers) [157–159]. Based on previous experience, several similarities can be observed to the previous section. First, the use of an IL stationary phase is the result of the incorrect peak shape, separation, efficiency and retention time obtained on traditional columns. As already mentioned, the same reasons concerned the use of ILs in the mobile phase. Secondly, research shows that both the cation and anion can be involved in the separation process. The imidazolium cation (single or multiple) is most commonly used to modify the stationary phase surface [160]. Furthermore, analytes were also separated on a column prepared using polymeric or chiral ILs [161,162]. There are many ways in the literature for obtaining IL-modified stationary phases based on various chemical reactions and substrates. However, their application is still not common. As mentioned, the number of publications is much smaller than the number of publications describing the suppression of free silanol groups by ILs present in the mobile phase.

This review focuses primarily on the determination of pharmaceuticals in biological and environmental samples, so it should be strongly highlighted here that the number of publications related to the determination of such analytes on IL columns by LC is negligible. Two such articles were published by Rahim et al. [163,164]. They prepared a stationary phase based on β-cyclodextrin and 3-benzylimidazolium tosylate as ILs for the enantioseparation of β-blockers and NSAIDs. The results confirmed an enhanced enantioseparation and better enantioresolution on the novelty stationary phase. Another publication in accordance with the criteria adopted in the review was published in 2019 by Xian et al. [165]. The stationary phase was prepared with photo-initiated thiol-ene click chemistry using the imidazolium cation and anion [Nf_2T]. Then, on the prepared column, the sulfonamides were separated by mixed-mode HPLC (MHPLC). The results confirmed good performance and separation selectivity, and additional research on commercial columns proved that the IL is responsible for a shorter separation time (Figure 8). Although the determination of drugs using IL column modifiers is very rare, their application in the determination of vitamins [166], flavonoids [167], amino acids [168] and many other compounds shows that perhaps in subsequent years these methods will be extended also for such analytes.

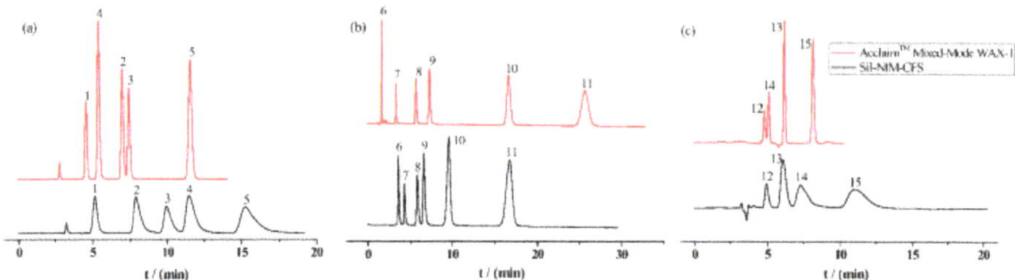

Figure 8. Separation of a mixture of nucleosides and nucleic bases, sulfonamides and inorganic anions on Sil-NIM-CFS and Acclaim™ Mixed-Mode WAX-1 columns. (1). uracil; (2). uridine; (3). cytosine; (4). adenine; (5). cytidine; (6). sulfanilamide; (7). sulfamethoxypyridazine; (8). sulfadiazine; (9). sulfathiazole; (10) sulfamethoxazole.; (11). sulfacetamide; (12). potassium bromate; (13). potassium bromide; (14). potassium iodate; (15). sodium iodide; (**a**) Mobile phase for Sil-NIM-CFS: ACN/10 mM ammonium formate (92:8, *v/v*); Mobile phase for Acclaim™ Mixed-Mode WAX-1 columns: ACN/10 mM ammonium formate (80:20, *v/v*); Ph = 5.6, flow rate: 0.6 mL/min, detection wavelength: 254 nm. (**b**) Mobile phase for Sil-NIM-CFS: ACN/H_2O (50:50, *v/v*), flow rate: 0.8 mL/min; Mobile phase for Acclaim™ Mixed-Mode WAX-1 columns: ACN/H_2O (60:40, *v/v*), flow rate: 1.0 mL/min; detection wavelength: 254 nm. (**c**) Mobile phase for Sil-NIM-CFS: ACN/5 mM Na_2SO_4 (5:95, *v/v*), pH = 4.28; Mobile phase for Acclaim™ Mixed-Mode WAX-1 columns: ACN/50 mM Na_3PO_4 (50:50, *v/v*), Ph = 6.0; flow rate: 0.6 mL/min; detection wavelength: 210 nm; Injection volume: 40 μL, column temperature: 25 °C. Figure adopted with permission from [165].

4.2. *Other Chromatographic Techniques*

4.2.1. Gas Chromatography

In gas chromatography (GC), ILs have found use as stationary phases. This is due to the fact that they have unique properties, such as a wide liquid phase range, low volatility (negligible vapour pressure), high viscosity, good thermal stability and variable polarities, which make them suitable for that purpose [169]. Research into using molten salts in GC started in the 1950s and now IL-based columns are used in the analysis of complex samples [29,170]. A characteristic property of ILs is that they display unusual dual nature retention behavior, separating both non-polar and polar compounds. On the one hand, ILs exhibit a similar behavior to polar stationary phases such as polyethylene glycol or cyanopropyl-substituted polysiloxanes due to their ability to display a high dipolar interaction and hydrogen bonding. On the other hand, they are able to retain non-polar solutes (i.e., alkanes and alkenes), similarly to the low polarity stationary phases such as phenyl substituted dimethyl polysiloxanes [171]. Another important feature of ILs is that varying the cation or anion might significantly affect their physical and chemical properties [169]. For example, imidazolium IL columns using the [Nf_2T] anion were the most efficient [29]. ILs formed by less coordinating or nucleophilic anions, such as [Nf_2T], tend to be more stable compared to those containing halide salts; in fact, the latter are characterized by a nucleophilic nature, and hence it is possible for them to undergo SN1 or SN2 reactions with the alkyl substituents of the cation. Phosphonium-based ILs, synthesized with a large alkyl chain substituent, have shown outstanding thermal stability; in particular, a dicationic phosphonium IL, namely the dicationic 1,12-di(tripropyl-phosphonium)dodecane bis(trifluoromethylsulfonyl)imide, was synthesized possessing a thermal stability of 425 °C [172]. In addition, it should be noted that dicationic and tricationic ILs exhibit significantly higher thermal stability compared to monocationic-based ILs [173].

At present, IL-coated capillary GC columns are commercially available under the trade name SLB-IL (Supelco–Sigma-Aldrich, Darmstadt, Germany) (Table 4). These columns are characterized by a different polarity and can work at temperatures passing 200 °C and approaching 300 °C [174]. Studies on the synthesis and properties of new IL-stationary phases are still continuing. Yu et al. [175] analyzed the application of a new triptycene-based amphiphilic material (TP-2IL) as the stationary phase for GC separations. In research, they showed that this column exhibited good performance for analytes from an apolar to a polar nature. Particularly, it has an outstanding capability for resolving critical pairs of anilines and phenols with good peak shapes, and shows distinct advantages over the typical conventional stationary phases. IL columns find many uses in the analysis of flavors and fragrances [176,177], fatty acids [178–182] and petrochemicals [183,184]. González Peredes et al. [185] evaluated different IL columns for the separation of chlorobenzenes and developed an analytical methodology based on the use of the IL stationary phase SLB-IL82 in GC with a microelectron capture detector for the determination of chlorobenzenes in soil samples.

Table 4. Selected commercial IL capillary GC columns with matrix active groups.

GC Capillary Column	Matrix Active Group
SLB-IL59	1,12-Di(tripropylphosphonium)dodecane bis(trifluoromethylsulfonyl)imide
SLB-IL60	1,12-Di(tripropylphosphonium)dodecane bis(trifluoromethylsulfonyl)imide
SLB-IL61	1,12-Di(tripropylphosphonium)dodecane bis(trifluoromethylsulfonyl)imide trifluoromethylsulfonate
SLB-IL76	Tri(tripropylphosphoniumhexanamido)triethylamine bis(trifluoromethylsulfonyl)imide
SLB-IL82	1,12-Di(2,3-dimethylimidazolium)dodecane bis(trifluoromethylsulfonyl)imide
SLB-IL110	1,9-Di(3-vinylimidazolium)nonane bis(trifluoromethylsulfonyl)imide
SLB-IL111	1,5-Di(2,3-dimethylimidazolium)pentane bis(trifluoromethylsulfonyl)imide
SLB-ILD3606	1,5-Di(2,3-dimethylimidazolium)pentane bis(trifluoromethylsulfonyl)imide

Do et al. [186] showed that the profiling of all 136 PCDD/Fs is greatly facilitated by using IL columns or combinations including such columns. Boczkaj et al. [187] tested three capillary columns (HP-5Ms, DB-624, SLB-IL 111) in the analysis of oxygenated volatile organic compounds (O-VOCs) in postoxidative effluents from the production of petroleum asphalt. Among the capillary columns investigated, a very polar column, SLB-IL 111, with an ionic liquid as the stationary phase was found to be superior for the separation of O-VOCs, as it has a high selectivity towards n-alkanes and oxygenated volatile organic compounds.

So far, IL stationary phases have not been widely applied in the field of bioanalysis [188]. Destaillats et al. [189] applied an IL-coated SLB-IL 111 column to identify the occurrence of petroselinic acid in human skin, hair and nails. They confirmed that this column can be used to obtain a baseline resolution between petroselinic acid and cis-8 18:1 acid methyl esters.

The use of IL-based stationary phases has extended to multidimensional gas chromatography (GCxGC), ensuring future applications. For example, Zapadlo et al. [190] investigated the use of GCxGC–TOFMS with highly polar IL-based columns for the analysis of polychlorobiphenyls (PCBs). They used a non-polar/ionic liquid column series consisting of poly (50%-n-octyl-50%-methyl) siloxane

(SPB-Octyl) and the ionic liquid SLB-IL59 in the first and second dimension, respectively. As a result, a total of 196 out of 209 PCBs congeners were resolved and all dioxin-like congeners were separated with no interferences from any PCB congener.

ILs have found another significant application as a solvent in headspace gas chromatography (HS-GC), ILs are ideal solvents for HS-GC, a more sensitive method of analysis compared to direct injection. HS-GC avoids direct liquid or solid probing and greatly decreases matrix interference [29]. ILs are excellent solvents and are now used for the analysis of residual solvents in a variety of pharmaceutical products. The detection and quantitation of residual solvents/impurities in drug substances or drug products is an important measure for pharmaceutical quality assurance/quality control, because the residual solvents/impurities that were not totally removed by practical manufacturing techniques always have a potential risk to human health from toxicity. Fink et al. [191] developed a rapid, accurate, IL-based HS-GC method for the determination of water in active pharmaceutical ingredients. The HS-GC method used an IL-based capillary GC column to increase the sensitivity and ruggedness of this method. ILs are also utilized as a headspace solvent. Studies have shown that the sensitivity of the HSGC method is 100 times greater than that of volumetric Karl Fischer titration (KFT) (which is the commonly used technique to determine water content), allowing very small sample sizes (e.g., 4 mg) to be accurately and reproducibly analyzed. In comparison, a typical sample size of 500–1000 mg is used in KFT. Liu and Jiang [192] applied [C_4MIM][BF_4] as the matrix medium in the analysis of six solvents utilized in the synthesis of Adefovir Dipivoxil: acetonitrile, dichloromethane, N-methyl-2-pyrrolidone ([NMPyrr]), toluene, dimethylformamide (DMF), n-butyl ether. The developed method proved accurate and linear (with $R \geq 0.9993$). All the RSDs were lower than 10%. Moreover, the comparison of [C_4MIM][BF_4] with DMSO as a matrix medium of headspace GC was also carried out in this study. In this research, it was indicated that DMSO, with its boiling point at 189 °C, has a higher vapor pressure, and the chromatographic peak of the DMSO matrix, with a much higher intensity, always occupies a wider baseline or interferes in the detection of analytes, especially at a higher equilibrium temperature. The impurities and the decomposed products of DMSO at a high equilibration temperature also became interfering substances for the detection of residual solvents.

A great analytical challenge for the pharmaceutical industry is the trace-level analysis of genotoxic impurities (GTIs) in drug substances. Ho et al. [193] used ILs (six compounds: ([C_4MPyrr][B(CN)4]), [C_4MIM][BF_4], [C_4MMIM][BF_4], [C_4MIM][Nf_2T], [C_4MMIM][Nf_2T] and ([$P_{6,6,6,14}{}^+$][Nf_2T])) as a new class of diluents for the analysis of two classes of genotoxic impurities (GTIs), namely, alkyl/aryl halides and nitro-aromatics, in small molecule drug substances by headspace gas chromatography coupled with electron capture detection (ECD) without the need for analyte derivatization. The low volatility and high thermal stability of ILs enables these compounds to be used at high headspace oven temperatures with the minimum chromatographic background. Studies have shown that increasing the headspace oven temperatures resulted in varying responses for alkyl/aryl halides and significant enhancements in the responses for all nitroaromatic GTIs. Furthermore, ILs with a conventional high-boiling organic diluent—DMSO, were compared. The chromatographic backgrounds from ILs are significantly lower than the backgrounds from DMSO. The LODs of all analytes obtained using the IL diluents were superior (5 to 500 ppb) to those obtained from pure DMSO. Research on organic solvent residues in drugs was also conducted by Ni et al. [194]. The main focus of this study was to investigate the relationship between analytes (organic solvents) and the matrix medium (ILs) by HS-GC in order to provide guidance in choosing a suitable matrix medium and next to determine the organic residual solvents in ketoconanzole to choose a suitable IL during the process of HS-GC. In research, [C_4MIM][PF_6] was chosen as the best headspace solvent, because an excellent separation of ethanol, dichloromethane, ethyl acetate, butyl alcohol, pyridine, DMF and DMSO was achieved. The evaluation of ILs for the analysis of residual solvents in pharmaceutical matrices was the subject of research by Laus et al. [195]. The authors chose ([C_4MIM][DMP]) as the most suitable ionic liquid as solvent for the HS-GC analysis of solvents with very low vapor pressure, such as dimethylsulfoxide,

N-methylpyrrolidone, sulfolane, tetralin, and ethylene glycol which can be found in pharmaceutical products. The limit of quantification (LOQ) of this method was from 59 (tetralin) to 113 μg/g (ethylene glycol), the accuracy was in the range of 96.6–103.7, and it showed high linearity in the tested range of 0.9890–0.9984. The developed method was applied for the detection of traces of sulfolane in a real sample of tablets containing the drug cefpodoxime proxetil. A trace of sulfolane was detected (estimated at 2 μg/g with respect to the tablet mass), which is safely below the regulatory limit of 160 μg/g. To the best of our knowledge, there are only a few reports on the use of ILs as stationary phases for GC separations in the analysis of environmental samples. One of them is the investigation performed by Reyes-Contreras et al. [171], who examined the suitability of ILs as stationary phases for GC–MS, and their application for the determination of nitrosamines and caffeine metabolites in wastewater samples. Studies have shown that the SLB-IL111 column enabled the baseline separation and quantification of 7 nitrosamines in a shorter analysis time compared with the commonly used cyanopropylphenyl polysiloxane. Furthermore, the SLB-IL59 column provided the elution of all the caffeine metabolites analyzed with the highest peak symmetry and an appropriate analysis time. Contaminants in wastewaters were also the subject of research by Domínguez et al. [169]. In this work, the application of IL stationary phases for the determination of benzothiazoles and benzotriazoles was examined. Among the IL columns evaluated, SLB-IL59 provided the total elution of all the target analytes with the highest peak symmetry and the lowest analysis time. Moreover, the lower stationary phase bleeding enabled positive identification and quantification.

In view of the properties of ILs, such as high thermal stability, high viscosity, and tunable selectivity through the modification of their chemical structure, their use, in particular as stationary phases in GC, will increase. New IL stationary phase chemistries that provide unique selectivity towards target analytes are needed to improve the separation performance and versatility of multidimensional GC [170].

4.2.2. Thin-Layer Chromatography

ILs are used in Thin-layer Chromatography (TLC) as stationary phase modifiers, especially in the separation of basic drug compounds. The separation of drug compounds is carried out normally with the use of silica-based stationary phases; however, it is often impossible because of the effect of free silanols on their chromatographic retention [196]. In order to remove this undesirable phenomenon, methods are used such as protonation, the addition of traditional amino quenchers, and changes in the mobile phase composition in order to increase ionic strength. The solution to this problem may be the application of ILs as silanol suppressing agents. The first study on improving separation in TLC by using ILs as mobile phase modifiers was presented by Kaliszan et al. [197]. The efficacy of using ILs as stationary phase modifiers in TLC has also been confirmed in research by Marszałł et al. [198]. The aim of this study was the application of imidazolium-based ILs to reduce the deleterious effects of free silanols on the LC separation of naphazoline nitrate. The authors used $[C_2MIM][BF_4]$ and $[C_6MIM][BF_4]$ as modifiers of the mobile phase. The results showed that ILs with short alkyl-chain lengths are efficient suppressors of free silanols, which are considered to be responsible for the troublesome and irreproducible chromatographic determinations of basic compounds. In the next study, Kaliszan et al. [199] also reported that ILs of the imidazolium tetrafluoroborate class when added to mobile phases blocked silanols and provided excellent TLC separations of strongly basic drugs which were otherwise not eluted, even with neat acetonitrile as the mobile phase. The ILs used by Marszałł et al. [198] as mobile phase modifiers were also tested in the studies reported by Mieszkowski et al. [196,200]. In the first study, 1-alkyl-3-methylimidazolium-based ILs (tetrafluoroborate $[C_2MIM][BF_4]$, L-(+)-lactate $[C_2MIM][LAC]$ and ethyl sulfate $[C_2MIM][ETOSO_3]$) were used as the mobile phase [196]. The subject of the research was the development of a new HPTLC method for the determination of perazine in oral tablets, and a comparative study between these three different ILs with the same cation but different counterions as additives to the mobile phase. In effect, among the selected ILs, the optimum distribution parameters, such as shape and quality of spots, high precision, and accuracy in qualitative and quantitative determination, characterize the system, with $[C_2MIM][BF_4]$ as the mobile phase modifier. Summarizing this study, it can be concluded that $[C_2MIM][BF_4]$ is a valuable and efficient suppressor of free silanols, which are responsible for unwanted interactions of chromatographic stationary phases in the determination of the above compounds. In the second study, the authors compared two TLC methods for the determination of haloperidol in oral drops—the pharmacopeia method (European Pharmacopeia 7.0) and an alternative with IL modifiers of the mobile phase. The addition of $[C_2MIM][BF_4]$ to the mobile phase gave similar separation and quantitative results with no peak tailing compared to the mobile phase suggested by the European Pharmacopeia 7.0 [200]. Besides the silanol-suppressing potency of $[C_2MIM][BF_4]$, a lack of interaction and interference with UV densitometric detection was observed. Research on the use of ILs in TLC was also conducted by Lu et al. [201], who used ILs as mobile and stationary phases of TLC to analyze berberine hydrochloride, tetrahydropalmatine and related Chinese patent medicine. In this study, the shape and value of target spots together with the developing duration were compared regarding four mobile phases which were a combination of the ILs ($[C_4MIM][OH]$), $[C_4MIM][BF_4]$, $[C_4MIM][Br]$, $[C_4MIM][PF_6]$) and methanol. Moreover, these IL mobile phases were compared with two traditional developing reagents, *n*-hexane-chloroform-methanol and *n*-butanol-acetic acid-water. As a result, it was found that $[C_4MIM][OH]$-methanol has a simpler composition and is more suitable for the simultaneous analysis of two target constituents in a plate. Besides any extra pH additives, the shape of spots was ideal and no tailing occured. $[C_4MIM][OH]$ was also used as the stationary phase, which was synthesized based on silica gel. The quantitative method for this kind of IL stationary phase showed a good correlation coefficient (R^2 = 0.9971–0.9976), good repeatability (%RSDs of berberine hydrochloride and tetrahydropalmatine were 0.88% and 0.79%, respectively) and method accuracy in terms of 95.91–104.85% (berberine hydrochloride) and 96.02–102.18% (tetrahydropalmatine). Research into the application of ILs as mobile phases in TLC was published by Tuzimski and Petruczynik [202]. The aim of the study was the separation of ten components of a mixture of isoquinoline alkaloids: allocryptopine, berberine, boldine, chelidonine, papaverine, emetine, columbamine, magnoflorine, palmatine and coptisine, using a 2D-TLC (two-dimensional TLC) method. The first dimension used an aqueous mobile phase (RP) (80% methanol–water–0.05 M/L−diethylamine), and in the second dimension a normal phase (NP) (75% methanol, 24.75% ethyl methyl ketone–0.25% IL $[C_4MIM][BF_4]$). The addition of ILs to conventional mobile phases caused a decrease in zone broadening and improved the chromatographic resolution. As shown in the results of the experiments, very symmetrical spots and peaks and high system efficiency were obtained. In conclusion, the authors proposed that mobile phase systems containing ionic liquids can be applied to the separation of isoquinoline alkaloids in other natural samples. The use of ILs as stationary phase modifiers can be an effective and more "green" alternative to classical mobile phases such as amines.

4.2.3. Supercritical Fluid Chromatography

Among the numerous applications of ILs, they can also be used in solvent systems composed of ILs and supercritical fluids with an emphasis on supercritical carbon dioxide ($scCO_2$). The specificity of IL–supercritical fluid biphasic systems follows from the availability of several mechanisms for tuning the solvent properties of such systems—apart from the wide selection of IL cations and IL anions to tailor the IL properties, the operating temperature and pressure are also available as variables to adjust the density and the solvent power of the supercritical fluid phase [203]. In an ILs-$scCO_2$ system the product recovery process is based on the principle that scCO2 is soluble in ILs, but ILs are not soluble in $scCO_2$. Since most organic compounds are soluble in $scCO_2$, with the high solubility of $scCO_2$ in ILs, these products are transferred from the IL to the supercritical phase [204]. Ji et al. [99] applied the IL $[C_8MIM][PF_6]$ and methanol as the extraction and dispersion solvents in a method for the determination of four NSAIDs—nabumetone, ibuprofen, naproxen and diclofenac—in tap water and drinks. The method was based on ultrasound-assisted ionic liquid dispersive liquid–liquid microextraction (US-ILDLLME) followed by ultra-high performance supercritical fluid chromatography

(UHPSFC) coupled to a photo-diode array detector (PDA). The developed method showed rapid separation (2.1 min), good recoveries (81.37–107.47%) and enrichment factors (126–132). The LODs for the analytes were from 0.62 (naproxen) to 7.69 (ibuprofen) ng/mL. This developed procedure was applied to real water samples, tap water, soda, lemon juice and green tea drink. In soda drink, ibuprofen was detected with detection levels of 16.43 ng/mL.

Because SFC can be performed with both polar and nonpolar stationary phases, columns that are marketed for HPLC can be used in SFC [205]. The application of immobilized ionic liquids (IILs) as a class of stationary phases for packed column SFC was studied by Smuts et al. [206]. The authors studied the cation and anion effect. The research was conducted on different IILs: tripropylphosphonium, tributylphosphonium, methylimidazolium, benzylimidazolium, triphenyl-phosphonium and 4,4′-bipyridyl while keeping the counteranion constant, and an immobilized tributylphosphonium with five different anions: acetate, trifluoroacetate (TFA), [Cl], perchlorate and [Nf_2T]. The best stationary phase in terms of low retention and good separation efficiency was the IIL tributylphosphonium with the TFA counter anion. Furthermore, the acetate anion exhibited the worst retention time and repeatability, and took the longest to reach baseline stability. [Nf_2T]$^-$ displayed poor efficiency in separations for tributylphosphonium-based stationary phases. Chou et al. [207] used covalently bonded 1-octyl-3-propylimidazolium chloride on a silica gel column for the simultaneous separation of acidic, basic and neutral compounds (fenoprofen, ibuprofen, acetaminophen, metoprolol, naphthalene and testosterone) using carbon dioxide subcritical/supercritical fluid chromatography. The data indicated that the IL-modified column, in terms of resolution, was clearly superior to commercial C18 columns. Also, the simultaneous separation of acidic, basic and neutral compounds via SFC was successful with a co-solvent content of 20% MeOH, a pressure of 110 bar, and a column temperature of 35 °C (Figure 9).

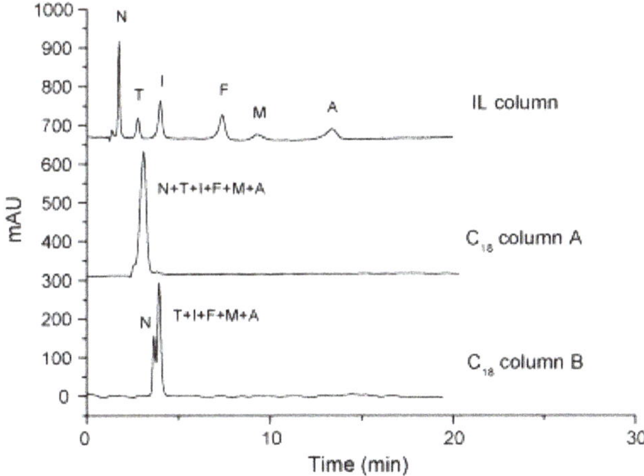

Figure 9. Separation of acidic, basic, and neutral compounds via SFC using the IL-modified column and a commercial C_{18} column. Figure adopted with copyright permission from [207].

In conclusion, it should be stated that ILs seem to be good replacements for volatile organic solvents, and the development of new applications utilizing ILs will increase. However, the high cost of ILs and lack of complete data on e.g., toxicity should be noted.

5. Electromigration Techniques

5.1. Capillary Electrophoresis

CE, which belongs to electromigration separation techniques, possesses many advantages, such as low sample and reagent consumption, high efficiency, simplicity, short analysis time, automation and inexpensive cost of capillaries in comparison to HPLC columns. CE separations are also extremely effective and allow substances with similar structures to be separated. These advantages mean that this technique has become an interesting alternative analytical tool to other chromatographic methods. Generally, CE analysis is carried out on fused-silica capillaries with silanol groups on the inner surface which are normally negatively charged. This results in the formation of an electroosmotic flow (EOF) that moves compounds toward the cathode when a voltage is applied across a tube filled with an electrolyte solution. Contrary to this effect, electrophoretic mobility exists, which moves a molecule to its opposite electrode. Each ion possesses a specific electrophoretic mobility resulting in a charge-to-mass ratio. However, the effect of EOF is generally predominant in respect to electrophoretic mobility, causing all the molecules to be moved at different speeds toward the cathode. A higher speed can be observed for cations, and neutral analytes take slightly longer to migrate, while negatively charged compounds take the longest to move because of their conflicting electrophoretic mobility. The fact that, simultaneously, both EOF and electrophoretic mobility occur, working on anions in opposite directions during electrophoretic separation, allows greater resolution to be obtained. The main parameters which can affect EOF mobility are the dielectric constant, the zeta potential value and the viscosity of the buffer. The values of these parameters can be regulated by the modification of the background electrolyte (BGE) and/or using different buffer additives, as well as when the physicochemical properties of the wall of the capillary are changed. ILs are considered as good EOF modifiers because of their good electrical conductivity and they are slightly more viscous than organic solvents. In effect, low IL concentrations can be enough for a significant improvement in the electrophoretic separation. According to the literature data, ILs have been applied as the BGE, as additives to the BGE and/or as covalent coating reagents of the capillary. However, taking into account the costs of these modifications, ILs were mainly used as electrolytes or additives to electrolytes to modify the capillary wall. It should be highlighted that both cations and anions of ILs may change the migration behavior of analytes, although the activity of IL cations have a major impact on the resolution in CE. The IL cations, by the modification of the ionic strength of the BGE, can change the EOF, which influences the migration times of the analytes and may improve separation efficiency. Other activity is related to the adsorption of IL cations on the capillary inner surface, which can reduce or even reverse the EOF as well as possibly correcting the peak tailing of some basic enantiomers. Both mechanisms mentioned above allow a significantly better resolution of analytes to be obtained [199,208].

For example, Qin et al. [209] used a 1-methylimidazolium-based IL for covalent bonding of the fused-silica capillary surface wall for reversing the EOF during the development of a CE-MS method for the determination of sildenafil (SL) and its metabolite UK-103,320 (UK) in human serum samples. The most effective separation was obtained with a BGE containing 10 mM of acetic acid (pH 4.5) and with a voltage of 25 kV. The sensitivity and resolution were significantly improved because this approach allowed the elimination of the adsorption of the compounds on the IL-coated capillary wall, which occurred on the bare fused-silica capillary wall. In effect, the analytes passed through the IL-coated capillary with a recovery of 98% and 100% for SL and UK, respectively. Moreover, the resolution between SL and UK was enhanced because of the modification of the EOF. The analytes were separated within 14 min with LODs of 14 and 17 ng/mL for SL and UK, respectively. El-Hady et al. [210] proposed a CE-UV method for the simultaneous determination of four anticancer drugs in human plasma and urine based on [C_4MIM][Br] as a component of the BGE. During the study, the parameters of CE separation were optimized. The best results were obtained when the analysis was carried out on a BGE containing a 12.5 mmol/L phosphate buffer at pH 7.4 and 0.1 μmol/L of [C_4MIM][Br] (IL), and 20 kV applied voltage. This approach allowed sensitivity to be increased 600 times over that observed in CE performed without the IL. The developed CE-UV method for the quantification of methotrexate, vinblastine, chlorambucil and dacarbazine in human plasma and urine allowed the analytes to be monitored with the LODs in the range of 0.01 to 0.05 μg/mL.

It should also be noted that excellent separation is particularly required for the analysis of racemic mixtures, including various groups of pharmaceuticals the enantiomers of which can possess significant different pharmacokinetic and pharmacodynamic properties and side effect profiles. The qualitative and quantitative analysis of the compounds in biological and environmental samples is necessary for better understanding the mechanism of their activity in live organisms and their influence on the environment. This issue was a predominant topic of many papers published in recent years in world scientific literature. In those studies, both achiral ILs and chiral ILs (CILs) were applied in combination with various types of chiral selectors (CS) like cyclodextrins (CDs) or their derivatives, antibiotics, polysaccharides or surfactants for the chiral separation of different pharmaceuticals. Typical achiral ILs applied in CE enantioseparation were tetraalkylammonium ILs, alkylimidazolium ILs and alkylpyridinium ILs with inorganic anions such as [OH], [Cl], [Br], [BF_4] and [PF_6]. Among them, tetraalkylammonium-based ILs are considered as more effective because of their relatively more hydrophilic character, which decreases the likelihood of entering the hydrophobic cavity of the CS. Moreover, their relatively lower conductivity and UV transparency in the wavelength ranges applied for enantiomer detection allow them to be used in higher concentration levels. These data are in accordance with the study reported by Huang et al. [211] who tested alkylpyridinium, tetraalkylammonium and alkylimidazolium-based ILs along with β-CDs for the chiral separation of five β-agonists. The results confirmed that tetraalkylammonium-based ILs were more effective because they could be used at much higher levels than the other tested ILs. Poor resolution was achieved when the long-chain IL, [C_8MPyrr][PF_6], was used as the BGE modifier. Moreover, the presence of ILs was required for the full enantioseparation of salbutamol, cimaterol and formoterol, which were not resolved using the BGE containing only β-CD as the CS.

Jiang et al. [212] used [C_2MIM][BF_4] for the coating of a silica capillary during the enantioseparation of ibuprofen, fenoprofen, naproxen and ketoprofen. It enabled the EOF to be modified, which provided the effective resolution of the enantiomers. The tested IL not only affected the EOF but also acted as a discriminator. Moreover, the interaction between hydrogen at the C–2 carbon of the IL and the acid drugs played an important role in the separation. The same type of IL was selected for the enantiorecognition of nine tricyclic antidepressants in the study reported by Tsai et al. [213]. The optimal simultaneous separation of all the tested pairs of enantiomers was achieved with 50 mM of [C_2MIM][BF_4] as the sole BGE at pH 3. Zhao et al. [214] used three ILs and hydroxypropyl-β-cyclodextrin (HP-β-CD) as the components of the BGE for the enantioseparation of itraconazole, ketoconazole, econazole and miconazole. Compared with [C_2MIM][L-lactate] or [C_2MPyrr][BF_4], [C_{12}MAmm][Cl] was the most effective. When this reagent was used along with HP-β-CD it allowed the resolutions of 3.8, 3.5, 2.8 and 2.5 for miconazole, econazole, ketoconazole and itraconazole, respectively, to be obtained.

In the paper published by Liu et al. [215], the effective chiral separation of racemic methyl-ephedrine hydrochloride, thebaine, codeine phosphate and acetylcodeine by capillary electrophoresis with electrochemical detection (CE-ECL) was observed when 0.6% [C_4MIM][BF_4] as the component of the BGE was applied (Figure 10).

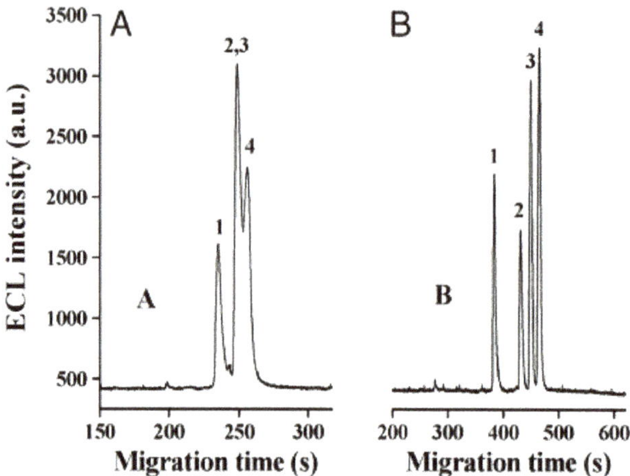

Figure 10. Electropherograms of four standard samples: (**A**) without IL in electrophoretic buffer; (**B**) with the use of 0.6% $BIMPF_4$ in the electrophoretic buffer. Peak: 1, 10 μmol/L methylephedrine hydrochloride; 2, 40 μmol/L of thebaine; 3, 25 μmol/L codeine phosphate; 4, 15 μmol/L acetylcodeine. Conditions: electrophoretic buffer, 14 mmol/L phosphate–borax at pH 7.4; electrokinetic injection, 10 s × 10 kV; separation voltage, 15 kV; detection potential, 1.2 V; ECL solution, 5 mmol/L $Ru(bpy)_3^{2+}$ with 50 mmol/L PBS at pH 8.2. Figure adopted from [215] with permission.

The developed method offered the quantification of four drug alkaloids in human urine samples with LODs from 1.4×10^{-7} to 6.3×10^{-8} mol/L. Jin et al. [216] reported the effective enantioseparation of propranolol, oxprenolol and pindolol by CE when a BGE containing the achiral IL—glycidyltrimethylammonium chloride ([GTMAmm][Cl]) as the modifier along with a dual CDs system based on 2,6-di-O-methyl-β-cyclodextrin (DM-β-CD) and 2,3,6-tri-O-methyl-β-cyclodextrin (TM-β-CD) was applied. The authors also used an on-line sample enrichment technique based on field-enhanced sample injection (FESI) for the improvement of sensitivity. The application of both approaches allowed the LODs of the enantiomers to be decreased from 0.10 to 0.65 nM. Finally, the developed CE method was successfully used for the analysis of spiked urine samples, with good recoveries.

Unfortunately, in many cases, the application of achiral ILs with a single chiral selector was not enough for the effective enantioseparation of the compounds of interest. An interesting alternative approach reported in the literature was using CILs which can possess either a chiral cation or achiral anion, or both. The application of these CILs in combination with traditional chiral selectors allows an extra "enantiorecognition" capability to be obtained while the capability of system modification is retained. In effect, a "synergistic system" occurs during electrophoretic separation, which can significantly improve the resolution of the analytes. The first paper reporting the use of this approach for the enantioseparation of pharmaceuticals was published by François et al. [217]. The authors developed and used two chiral choline-based ILs—ethylcholine bis(trifluoromethylsulfonyl)imide ([EtChol][Nf_2T]) and phenylcholine bis(trifluoromethylsulfonyl)imide ([PhChol][Nf_2T]) alone or in combination with DM-β-CD or TM-β-CD for the analysis of the anti-inflammatory drugs, 2-arylpropionic acids, as model compounds. The developed CILs were applied as BGE additives, chiral ligands and CSs. Moreover, the enantioseparation efficiency in respect to the type and concentrations of tested CILs and CDs, as well as the methanol addition to the BGE, were evaluated. The results indicated that the effective separation of the analytes was achieved only upon adding one of the CILs containing DM-β-CD or TM-β-CD and methanol to the BGE. Thus, the synergistic effect between the tested chiral choline-based ILs and CDs in the dual separation system was confirmed. In another study, two chiral synergistic systems based on tetramethylammonium-L-arginine (TMA-L-Arg)/glycogen and tetramethylammonium-L-aspartic acid

(TMA-L-Asp)/glycogen were compared with the system containing achiral tetramethylammonium hydroxide (TMA-OH)/glycogen for the chiral separation of nefopam, citalopram and duloxetine [218]. Each tested IL/glycogen synergistic system gave better resolutions of the tested enantiomers compared to those observed for the separation using glycogen alone. However, the addition of TMA-L-Arg to the BGE composition was more effective than TMA-L-Asp, while the TMA-OH/glycogen separation system gave poorer resolution. Zhang et al. [219] tested tetramethylammonium-L-arginine (TMA-L-Arg), tetramethyl-ammonium-L-hydroxyproline (TMA-L-Hyp) and tetramethylammonium-L-isoleucine (TMA-L-Ile) as BGE additives in combination with HP-β-CD for the enantioseparation of amlodipine, nefopam, duloxetine and propranolol. The highest signals of the tested analytes and the best resolution was achieved using a 40 mM Tris/H_3PO_4 buffer solution (pH 2.6) containing 20 mM of HP-β-CD and 30 mM of TMA-L-Arg (Figure 11). Zuo et al. [220] reported the enantioseparation of twelve pharmaceuticals using 1-ethyl-3-methylimidazolium-L-lactate ([C_2MIM][L-lactate]) and 1-butyl-3-methylimidazolium-L-lactate ([C_4MIM][L-lactate]) in combination with β-CD in a BGE. The resolution was better in a dual system based on one of the tested CILs and β-CD compared to the β-CD alone, although the addition of [C_2MIM][L-lactate] was more effective. Finally, the BGE composed of 20 mM of [C_2MIM][L-lactate] and 10 mM of β-CD at pH 2.5 was selected as optimal for the separation of most analytes.

Only the analysis of homatropine methylbromide was carried out on 30 mM of Tris-H_3PO_4 at pH 2.0 (more effective separation), while the enantiomers of venlafaxine and sibutramine were not baseline resolved. Kolobova et al. [221] confirmed that 1-butyl-3-methylimidazolium L-prolinate [C_4MIM][L-Pro] as a CS in combination with 2-hydroxypropyl-β-cyclodextrin (2HP-β-CD) allowed a significant improvement in the chiral separation of carvedilol and propranolol.

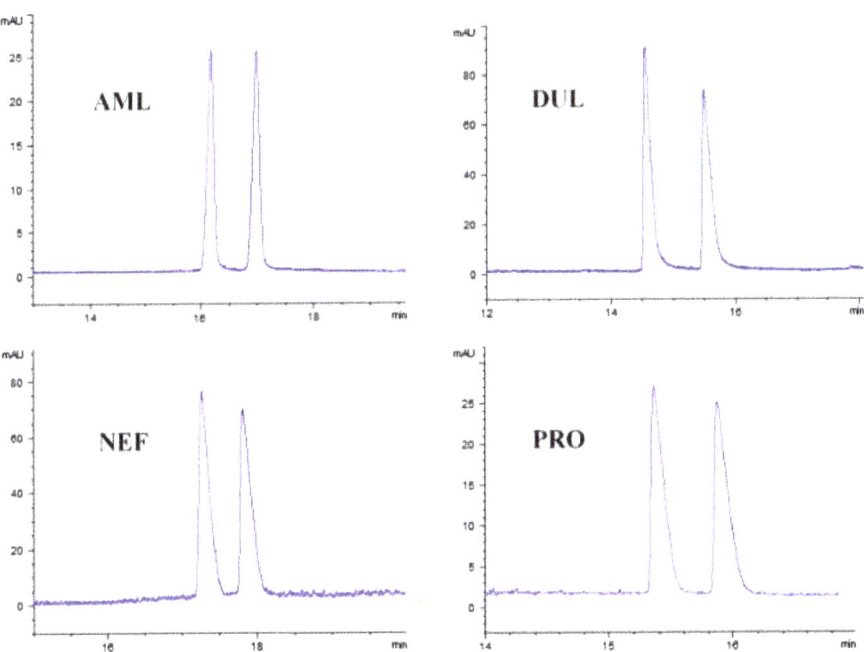

Figure 11. Chiral separation of all drug enantiomers in the optimized HP-β-CD/TMA-L-Arg synergic system. Conditions: focused-silica capillary, 50 cm (41.5 cm effective length) × 50 μm i.d; applied voltage, 20 kV; capillary temperature, 15 °C; BGE, 40 mM Tris/H_3PO_4 buffer solution (Ph 2.6) containing 20 mM HP-β-CD/TMA-L-Arg. Figure adopted from [219] with copyright permission.

Zhang et al. [222] designed a lactobionic acid LA-based IL, namely tetramethylammonium-lactobionate (TMA-LA), which was applied for the chiral separation of atenolol, metoprolol, propranolol, nefopam and duloxetine. In the study, three combinations, namely a single LA system, β-TMA chloride (TMA-Cl) system and TMA-LA IL system, were tested. The best results were achieved when the IL TMA-LA as the CS was applied. Finally, the BGE containing 40 mM of borax buffer, pH 7.6, 40% *v/v* methanol, 200 mM of TMA-LA and 20 kV applied voltage was selected as the most effective. Zhang et al. [223] tested L-alanine tert-butyl ester bis (trifluoromethane) sulfonamide (L-AlaC$_4$Nf$_2$T) and L-valine *tert*-butyl ester bis (trifluoromethane) sulfonamide (L-ValC$_4$Nf$_2$T) as additives to the BGE in combination with M-β-CD, HP-β-CD and glucose-β-CDs (Glu-β-CD) for the enantioseparation of naproxen, pranoprofen and warfarin. Compared to CDs alone, significantly better chiral recognitions of all analytes were obtained, although the resolutions of these dual systems were different. Moreover, the addition of organic modifiers to the BGE additionally improved selectivity. This was probably related to decreasing the EOF, which allowed interactions to be increased between AAILs, M-β-CD and the racemates. The best separations of the analytes were observed when 15 mM of CILs was introduced into the 30 mM sodium citrate/citric acid buffer solution at pH 5.0 containing 20 mM of M-β-CD and 20% ethanol as the organic modifier with a 20 kV applied voltage. The potential synergistic effects of L-AlaC$_4$Nf$_2$T and L-ValC$_4$Nf$_2$T were also checked in combination with vancomycin during the enantioseparation of naproxen, carprofen, ibuprofen, ketoprofen and pranoprofen [224]. Both dual synergic separation systems were also able more effectively to separate the enantiomers compared to the vancomycin-alone case. Xu et al. [225] applied tetramethylammonium-L-hydroxyproline (TMA-L-Hyp) with clindamycin phosphate (CP) for the separation of a racemic mixture of propranolol, nefopam, citalopram and chlorphenamine. The authors optimized the electrophoretic conditions in terms of the BGE composition, pH, voltage, temperature and UV parameters. The best results were obtained when the CE separation was carried out on an uncoated fused-silica capillary (50 cm total and 41.5 cm effective length × 50 μm i.d.) with a 40 mM borax buffer (pH 7.6) containing 80 mM of CP and 30 mM of TMA-L-Hyp and methanol (20% *v/v*). A voltage of 20 kV and a temperature of 20 °C were used. Nefopam, citalopram, chlorphenamine and propranolol were monitored at 289, 230, 265 and 237 nm, respectively. AAILs based on a tetramethylammonium cation were also tested with maltodextrin for the enantio-separation of pharmaceuticals belonging to different classes. For example, Yang et al. [226] used tetramethylammonium-D-pantothenate (TMA-D-PAN) and tetramethylammonium-D-quinate (TMA-D-QUI) as additives to the maltodextrin-based synergistic systems in a CE method developed for the analysis of racemic mixtures of nefopam, ketoconazole, econazole and voriconazole. For both of the CIL/maltodextrin systems, significantly improved R_s were observed for all the tested enantiomers, although TMA-D-PAN offered better separation results. This synergistic effect was probably related to a decrease in the density of the negative charge as an effect of the adsorption of the CIL cations on the surface of the capillary. This caused increasing complexation between the racemates and the CIL, which improved the resolution for all analytes.

Chen et al. [227] used tetramethylammonium-L-arginine (TMA-L-Arg) and tetramethyl-ammonium-L-aspartic acid (TMA-L-Asp) in combination with maltodextrin for the enantioseparation of nefopam, citalopram, cetirizine, duloxetine and ketoconazole. The most effective chiral separation was observed when a BGE composed of 60 mM of TMA-L-Arg, 7.0% maltodextrin in 50 mM of Tris-H_3PO_4 (pH 3.0) and with a voltage of 18.0 kV was applied. Zhang et al. compared the separation systems based on 1-butyl-3-methylimidazolium (T-4)-bis[(2S)-2-(hydroxy-κO)-3-methylbutanoato-κO]borate ([C$_4$MIM][BLHvB]) and 1-butyl-3-methylimidazolium (T-4)-bis[(αS)-α-(hydroxy-κO)-4-methylbenzeneacetato-κO]borate ([C$_4$MIM][BSMB]) along with HP-β-CD [228] as well as dextrin [229] as the CS in CE enantioseparations. In both studies, the addition of the CIL enabled the synergistic effect to occur between them and the used CS, which allowed better resolutions to be obtained and higher peak efficiencies compared to those calculated for the HP-β-CD or the dextrin alone. On the other hand, [C$_4$MIM][BLHvB] was more effective than

$[C_4MIM][BSMB]$. This was probably related to the structure of the $[C_4MIM][BSMB]$ anion whose aromatic ring substituent could disturb chiral recognition.

An interesting approach was presented by Zhang et al. [230] who employed IL-dispersed NPs as buffer modifiers for the chiral separation of laudanosine, propranolol, amlodipine, citalopram and nefopam in CE. In the study, $[C_4MIM]BF_4]$, ($[C_4MIM][PF_6]$), 1-dodecyl-3-methylimidazolium chloride ($[C_{12}MIM][Cl]$) and 1-aminoethyl-3-methylimidazolium bromide ($[C_2NH_2MIM][Br]$) ILs were dispersed in multi-walled carbon nanotubes (ILs-MWNTs) and applied as the BGE modifier in combination with chondroitin sulfate E (CSE), as the CS. The obtained results indicated that significantly better separation, selectivity and peak shapes were achieved in the ILs-MWNTs modified system compared to that observed in CSE alone. The parameters affecting the electrophoretic separation were also investigated and optimized. The best results were obtained when CE was carried out on a 20 mM Tris/H_3PO_4 buffer solution containing 2.5% CSE and 2.4 μg/mL of ILs-MWNTs at pH 2.8–3.4 and with 15 kV applied voltage.

It should be highlighted that most of the studies described above indicated that the application of CILs alone as BGE modifiers was not able to effectively to separate the enantiomers. However, in the literature there are also a few reports describing the synthesis of novel CIL structures the activity of which was enough to achieve full resolution of drug enantiomers. For example, Yu et al. [231] synthesized a β-CDs-based CIL, 6-O-2-hydroxypropyltrimethylammonium-β-cyclodextrin tetra-fluoroborate ([HPTMA-β-CD][BF_4]), and used it as a CS for the enantioseparation of eight pairs of drug enantiomers. The novel CIL offered higher solubility of the analytes in the BGE and gave better stabilization of reversed EOF in CE compared to the parent β-CDs, which allowed a higher intensity of the signals and a more effective resolution to be obtained. The results confirmed that the enantiomers of chlorpheniramine, brompheniramine, promethazine, liarozole, tropicamide, warfarin, pheniramine and bifonazole were more effectively separated with [HPTMA-β-CD][BF_4] as the CS than with β-CDs. Recently, a report describing the synthesis of mono-6-deoxy-6-(3-methylimidazolium)-β-cyclodextrin tosylate (β-CDMIMOTs) CIL was also published by Zhou et al. [63]. The authors applied this new CIL as a coating material to modify the EOF in the CE method for the enantioseparation of oxytetracycline, tetracycline, chlortetracycline and doxycycline in environmental samples. The researchers achieved good separation of the analytes due to the multiple functions of β-CD-IL, which enabled the tetracyclines to be entrapped to form an inclusion complex (Figure 12).

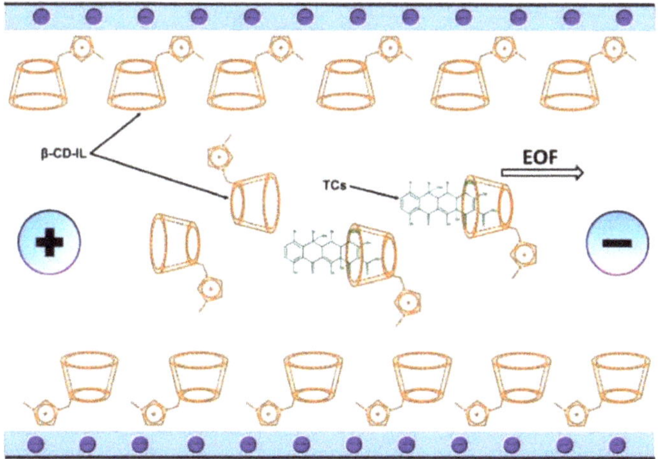

Figure 12. Mechanism of separation of four TCs using β-CD–IL as dynamic coating material. Figure adopted with permission from [63].

Compared to β-CD alone, β-CD-IL offered better solubility in an aqueous buffer. A stable suppressed EOF in the capillary was also generated as the effect of the occurrence of hydrogen bonding and the electrostatic interaction with the capillary inner wall. The authors selected the best CE conditions for tetracycline separation, which were achieved when a BGE composed of 10 mmol/L, a pH 7.2 phosphate buffer and 20 mmol/L of β-CD-IL and electrochemical detection at 1 V was used. The developed CE method allowed the compounds of interest to be monitored in environmental water samples with LODs from 0.33 to 0.67 μmol/L.

5.2. Micellar Electrokinetic Chromatography

Considered as a mode of CE, micellar electrokinetic chromatography (MEKC) allows both neutral and charged analytes to be separated. In MEKC, the surfactant monomers are added to the run separation buffer above the critical micelle concentration (CMC), which allows aggregates called micelles to form as a pseudostationary phase. The separation process is based on differences between the analytes partitioning in a micellar stationary phase, and is related to the electrophoretic mobility of the compounds. Therefore, the neutral and hydrophobic analytes incorporated into the micelles gain an apparent electrophoretic mobility and will move at the same velocity as the micelle under electrophoretic conditions. This allows the neutral and charged compounds with the same charge-to-mass ratio to be separated because the migration time in MEKC is dependent on the electrophoretic velocity of the micelle, the distribution ratio and the EOF velocity. The use of additional BGE modifiers can increase efficiency and selectivity. ILs as BGE additions have become interesting alternatives because the long-chain part of the AAILs can act as a surfactant to form a micelle in the BGE when the level of ILs exceeds the CMC. Moreover, the electrostatic interaction between the acidic analyte and the cationic micelle (AAILs) offered a more effective enantiorecognition of the analytes. Higher concentrations of ILs may also be used compared to organic solvent surfactants because of higher conductivity, hydrophobicity and solvation, which decreases the risk of destroying the micellar system in MEKC. In the literature, there are a few papers reporting the use of ILs in MECK. For example, Wang et al. published two consecutive papers [232,233] demonstrating the combination of TM-β-CD with N-undecenoxycarbonyl-L-leucinol bromide (L-UCLB) CIL as a dual chiral selector for the enantiodiscrimination of fenoprofen, indoprofen, ketoprofen, suprofen and ibuprofen. In the study, different levels of CILs and TM-β-CD were tested. The results indicated that TM-β-CD alone could not resolve the enantioseparation of the racemates, whereas the addition of L-UCLB at a concentration of 1.5 to 2.0 mM to the BGE with TM-β-CD provided an excellent resolution. This was related to the competitive inhibition of the interaction between the CIL and the capillary wall in the presence of TM-β-CD. Cui et al. [234] used L-ethyl-3-methylimidazolium-L-lactate, [C_2MIM][L-lactate] and 1-ethyl-3-methylimidazolium-L-(β)-lactate [C_2MIM][DL-lactate] alone or in combination with HP-β-CD for the chiral resolution of ten analytes belonging to different classes of pharmaceuticals. The results confirmed that the best enantiorecognition was obtained when a BGE composed of 40 mM of HP-β-CD, 50 mM of NaH_2PO_4-H_3PO_4, pH 2.75, and 30 mM of [C_2MIM][L-lactate] was used during the enantiomeric separation. Moreover, this effect was mainly correlated with the cationic activity of the IL, which played an important role in the increased resolution, whereas the anionic part of the CIL possessed a low influence on the chirality and nature of the enantioseparation. Su et al. [235] tested the addition of [C_4MIM][Cl], [C_4MIM][PF_6], [C_4MIM][Nf_2T] and SDS as modifiers in the BGE during the optimization of MEKC conditions for the separation of seven benzodiazepines. The results confirmed that the BGE containing 170 mM of [C_4MIM][Nf_2T] and 10 mM of SDS offered the most effective selectivity and resolution of the compounds of interest. This was related to different degrees of association of the tested analytes, which gave a more satisfactory separation compared to the results observed using the IL or SDS alone. The anionic moiety of [C_4MIM][Nf_2T] probably played a dominant function during the separation process as a heteroassociation site for the benzodiazepines, while the SDS improved the resolution. The developed MEKC method allowed the analytes to be detected in human urine samples with LODs in the range of 2.74 to 4.42 μg/mL.

5.3. Non-Aqueous Capillary Electrophoresis

In recent years, non-aqueous capillary electrophoresis (NACE) has become an interesting separation technique because it allows the detection of water-insoluble analytes which cannot be measured in traditional aqueous CE. Additionally, the analysis time in NACE can be shortened because of the lower viscosity of the buffer solution and the higher EOF as well as the reduction of the electrophoretic current. Moreover, the application of organic solvents allows the analytes to be detected online by MS. As it was earlier mentioned, ILs possess some advantages over conventional organic solvent modifiers, such as good conductivity. Hence, using ILs in NACE can give a better separation effect. These possibilities were confirmed by Ma et al. [236] who applied an ephedrine-based CIL as the CS for the enantiomeric resolution of omeprazole and rabeprazole by NACE. A reversed EOF (anodic flow), probably caused by the adsorption of the cations onto the capillary wall, was observed when (+)-N,N-dimethylephedrinium-bis(trifluoromethanesulfon)imidate ($[DMP]^+[Nf_2T]^-$) was added to the BGE. The best resolution was achieved with the BGE containing an acetonitrile-methanol mixture (60:40, *v/v*) and 60 mM of $[DMP]^+[Nf_2T]^-$. The authors found that the enantioseparation was related to ion-pair interactions dependent on equilibrium constants between the negatively charged enantiomers and DMP cations. Moreover, hydrogen-bonding between the hydroxyl group of DMP^+ and the sulfoxide group of the analytes as well as π–π and dipole–dipole interactions were responsible for the separation mechanism.

Summarizing, the application of ILs in electromigration techniques offers new opportunities to solve many analytical problems in the separation field. One of them is the chiral recognition of racemic mixtures of pharmaceuticals having different chemical structures and biological activity. The results of numerous studies based on drug standards confirmed the great potential of ILs in CE applications. On the other hand, there are relatively few reports describing the separations of drugs in real biological and environmental samples. This seems to be caused by the relatively low sensitivity of CE-based methods compared to LC and GC techniques, which may be not enough for many pharmaceutical, clinical and environmental applications. On the other hand, lower LOD values can be obtained in electromigration techniques supported by ILs, which allows a partial resolution for this analytical problem. Moreover, intensive progress is continuing systematically in developing new approaches for improving sensitivity in electromigration techniques based on techniques such as field-enhanced sample injection (FESI), field-amplified sample injection (FASI), field-amplified sample stacking (FASS) or a combination of simultaneous electrokinetic and hydrodynamic injection (SEHI) and field-enhanced sample injection in conjunction with a sweeping technique known as sequential stacking featuring sweeping (SSFS) [237,238]. Probably, when scientists apply both ILs and new technical resolutions in CE, it will allow the required sensitivity to be obtained for clinical and environmental studies. These studies are very important because both CE-based techniques and ILs are environmentally-friendly, so connecting them in one analytical tool could be an important factor supporting the protection of nature.

6. Current Trends and Future Perspectives

Pharmaceuticals possess high biological activity and they can take part in various types of interactions, which means that these substances have a huge influence on the functioning of both live organisms as well as whole ecosystems. Therefore, as it was mentioned in Section 1, it is very important to develop sensitive, selective, accurate and precise methods for reliable drug determination in biological and environmental samples. An interesting approach is the application of ILs during method development. According to the data presented in this review, there are several interesting trends in the application of ILs for the determination of pharmaceuticals. First of all, ILs are most often applied at the stage of sample preparation (Table 2). The vast majority of studies concerned the extraction (or actually microextraction) of biological and environmental samples. Moreover, the most common type of analyte extraction from both these matrices was DLLME. Researchers pay a lot of attention to improving these methods by introducing modifications using physical and chemical factors. As a result, they promote the development of environmentally-friendly solutions in the field of analytical chemistry and the improvement of validation parameters. Unfortunately, it should be noted that despite the development of various IL-based methods, the majority of procedures are still supported by organic solvents. In DLLME, their basic function is the dispersion of ionic liquids. In turn, because of the high viscosity of ILs, sample detection is only possible after dissolving the sample in MeOH, ACN and others. Thus, the application of ILs leads to improved validation parameters, but the developed methods are not completely eco-friendly. The results prove that despite moving in the right direction, this area requires further development. Improving the results is possible not only by proper sample preparation, but also by the application of ILs in chromatographic and electrophoretic techniques. The addition of ILs to mobile phases is the main way of using them for the determination of pharmaceutical drugs by chromatographic techniques. As the results show, the suppression of the interaction of silanol by use of ILs is a huge advance in the problematic analysis of basic drugs (Table 3). The use of ILs in the BGE in electrophoretic techniques, which in many respects are compatible with green chemistry, although their sensitivity still remains a challenging task for the analyst, seems to be promising. It may be surprising that despite the existence of commercially available and described methods for the self-preparation of IL-based chromatographic columns and capillaries for electrophoresis, such methods of their use is very rare for pharmaceuticals. If the huge potential of ILs is to be discovered, it should also be noted that in addition to the above detection methods, researchers are trying to use them with other chromatographic techniques. Although such applications are not yet widespread in the analysis of pharmaceuticals, their dynamic development may cause such experiments to be performed in the future. In addition to trends in the design of analytical methods, the qualification of ILs with similar structures to a specific stage of analysis is the constant rule. In many works, optimization concerns the selection of a specific IL from a large diverse group of IL molecules. However, according to the data presented in different reports, the final optimization effect leads to the selection of the same IL. For example, an IL with hydrophobic properties was sought for liquid-phase extraction and the best results were often achieved for the imidazolium cation and anion [PF_6]. In turn, as an addition to the mobile phase, the selection of the IL [PF_6] was not suitable due to too strong adsorption on the column and was replaced by [BF_4]. It must be highlighted that these are trends for most, but not all papers (detailed in Tables 2 and 3). However, the fact is that despite access to a vast amount of ILs, only a few have been tested in experiments, and the final selection focuses on a small number. As mentioned, the samples are analyzed by various chromatographic and electrophoretic techniques, while a UV detector is almost always used for analyte detection, rarely FL and almost never MS/MS.

The above examples confirm that there are no ideal solutions in the design of analytical methods for the determination of pharmaceuticals in biological and environmental samples. However, in the case of ILs, their advantages over disadvantages and also the incomplete data on them prove the need for continuous interest and development in this area.

7. Conclusions

ILs as molecules with unique properties have been the subject of increased interest in recent years. Undoubtedly, the key issue is "green chemistry", which has set the direction of current research. Due to their huge potential, it is natural to use ILs in the search for solutions to many problems in modern laboratories, including their participation in analyzing pharmaceuticals in real samples. The use of analytical methods at various stages confirms the universality and enormous potential of this "solvent design". The application of various chromatographic and electrophoretic techniques and extraction methods together with the possibility of the use of ILs for a wide range of analytes prove that their contribution to the development of analytical methods is not overestimated. At the same time, the limitations that appear during their use show that success in experiments is not easy and this field of research requires further development.

Funding: The study was supported by the project POWR.03.02.00-00-I014/17-00 co-financed by the European Union through the European Social Fund under the Operational Programme Knowledge Education Development 2014–2020. This research was funded by the National Science Centre in Poland, project No. 2017/01/X/ST4/00225 and by the National Centre for Research and Development in Poland within V4-Korea Joint Research Program, project MTB No. DZP/V4-Korea-I/20/2018.

Conflicts of Interest: The authors declare that there are no conflict of interest.

References

1. Ma, W.; Row, K.H. Simultaneous determination of levofloxacin and ciprofloxacin in human urine by ionic-liquid-based, dual-template molecularly imprinted coated graphene oxide monolithic solid-phase extraction. *J. Sep. Sci.* **2019**, *42*, 642–649. [CrossRef]
2. Gorynski, K. A critical review of solid-phase microextraction applied in drugs of abuse determinations and potential applications for targeted doping testing. *TrAC Trends Anal. Chem.* **2019**, *112*, 135–146. [CrossRef]
3. Boyacı, E.; Rodríguez-Lafuente, A.; Gorynski, K.; Mirnaghia, F.; Souza-Silva, E.A.; Heind, D.; Pawliszyn, J. Sample preparation with solid phase microextraction and exhaustive extraction approaches: Comparison for challenging cases. *Anal. Chim. Acta* **2015**, *873*, 14–30. [CrossRef]
4. Kokosa, J.M. Recent trends in using single-drop microextraction and related techniques in green analytical methods. *TrAC Trends Anal. Chem.* **2015**, *71*, 194–204. [CrossRef]
5. Yamini, Y.; Safari, M.; Shamsayei, M. Simultaneous determination of steroid drugs in the ointment via magnetic solid phase extraction followed by HPLC-UV. *J. Pharm. Anal.* **2018**, *8*, 250–257. [CrossRef] [PubMed]
6. Carda-Broch, S.; García-Alvarez-Coque, M.C.; Ruiz-Angel, M.J. Extent of the influence of phosphate buffer and ionic liquids on the reduction of the silanol effect in a C18 stationary phase. *J. Chromatogr. A* **2018**, *1559*, 112–117. [CrossRef] [PubMed]
7. Peris-García, E.; Alvarez-Coque, M.C.G.; Carda-Broch, S.; Ruiz-Angel, M.J. Effect of buffer nature and concentration on the chromatographic performance of basic compounds in the absence and presence of 1-hexyl-3-methylimidazolium chloride. *J. Chromatogr. A* **2019**, *1602*, 397–408. [CrossRef]
8. Ren, Y.M.; Yang, R.C. Efficient method for the synthesis of 3-arylbenzoquinoline, pyranoquinoline, and thiopyranoquinoline derivatives using PEG1000-based dicationic acidic ionic liquid as recyclable catalyst via a one-pot multicomponent reaction. *Synth. Commun.* **2016**, *46*, 1318–1325. [CrossRef]
9. Halayqa, M.; Zawadzki, M.; Domańska, U.; Plichta, A. Polymer—Ionic liquid—Pharmaceutical conjugates as drug delivery systems. *J. Mol. Struct.* **2019**, *1180*, 573–584. [CrossRef]
10. Abbott, A.P.; Frisch, G.; Ryder, K.S. Electroplating Using Ionic Liquids. *Annu. Rev. Mater. Res.* **2013**, *43*, 335–358. [CrossRef]
11. Pei, Y.C.; Wang, J.J.; Xuan, X.P.; Fan, J.; Fan, M. Factors Affecting Ionic Liquids Based Removal of Anionic Dyes from Water. *Environ. Sci. Technol.* **2007**, *41*, 5090–5095. [CrossRef] [PubMed]
12. Kobylis, P.; Lis, H.; Stepnowski, P.; Caban, M. Spectroscopic verification of ionic matrices for MALDI analysis. *J. Mol. Liq.* **2019**, *284*, 328–342. [CrossRef]
13. Shekaari, H.; Zafarani-Moattar, M.T.; Mirheydari, S.N. Conductometric analysis of 1-butyl-3-methylimidazolium ibuprofenate as an active pharmaceutical ingredient ionic liquid (API-IL) in the aqueous amino acids solutions. *J. Chem. Thermodyn.* **2016**, *103*, 165–175. [CrossRef]
14. Nacham, O.; Ho, T.D.; Anderson, J.L.; Webster, G.K. Use of ionic liquids as headspace gas chromatography diluents for the analysis of residual solvents in pharmaceuticals. *J. Pharm. Biomed. Anal.* **2017**, *145*, 879–886. [CrossRef]
15. De Boeck, M.; Damilano, G.; Dehaen, W.; Tytgat, J.; Cuypers, E. Evaluation of 11 ionic liquids as potential extraction solvents for benzodiazepines from whole blood using liquid-liquid microextraction combined with LC-MS/MS. *Talanta* **2018**, *184*, 369–374. [CrossRef]
16. Hamed Mosavian, M.T.; Es'haghi, Z.; Razavi, N.; Banihashemi, S. Pre-concentration and determination of amitriptyline residues in waste water by ionic liquid based immersed droplet microextraction and HPLC. *J. Pharm. Anal.* **2012**, *2*, 361–365. [CrossRef]
17. Horváth, I.T.; Anastas, P.T. Introduction: Green Chemistry. *Chem. Rev.* **1999**, *107*, 2167–2168. [CrossRef]
18. Koel, M. General review of ionic liquids and their properties. In *Analytical Applications of Ionic Liquids*; Schröder, C., Ed.; World Scientific Publishing Europe: London, UK, 2016.

19. Kantae, V.; Krekels, E.H.; Van Esdonk, M.J.; Lindenburg, P.; Harms, A.C.; Knibbe, C.A.J.; Van der Graaf, P.H.; Hankemeier, T. Integration of pharmacometabolomics with pharmacokinetics and pharmacodynamics: Towards personalized drug therapy. *Metabolomics* **2017**, *13*, 9. [CrossRef]
20. González-Mariño, I.; Castro, V.; Montes, R.; Rodil, R.; Lores, A.; Cela, R.; Quintana, J.B. Multi-residue determination of psychoactive pharmaceuticals, illicit drugs and related metabolites in wastewater by ultra-high performance liquid chromatography-tandem mass spectrometry. *J. Chromatogr. A* **2018**, *1569*, 91–100. [CrossRef]
21. Ekwanzala, M.D.; Dewar, J.B.; Kamika, I.; Momba, M.N.B. Tracking the environmental dissemination of carbapenem-resistant Klebsiella pneumoniae using whole genome sequencing. *Sci. Total. Environ.* **2019**, *691*, 80–92. [CrossRef]
22. Gustavsson, P.; Andersson, T.; Gustavsson, A.; Reuterwall, C. Cancer incidence in female laboratory employees: Extended follow-up of a Swedish cohort study. *Occup. Environ. Med.* **2017**, *74*, 823–826. [CrossRef] [PubMed]
23. Koel, M. Do we need Green Analytical Chemistry? *Green Chem.* **2016**, *18*, 923–931. [CrossRef]
24. Kaljurand, M.; Koel, M. Recent advancements on greening analytical separation. *Crit. Rev. Anal. Chem.* **2005**, *35*, 177–192. [CrossRef]
25. Yu, H.; Ho, T.D.; Anderson, J.L. Ionic liquid and polymeric ionic liquid coatings in solid-phase microextraction. *Trends Anal. Chem.* **2013**, *45*, 219–232. [CrossRef]
26. Patinha, D.J.S.; Silvestre, A.J.D.; Marrucho, I.M. Poly(ionic liquids) in solid phase microextraction: Recent advances and perspectives. *Prog. Polym. Sci.* **2019**, *98*, 101148. [CrossRef]
27. Xu, J.; Zheng, J.; Tian, J.; Zhu, F.; Zeng, F.; Su, Ch.; Ouyang, G. New materials in solid-phase microextraction. *TrAC Trends Anal. Chem.* **2013**, *47*, 68–83. [CrossRef]
28. Mei, M.X.; Huang, X.L.; Chen, L. Recent development and applications of poly (ionic liquids) in microextraction techniques. *Trends Anal. Chem.* **2019**, *112*, 123–134. [CrossRef]
29. Sun, P.; Armstrong, D.W. Ionic liquids in analytical chemistry. *Anal. Chim. Acta* **2010**, *661*, 1–16. [CrossRef]
30. Soares, B.; Passos, H.; Freire, C.S.R.; Coutinho, J.A.P.; Silvestre, A.J.D.; Freire, M.G. Ionic liquids in chromatographic and electrophoretic techniques: Toward additional improvements in the separation of natural compounds. *Green Chem.* **2016**, *17*, 4582–4604. [CrossRef]
31. Ho, T.D.; Zhang, C.; Hantao, L.W.; Anderson, J.L. Ionic Liquids in Analytical Chemistry: Fundamentals, Advances, and Perspectives. *Anal. Chem.* **2014**, *86*, 262–285. [CrossRef]
32. Koel, M. Ionic liquids in chemical analysis. *Crit. Rev. Anal. Chem.* **2005**, *35*, 177–192. [CrossRef]
33. Wilkes, J.S. A short history of ionic liquids—From molten salts to neoteric solvents. *Green Chem.* **2002**, *4*, 73–80. [CrossRef]
34. Gabriel, S.; Weiner, J. Ueber einige Abkömmlinge des Propylamins. *Ber. Dtsch. Chem. Ges.* **1888**, *21*, 2669–2679. [CrossRef]
35. Walden, P. Über die Molekulargrösse und elektrische Leitfähigkeit einiger geschmolzener Salze. *Bull. l'Académie Impériale Sci. St. Pétersbourg* **1914**, *8*, 405–422.
36. Boeck, G. Paul Walden (1863–1957): The man behind the Walden inversion, the Walden rule, the Ostwald-Walden-Bredig rule and Ionic liquids. *ChemTexts* **2019**, *5*, 6. [CrossRef]
37. Yoke, J.T.; Weiss, J.F.; Tollin, G. Reactions of Triethylamine with Copper (I) and Copper (II) Halides. *Inorg. Chem.* **1963**, *2*, 1210–1216. [CrossRef]
38. Koch, V.R.L.; Miller, L.; Osteryoung, R.A. Electroinitiated Friedel-Crafts transalkylations in a room-temperature molten-salt medium. *J. Am. Chem. Soc.* **1976**, *98*, 5277–5284. [CrossRef]
39. Pacholec, F.; Butler, H.T.; Poole, C.F. Molten organic salt phase for gas-liquid chromatography. *Anal. Chem.* **1982**, *54*, 1938–1941. [CrossRef]
40. Wilkes, J.S.; Levisky, J.A.; Wilson, R.A.; Hussey, C.L. Dialkylimidazolium chloroaluminate melts: A new class of room-temperature ionic liquids for electrochemistry, spectroscopy and synthesis. *Inorg. Chem.* **1982**, *21*, 1263–1264. [CrossRef]
41. Visser, A.E.; Swatloski, R.P.; Reichert, W.M.; Rebecca Mayton, R.; Sheff, S.A.; Davis, J.H.; Rogers, R.D. Task-specific ionic liquids for the extraction of metal ions from aqueous solutions. *Chem. Commun* **2001**, *135–136*. [CrossRef]
42. Ludwig, R. *Ionic Liquids in Synthesis*; Wasserscheid, P., Welton, T., Eds.; ohn Wiley & Sons: Weinheim, Germany, 2008; Volume 1, pp. 863–864. [CrossRef]

43. Huddleston, J.G.; Willauer, H.D.; Swatloski, R.P.; Visser, A.E.; Rogers, R.D. Room temperature ionic liquids as novel media for 'clean' liquid-liquid extraction—Chemical Communications (RSC Publishing). *Chem. Commun.* **1998**, *16*, 1765–1766. [CrossRef]
44. Liu, J.; Li, N.; Jiang, G.; Liu, J.; Jönsson, J.Å.; Wen, M. Disposable ionic liquid coating for headspace solid-phase microextraction of benzene, toluene, ethylbenzene, and xylenes in paints followed by gas chromatography–flame ionization detection. *J. Chromatogr. A* **2005**, *1066*, 27–32. [CrossRef]
45. Moschovi, A.M.; Dracopoulos, V.; Nikolakis, V. Inter- and Intramolecular Interactions in Imidazolium Protic Ionic Liquids. *J. Phys. Chem. B* **2014**, *118*, 8673–8683. [CrossRef] [PubMed]
46. Shi, B.; Wang, Z.; Wen, H. Research on the strengths of electrostatic and van der Waals interactions in ionic liquids. *J. Mol. Liq.* **2017**, *241*, 486–488. [CrossRef]
47. Seddon, K.R.; Carmichael, A.J. Polarity study of some 1-alkyl-3-methylimidazolium ambient-temperature ionic liquids with the solvatochromic dye, Nile Red. *J. Phys. Org. Chem.* **2000**, *13*, 591–595. [CrossRef]
48. Mondal, S.S.; Müller, H.; Junginger, M.; Kelling, A.; Schilde, U.; Strehmel, V.; Holdt, H.J. Imidazolium 2-Substituted 4,5-Dicyanoimidazolate Ionic Liquids: Synthesis, Crystal Structures and Structure-Thermal Property Relationships. *Chem. Eur. J.* **2014**, *20*, 8170–8181. [CrossRef]
49. Marcinkowski, Ł.; Szepiński, E.; Milewska, M.J.; Kloskowski, A. Density, sound velocity, viscosity, and refractive index of new morpholinium ionic liquids with amino acid-based anions: Effect of temperature, alkyl chain length, and anion. *J. Mol. Liq.* **2019**, *284*, 557–568. [CrossRef]
50. Haghbakhsh, R.; Raeissi, S. Estimation of viscosities of 1-alkyl-3-methylimidazolium ionic liquids over a range of temperatures using a simple correlation. *Phys. Chem. Liq.* **2019**, *57*, 401–421. [CrossRef]
51. Sun, L.; Morales-Collazo, O.; Xia, H.; Brennecke, J.F. Effect of Structure on Transport Properties (Viscosity, Ionic Conductivity, and Self-Diffusion Coefficient) of Aprotic Heterocyclic Anion (AHA) Room Temperature Ionic Liquids. 2. Variation of Alkyl Chain Length in the Phosphonium Cation. *J. Psych. Chem. B* **2016**, *120*, 5767–5776. [CrossRef]
52. Rashid, T.; Kait, C.F.; Murugesan, T. Effect of alkyl chain length on the thermophysical properties of pyridinium carboxylates. *Chin. J. Chem. Eng.* **2016**, *25*, 1266–1272. [CrossRef]
53. Yuan, W.L.; Yang, X.; He, L.; Xue, Y.; Qin, S.; Tao, G.H. Viscosity, Conductivity, and Electrochemical Property of Dicyanamide Ionic Liquids. *Front. Chem.* **2018**, *6*, 59. [CrossRef] [PubMed]
54. Karpińska, M.; Królikowska, M. The influence of structure of ionic liquid containing cyano group on mutual solubility with water. *J. Chem. Thermodyn.* **2019**, *137*, 56–61. [CrossRef]
55. Handy, S.T.; Okello, M. The 2-Position of Imidazolium Ionic Liquids: Substitution and Exchange. *J. Org. Chem.* **2005**, *13902*, 4673–4676. [CrossRef] [PubMed]
56. Cao, Y.; Mu, T. Comprehensive Investigation on the Thermal Stability of 66 Ionic Liquids by Thermogravimetric Analysis. *Ind. Eng. Chem. Res.* **2014**, *53*, 8651–8664. [CrossRef]
57. Efimova, A.; Varga, J.; Matuschek, G.; Saraji-Bozorgzad, M.R.; Denner, T.; Zimmermann, R.; Schmidt, P. Thermal Resilience of Imidazolium-Based Ionic Liquids—Studies on Short- and Long-Term Thermal Stability and Decomposition Mechanism of 1-Alkyl-3-methylimidazolium Halides by Thermal Analysis and Single-Photon Ionization Time-of-Flight Mass Spectrometry. *J. Phys. Chem. B* **2018**, *122*, 8738–8749. [CrossRef]
58. Götz, M.; Reimert, R.; Bajohr, S.; Schnetzer, H.; Wimberg, J.; Schubert, T.J.S. Long-term thermal stability of selected ionic liquids in nitrogen and hydrogen atmosphere. *Thermochim. Acta* **2015**, *600*, 82–88. [CrossRef]
59. Zaitsau, D.H.; Yermalayeu, A.V.; Schubert, T.J.S.; Verevkin, S.P. Alkyl-imidazolium tetrafluoroborates: Vapor pressure, thermodynamics of vaporization, and enthalpies of formation. *J. Mol. Liq.* **2017**, *242*, 951–957. [CrossRef]
60. Aschenbrenner, O.; Supasitmongkol, S.; Taylor, M.; Styring, P. Measurement of vapour pressures of ionic liquids and other low vapour pressure solvents. *Green Chem.* **2009**, *11*, 1217–1221. [CrossRef]
61. Huo, Y.; Xia, S.; Yi, S.; Ma, P. Measurement and correlation of vapor pressure of benzene and thiophene with [BMIM][PF6] and [BMIM][BF4] ionic liquids. *Fluid Phase Equilibria* **2009**, *276*, 46–52. [CrossRef]
62. Hough, W.L.; Smiglak, M.; Rodríguez, H.; Swatloski, R.P.; Spear, S.C.; Daly, D.T.; Pernak, J.; Grisel, J.E.; Carliss, R.D.; Soutullo, M.D.; et al. The third evolution of ionic liquids: Active pharmaceutical ingredients. *New J. Chem.* **2007**, *31*, 1429–1436. [CrossRef]

63. Zhou, C.; Deng, J.; Shi, G.; Zhou, T. β-cyclodextrin-ionic liquid polymer based dynamically coating for simultaneous determination of tetracyclines by capillary electrophoresis. *Electrophoresis* **2017**, *38*, 1060–1067. [CrossRef] [PubMed]
64. Chatzimitakos, T.; Binellas, C.; Maidatsi, K.; Stalikas, C. Magnetic ionic liquid in stirring-assisted drop-breakup microextraction: Proof-of-concept extraction of phenolic endocrine disrupters and acidic pharmaceuticals. *Anal. Chim. Acta* **2016**, *910*, 53–59. [CrossRef] [PubMed]
65. Yu, H.; Merib, J.; Anderson, J.L. Crosslinked polymeric ionic liquids as solid-phase microextraction sorbent coatings for high performance liquid chromatography. *J. Chromatogr. A* **2016**, *1438*, 10–21. [CrossRef]
66. Casado, N.; Salgado, A.; Castro-Puyana, M.; Gacia, M.A.; Marina, M.L. Enantiomeric separation of ivabradine by cyclodextrin-electrokinetic chromatography. Effect of amino acid chiral ionic liquids. *J. Chromatogr. A* **2019**, *1608*, 460407. [CrossRef] [PubMed]
67. Yu, W.; Li, K.; Liu, Z.; Zhang, H.; Jin, X. Novelty aqueous two-phase extraction system based on ionic liquid for determination of sulfonamides in blood coupled with high-performance liquid chromatography. *Microchem. J.* **2018**, *136*, 263–269. [CrossRef]
68. Kiszkiel, I.; Starczewska, B.; Leśniewska, B.; Poźniak, P. Extraction of ranitidine and nizatidine with using imidazolium ionic liquids prior spectrophotometric and chromatographic detection. *J. Pharm. Biomed. Anal.* **2015**, *106*, 85–91. [CrossRef] [PubMed]
69. Rezaee, M.; Assadi, Y.; Milani Hosseini, M.R.; Aghaee, E.; Ahmadi, F.; Berijani, S. Determination of organic compounds in water using dispersive liquid-liquid microextraction. *J. Chromatogr. A* **2006**, *1116*, 1–9. [CrossRef]
70. Cruz-Vera, M.; Lucena, R.; Cárdenas, S.; Valcárcel, M. One-step in-syringe ionic liquid-based dispersive liquid-liquid microextraction. *J. Chromatogr. A* **2009**, *1216*, 6459–6465. [CrossRef]
71. Hatami, M.; Karimnia, E.; Farhadi, K. Determination of salmeterol in dried blood spot using an ionic liquid based dispersive liquid-liquid microextraction coupled with HPLC. *J. Pharm. Biomed. Anal.* **2013**, *85*, 283–287. [CrossRef]
72. Padró, J.M.; Marsón, M.E.; Mastrantonio, G.E.; Altcheh, J.; García-Bournissen, F.; Reta, M. Development of an ionic liquid-based dispersive liquid-liquid microextraction method for the determination of nifurtimox and benznidazole in human plasma. *Talanta* **2013**, *107*, 95–102. [CrossRef]
73. Padró, J.M.; Pellegrino Vidal, R.B.; Echevarria, R.N.; Califano, A.N.; Reta, M.R. Development of an ionic-liquid-based dispersive liquid-liquid microextraction method for the determination of antichagasic drugs in human breast milk: Optimization by central composite design. *J. Sep. Sci.* **2015**, *38*, 1591–1600. [CrossRef] [PubMed]
74. Xiao, C.; Tang, M.; Li, J.; Yin, C.; Xiang, G.; Xu, L. Determination of sildenafil, vardenafil and aildenafil in human plasma by dispersive liquid-liquid microextraction-back extraction based on ionic liquid and high performance liquid chromatography-ultraviolet detection. *J. Chromatogr. B* **2013**, *931*, 111–116. [CrossRef]
75. Gong, A.; Zhu, X. Dispersve solvent-free ultrasound-assisted ionic liquid dispersive liquid-liquid microextraction coupled with HPLC for determination of ulipriststal acetate. *Talanta* **2015**, *131*, 603–608. [CrossRef] [PubMed]
76. De Boeck, M.; Missotten, S.; Dehaen, W.; Tytgat, J.; Cuypers, E. Development and validation of a fast ionic liquid-based dispersive liquid-liquid microextraction procedure combined with LC-MS/MS analysis for the quantification of benzodiazepines and benzodiazepine-like hypnotics in whole blood. *Forensic Sci. Int.* **2017**, *274*, 44–54. [CrossRef] [PubMed]
77. De Boeck, M.; Dehaen, W.; Tytgat, J.; Cuypers, E. Ionic Liquid-Based Liquid-Liquid Microextraction for Benzodiazepine Analysis in Postmortem Blood Samples. *J. Forensic Sci.* **2018**, *63*, 1875–1879. [CrossRef]
78. De Boeck, M.; Dubrulle, L.; Dehaen, W.; Tytgat, J.; Cuypers, E. Fast and easy extraction of antidepressants from whole blood using ionic liquids as extraction solvent. *Talanta* **2018**, *180*, 292–299. [CrossRef]
79. Vaghar-Lahijani, G.; Aberoomand-Azar, P.; Saber-Tehrani, M.; Soleimani, M. Application of ionic liquid-based ultrasonic-assisted microextraction coupled with HPLC for determination of citalopram and nortriptyline in human plasma. *J. Liq. Chromatogr. Relat. Technol.* **2017**, *40*, 1–7. [CrossRef]
80. Rao, R.N.; Raju, S.S.; Vali, R.M. Ionic-liquid based dispersive liquid-liquid microextraction followed by high performance liquid chromatographic determination of anti-hypertensives in rat seru. *J. Chromatogr. B* **2013**, *931*, 174–180. [CrossRef]

81. Rao, R.N.; Naidu, C.G.; Suresh, C.V.; Srinath, N.; Padiya, R. Ionic liquid based dispersive liquid-liquid microextraction followed by RP-HPLC determination of balofloxacin in rat serum. *Anal. Methods* **2014**, *6*, 1674–1683. [CrossRef]
82. Rao, R.N.; Mastan Vali, R.; Vara Prasada Rao, A. Determination of rifaximin in rat serum by ionic liquid based dispersive liquid-liquid microextraction combined with RP-HPLC. *J. Sep. Sci.* **2012**, *35*, 1945–1952. [CrossRef]
83. Vaghar-Lahijani, G.; Saber-Tehrani, M.; Aberoomand-Azar, P.; Soleimani, M. Extraction and Determination of Two Antidepressant Drugs in Human Plasma by Dispersive Liquid-Liquid Microextraction—HPLC. *J. Anal. Chem.* **2018**, *73*, 145–151. [CrossRef]
84. Gong, A.; Zhu, X. Miniaturized ionic liquid dispersive liquid-liquid microextraction in a coupled-syringe system combined with UV for extraction and determination of danazol in danazol capsule and mice serum. *Spectrochim. Acta A* **2016**, *159*, 163–168. [CrossRef] [PubMed]
85. Zeeb, M.; Jamil, P.T.; Berenjian, A.; Ganjali, M.R.; Olyai, M.R.T.B. Quantitative analysis of piroxicam using temperature-controlled ionic liquid dispersive liquid phase microextraction followed by stopped-flow injection spectrofluorimetry. *DARU. J. Pharm. Sci.* **2013**, *21*, 63. [CrossRef]
86. Zeeb, M.; Ganjali, M.R.; Norouzi, P. Modified ionic liquid cold-induced aggregation dispersive liquid-liquid microextraction combined with spectrofluorimetry for trace determination of ofloxacin in pharmaceutical and biological samples. *DARU. J. Pharm. Sci.* **2011**, *19*, 446–454.
87. Song, J.; Zhang, Z.H.; Zhang, Y.Q.; Feng, C.; Wang, G.N.; Wang, J.P. Ionic liquid dispersive liquid-liquid microextraction combined with high performance liquid chromatography for determination of tetracycline drugs in eggs. *Anal. Methods* **2014**, *6*, 6459–6466. [CrossRef]
88. Wang, G.N.; Feng, C.; Zhang, H.C.; Zhang, Y.Q.; Zhang, L.; Wang, J.P. Determination of fluoroquinolone drugs in meat by ionic-liquid-based dispersive liquid-liquid microextraction-high performance liquid chromatography. *Anal. Methods* **2015**, *7*, 1046–1052. [CrossRef]
89. Liu, X.; Fu, F.; Li, M.; Guo, L.P.; Yang, L. Ionic Liquid-Based Dispersive Liquid-Liquid Microextraction Coupled with Capillary Electrophoresis to Determine Drugs of Abuse in Urine. *Chin. J. Anal. Chem.* **2013**, *41*, 1919–1922. [CrossRef]
90. Rao, T.S.; Sridevi, M.; Naidu, C.G.; Nagaraju, B. Ionic liquid-based vortex-assisted DLLME followed by RP-LC-PDA method for bioassay of daclatasvir in rat serum: Application to pharmacokinetics. *J. Anal. Sci. Technol.* **2019**, *10*, 20. [CrossRef]
91. Cruz-Vera, M.; Lucena, R.; Cárdenas, S.; Valcárcel, M. Ionic liquid-based dynamic liquid-phase microextraction: Application to the determination of anti-inflammatory drugs in urine samples. *J. Chromatogr. A* **2008**, *1202*, 1–7. [CrossRef]
92. Yao, C.; Li, T.; Twu, P.; Pitner, W.R.; Anderson, J.L. Selective extraction of emerging contaminants from water samples by dispersive liquid-liquid microextraction using functionalized ionic liquids. *J. Chromatogr. A* **2011**, *1218*, 1556–1566. [CrossRef]
93. Zhao, R.S.; Wang, X.; Sun, J.; Wang, S.S.; Yuan, J.P.; Wang, X.K. Trace determination of triclosan and triclocarban in environmental water samples with ionic liquid dispersive liquid-phase microextraction prior to HPLC-ESI-MS-MS. *Anal. Bioanal. Chem.* **2010**, *397*, 1627–1633. [CrossRef] [PubMed]
94. Yu, H.; Merib, J.; Anderson, J.L. Faster dispersive liquid-liquid microextraction methods using magnetic ionic liquids as solvents. *J. Chromatogr. A* **2016**, *1463*, 11–19. [CrossRef] [PubMed]
95. Mao, T.; Hao, B.; He, J.; Li, W.; Li, S.; Yu, Z. Ultrasound assisted ionic liquid dispersive liquid phase extraction of lovastatin and simvastatin: A new pretreatment procedure. *J. Sep. Sci.* **2009**, *32*, 3029–3033. [CrossRef] [PubMed]
96. Hou, D.; Guan, Y.; Di, X. Temperature-Induced Ionic Liquids Dispersive Liquid-Liquid Microextraction of Tetracycline Antibiotics in Environmental Water Samples Assisted by Complexation. *Chromatographia* **2011**, *73*, 1057–1064. [CrossRef]
97. Parrilla Vázquez, M.M.; Parrilla Vázquez, P.; Martínez Galera, M.; Gil García, M.D.; Uclés, A. Ultrasound-assisted ionic liquid dispersive liquid-liquid microextraction coupled with liquid chromatography-quadrupole-linear ion trap-mass spectrometry for simultaneous analysis of pharmaceuticals in wastewaters. *J. Chromatogr. A* **2013**, *1291*, 19–26. [CrossRef]

98. Vázquez, M.P.; Vázquez, P.P.; Galera, M.M.; García, M.G. Determination of eight fluoroquinolones in groundwater samples with ultrasound-assisted ionic liquid dispersive liquid-liquid microextraction prior to high-performance liquid chromatography and fluorescence detection. *Anal. Chim. Acta* **2012**, *748*, 20–27. [CrossRef]
99. Ji, Y.; Du, Z.; Zhang, H.; Zhang, Y. Rapid analysis of non-steroidal anti-inflammatory drugs in tap water and drinks by ionic liquid dispersive liquid-liquid microextraction coupled to ultra-high performance supercritical fluid chromatography. *Anal. Methods* **2014**, *6*, 7294–7304. [CrossRef]
100. Xu, X.; Su, R.; Zhao, X.; Liu, Z.; Zhang, Y.; Li, D.; Li, X.; Zhang, H.; Wang, Z. Ionic liquid-based microwave-assisted dispersive liquid-liquid microextraction and derivatization of sulfonamides in river water, honey, milk, and animal plasma. *Anal. Chim. Acta* **2011**, *707*, 92–99. [CrossRef]
101. Ge, D.; Lee, H.K. Ionic liquid based dispersive liquid-liquid microextraction coupled with micro-solid phase extraction of antidepressant drugs from environmental water samples. *J. Chromatogr. A* **2013**, *1317*, 217–222. [CrossRef]
102. Zhao, R.S.; Wang, X.; Sun, J.; Hu, C.; Wang, X.K. Determination of triclosan and triclocarban in environmental water samples with ionic liquid/ionic liquid dispersive liquid-liquid microextraction prior to HPLC-ESI-MS/MS. *Microchim. Acta* **2011**, *174*, 145–151. [CrossRef]
103. Toledo-Neira, C.; Álvarez-Lueje, A. Ionic liquids for improving the extraction of NSAIDs in water samples using dispersive liquid-liquid microextraction by high performance liquid chromatography-diode array—Fluorescence detection. *Talanta* **2015**, *134*, 619–626. [CrossRef] [PubMed]
104. Zare, F.; Ghaedi, M.; Daneshfar, A. Ionic-liquid-based surfactant-emulsified microextraction procedure accelerated by ultrasound radiation followed by high-performance liquid chromatography for the simultaneous determination of antidepressant and antipsychotic drugs. *J. Sep. Sci.* **2015**, *38*, 844–851. [CrossRef] [PubMed]
105. Liu, Z.; Yu, W.; Zhang, H.; Gu, F.; Jin, X. Salting-out homogenous extraction followed by ionic liquid/ionic liquid liquid-liquid micro-extraction for determination of sulfonamides in blood by high performance liquid chromatography. *Talanta* **2016**, *161*, 748–754. [CrossRef] [PubMed]
106. Cruz-Vera, M.; Lucena, R.; Cárdenas, S.; Valcárcel, M. Determination of phenothiazine derivatives in human urine by using ionic liquid-based dynamic liquid-phase microextraction coupled with liquid chromatography. *J. Chromatogr. B* **2009**, *877*, 37–42. [CrossRef] [PubMed]
107. Qin, H.; Li, B.; Liu, M.S.; Yang, Y.L. Separation and pre-concentration of glucocorticoids in water samples by ionic liquid supported vortex-assisted synergic microextraction and HPLC determination. *J. Sep Sci.* **2013**, *36*, 1463–1469. [CrossRef] [PubMed]
108. Song, Y.; Wu, L.; Lu, C.; Li, N.; Hu, M.; Wang, Z. Microwave-assisted liquid-liquid microextraction based on solidification of ionic liquid for the determination of sulfonamides in environmental water samples. *J. Sep. Sci.* **2014**, *37*, 3533–3538. [CrossRef]
109. Tao, Y.; Liu, J.F.; Hu, X.L.; Li, H.C.; Wang, T.; Jiang, G.B. Hollow fiber supported ionic liquid membrane microextraction for determination of sulfonamides in environmental water samples by high-performance liquid chromatography. *J. Chromatogr. A* **2009**, *1216*, 6259–6266. [CrossRef]
110. Goh, S.X.L.; Goh, H.A.; Lee, H.K. Automation of ionic liquid enhanced membrane bag-assisted-liquid-phase microextraction with liquid chromatography-tandem mass spectrometry for determination of glucocorticoids in water. *Anal. Chim. Acta* **2018**, *1035*, 77–86. [CrossRef]
111. Hanapi, N.S.M.; Sanagi, M.M.; Ismail, A.K.; Ibrahim, W.A.W.; Saim, N.; Ibrahim, W.N.W. Ionic liquid-impregnated agarose film two-phase micro-electrodriven membrane extraction (IL-AF-μ-EME) for the analysis of antidepressants in water samples. *J. Chromatogr. B* **2017**, *1046*, 73–80. [CrossRef]
112. Shao, M.; Zhang, X.; Li, N.; Shi, Ji.; Zhang, H.; Wang, Z.; Zhang, H.; Yu, A.; Yu, Y. Ionic liquid-based aqueous two-phase system extraction of sulfonamides in milk. *J. Chromatogr. B* **2014**, *961*, 5–12. [CrossRef] [PubMed]
113. Han, J.; Wang, Y.; Yu, C.; Li, C.; Yan, Y.; Liu, Y.; Wang, L. Separation, concentration and determination of chloramphenicol in environment and food using an ionic liquid/salt aqueous two-phase flotation system coupled with high-performance liquid chromatography. *Anal. Chim. Acta* **2011**, *685*, 138–145. [CrossRef] [PubMed]
114. Yao, T.; Yao, S. Magnetic ionic liquid aqueous two-phase system coupled with high performance liquid chromatography: A rapid approach for determination of chloramphenicol in water environment. *J. Chromatogr. A* **2017**, *1481*, 12–22. [CrossRef] [PubMed]

115. Wen, Y.; Chen, L.; Li, J.; Liu, D.; Chen, L. Recent advances in solid-phase sorbents for sample preparation prior to chromatographic analysis. *TrAC Trends Anal. Chem.* **2014**, *59*, 26–41. [CrossRef]
116. Pang, X.; Bai, L.; Lan, D.; Guo, B.; Wang, H.; Liu, H.; Ma, Z. Development of an SPE Method Using Ionic Liquid-Based Monolithic Column for On-Line Cleanup of Human Plasma and Simultaneous Determination of Five Steroid Drugs. *Chromatographia* **2018**, *81*, 1391–1400. [CrossRef]
117. Liu, S.; Wang, C.; He, S.; Bai, L.; Liu, H. On-line SPE Using Ionic Liquid-Based Monolithic Column for the Determination of Antihypertensives in Human Plasma. *Chromatographia* **2016**, *79*, 441–449. [CrossRef]
118. Ferreira, D.C.; De Toffoli, A.L.; Maciel, E.V.S.; Lanças, F.M. Online fully automated SPE-HPLC-MS/MS determination of ceftiofur in bovine milk samples employing a silica-anchored ionic liquid as sorbent. *Electrophoresis* **2018**, *39*, 2210–2217. [CrossRef]
119. Da Silva, M.R.; Lanças, M.F. Evaluation of ionic liquids supported on silica as a sorbent for fully automated online solid-phase extraction with LC-MS determination of sulfonamides in bovine milk samples. *J. Sep. Sci.* **2018**, *41*, 2237–2244. [CrossRef]
120. Yan, H.; Liu, S.; Gao, M.; Sun, N. Ionic liquids modified dummy molecularly imprinted microspheres as solid phase extraction materials for the determination of clenbuterol and clorprenaline in urine. *J. Chromatogr. A* **2013**, *1294*, 10–16. [CrossRef]
121. Wu, J.; Zhao, H.; Xiao, D.; Chuong, P.H.; He, J.; He, H. Mixed hemimicelles solid-phase extraction of cephalosporins in biological samples with ionic liquid-coated magnetic graphene oxide nanoparticles coupled with high-performance liquid chromatographic analysis. *J. Chromatogr. A.* **2016**, *1454*, 1–8. [CrossRef]
122. Taghvimi, A.; Hamidi, S.; Javadzadeh, Y. Mixed hemimicelle magnetic dispersive solid-phase extraction using carbon-coated magnetic nanoparticles for the determination of tramadol in urine samples. *J. Sep. Sci.* **2019**, *42*, 582–590. [CrossRef]
123. Yan, H.; Gao, M.; Yang, C.; Qiu, M. Ionic liquid-modified magnetic polymeric microspheres as dispersive solid phase extraction adsorbent: A separation strategy applied to the screening of sulfamonomethoxine and sulfachloropyrazine from urine. *Anal. Bioanal. Chem.* **2014**, *406*, 2669–2677. [CrossRef] [PubMed]
124. Wang, Z.; He, M.; Jiang, Ch.; Zhang, F.; Du, S.; Feng, W.; Zhang, H. Matrix solid-phase dispersion coupled with homogeneous ionic liquid microextraction for the determination of sulfonamides in animal tissues using high-performance liquid chromatography. *J. Sep. Sci.* **2015**, *38*, 4127–4135. [CrossRef] [PubMed]
125. Gao, S.; Yu, W.; Yang, X.; Liu, Z.; Jia, Y.; Zhang, H. On-line ionic liquid-based dynamic microwave-assisted extraction-high performance liquid chromatography for the determination of lipophilic constituents in root of Salvia miltiorrhiza Bunge. *J. Sep. Sci.* **2012**, *35*, 2813–2821. [CrossRef] [PubMed]
126. Amiri, M.; Yamini, Y.; Safari, M.; Asiabi, H. Magnetite nanoparticles coated with covalently immobilized ionic liquids as a sorbent for extraction of non-steroidal anti-inflammatory drugs from biological fluids. *Microchim. Acta* **2016**, *183*, 2297–2305. [CrossRef]
127. Fontanals, N.; Ronka, S.; Borrull, F.; Trochimczuk, A.W.; Marcé, R.M. Supported imidazolium ionic liquid phases: A new material for solid-phase extraction. *Talanta* **2009**, *80*, 250–256. [CrossRef]
128. Bratkowska, D.; Fontanals, N.; Ronka, S.; Trochimczuk, A.W.; Borrull, F.; Marcé, R.M. Comparison of different imidazolium supported ionic liquid polymeric phases with strong anion-exchange character for the extraction of acidic pharmaceuticals from complex environmental samples. *J. Sep. Sci.* **2012**, *35*, 1953–1958. [CrossRef]
129. Zhu, G.; Cheng, G.; Wang, P.; Li, W.; Wang, Y.; Fan, J. Water compatible imprinted polymer prepared in water for selective solid phase extraction and determination of ciprofloxacin in real samples. *Talanta* **2019**, *200*, 307–315. [CrossRef]
130. Arthur, C.L.; Pawliszyn, J. Solid phase microextraction with thermal desorption using fused silica optical fibers. *Anal. Chem.* **1990**, *62*, 2145–2148. [CrossRef]
131. Jamshidi, S.; Rofouei, M.K.; Thorsen, G. Using magnetic core-shell nanoparticles coated with an ionic liquid dispersion assisted by effervescence powder for the micro-solid-phase extraction of four beta blockers from human plasma by ultra high performance liquid chromatography with mass. *J. Sep. Sci.* **2019**, *42*, 698–705. [CrossRef]
132. Yang, M.; Wu, X.; Jia, Y.; Xi, X.; Yang, X.; Lu, R.; Zhang, R.; Gao, H.; Zhou, W. Use of magnetic effervescent tablet-assisted ionic liquid dispersive liquid-liquid microextraction to extract fungicides from environmental waters with the aid of experimental design methodology. *Anal. Chim. Acta* **2016**, *906*, 118–127. [CrossRef]

133. Serrano, M.; Chatzimitakos, T.; Gallego, M.; Stalikas, C.D. 1-Butyl-3-aminopropyl imidazolium-functionalized graphene oxide as a nanoadsorbent for the simultaneous extraction of steroids and β-blockers via dispersive solid-phase microextraction. *J. Chromatogr. A* **2016**, *1436*, 9–18. [CrossRef] [PubMed]
134. Pacheco-Fernández, I.; Najafi, A.; Pino, V.; Anderson, J.L.; Ayala, J.H.; Afonso, A.M. Utilization of highly robust and selective crosslinked polymeric ionic liquid-based sorbent coatings in direct-immersion solid-phase microextraction and high-performance liquid chromatography for determining polar organic pollutants in waters. *Talanta* **2016**, *158*, 125–133. [CrossRef] [PubMed]
135. Talebpour, Z.; Taraji, M.; Adib, N. Stir bar sorptive extraction and high performance liquid chromatographic determination of carvedilol in human serum using two different polymeric phases and an ionic liquid as desorption solvent. *J. Chromatogr. A* **2012**, *1236*, 1–6. [CrossRef] [PubMed]
136. Męczykowska, H.; Kobylis, P.; Stepnowski, P.; Caban, M. Calibration of Passive Samplers for the Monitoring of Pharmaceuticals in Water-Sampling Rate Variation. *Crit. Rev. Anal. Chem.* **2017**, *47*, 204–222. [CrossRef]
137. Caban, M.; Męczykowska, H.; Stepnowski, P. Application of the PASSIL technique for the passive sampling of exemplary polar contaminants (pharmaceuticals and phenolic derivatives) from water. *Talanta* **2016**, *155*, 185–192. [CrossRef]
138. Męczykowska, H.; Stepnowski, P.; Caban, M. Effect of salinity and pH on the calibration of the extraction of pharmaceuticals from water by PASSIL. *Talanta* **2018**, *179*, 271–278. [CrossRef]
139. Męczykowska, H.; Stepnowski, P.; Caban, M. Impact of humic acids, temperature and stirring on passive extraction of pharmaceuticals from water by trihexyl (tetradecyl) phosphonium dicyanamide. *Microchem. J.* **2019**, *144*, 500–505. [CrossRef]
140. Petruczynik, A.; Wróblewski, K.; Waksmundzka-Hajnos, M. Application of mobile phases containing ionic liquid for the analysis of selected psychotropic drugs by HPLC-DAD and HPLC-MS. *Acta Chromatogr.* **2018**, *31*, 255–261. [CrossRef]
141. Herrera-Herrera, A.V.; Hernández-Borges, J.; Rodríguez-Delgado, M.Á. Fluoroquinolone antibiotic determination in bovine, ovine and caprine milk using solid-phase extraction and high-performance liquid chromatography-fluorescence detection with ionic liquids as mobile phase additives. *J. Chromatogr. A* **2009**, *1216*, 7281–7287. [CrossRef]
142. Herrera-Herrera, A.V.; Hernández-Borges, J.; Rodríguez-Delgado, M.Á. Ionic liquids as mobile phase additives for the high-performance liquid chromatographic analysis of fluoroquinolone antibiotics in water samples. *Anal. Bioanal. Chem.* **2008**, *392*, 1439–1446. [CrossRef]
143. Marszałł, M.P.; Baczek, T.; Kaliszan, R. Reduction of silanophilic interactions in liquid chromatography with the use of ionic liquids. *Anal. Chim. Acta* **2005**, *547*, 172–178. [CrossRef]
144. Nageswara Rao, R.; Ramachandra, B.; Mastan Vali, R. Reversed-phase liquid chromatographic separation of antiretroviral drugs on a monolithic column using ionic liquids as mobile phase additives. *J. Sep. Sci.* **2011**, *34*, 500–507. [CrossRef] [PubMed]
145. Cruz-Vera, M.; Lucena, R.; Cárdenas, S.; Valcárcel, M. Combined use of carbon nanotubes and ionic liquid to improve the determination of antidepressants in urine samples by liquid chromatography. *Anal. Bioanal. Chem.* **2008**, *391*, 1139–1145. [CrossRef] [PubMed]
146. Somova, V.D.; Bessonova, E.A.; Kartsova, L.A. A new version of hydrophilic interaction liquid chromatography with the use of ionic liquids based on imidazole for the determination of highly polar drugs in body fluids. *Аналитика И Контроль* **2017**, *21*, 241–250. [CrossRef]
147. Calabuig-Hernández, S.; Peris-García, E.; García-Alvarez-Coque, M.C.; Ruiz-Angel, M.J. Suitability of 1-hexyl-3-methylimidazolium ionic liquids for the analysis of pharmaceutical formulations containing tricyclic antidepressants. *J. Chromatogr. A* **2018**, *1559*, 118–127. [CrossRef]
148. Fernández-Navarro, J.J.; Torres-Lapasió, J.R.; Ruiz-Ángel, M.J.; García-Álvarez-Coque, M.C. Silanol suppressing potency of alkyl-imidazolium ionic liquids on C18 stationary phases. *J. Chromatogr. A* **2012**, *1232*, 166–175. [CrossRef]
149. Ubeda-Torres, M.T.; Ortiz-Bolsico, C.; García-Alvarez-Coque, M.C.; Ruiz-Angel, M.J. Gaining insight in the behaviour of imidazolium-based ionic liquids as additives in reversed-phase liquid chromatography for the analysis of basic compounds. *J. Chromatogr. A* **2015**, *1380*, 96–103. [CrossRef]
150. Fernández-Navarro, J.J.; García-Álvarez-Coque, M.C.; Ruiz-Ángel, M.J. The role of the dual nature of ionic liquids in the reversed-phase liquid chromatographic separation of basic drugs. *J. Chromatogr. A* **2011**, *1218*, 398–407. [CrossRef]

151. Caban, M.; Stepnowski, P. The antagonistic role of chaotropic hexafluorophosphate anions and imidazolium cations composing ionic liquids applied as phase additives in the separation of tri-cyclic antidepressants. *Anal. Chim. Acta* **2017**, *967*, 102–110. [CrossRef]
152. Han, D.; Wang, Y.; Jin, Y.; Row, K.H. Analysis of some β-Lactam antibiotics using ionic liquids as mobile phase additives by RP-HPLC. *J. Chromatogr. Sci.* **2011**, *49*, 63–66. [CrossRef]
153. Ruiz-Angel, M.J.; Carda-Broch, S.; Berthod, A. Ionic liquids versus triethylamine as mobile phase additives in the analysis of β-blockers. *J. Chromatogr. A* **2006**, *1119*, 202–208. [CrossRef]
154. Flieger, J. Effect of Ionic Liquids as Mobile-Phase Additives on Chromatographic Parameters of Neuroleptic Drugs in Reversed-Phase High-Performance Liquid Chromatography. *Anal. Lett.* **2009**, *42*, 1632–1649. [CrossRef]
155. Marszałł, M.P.; Baczek, T.; Kaliszan, R. Evaluation of the silanol-suppressing potency of ionic liquids. *J. Sep. Sci.* **2006**, *29*, 1138–1145. [CrossRef]
156. Shu, X.H.J.; Liu, F.; Zhou, X.; Jiang, S.X.; Liu, X.; Zhao, L. Surface confined ionic liquid—A new stationary phase for the separation of ephedrines in high-performance liquid chromatography. *Chin. Chem. Lett.* **2004**, *15*, 1060–1062.
157. Zhang, M.; Chen, J.; Gu, T.; Qiu, H.; Jiang, S. Novel imidazolium-embedded and imidazolium-spaced octadecyl stationary phases for reversed phase liquid chromatography. *Talanta* **2014**, *126*, 177–184. [CrossRef] [PubMed]
158. Hong, X.; Huanjun, P.; Xiang, W.; Dengying, L.; Ranxi, N.; Jun, C.; Shiyu, L.; Zhongying, Z.; Jingdong, P. Development and evaluation of new imidazolium-based zwitterionic stationary phases for hydrophilic interaction chromatography. *J. Chromatogr. A* **2013**, *1286*, 137–145. [CrossRef]
159. Qiu, H.; Zhang, M.; Chen, J.; Gu, T.; Takafuji, M.; Ihara, H. Anionic and cationic copolymerized ionic liquid-grafted silica as a multifunctional stationary phase for reversed-phase chromatography. *Anal. Methods* **2014**, *6*, 469–475. [CrossRef]
160. Sun, M.; Feng, J.; Wang, X.; Duan, H.; Li, L.; Luo, C. Dicationic imidazolium ionic liquid modified silica as a novel reversed-phase/anion-exchange mixed-mode stationary phase for high-performance liquid chromatography. *J. Sep. Sci.* **2014**, *37*, 2153–2159. [CrossRef] [PubMed]
161. He, S.; He, Y.; Cheng, L.; Wu, Y.; Ke, Y. Novel chiral ionic liquids stationary phases for the enantiomer separation of chiral acid by high-performance liquid chromatography. *Chirality* **2018**, *30*, 670–679. [CrossRef] [PubMed]
162. Qiu, H.; Mallik, A.K.; Takafuji, M.; Jiang, S.; Ihara, H. New poly(ionic liquid)-grafted silica multi-mode stationary phase for anion-exchange/reversed-phase/hydrophilic interaction liquid chromatography. *Analyst* **2012**, *137*, 2553. [CrossRef]
163. Rahim, N.Y.; Tay, K.S.; Mohamad, S. Chromatographic and spectroscopic studies on β-cyclodextrin functionalized ionic liquid as chiral stationary phase: Enantioseparation of NSAIDs. *Adsorpt. Sci. Technol.* **2018**, *36*, 130–148. [CrossRef]
164. Rahim, N.Y.; Tay, K.S.; Mohamad, S. β-Cyclodextrin functionalized ionic liquid as chiral stationary phase of high performance liquid chromatography for enantioseparation of β-blockers. *J. Incl. Phenom. Macrocl. Chem.* **2016**, *85*, 303–315. [CrossRef]
165. Xian, H.; Peng, H.; Wang, X.; Long, D.; Ni, R.; Chen, J.; Li, S.; Zhang, Z.; Peng, Z. Preparation and evaluation a mixed-mode stationary phase with imidazolium and carboxyl group for high performance liquid chromatography. *Microchem. J.* **2019**, *150*, 104131. [CrossRef]
166. Dai, Q.; Ma, J.; Ma, S.; Wang, S.; Li, L.; Zhu, X.; Qiao, X. Cationic Ionic Liquids Organic Ligands Based Metal-Organic Frameworks for Fabrication of Core-Shell Microspheres for Hydrophilic Interaction Liquid Chromatography. *ACS Appl. Mater. Inter.* **2016**, *8*, 21632–21639. [CrossRef] [PubMed]
167. Qiao, L.; Lv, W.; Chang, M.; Shi, X.; Xu, G. Surface-bonded amide-functionalized imidazolium ionic liquid as stationary phase for hydrophilic interaction liquid chromatography. *J. Chromatogr. A* **2018**, *1559*, 141–148. [CrossRef] [PubMed]
168. Wu, Q.; Sun, Y.; Gao, J.; Chen, L.; Dong, S.; Luo, G.; Li, H.; Wang, L.; Zhao, L. Ionic liquid-functionalized graphene quantum dot-bonded silica as multi-mode HPLC stationary phase with enhanced selectivity for acid compounds. *New J. Chem.* **2018**, *42*, 8672–8680. [CrossRef]

169. Domínguez, C.; Reyes-Contreras, C.; Bayona, J.M. Determination of benzothiazoles and benzotriazoles by using ionic liquid stationary phases in gas chromatography mass spectrometry. Application to their characterization in wastewaters. *J. Chromatogr. A* **2012**, *1230*, 117–122. [CrossRef]
170. Nan, H.; Anderson, J.L. Ionic liquid stationary phases for multidimensional gas chromatography. *TrAC Trends Anal. Chem.* **2018**, *105*, 367–379. [CrossRef]
171. Reyes-Contreras, C.; Domínguez, C.; Bayona, J.M. Determination of nitrosamines and caffeine metabolites in wastewaters using gas chromatography mass spectrometry and ionic liquid stationary phases. *J. Chromatogr. A* **2012**, *1261*, 164–170. [CrossRef]
172. Ragonese, C.; Sciarrone, D.; Tranchida, P.Q.; Dugo, P.; Mondello, L. Use of ionic liquids as stationary phases in hyphenated gas chromatography techniques. *J. Chromatogr. A* **2012**, *1255*, 130–144. [CrossRef]
173. Trujillo-Rodríguez, V.; Afonso, M.J.; Pino, A.M. Analytical Applications of Ionic Liquids in Chromatographic and Electrophoretic Separation Techniques. In *Ionic Liquids for Better Separation Processes*; Springer: Berlin/Heidelberg, Germany, 2016; pp. 193–233. [CrossRef]
174. Berthod, A.; Ruiz-Ángel, M.J.; Carda-Broch, S. Recent advances on ionic liquid uses in separation techniques. *J. Chromatogr. A* **2018**, *1559*, 2–16. [CrossRef] [PubMed]
175. Yu, L.; He, J.; Qi, M.; Huang, X. Amphiphilic triptycene-based stationary phase for high-resolution gas chromatographic separations. *J. Chromatogr. A* **2019**, *1599*, 239–246. [CrossRef] [PubMed]
176. Cagliero, C.; Bicchi, C.; Cordero, C.; Liberto, E.; Sgorbini, B.; Rubiolo, P. Room temperature ionic liquids: New GC stationary phases with a novel selectivity for flavor and fragrance analyses. *J. Chromatogr. A* **2012**, *1268*, 130–138. [CrossRef] [PubMed]
177. Cagliero, C.; Bicchi, C.; Cordero, C.; Liberto, E.; Rubiolo, P.; Sgorbini, B. Analysis of essential oils and fragrances with a new generation of highly inert gas chromatographic columns coated with ionic liquids. *J. Chromatogr. A* **2017**, *1495*, 64–75. [CrossRef]
178. Zeng, A.X.; Chin, S.T.; Nolvachai, Y.; Kulsing, C.; Sidisky, L.M.; Marriott, P.J. Characterisation of capillary ionic liquid columns for gas chromatography–mass spectrometry analysis of fatty acid methyl esters. *Anal. Chim. Acta* **2013**, *803*, 166–173. [CrossRef]
179. Gómez-Cortés, P.; Rodríguez-Pino, V.; Juárez, M.; De la Fuente, M.A. Optimization of milk odd and branched-chain fatty acids analysis by gas chromatography using an extremely polar stationary phase. *Food Chem.* **2017**, *231*, 11–18. [CrossRef]
180. Talebi, M.; Patil, R.A.; Sidisky, L.M.; Berthod, A.; Armstrong, D.W. Branched-chain dicationic ionic liquids for fatty acid methyl ester assessment by gas chromatography. *Anal. Bioanal. Chem.* **2018**, *410*, 4633–4643. [CrossRef]
181. Hammann, S.; Vetter, W. Gas chromatographic separation of fatty acid esters of cholesterol and phytosterols on an ionic liquid capillary column. *J. Chromatogr. B* **2015**, *1007*, 67–71. [CrossRef]
182. Lin, C.C.; Wasta, Z.; Mjøs, S.A. Evaluation of the retention pattern on ionic liquid columns for gas chromatographic analyses of fatty acid methyl esters. *J. Chromatogr. A* **2014**, *1350*, 83–91. [CrossRef]
183. Krupčík, J.; Gorovenko, R.; Špánik, I.; Bočková, I.; Sandra, P.; Armstrong, D.W. On the use of ionic liquid capillary columns for analysis of aromatic hydrocarbons in low-boiling petrochemical products by one-dimensional and comprehensive two-dimensional gas chromatography. *J. Chromatogr. A* **2013**, *1301*, 225–236. [CrossRef]
184. Frink, L.A.; Armstrong, D.W. Determination of Trace Water Content in Petroleum and Petroleum Products. *Anal. Chem.* **2016**, *88*, 8194–8201. [CrossRef] [PubMed]
185. González Paredes, R.M.; García Pinto, C.; Pavón, J.L.P.; Moreno Cordero, B. Ionic liquids as stationary phases in gas chromatography: Determination of chlorobenzenes in soils. *J. Sep. Sci.* **2014**, *37*, 1448–1455. [CrossRef] [PubMed]
186. Do, L.; Liljelind, P.; Zhang, J.; Haglund, P. Comprehensive profiling of 136 tetra- to octa-polychlorinated dibenzo-p-dioxins and dibenzofurans using ionic liquid columns and column combinations. *J. Chromatogr. A* **2013**, *1311*, 157–169. [CrossRef] [PubMed]
187. Boczkaj, G.; Makoś, P.; Przyjazny, A. Application of dynamic headspace and gas chromatography coupled to mass spectrometry (DHS-GC-MS) for the determination of oxygenated volatile organic compounds in refinery effluents. *Anal. Methods* **2016**, *8*, 3570–3577. [CrossRef]
188. Escudero, L.B.; Grijalba, C.A.; Martinis, E.M.; Wuilloud, R.G. Bioanalytical separation and preconcentration using ionic liquids. *Anal. Bioanal. Chem.* **2013**, *405*, 7597–7613. [CrossRef] [PubMed]

189. Destaillats, F.; Guitard, M.; Cruz-Hernandez, C. Identification of Δ6-monounsaturated fatty acids in human hair and nail samples by gas-chromatography-mass-spectrometry using ionic-liquid coated capillary column. *J. Chromatogr. A* **2011**, *1218*, 9384–9389. [CrossRef]
190. Zapadlo, M.; Krupčík, J.; Kovalczuk, T.; Májek, P.; Špánik, I.; Armstrong, D.W.; Sandrae, P. Enhanced comprehensive two-dimensional gas chromatographic resolution of polychlorinated biphenyls on a non-polar polysiloxane and an ionic liquid column series. *J. Chromatogr. A* **2011**, *1218*, 746–751. [CrossRef]
191. Frink, L.A.; Weatherly, C.A.; Armstrong, D.W. Water determination in active pharmaceutical ingredients using ionic liquid headspace gas chromatography and two different detection protocols. *J. Pharm. Biomed. Anal.* **2014**, *94*, 111–117. [CrossRef]
192. Liu, F.; Jiang, Y. Room temperature ionic liquid as matrix medium for the determination of residual solvents in pharmaceuticals by static headspace gas chromatography. *J. Chromatogr. A* **2007**, *1167*, 116–119. [CrossRef]
193. Ho, T.D.; Yehl, P.M.; Chetwyn, N.P.; Wang, J.; Anderson, J.L.; Zhong, Q. Determination of trace level genotoxic impurities in small molecule drug substances using conventional headspace gas chromatography with contemporary ionic liquid diluents and electron capture detection. *J. Chromatogr. A* **2014**, *1361*, 217–228. [CrossRef]
194. Ni, M.; Sun, T.; Zhang, L.; Liu, Y.; Xu, M.; Jiang, Y. Relationship study of partition coefficients between ionic liquid and headspace for organic solvents by HS-GC. *J. Chromatogr. B* **2014**, *945–946*, 60–67. [CrossRef] [PubMed]
195. Laus, G.; Andre, M.; Bentivoglio, G.; Schottenberger, H. Ionic liquids as superior solvents for headspace gas chromatography of residual solvents with very low vapor pressure, relevant for pharmaceutical final dosage forms. *J. Chromatogr. A* **2009**, *1216*, 6020–6023. [CrossRef] [PubMed]
196. Mieszkowski, D.; Sroka, W.D.; Marszałł, M.P. Influence of the anionic part of 1-Alkyl-3-methylimidazolium-based ionic liquids on the chromatographic behavior of perazine in RP-HPTLC. *J. Liq. Chromatogr. Relat. Technol.* **2015**, *38*, 1499–1506. [CrossRef]
197. Buszewska-Forajta, M.; Markuszewski, M.J.; Kaliszan, R. Free silanols and ionic liquids as their suppressors in liquid chromatography. *J. Chromatogr. A* **2018**, *1559*, 17–43. [CrossRef] [PubMed]
198. Marszall, M.P.; Sroka, W.D.; Balinowska, A.; Mieszkowski, D.; Koba, M.; Kaliszan, R. Ionic Liquids as Mobile Phase Additives for Feasible Assay of Naphazoline in Pharmaceutical Formulation by HPTLC-UV-Densitometric Method. *J. Chromatogr. Sci.* **2013**, *51*, 560–565. [CrossRef]
199. Kaliszan, R.; Marszałł, M.P.; Markuszewski, M.J.; Bączek, T.; Pernak, J. Suppression of deleterious effects of free silanols in liquid chromatography by imidazolium tetrafluoroborate ionic liquids. *J. Chromatogr. A* **2004**, *1030*, 263–271. [CrossRef]
200. Mieszkowski, D.; Siódmiak, T.; Marszałł, M.P. 1-alkyl-3-methylimidazolium tetrafluoroborate as an alternative mobile phase additives for determination of haloperidol in pharmaceutical formulation by HPTLC UV densitometric method. *J. Liq. Chromatogr. Relat. Technol.* **2014**, *37*, 1524–1534. [CrossRef]
201. Lu, J.; Ma, H.Y.; Zhang, W.; Ma, Z.G.; Yao, S. Separation of Berberine Hydrochloride and Tetrahydropalmatine and Their Quantitative Analysis with Thin Layer Chromatography Involved with Ionic Liquids. *J. Anal. Methods Chem.* **2015**, *2015*, 642401. [CrossRef]
202. Tuzimski, T.; Petruczynik, A. Application of mobile phases containing ionic liquid for the separation of a mixture of ten selected isoquinoline alkaloids by 2D-TLC and identification of analytes in *Rhizoma Coptidis* (Huang Lian) Extract by TLC and HPLC—DAD. *J. Planar Chromatogr. Mod. TLC* **2017**, *30*, 245–250. [CrossRef]
203. Roth, M. Partitioning behaviour of organic compounds between ionic liquids and supercritical fluids. *J. Chromatogr. A* **2009**, *1216*, 1861–1880. [CrossRef]
204. Keskin, S.; Kayrak-Talay, D.; Akman, U.; Hortaçsu, Ö. A review of ionic liquids towards supercritical fluid applications. *J. Supercrit. Fluids* **2007**, *43*, 150–180. [CrossRef]
205. Lemasson, E.; Bertin, S.; Hennig, P.; Boiteux, H.; Lesellier, E.; West, C. Development of an achiral supercritical fluid chromatography method with ultraviolet absorbance and mass spectrometric detection for impurity profiling of drug candidates. Part I: Optimization of mobile phase composition. *J. Chromatogr. A* **2015**, *408*, 217–226. [CrossRef] [PubMed]
206. Smuts, J.; Wanigasekara, E.; Armstrong, D.W. Comparison of stationary phases for packed column supercritical fluid chromatography based upon ionic liquid motifs: A study of cation and anion effects. *Anal. Bioanal. Chem.* **2011**, *400*, 435–447. [CrossRef]

207. Chou, F.M.; Wang, W.T.; Wei, G.T. Using subcritical/supercritical fluid chromatography to separate acidic, basic, and neutral compounds over an ionic liquid-functionalized stationary phase. *J. Chromatogr. A* **2009**, *1216*, 3594–3599. [CrossRef]
208. Marszałł, M.P.; Markuszewski, M.J.; Kaliszan, R. Separation of nicotinic acid and its structural isomers using 1-ethyl-3-methylimidazolium ionic liquid as a buffer additive by capillary electrophoresis. *J. Pharm. Biomed. Anal.* **2006**, *41*, 329–332. [CrossRef]
209. Qin, W.; Li, S.F.Y. An ionic liquid coating for determination of sildenafil and UK-103,320 in human serum by capillary zone electrophoresis-ion trap mass spectrometry. *Electrophoresis* **2002**, *23*, 4110–4116. [CrossRef]
210. Abd El-Hady, D.; Albishri, H.M.; Rengarajan, R. Eco-friendly ionic liquid assisted capillary electrophoresis and α -acid glycoprotein-assisted liquid chromatography for simultaneous determination of anticancer drugs in human fluids. *Biomed. Chromatogr.* **2015**, *29*, 925–934. [CrossRef]
211. Huang, L.; Lin, J.M.; Yu, L.; Xu, L.; Chen, G. Improved simultaneous enantioseparation of β-agonists in CE using β-CD and ionic liquids. *Electrophoresis* **2009**, *30*, 1030–1036. [CrossRef]
212. Jiang, T.F.; Wang, Y.H.; Lv, Z.H. Dynamic coating of a capillary with room-temperature ionic liquids for the separation of amino acids and acid drugs by capillary electrophoresis. *J. Anal. Chem.* **2006**, *61*, 1108–1112. [CrossRef]
213. Tsai, C.Y.; Yang, C.F.; Whang, C.W. Capillary electrophoretic separation of tricyclic antidepressants using 1-Alkyl-3-methylimidazolium-based ionic liquids as background electrolyte. *J Chin. Chem. Soc.* **2009**, *56*, 351–359. [CrossRef]
214. Zhao, M.; Cui, Y.; Yu, J.; Xu, S.; Guo, X. Combined use of hydroxypropyl-β-cyclodextrin and ionic liquids for the simultaneous enantioseparation of four azole antifungals by CE and a study of the synergistic effect. *J. Sep. Sci.* **2014**, *37*, 151–157. [CrossRef] [PubMed]
215. Liu, Y.M.; Tian, W.; Jia, Y.X.; Yue, H.Y. The use of CE ECL with ionic liquid for the determination of drug alkaloids and applications in human urine. *Electrophoresis* **2009**, *30*, 1406–1411. [CrossRef] [PubMed]
216. Jin, Y.; Chen, C.; Meng, L.; Chen, J.; Li, M.; Zhu, Z. Simultaneous and sensitive capillary electrophoretic enantioseparation of three β-blockers with the combination of achiral ionic liquid and dual CD derivatives. *Talanta* **2012**, *89*, 149–154. [CrossRef]
217. François, Y.; Varenne, A.; Juillerat, E.; Villemin, D.; Gareil, P. Evaluation of chiral ionic liquids as additives to cyclodextrins for enantiomeric separations by capillary electrophoresis. *J. Chromatogr. A* **2007**, *1155*, 134–141. [CrossRef] [PubMed]
218. Zhang, Q.; Du, Y. Evaluation of the enantioselectivity of glycogen-based synergistic system with amino acid chiral ionic liquids as additives in capillary electrophoresis. *J. Chromatogr. A* **2013**, *1306*, 97–103. [CrossRef]
219. Zhang, Q.; Qi, X.; Feng, C.; Tong, S.; Rui, M. Three chiral ionic liquids as additives for enantioseparation in capillary electrophoresis and their comparison with conventiona modifiers. *J. Chromatogr. A* **2016**, *1462*, 146–152. [CrossRef]
220. Zuo, L.; Meng, H.; Wu, J.; Jiang, Z.; Xu, S.; Guo, X. Combined use of ionic liquid and β-CD for enantioseparation of 12 pharmaceuticals using CE. *J. Sep Sci.* **2013**, *36*, 517–523. [CrossRef]
221. Kolobova, E.; Kartsova, L.; Alopina, E.; Smirnova, N. Separation of amino acids and β-blockers enantiomers by capillary electrophoresis with 1-butyl-3-methylimidazolium *L*-prolinate [C4MIm][*L*-Pro] as a chiral selector. *Anal. I Kontrol* **2018**, *22*, 51–60. [CrossRef]
222. Zhang, Q.; Du, Y.; Du, S.; Zhang, J.; Feng, Z.; Zhang, Y.; Li, X. Tetramethylammonium-lactobionate: A novel ionic liquid chiral selector based on saccharides in capillary electrophoresis. *Electrophoresis* **2015**, *36*, 1216–1223. [CrossRef]
223. Zhang, J.; Du, Y.; Zhang, Q.; Chen, J.; Xu, G.; Yu, T.; Hua, X.J. Investigation of the synergistic effect with amino acid-derived chiral ionic liquids as additives for enantiomeric separation in capillary electrophoresis. *J. Chromatogr. A* **2013**, *1316*, 119–126. [CrossRef] [PubMed]
224. Zhang, J.; Du, Y.; Zhang, Q.; Lei, Y. Evaluation of vancomycin-based synergistic system with amino acid ester chiral ionic liquids as additives for enantioseparation of non-steroidal anti-inflamatory drugs by capillary electrophoresis. *Talanta* **2014**, *119*, 193–201. [CrossRef] [PubMed]
225. Xu, G.; Du, Y.; Du, F.; Chen, J.; Yu, T.; Zhang, Q.; Zhang, J.; Du, S.; Feng, Z. Establishment and Evaluation of the Novel Tetramethylammonium-L-Hydroxyproline Chiral Ionic Liquid Synergistic System Based on Clindamycin Phosphate for Enantioseparation by Capillary Electrophoresis. *Chirality* **2015**, *27*, 598–604. [CrossRef] [PubMed]

226. Yang, X.; Du, Y.; Feng, Z.; Liu, Z.; Li, J. Establishment and molecular modeling study of maltodextrin-based synergistic enantioseparation systems with two new hydroxy acid chiral ionic liquids as additives in capillary electrophoresis. *J. Chromatogr. A* **2018**, *1559*, 170–177. [CrossRef] [PubMed]
227. Chen, J.; Du, Y.; Sun, X. Investigation of maltodextrin-based synergistic system with amino acid chiral ionic liquid as additive for enantioseparation in capillary electrophoresis. *Chirality* **2017**, *29*, 824–835. [CrossRef] [PubMed]
228. Zhang, Y.; Du, S.; Feng, Z.; Du, Y.; Yan, Z. Evaluation of synergistic enantioseparation systems with chiral spirocyclic ionic liquids as additives by capillary electrophoresis. *Anal. Bioanal. Chem.* **2016**, *408*, 2543–2555. [CrossRef]
229. Zhang, Y.; Du, Y.; Yu, T.; Feng, Z.; Chen, J. Investigation of dextrin-based synergistic system with chiral ionic liquids as additives for enantiomeric separation in capillary electrophoresis. *J. Pharm. Biomed. Anal.* **2019**, *164*, 413–420. [CrossRef]
230. Zhang, Q.; Du, Y.; Du, S. Evaluation of ionic liquids-coated carbon nanotubes modified chiral separation system with chondroitin sulfate E as chiral selector in capillary electrophoresis. *J. Chromatogr. A* **2014**, *1339*, 185–191. [CrossRef] [PubMed]
231. Yu, J.; Zuo, L.; Liu, H.; Zhang, L.; Guo, X. Synthesis and application of a chiral ionic liquid functionalized β-cyclodextrin as a chiral selector in capillary electrophoresis. *Biomed. Chromatogr.* **2013**, *27*, 1027–1033. [CrossRef]
232. Wang, B.; He, J.; Bianchi, V.; Shamsi, S.A. Combined use of chiral ionic liquid and cyclodextrin for MEKC: Part I. Simultaneous enantioseparation of anionic profens. *Electrophoresis* **2009**, *30*, 2812–2819. [CrossRef]
233. Wang, B.; He, J.; Bianchi, V.; Shamsi, S.A. Combined use of chiral ionic liquid and CD for MEKC: Part II. Determination of binding constants. *Electrophoresis* **2009**, *30*, 2820–2828. [CrossRef] [PubMed]
234. Cui, Y.; Ma, X.; Zhao, M.; Jiang, Z.; Xu, S.; Guo, X. Combined Use of Ionic Liquid and Hydroxypropyl-β-Cyclodextrin for the Enantioseparation of Ten Drugs by Capillary Electrophoresis. *Chirality* **2013**, *25*, 409–414. [CrossRef]
235. Su, H.L.; Kao, W.C.; Lin, K.W.; Lee, C.; Hsieh, Y.Z. 1-Butyl-3-methylimidazolium-based ionic liquids and an anionic surfactant: Excellent background electrolyte modifiers for the analysis of benzodiazepines through capillary electrophoresis. *J. Chromatogr. A* **2010**, *1217*, 2973–2979. [CrossRef] [PubMed]
236. Ma, Z.; Zhang, L.; Lin, L.; Ji, P.; Guo, X. Enantioseparation of rabeprazole and omeprazole by nonaqueous capillary electrophoresis with an ephedrine-based ionic liquid as the chiral selector. *Biomed. Chromatogr.* **2010**, *24*, 1332–1337. [CrossRef] [PubMed]
237. Olędzka, I.; Kowalski, P.; Plenis, A.; Miękus, N.; Grabow, N.; Eickner, T.; Bączek, T. Simultaneous electrokinetic and hydrodynamic injection and sequential stacking featuring sweeping for signal amplification following MEKC during the analysis of rapamycin (sirolimus) in serum samples. *Electrophoresis* **2018**, *39*, 2590–2597. [CrossRef]
238. Kowalski, P.; Olędzka, I.; Plenis, A.; Miękus, N.; Pieckowski, M.; Bączek, T. Combination of field amplified sample injection and hydrophobic interaction electrokinetic chromatography (FASI-HIEKC) as a signal amplification method for the determination of selected macrocyclic antibiotics. *Anal. Chim. Acta* **2019**, *1046*, 192–198. [CrossRef]

MDPI
St. Alban-Anlage 66
4052 Basel
Switzerland
Tel. +41 61 683 77 34
Fax +41 61 302 89 18
www.mdpi.com

Molecules Editorial Office
E-mail: molecules@mdpi.com
www.mdpi.com/journal/molecules

www.ingramcontent.com/pod-product-compliance
Lightning Source LLC
LaVergne TN
LVHW070203170726
843515LV00007B/1987

* 9 7 8 3 0 3 9 4 3 4 8 5 5 *